BEAM COOLING AND RELATED TOPICS

International Workshop on Beam Cooling
and Related Topics - COOL05

Galena, Illinois, U.S.A.   18 – 23 September 2005

EDITORS
Sergei Nagaitsev
Ralph J. Pasquinelli
Fermi National Accelerator Laboratory
Batavia, Illinois

SPONSORING ORGANIZATIONS
Fermi National Accelerator Laboratory
Argonne National Laboratory
Brookhaven National Laboratory
Lawrence Berkeley National Laboratory
United States Department of Energy

CD-ROM INCLUDED

Melville, New York, 2006
AIP CONFERENCE PROCEEDINGS ■ VOLUME 821

Editors:

Sergei Nagaitsev
Fermi National Accelerator Laboratory
P.O. Box 500
MS 323
Batavia, IL 60510
E-mail: nsergei@fnal.gov

Ralph J. Pasquinelli
Fermi National Accelerator Laboratory
P.O. Box 500
MS 341
Batavia, IL 60510
E-mail: pasquin@fnal.gov

Authorization to photocopy items for internal or personal use, beyond the free copying permitted under the 1978 U.S. Copyright Law (see statement below), is granted by the American Institute of Physics for users registered with the Copyright Clearance Center (CCC) Transactional Reporting Service, provided that the base fee of $23.00 per copy is paid directly to CCC, 222 Rosewood Drive, Danvers, MA 01923, USA. For those organizations that have been granted a photocopy license by CCC, a separate system of payment has been arranged. The fee code for users of the Transactional Reporting Services is: 0-7354-0314-7/06/$23.00

© 2006 American Institute of Physics

Permission is granted to quote from the AIP Conference Proceedings with the customary acknowledgment of the source. Republication of an article or portions thereof (e.g., extensive excerpts, figures, tables, etc.) in original form or in translation, as well as other types of reuse (e.g., in course packs) require formal permission from AIP and may be subject to fees. As a courtesy, the author of the original proceedings article should be informed of any request for republication/reuse. Permission may be obtained online using Rightslink. Locate the article online at http://proceedings.aip.org, then simply click on the Rightslink icon/"Permission for Reuse" link found in the article abstract. You may also address requests to: AIP Office of Rights and Permissions, Suite 1NO1, 2 Huntington Quadrangle, Melville, NY 11747-4502, USA; Fax: 516-576-2450; Tel.: 516-576-2268; E-mail: rights@aip.org.

L.C. Catalog Card No. 2006921301
ISBN 0-7354-0314-7
ISSN 0094-243X

Printed in the United States of America

BEAM COOLING AND RELATED TOPICS

To learn more about the AIP Conference Proceedings, including the Conference Proceedings Series, please visit the webpage
http://proceedings.aip.org/proceedings

CONTENTS

Preface..xi
Committees...xii
Group Photograph..xiii

OVERVIEW OF BEAM COOLING

Comment on the Word "Cooling" as it is Used in Beam Physics...............3
    A. M. Sessler
The Reason for Beam Cooling: Some of the Physics that
Cooling Allows..6
    W. Oelert
Overview of Recent Trends in Beam Cooling Methods
and Technology...16
    I. Meshkov and D. Möhl

REPORTS ON OPERATING AND NEW PROJECTS

Status of the FAIR Facility.....................................29
    M. Steck
Antiproton Cooling in the Fermilab Recycler Ring.........................39
    S. Nagaitsev, A. Bolshakov, D. Broemmelsiek, A. Burov, K. Carlson,
    C. Gattuso, M. Hu, G. Kazakevich, B. Kramper, T. Kroc, J. Leibfritz,
    L. Prost, S. Pruss, G. Saewert, C. W. Schmidt, S. Seletskiy, A. Shemyakin,
    M. Sutherland, V. Tupikov, A. Warner, and P. Zenkevich
Report on Operation of Antiproton Decelerator...........................48
    P. Belochitskii on behalf of AD team
LEIR Cooler Status..57
    G. Tranquille
Commissioning of HIRFL-CSR and its Electron Coolers....................65
    X. Yang, V. Parkhomchuk, W. Zhan, J. Xia, H. Zhao, Y. Yuan, M. Song,
    J. Li, L. Mao, W. Lu, Z. Wang, and BINP Electron Cooler Group
High-Current ERL-Based Electron Cooling System for RHIC................75
    I. Ben-Zvi
FLAIR Project at GSI..85
    C. P. Welsch, M. Grieser, J. Ullrich, and
    A. Wolf for the FLAIR Collaboration
Status of the LEPTA Project......................................95
    A. Kobets, Y. Korotaev, V. Malakhov, I. Meshkov, V. Pavlov, R. Pivin,
    I. Seleznev, A. Sidorin, A. Smirnov, G. Trubnikov, and S. Yakovenko
S-LSR, Cooler Ring Development at Kyoto University......................103
    T. Shirai, S. Fujimoto, M. Ikegami, A. Noda, H. Souda, M. Tanabe,
    H. Tongu, K. Noda, S. Shibuya, T. Takeuchi, T. Fujimoto, S. Iwata,
    A. Takubo, H. Okamoto, Y. Yuri, M. Grieser, and E. M. Syresin

Antiproton—Ion Collider for FAIR Project 108
    P. Beller, B. Franzke, P. Kienle, R. Kruecken, I. Koop, V. Parkhomchuk,
    Y. Shatunov, A. Skrinsky, V. Vostrikov, and E. Widmann

GENERAL TOPICS

Transverse-Longitudinal Phase-Space Manipulations and Correlations 115
    K.-J. Kim and A. Sessler
Optics of Electron Beam in the Recycler 139
    A. Burov, G. Kazakevich, T. Kroc, V. Lebedev, S. Nagaitsev, L. Prost,
    S. Pruss, A. Shemyakin, M. Sutherland, M. Tiunov, and A. Warner
Experimental Study of Dispersion Control Utilizing both Magnetic
and Electric Fields 144
    M. Tanabe, M. Ikegami, A. Noda, T. Shirai, H. Souda, H. Tongu,
    S. Shibuya, and K. Noda
Transverse Echo Measurements in RHIC 149
    W. Fischer
Studies of Beam Dynamics in Cooler Rings 154
    J. Dietrich, I. Meshkov, A. Sidorin, A. Smirnov, and J. Stein
IBS in a CAM-Dominated Electron Beam 159
    A. Burov, I. Gusachenko, S. Nagaitsev, and A. Shemyakin
Stability Studies under Dipole Oscillation Model for RHIC E-Cooling 164
    G. Wang
Hamiltonian Analysis of the Particle Motion in an Accelerator with
the Longitudinal Magnetic Field 169
    V. B. Reva

STOCHASTIC COOLING

Stochastic Cooling Developments at GSI. 177
    F. Nolden, K. Beckert, P. Beller, A. Dolinskii, B. Franzke, U. Jandewerth,
    I. Nesmiyan, C. Peschke, P. Petri, M. Steck, F. Caspers, D. Möhl, and
    L. Thorndal
Bunched Beam Stochastic Cooling Project for RHIC 185
    J. M. Brennan and M. Blaskiewicz
Cooling Scenario for the HESR Complex 190
    H. Stockhorst, D. Prasuhn, R. Maier, and B. Lorentz
Stacking of 3 GeV Antiprotons with a Moving Barrier Bucket
Method at the GSI-RESR 196
    T. Katayama, P. Beller, B. Franzke, I. Nesmiyan, F. Nolden, M. Steck,
    D. Möhl, and T. Kikuchi
Bunched Beam Stochastic Cooling and Coherent Lines 206
    M. Blaskiewicz and J. M. Brennan

Applications of Schottky Spectroscopy at the Storage Ring ESR of GSI 211
    F. Nolden, K. Beckert, P. Beller, B. Franzke, V. Gostishchev,
    C. Kozhuhzrov, Y. A. Litvinov, A. Schwinn, and M. Steck

Pick-Up and Kicker Electrodes for the CR 221
    C. Peschke, F. Nolden, and L. Thorndal

Fermilab Recycler Stochastic Cooling for Luminosity Production 226
    D. Broemmelsiek and C. Gattuso

Stochastic Cooling with Schottky Band Overlap 231
    V. Lebedev

Debuncher Cooling Performance 237
    P. F. Derwent, D. McGinnis, R. Pasquinelli, D. Vander Meulen, and
    S. Werkema

Performance and Upgrades of the Fermilab Accumulator Stacktail Stochastic Cooling 242
    P. F. Derwent, E. Cullerton, D. McGinnis, R. Pasquinelli, D. Sun, and
    D. Tinsley

ELECTRON COOLING

Development of a New Generation of Coolers with a Hollow Electron Beam and Electrostatic Bending 249
    V. V. Parkhomchuk

Cooling Force Measurements at CELSIUS 259
    B. Gålnander, A. V. Fedotov, V. N. Litvinenko, T. Lofnes, A. O. Sidorin,
    A. V. Smirnov, and V. Ziemann

Experimental Benchmarking of the Magnetized Friction Force 265
    A. V. Fedotov, B. Galnander, V. N. Litvinenko, T. Lofnes, A. O. Sidorin,
    A. V. Smirnov, and V. Ziemann

Electron Cooling of Intense Ion Beam 270
    J. Dietrich, V. Kamerdjiev, Yu. Korotaev, R. Maier, I. Meshkov, D. Prasuhn,
    A. Sidorin, A. Smirnov, J. Stein, and H. Stockhorst

Attainment of a High-Quality Electron Beam for Fermilab's 4.3 MeV Cooler 280
    A. Shemyakin, A. Burov, K. Carlson, M. Hu, G. Kazakevich, B. Kramper,
    T. Kroc, J. Leibfritz, S. Nagaitsev, L. Prost, S. Pruss, G. Saewert,
    C. W. Schmidt, S. Seletskiy, M. Sutherland, V. Tupikov, and A. Warner

The HESR Electron Cooling Proposal 289
    D. Reistad

The Proposed 2 MeV Electron Cooler for COSY 299
    J. Dietrich, V. V. Parkhomchuk, V. B. Reva, and M. A. Vedenev

Budker INP Proposals for HESR and COSY Electron Cooler Systems 308
    V. Bocharov, M. Bryzgunov, A. Bubley, V. Gosteev, I. Kazarezov,
    A. Kryuchkov, V. Panasyuk, V. Parkhomchuk, V. Pavlov, D. Pestrikov,
    V. Reva, V. Shamovskij, A. Skrinsky, B. Sukhina, M. Vedenev, and
    V. Vostrikov

Summary Report: Working Group on COSY 2 MV Cooler 317
    S. Nagaitsev and I. Meshkov

Detailed Studies of Electron Cooling Friction Force 319
    A. V. Fedotov, D. L. Bruhwiler, D. T. Abell, and A. O. Sidorin

Simulations of Dynamical Friction Including Spatially-Varying Magnetic Fields 329
    G. I. Bell, D. L. Bruhwiler, V. N. Litvinenko, R. Busby, D. T. Abell,
    P. Messmer, S. Veitzer, and J. R. Cary

Comission of Electron Cooler EC-300 for HIRFL-CSR 334
    E. Behtenev, V. Bocharov, V. Bubley, M. Vedenev, R. Voskoboinikov,
    A. Goncharov, Yu. Evtushenko, N. Zapiatkin, M. Zakhvatkin, A. Ivanov,
    V. Kokoulin, V. Kolmogorov, M. Kondaurov, S. Konstantinov, G. Krainov,
    V. Kozak, A. Kruchkov, E. Kuper, A. Medvedko, L. Mironenko,
    V. Panasiuk, V. Parkhomchuk, V. Reva, A. Skrinsky, B. Smirnov,
    B. Skarbo, B. Sukhina, K. Shrainer, X. D. Yang, H. W. Zhao, J. Li, W. Lu,
    L. J. Mao, Z. X. Wang, H. B. Yan, W. Zhang, and J. H. Zhang,

Recuperation of Electron Beam in the Coolers with Electrostatic Bending 341
    M. Bryzgunov, V. Panasyuk, V. Parkhomchuk, V. Reva, and M. Vedenev

Low Energy Electron Cooling and Accelerator Physics for the Heidelberg CSR 346
    H. Fadil, M. Grieser, R. von Hahn, D. Orlov, D. Schwalm, A. Wolf, and
    D. Zajfman

Electron Cooling of Bunched Beams 351
    T. Uesugi, K. Noda, E. Syresin, I. Meshkov, and S. Shibuya

First Tests of LEIR—Cooler at BINP 355
    V. Bocharov, M. Brizgunov, A. Bubley, V. Ershov, A. Goncharov,
    S. Konstantinov, A. Lomakin, V. Panasyuk, V. Parkhomchuk, V. Polukhin,
    V. Reva, B. Skarbo, B. Sukhina, M. Vedenev, M. Zakhvatkin, and
    N. Zapiatkin

Precise Measurements of a Magnetic Field at the Solenoids for Low Energy Coolers 360
    V. Bocharov, A. Bubley, S. Konstantinov, V. Panasyuk, and
    V. Parkhomchuk

Electron Cooling for Cold Beam Synchrotron for Cancer Therapy 365
    B. Grishanov, M. Kumada, V. Parkhomchuk, S. Rastigeev, V. Reva, and
    V. Vostrikov

Electron Beam Size Measurements in the Fermilab Electron Cooling System 370
    T. K. Kroc, A. V. Burov, T. B. Bolshakov, A. Shemyakin, and
    S. M. Seletskiy

Magnetic Field Measurement and Compensation in the Recycler Electron Cooler 375
    V. Tupikov, G. Kazakevich, T. K. Kroc, S. Nagaitsev, L. Prost,
    A. Shemyakin, C. W. Schmidt, M. Sutherland, and A. Warner

OTR Measurements and Modeling of the Electron Beam Optics at
the E-Cooling Facility 380
      A. Warner, A. Burov, K. Carlson, G. Kazakevich, S. Nagaitsev, L. Prost,
      M. Sutherland, and M. Tiunov

Beam-Based Alignment of Magnetic Field in the Fermilab Electron
Cooler Cooling Section... 386
      S. M. Seletskiy and V. Tupikov

Full Discharges in Fermilab's Electron Cooler......................... 391
      L. R. Prost and A. Shemyakin

Cooling Rates of the USR as Calculated with BETACOOL................. 397
      C. P. Welsch and A. Smirnov

MUON COOLING

Recent Innovations in Muon Beam Cooling.............................. 405
      R. P. Johnson, M. Alsharo'a, C. Ankenbrandt, E. Barzi, K. Beard,
      S. A. Bogacz, Y. Derbenev, L. Del Frate, I. Gonin, P. M. Hanlet,
      R. Hartline, D. M. Kaplan, M. Kuchnir, A. Moretti, D. Neuffer, K. Paul,
      M. Popovic, T. J. Roberts, G. Romanov, D. Turrioni, V. Yarba, and
      K. Yonehara

6D Cooling of a Circulating Muon Beam 415
      A. Garren, D. Cline, S. Kahn, H. Kirk, and F. Mills

Parametric-Resonance Ionization Cooling and Reverse Emittance
Exchange for Muon Colliders.. 420
      Y. Derbenev and R. P. Johnson

MICE: The International Muon Ionization Cooling Experiment 427
      D. M. Kaplan

6D Muon Ionization Cooling with an Inverse Cyclotron 432
      D. J. Summers, S. B. Bracker, L. M. Cremaldi, R. Godang, and
      R. B. Palmer

The Muon Cooling RF R&D Program 437
      Y. Torun, A. Bross, D. Li, A. Moretti, J. Norem, Z. Qian, R. A. Rimmer,
      and M. S. Zisman

Mucool Hydrogen Absorber R&D 442
      M. A. Cummings on behalf of the Muon Collaboration

Cryogenics for the MuCool Test Area (MTA) 448
      C. Darve, B. Norris, and L. Pei

g4 Beamline Simulations of Parametric Resonance Ionization Cooling
of Muon Beams.. 453
      K. Beard, S. A. Bogacz, Y. Derbenev, K. Yonehara, R. P. Johnson, K. Paul,
      and T. J. Roberts

Simulations of MANX, A Practical Six Dimensional Muon Beam
Cooling Experiment 458
      K. Yonehara, K. Beard, A. Bogacz, Y. Derbenev, R. P. Johnson, Kaplan,
      K. Paul, and T. Roberts

ELECTROSTATIC RINGS

DESIREE—A Double Electrostatic Storage Ring for Merged-Beam Experiments 465
    H. Danared, L. Liljeby, G. Andler, L. Bagge, M. Blom, A. Källberg,
    S. Leontein, P. Löfgren, A. Paál, K.-G. Rensfelt, A. Simonsson,
    H. T. Schmidt, H. Cederquist, M. Larsson, S. Rosén, and K. Schmidt

The Heidelberg CSR: Stored Ion Beams in a Cryogenic Environment 473
    A. Wolf, R. von Hahn, M. Grieser, D. A. Orlov, H. Fadil,
    C. P. Welsch, V. Andrianarijaona, A. Diehl, C. D. Schröter,
    J. R. Crespo López-Urrutia, M. Rappaport, X. Urbain, T. Weber,
    V. Mallinger, C. Haberstroh, H. Quack, D. Schwalm, J. Ullrich,
    and D. Zajfman

Ultra-Cold Electron Beams for the Heidelberg TSR and CSR 478
    D. A. Orlov, M. Lestinsky, F. Sprenger, D. Schwalm, A. S. Terekhov, and
    A. Wolf

LASER COOLING

Laser Cooling for 3-D Crystalline State at S-LSR 491
    A. Noda, S. Fujimoto, M. Ikegami, T. Shirai, H. Souda, M. Tanabe,
    H. Tongu, K. Noda, S. Yamada, S. Shibuya, T. Takeuchi, H. Okamoto, and
    M. Grieser

Combined Laser and Electron Cooling of Bunched C3+ Ion Beams at the Storage Ring ESR 501
    U. Schramm, M. Bussmann, D. Habs, T. Kühl, P. Beller, B. Franzke,
    F. Nolden, M. Steck, G. Saathoff, S. Reinhardt, and S. Karpuk

TRAPS

Electron Cooling of Ions and Antiprotons in Traps 513
    G. Zwicknagel

Aspects of Cooling at the TRIμP Facility 523
    L. Willmann, G. P. Berg, U. Dammalapati, S. De, P. Dendooven,
    O. Dermois, K. Jungmann, A. Mol, C. J. G. Onderwater, A. Rogachevskiy,
    M. Sohani, E. Traykov, and H. W. Wilschut

Photographs 528
Program 531
List of Participants 535
Author Index 538

Preface

This was the eighth meeting in the series of beam cooling workshops and the first to be held in the United States since the early 1980's. Participation included 87 engineers and scientists from 30 institutions around the globe. Presentations highlighted beam cooling as a mainstream part of accelerator facilities worldwide. In particular, electron cooling was extensively covered in the program. Normally used at low beam energies, the Fermilab Electron Cooling project for the Recycler was successfully commissioned shortly before this meeting. Eight GeV antiprotons are routinely being cooled with a 4.3 MeV electron beam. The future of beam cooling also looks promising as the FAIR project at GSI in Darmstadt, Germany is in the approval stages. Ionization cooling experiments are taking place at a number of facilities in support of Muon colliders and neutrino sources. The future of beam cooling looks to be bright.

Sergei Nagaitsev
Ralph J. Pasquinelli
COOL05 Co-chairmen

Committees

Local Organizing Committee

Sergei Nagaitsev (co-chair), *FNAL*
Ralph Pasquinelli (co-chair), *FNAL*
Cynthia Sazama, *FNAL*
Suzanne Weber, *FNAL*
Monica Sasse, *FNAL*
Sara Webber, *FNAL*

International Program Committee

Ilan Ben-Zvi, *BNL*
Håkan Danared, *MSL-Stockholm*
Yaroslav Derbenev, *Jefferson Lab*
Dan Kaplan, *IIT*
Kwang-Je Kim, *ANL*
Igor Meshkov, *JINR-Dubna*
Dieter Möhl, *CERN*
Yoshiharu Mori, *KEK*
Akira Noda, *Kyoto University*
Vasily Parkhomchuk, *BINP-Novosibirsk*
Dieter Prasuhn, *IKP-Jülich*
Tor Raubenheimer, *SLAC*
Andrew Sessler, *LBNL*
Markus Steck, *GSI*
Gerard Tranquille, *CERN*
Yasunori Yamazaki, *RIKEN & University of Tokyo*

Sponsored by:

OVERVIEW OF BEAM COOLING

Comment on the Word "Cooling" as it is Used in Beam Physics

Andrew M. Sessler

Lawrence Berkeley National Laboratory
1 Cyclotron Road, BLDG: 71R0259, Berkeley, CA 94720

Abstract. Beam physicists use the word "cooling" differently than it is used by the general public or even by other physicists. It is recommended that we no longer use this term, but replace it with some other term such as: "Phase Density Cooling" (PDF) or "damping", or alternatively "Liouville Cooling", which would make our field more easily understood by outsiders.

Keywords: Beam cooling, phase space, Liouville, beam damping
PACS: 29.27.-a, 29.27.Eg. 41.75.-i, 41,75,Lx

COOLING: HISTORY AND USAGE

Many millions of years ago, our ancestors understood that there were hot objects and cold objects. Surely they also understood when something was heating up (say over a fire or sitting in the sun) and when something was cooling down (say due to shade or wind). They must have developed signs, and later words, to describe the phenomena. In English, the word "cooling" has been in use for thousands of years.

This concept is very much with us, for it is deeply ingrained in our psyche as a result of common experience. Each of us, from early childhood on, knows about "hot" and "cold" (perhaps the very second, or third, word that a child learns). Throughout our lives we have a deep emotional understanding of hot, cold, and cooling.

The development of thermometers took this basic understanding and made it quantitative. Thermometers allowed a rather accurate measurement of temperature and, therefore, a rather accurate meaning to the concept of "cooling". It was now possible to ascribe units to the concept and to numerically evaluate it experimentally. With the advent of thermodynamics in the 19th century, physicists were for the first time able to make the concept of temperature very precise with the use of ideal Carnot cycles, (introduced in 1824) and so, therefore, made very precise the concept of "cooling". [1, 2, 3]

The development of statistical mechanics, again in the 19th century, provided a molecular basis for thermodynamics. If the distribution function in velocity space was Maxwellian (concept introduced in 1860) then the results of thermodynamics could be explained on a molecular level. [4] For example, if a container of gas was put in a refrigerator, the physical size (except for negligible contraction) remains the same, but the velocity of molecules is reduced (although the distribution is still Maxwellian). In this case, statistical mechanics would say that the Maxwellian describing velocity space is narrower, that the density in phase space is increased and the gas is cooler. [5]

Another example might be the adiabatic expansion of a gas leading to cooling of the gas according to the well known law that PV^γ remains a constant. Statistical mechanics describes this process as a narrowing of the Maxwellian and an increase or decrease of phase density to a degree that depends upon the value of γ which depends on the gas in question. (Only for $\gamma=3$ is phase density preserved.)[6]

Even at the forefront of physics; namely while making Bose-Einstein condensates down to tiny fractions of a degree, a thermodynamic concept is employed; namely evaporative cooling. [7] In this case, just as in the evaporation of water, the more energetic molecules leave the condensate and so the temperature of the condensate is reduced. Statistical mechanics simply describes this process as a narrowing of the Maxwellian with a reduction in phase volume, but not as an increase in phase density.

In beam physics it is very different. We only consider a beam as being cooled if the density in six dimensional phase space is increased. Let me give some examples of what we consider *non-cooling*.

Consider the case where there are beam scrapers so particles of larger transverse oscillations are scraped off. The transverse average energy certainly goes down, just as in evaporative cooling. We beam physicists would certainly *not* call this "cooling".

Consider a transport channel in which the beam is rather dilute (so space charge effects can be neglected) and the amplitudes are small (so a linear approximation is valid). Neither of these approximations is necessary, but they make the discussion simpler. When the beam goes through a region in which it expands, and because collisions are negligible, the transverse velocity, even energy, is reduced. Everyone else would call this "cooling" as the transverse temperature has certainly gone down. We beam physicists would certainly *not* call this "cooling".

Consider a system that transfers phase volume from one degree of freedom to the other, for example, a coupling resonance. If the dynamics can be described by a Hamiltonian, and in most cases this is true, then, again, we would *not* consider this "cooling", but simply a transfer of phase space.

We beam physicists carefully reserve the term "cooling" for non-Hamiltonian processes where Liouville's theorem (1838) is violated; i.e., where there is an increase in phase space density. [8,9] There are a good number of methods for "cooling" such as: Stochastic Cooling, Electron Cooling, Radiation Damping, Laser Doppler Cooling, Energy Loss Foils, etc. [10] I suggest that we call these processes for what they are;

namely "Phase Density Cooling (PDC)", or more simply, "damping", or alternatively "Liouville Cooling".

Then, reserving the word "damping", or "Liouville Cooling" for violations of Liouville's Theorem, we might use the term "cooling" in the same way as all the other physicists use it; namely for a simple reduction in transverse (or longitudinal) oscillation energy. This change would remove some confusing terms from our field and, consequently, make our work more accessible to other physicists and students.

ACKNOWLEDGMENTS

Work supported by the U.S. Department of Energy, Office of Basic Energy Sciences, under Contract No. DE-AC02-05CH11231.

REFERENCES

1. S. Carnot, "R´eflexions sur la puissance motrice du feu et sur les machines propres a d´evelopper cette puissance", Bachelier, Paris (1824).
2. R.H. Fowler and E.A. Guggenheim, "Statistical Thermodynamics", p. 57 Cambridge University Press , Cambridge (1960).
3. A.H. Wilson, "Thermodynamics and Statistical Mechanics", p. 21. Cambridge University Press, Cambridge (1960).
4. J.C. Maxwell, "Illustration of the Dynamical Theory of Gases", Phil. Mag. 1860, in The Scientific Papers of James Clerk Maxwell, Dover Publications, Inc., New York (1965).
5. L. Landau and E.M. Lifshitz, "Statistical Physics", p.81, Pergamon Press, London (1958).
6. F. Reif, "Fundamentals of Statistical and Thermal Physics", p.159, McGraw-Hill, New York (1965).
7. E.A. Cornell and C. Weiman, W. Ketterle, Two Nobel Lectures of 2001, Les Prix Nobel, Ed.T. Frangsmyr, Nobel Foundation, Stockholm (2002).
8. J. Liouville, J. de Math.3, 349 (1838).
9. D. Mohl and A.M. Sessler, "Beam cooling: principles and achievements", Proceedings of COOL03, Nucl. Instr. and Meth. A532 [Nos. 1 and 2], p.1 (2004).
10. A.W. Chao and M. Tigner, "Handbook of Accelerator Physics and Engineering", 2nd printing, Sect. 2.8, p.168-182 (2002).

The Reason for Beam Cooling: Some of the Physics that Cooling Allows

W. Oelert

IKP – Research Centre Jülich, D–52425 Jülich, Germany

Abstract. There are many examples of achievements in physics which would not be possible without cooling. Different mechanisms for cooling exist and some will be presented in this introductory talk where we distinguish between "relative" and "absolute" cooling. A short reminder to high and medium energy physics with antiprotons as performed at the accelerators of CERN will be delineated. The success in applying cooling of beams in hadron physics at the internal COSY–11 experiment installed at the cooler synchrotron COSY will be presented. COSY–11 aims for meson production investigations at threshold in nucleon–nucleon collisions. Again, such investigations would not be feasible without cooling especially regarding the precision required and obtained. The need of cooling for the production and trapping of antihydrogen atoms is demonstrated – as an example – by the ATRAP experiment at the CERN antiproton decelerator AD aiming for a comparison of hydrogen (H^0) to antihydrogen ($\bar{H}^0$) atom spectroscopy.

Keywords: Cooling, Antihydrogen, Hadron Physics Experiments

INTRODUCTION

Temperature and especially the control of temperature is one of the most important features of our existence. If the average temperature on earth would be a few percent higher or lower as compared to the long range average it would turn into a dessert planet. If the temperature of elements would be a few thousand degree and there would be no way of cooling, we would experience the total chaos or rather not experience it. Thus, without any doubt it is obvious that cooling and control of cooling is the essential reason for our existence and gives us the opportunity to talk about some of the physics that cooling allows.
There is a very good reason to do experiments with cooling since at least 22 Nobel prices from the 98 times it was awarded since 1901 the 'nobel' research was done using cooling in a direct way. And thus it can make people happy as demonstrated by the famous picture of C. Rubbia and S. van der Meer, shown in figure 1, when awarded the Nobel price in 1984 for their work in physics of weak interaction which certainly only ever was possible with beam cooling.
 The cooling applications can be separated in relative and absolute cooling which are qualitatively not that much different. The relative cooling of particles, which we might define as a process keeping the same mean momentum but reducing ΔP such that the individual particles are running parallel to each other with ε distinction in their velocity. The beam cooling reduces the phase space of the beam around a central momentum P_0 (in the z–direction) with small values of $x, \dot{x}, y, \dot{y}$. Such cooling is needed for several reasons when particles are used to produce secondary particles like the field particles of the weak interaction at CERN, meson production at threshold at COSY or hyperon produc-

FIGURE 1. The Nobel price Laureates Carlo Rubbia and Simon van der Meer at CERN 1984.

tion with antiprotons at LEAR only to name a very limited selection of the thousands of applications used in high energy or hadron physics.

The second cooling phrase concerns the absolute cooling with the central momentum P_0 approaching zero, where a temperature close to the absolute minimum is needed since otherwise the physics just could not be done. Here I would like to only mention the Bose–Einstein–Condensation and to present steps towards the aim of studying properties of antihydrogen in relation to its counterpart the hydrogen atom for investigating CPT symmetry breaking effects.

Thus we are facing during this conference the interesting and exciting fields of beam cooling techniques, necessary to perform experiments which require at least a certain precision. We will learn about intense cooled beams, intra beam scattering, space charge limits and beam stability, cooling applications and phase space manipulation, and especially correlations with cold beams or cold clouds of charged particles.

The different techniques used and applied include electron cooling, stochastic cooling, muon cooling, laser cooling, synchrotron cooling, ion cooling, evaporative cooling, sympathetic cooling, side band cooling, and so on. Here I only can concentrate on a few of them giving some typical examples.

For high energy particles in storage rings both electron and stochastic cooling were proven to really work experimentally in 1974 in Novosibirsk [1] and at the Intersecting Storage Rings (ISR) of CERN [2], respectively. Finally, for the development of the high energy antiproton-proton collider experiments at CERN, the Initial Cooling Experiment (ICE) [3] used the magnets from the g-2 experiment – which later were transferred to the TSL in Uppsala for the CELSIUS ring – and proved extremely successfully the stochastic cooling, as demonstrated by the famous picture shown in the left part of figure 2 where $5 \cdot 10^7$ circulating protons were first spread out by applying noise on an RF-cavity and then the stochastic cooling was turned on. After 4 minutes the phase space volume of the protons shrank and a narrow momentum distribution was seen.

Very recently the first cooling of 8.9 GeV/c $\bar{p}$'s were reported achieved in the recycler ring at Fermilab. This is a very good and trend-setting result for high energy electron cooling in accelerator physics, congratulations.

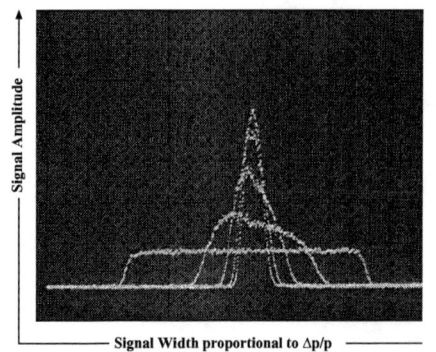
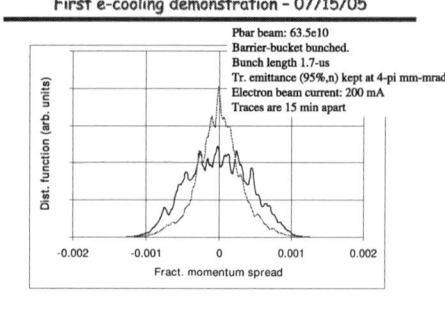

FIGURE 2. Impressive demonstration of stochastic cooling of the proton momentum at the 2 GeV (2.8 GeV/c) ICE ring of CERN in four minutes. Right: Pioneering electron cooling of 8.9 GeV/c $\bar{p}$ in the recycler ring at Fermilab. The two $\bar{p}$ longitudinal (1.75 GHz) Schottky spectra, taken 15 minutes apart, were observed on 07/15/05.

BEAM COOLING AT COSY FOR THE EXPERIMENT COSY–11

COSY–11 is an experiment [4] for medium energy hadron physics at the **Co**oler **S**ynchrotron COSY [5] of the Research Centre Jülich. It uses the internal accelerator beam for precision studies to understand the interaction strengths in the various hadronic systems as well as the production and reaction mechanisms of the creation of mesons with and without strangeness. The COSY–11 collaboration concentrates on

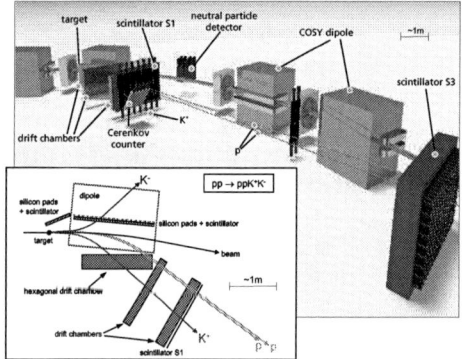
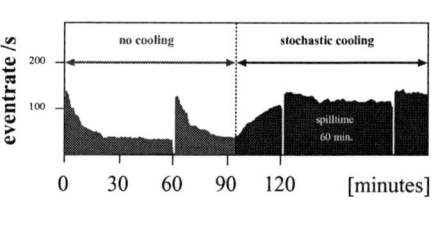

FIGURE 3. Left: Schematical view of the zero-degree spectrometer COSY-11. Right: Luminosity with and without stochastic beam cooling at COSY–11 at a cycle length of 60 minutes.

the threshold region for production of mesons and meson-pairs using a regular internal COSY dipole as a zero degree spectrometer, shown in the left part of figure 3, since at threshold the number of participating partial waves is small and thus the interpretation of reaction mechanisms and production processes gets simplest.

A procedure of beam cooling may have the sequence: injection of the beam into the COSY ring, electron cooling at injection energy, beam ramping to the desired particle momentum, stochastic cooling operated during the measurement. (For the COSY-11 experiments mostly the electron cooling at injection energy is not used.) At the right side of figure 3, the coincident counting rate of the detectors installed for measuring the elastic scattering demonstrate the strong losses within the first 10 minutes of effective beam target interaction without cooling and the impressive improvement of the luminosity when cooling is turned on. Cooling prevents the beam from blowing up and from moving out of the target range, because the COSY-11 target in front of a dipole is located in a dispersive region and the stochastic cooling compensates the energy loss and keeps the beam momentum at a fixed value.

More important for the physics is of course the reduction of the momentum resolution. Beam cooling is essential for threshold measurements as it was demonstrated clearly first by the pioneering IUCF experiments of pion production [6], see figure 4 left part, and is illustrated for the COSY experiments further in figure 4 where it is shown that the

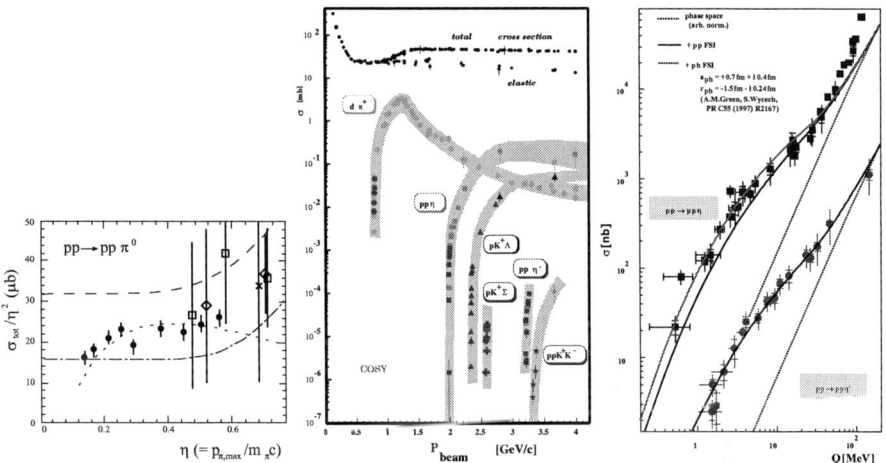

FIGURE 4. Left: Total cross section divided by η^2 for the $pp \to pp\pi^0$ reaction vs η, see Ref. [6] for more details. Center: Excitation function for meson production in $p-p$ scattering close to threshold. Right: Cross sections for the η and η' production in $p-p$ scattering.

cross sections increase drastically within a few Mev above threshold when increasing the beam momentum. Thus the knowledge of the precise beam momentum as well as a small $\Delta p/p$ value are mandatory for such kind of investigations.

As an example of the experimental results [7] figure 5 shows the missing mass spectrum of the $pp \to ppX$ reaction in a range from 0.6 - 1 GeV/c^2 (left side) and in the range close to the mass of the η' meson (right side). A statistically significant sharp signal due to the creation of the η' meson with a mass resolution of less than $1 MeV/c^2$ is clearly seen over the continuous spectrum resulting from multi-pion production. The COSY-11 detection system permits to measure the momentum vectors of both outgoing protons only. In order to calculate the missing mass of an unobserved neutral system it is necessary to know the momentum of the reacting beam protons. The accuracy of

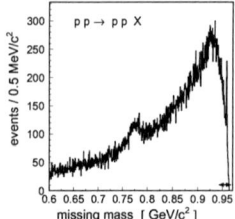 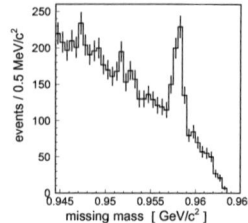

FIGURE 5. Missing mass spectrum of the reaction $pp \to ppX$ and a blow up close to the threshold range of the η production.

the missing mass reconstruction depends on the spread of the beam momentum and on the precision of the measured momentum of the outgoing protons. The latter is predominantly due to the finite dimensions of the interacting region. Thus, both main sources contributing to the inaccuracy of the missing mass determination are due to the geometrical and momentum spread of the proton beam. That is why beam cooling is of great importance for high resolution meson production studies close to threshold. In this special example of η' production the beam cooling resulted in a reduction of $\Delta p/p$ by a factor of 5. A smearing of the η' signal by a factor 5 would give a barely visible bump instead of a clear peak.

COOLING OF CHARGED PARTICLE CLOUDS FOR THE PRODUCTION OF ANTIHYDROGEN

Fortunately, accelerator physicists need the H^0 diagnostics in electron cooling experiments with proton beams. This circumstance and the circumspection of our colleagues H. Herr, D. Möhl and A. Winnacker [8], who thought already in 1982 about the antihydrogen production in a system of positron–antiproton interaction, similar to the electron cooling, allowed about ten years ago for the first observation of a few atoms of antihydrogen at LEAR [9], see left part of figure 6. The main success of this experiment was the possibility to continue after the closing of LEAR physics with low energy antiprotons at AD/CERN. These first antimatter atoms with velocities of about 95% of the speed of light should be considered as a proof of principle to synthesize antihydrogen because high precision spectroscopic studies need cold antihydrogen where trap experiments are the right tools. The fraction of antihydrogen atoms trappable as a function of trap depth and temperature is shown in the right part of figure 6 which demonstrates drastically the need of very cold antihydrogen atoms to be useful for studying its properties. With superconducting magnets field gradients of about 1 Tesla are achievable corresponding to trap depths of 0.67 K.

For the trapping of ions as well as the production and trapping of antimatter atoms electro–magnetic traps (penning traps, nested traps, Ioffe traps) are commonly used which act as very small but still macroscopic storage rings.

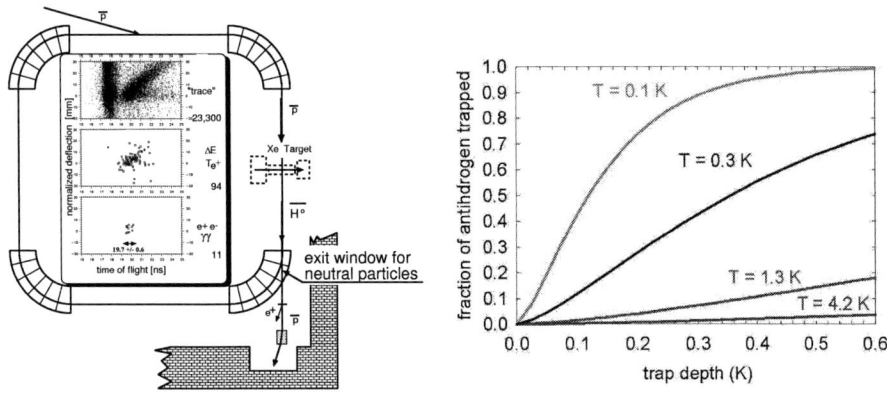

FIGURE 6. Left: Layout of the experiment which first observed the production of a few antihydrogen atoms, with the exit window which was already installed due to suggestions described in [8] Right: Trapping efficiency as a function of trap depth in K and temperature.

To trap antiprotons an energy reduction from the production energy by several orders of magnitude is needed. Figure 7 shows the AD preparation sequence for an extracted antiproton beam. The antiprotons are injected into the AD at 3.5 GeV/c and decelerated to 100 MeV/c with stochastic cooling at 2 GeV/c and electron cooling at 300 MeV/c and 100 MeV/c to minimize the beam losses. Each 85 seconds, close to the design value of 60 seconds) a bunch of fast extracted $\bar{p}$'s ($\approx 3 \times 10^7 \bar{p}$'s in ≈ 80 ns) is delivered to the experiments.

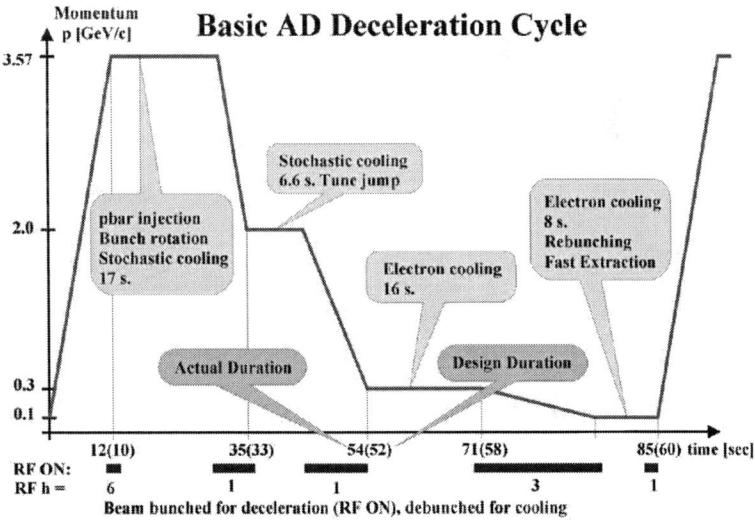

FIGURE 7. Cycle of the antiproton preparation for the experiments at the AD.

In the next step the antiprotons enter into the trap via a degrader of about 100 μ Beryllium to reduce the 5 MeV particles to energies which are optimized for trapping them in a 4 keV potential. This step includes no cooling mechanism and a very large amount of $\bar{p}$'s is lost due to energy smearing and mutiple scattering and via annihilation in the degrader material. Only up to about 20 000 $\bar{p}$'s i.e. less than 0.1 % can be trapped. That is why the community at the AD is preparing a proposal for a further deceleration ring ELENA [10], which is depicted in figure 8, to minimize such losses. If ever reality such a ring should be placed between the AD and the experiments.

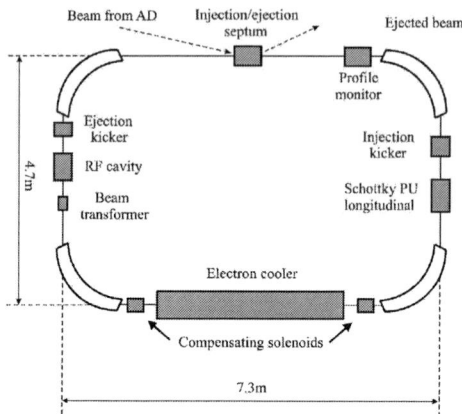

FIGURE 8. Proposed layout of the ELENA ring to be placed between AD and the experiments.

As a first step towards $\bar{H}^0$ production antiprotons are captured and cooled with electrons within the Penning trap, where a 5.4 Tesla magnetic field is directed along the vertical axis of a stack of copper ring electrodes cooled to liquid helium temperature. When a bunch of $\bar{p}$'s from the AD enters the trap is opened by grounding the degrader as indicated in figure 9 (right). Before the $\bar{p}$'s can return to the entrance, reflected from the potential barrier at the top of the trap, the potential at the window degrader is turned on to negative potential and thus the $\bar{p}$'s are trapped.

The cyclotron motion of electrons cools via spontaneous emission of synchrotron radiation with a 0.1 s time constant and therefore for electron cooling of the $\bar{p}$'s electrons were loaded before in short wells. After the trapped $\bar{p}$'s are cooled into the electron wells the potential barrier at the degrader is reduced again for the next injection of a stack of new $\bar{p}$'s from the AD. The left hand side of figure 9 [11] demonstrates that a large number of $\bar{p}$'s can be accumulated by this stacking method showing a linear increase with the number of $\bar{p}$ bunches injected, here up to 32.

After the required number of $\bar{p}$'s are trapped the electrons are released from the trap and the $\bar{p}$'s are launched through a positron cloud in a nested Penning trap configuration which cool the $\bar{p}$'s like the electrons.

Figure 9 shows a systematic study of the time dependence of $\bar{p}$ cooling with positrons [12] in a potential structure as given in the right part of the figure. When the two species $\bar{p}$ and e^+ were interacting, antihydrogen atoms were produced routinely at ATHENA [13] and ATRAP [14] during the last two years.

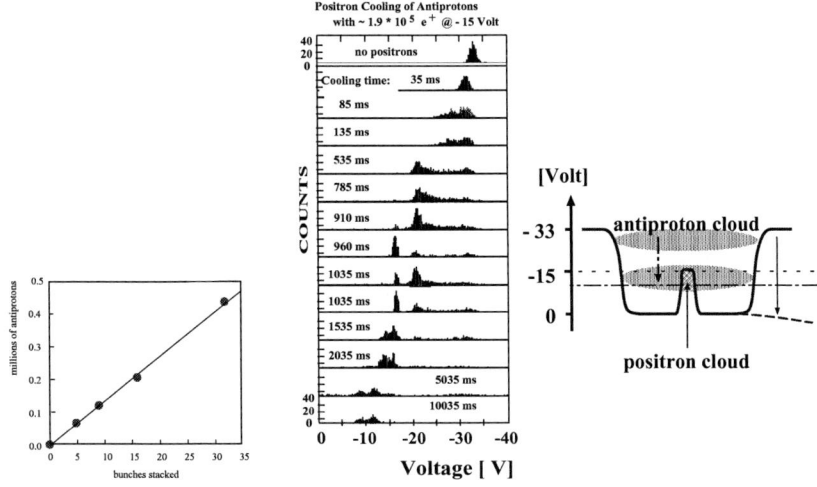

FIGURE 9. Left: Linear increase of trapped $\bar{p}$'s with number of stacked bunches. Right: Time dependence of $\bar{p}$ cooling with positrons.

Detailed studies [15] concerning shape parameters of the antiproton ($\bar{p}$) and positron (e^+) clouds, N-state distribution of the produced Rydberg $\bar{H}^0$ and $\bar{H}^0$ velocity have been performed to improve the production efficiency of useful $\bar{H}^0$ atoms.

As schown in figure 10, the velocity measurement [16] indicates a kinetic energy of the $\bar{H}$ atoms of $\approx$ 200 meV (2400 K in temperature units) which is much too high for $\bar{H}$ trapping experiments and has to be reduced. Studies in this direction were performed and will be continued with the next beam time.

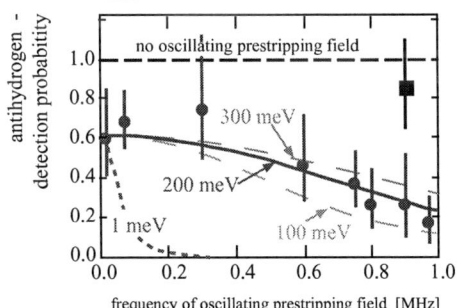

FIGURE 10. The fraction of antihydrogen atoms detected in a detection well decreases as the frequency of an oscillating electric field, which ionizes the Rydberg antihydrogen atom, is increased.

ATRAP demonstrated also a second, entirely new method for producing slow $\bar{H}^0$ atoms [17]. Two lasers excite Cs atoms to high Rydberg states Cs^*. Two resonant charge exchange collisions [18] transfer the laser–selected Cs^* binding energy to an excited positronium – (Ps^*) and then to an excited $\bar{H}^{0*}$ atom. Very slow $\bar{H}^{0*}$ atoms are expected since a Ps^* transfers little kinetic energy to a $\bar{p}$ (which is at liquid He temperature) as $\bar{H}^{0*}$ forms. As soon as $\bar{H}^0$ are trapped another cooling mechanism, the laser cooling, will be applied to further reduce the velocity of the antihydrogen atoms.

APPLICATION AND OUTLOOK

Finally figure 11 shows an impressive application of a technique which has in principle been developed for stochastic cooling purposes and which was installed at the unique combination of the fragment separator and the storage–cooler ring facilities at GSI/Darmstadt, see reference [19].

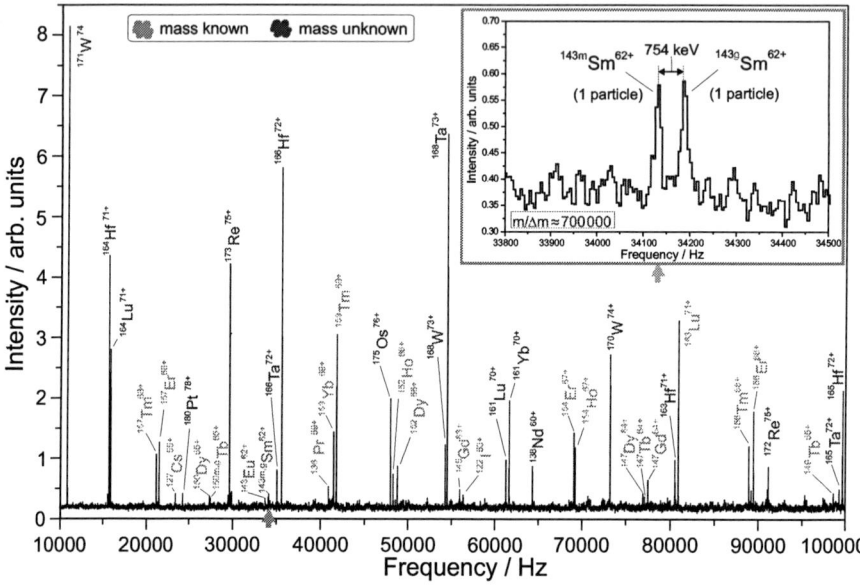

FIGURE 11. Example of a Schottky frequence spectrum with lage bandwidth. The prominent lines are labelled.

Here masses were measured with time-resolved Schottky mass spectrometry and many masses were obtained for the first time. A typical mass accuracy of 30 μu was achieved. With the new developed technique it became possible to trace the properties of stored ions in time. The frequency-peak shifts due to Coulomb interaction between the ions with similar mass-to-charge ratios could be detected and analyzed. This method proved to be so successful that a new generation of Schottky mass measurements will be opened up with the construction of the new Super-fragment-separator [20] and the new CR-RESR-NESR [21] ring complex within the FAIR project [22].

I hope I could demonstrate that cooling is a very essential tool in all kind of physics, here shown for the production experiments in hadron physics serving precision and intensity as well as for trapping the very expensive antihydrogen atoms. Since traps with a magnetic gradient are rather shallow, extremely cold atoms are needed. Laser cooling of hydrogen atoms in a magnetic trap has been done using a pulsed Lyman–α source [23]. A continuous source presently under development at the MPQ in Garching should be favourable, since it will provide a larger rate for laser cooling. Experiments to measure antimatter gravity can most likely only be done with temperatures < mK. In ref [24, 25] novel schemes have been outlined to cool antihydrogen atoms at these ultra low temperatures.

There certainly is an exciting future ahead of us, now in short terms during this conference and on a longer time scale in physics that cooling allows.

ACKNOWLEDGMENTS

I would like to thank the organizers of the COOL–05 conference for giving me the opportunity to participate in this exciting event.

The help of many people in performing our experiments is deeply acknowledged.

I would like to thank Dieter Möhl for inspiring discussions and Dieter Grzonka for a careful reading of the manuscript and many helpful suggestions.

REFERENCES

1. G.I. Budker, et al., IEEE Trans. Nucl. Sci. **NS-22** (1975) 2093.
2. P. Bramham, et al., Nucl. Inst. and Meth. **125** (1975) 201.
3. G. Carron, et al., Phys. Lett. **B 77** (1978) 353.
4. S. Brauksiepe et al., Nucl. Inst. and Meth. **A376** (1996) 397.
5. R. Maier et al., Nucl. Inst. and Meth. **A390** (1997) 1.
6. H. O. Meyer etal., Phys. Rev. Lett. **65** (1990) 2846-2849.
7. P. Moskal et al., Phys. Lett. **B 474** (2000) 416.
8. H. Herr, D. Möhl and A. Winnacker, Proc. 2nd Workshop on Physics at LEAR, Erice (1982) 659.
9. G. Baur et al., Phys. Lett. **B 368** (1996) 251.
10. P. Belochitskii, Talk presented at the Villars Meeting, Fall 2004.
11. G. Gabrielse, Phys. Lett. **B 548** (2002) 140.
12. G. Gabrielse, et al., Phys. Lett., **B 507** (2001) 1.
13. http://athena.web.cern.ch/athena
14. http://hussle.harvard.edu/~atrap
15. G. Gabrielse, et al., Phys. Rev. Lett., **89 (2002)** 213401.
    G. Gabrielse, et al., Phys. Rev. Lett., **89 (2002)** 233401.
16. G. Gabrielse, et al., Phys. Rev. Lett., **93** (2004) 073401.
17. C.H. Storry, et al., Phys. Rev. Lett., **93** (2004) 263401.
18. E.A. Hessels et al., Phys. Rev. A **57** (1998) 1668.
19. Yu.A. Litvinov et al., Nucl. Phys. **A756** (2005) 3.
20. H Geissel et al., Nucl. Inst. & Meth, **B204** (2003) 71.
21. P. Beller et al., Proceedins EPAC, (2004) 1174.
22. see homepage: http://www.gsi.de/GSI-Future/CDR.
23. I.D. Setija, et al., Phys. Rev. Lett. **70** (1993) 2257.
24. J. Walz and A. Kellerbauer, Proceedings LEAP-05, Bonn May 2005, to be published by AIP.
25. J. Walz and T.W. Hänsch, General Relativity and Gravitation, **36** (2004) 361.

Overview of Recent Trends in Beam Cooling Methods and Technology

Igor Meshkov[1] and Dieter Möhl[2]

[1]*JINR, Dubna, Russia*
[2]*CERN, Geneva, Switzerland*

Abstract. In this introductory paper, we try to give an idea of new developments in beam cooling since COOL03. We will concentrate on trends in electron cooling, stochastic cooling, muon cooling and beam crystallization; trends, which we think, will mark the future. We hope to touch upon some of the major ideas and topics that will be developed in detail at this workshop.

Keywords: particle storage rings, cooling methods, electron beam, Schottky noise.
PACS: 29.20.C, 29.20.Dh; 29.27.Bd

INTRODUCTION

A variety of remarkable events in the field of beam cooling have occurred since our last workshop (COOL03) in May 2003:
- The first demonstration of electron cooling at intermediate energy (8 GeV antiprotons in the FERMILAB recycler) [1].
- The start of commissioning of two state-of-the-art low energy electron coolers (LANZHOU and LEIR) built in Novosibirsk [2].
- The commissioning of the storage ring LEPTA aimed for "electron cooling all around" (at JINR Dubna) [3].
- The start of the construction of a special dispersion-free ring for laser cooling/ beam ordering experiments (at Kyoto University) [4].
- The approval of the international Muon Ionisation Cooling Experiment MICE (at Rutherford Appleton Laboratory) [5].

In addition there has been considerable advancement both in the understanding and the scope of beam cooling:
- An international effort has lead to a big step forwards in the conception, modeling, benchmarking and hardware design for various medium and high-energy (both stochastic and electron) coolers (e.g. for RHIC, FAIR, TEVATRON...) [6, 7].
- New proposals have emerged for the use of cooled beams (e.g. very small aperture machines for medical and particle physics applications) [8].
- There has been great progress concerning the conditions for and the potential use of ordered (crystalline) beams.

All this, and much more we believe, has been presented in detail at this Workshop. In this introductory talk, we try to give a preview of the trends and refer to the presentations at the Workshop.

TRENDS IN ELECTRON COOLING

Medium and High Energy

All proposals (FERMILAB, RHIC, FAIR ...) are based on a very long interaction region (15–20 m). To sustain low temperature, the electrons are accelerated in a linear device all the way from the cathode to the interaction energy. They make a single traversal of the cooling region and are then decelerated to recuperate their energy. Thus the arrangement is similar to low emittance linacs with energy recovery as proposed e.g. for advanced synchrotron light sources.

Different schemes of acceleration/deceleration have been proposed using either electrostatic (continuous beam) or RF (bunched beam) acceleration. The FERMILAB scheme ([1]) uses a pelletron high voltage device to generate the 4.3 MeV electrostatic acceleration potential. The BNL proposal for RHIC ([9]) is based on an linac with electron bunches matching the RHIC bunch structure. A Novosibirsk proposal [10] for FLAIR uses a proton or H⁻ beam from a cyclotron to charge up a high voltage platform. A question of particular importance is the magnetisation (i.e. the immersion into a strong longitudinal B-field) of the electron beam. This can increase the cooling speed, ideally without augmenting, in the case of heavy ions, the electron-ion recombination rate.

Obviously the generation of a strong magnetic field along the orbit in a high voltage device is a challenge. In fact the FERMILAB device goes without strong magnetisation whereas in the FAIR proposal it is an essential ingredient. The Novosibirsk team proposes to solve the problem using isolated multiple coils fed by individual generators on high voltage ([10]). The technology will be tested in a new cooler at COSY ([11])

The basis of non-magnetized medium energy cooling has been demonstrated by the pioneering work at FERMILAB. In the future we will see much work directed towards magnetized electron cooling at medium and high energy.

Low-Energy

The design for low energy (2–200 KeV electron energy) has only relatively little changed, evolving from the pioneering Novosibirsk concept in the early 1970s. After the construction and the use of more than a dozen of coolers all over the world, the latest generation (Lanzhou, LEIR, [2, 12, 13]) has the following new features:

1) Very precise magnetic field with a great number of trim coils (to allow fast cooling);

2) "Hollow" e-beam (to avoid "overcooling" in center and also to reduce ion-electron recombination). However ion-electron instability may become especially critical due to non-linear beam-beam tuneshift;

3) Electrostatic bends (to reduce trapping of secondary particles);

4) Magnetic expansion (to adjust beam size);

5) Magnetisation of an e-beam of a relatively high transverse "temperature" (to decouple "temperature" for cooling from "temperature" for ion-electron recombination);

6) High perveance together with elaborate measures to stabilize the beams (to reach fast cooling). As an example: a beam of 1 A at 25 KeV (45 MeV proton energy) was stably reached at COSY, in August 2005 ([14]).

Such cooling devices can now be bought "off the shelf" (from Novosibirsk). However in the past, every new design has had its own surprises. Results from Lanzhou and LEIR with the new state-of-the-art coolers are therefore impatiently awaited.

Very Low-Energy

The ELENA and FLAIR [15] proposals of post-deceleration/cooling rings after the AD and the FAIR–RESR [16] respectively require efficient cooling of antiprotons with an energy as low as 100 KeV ($v/c \approx 1.5 \cdot 10^{-2}$). The FLAIR project [15] calls for cooling also of ions with very low velocity. In addition cooling rings (using electrostatic bending and focusing) for molecules with v/c in the few percent range have been constructed or are being planned [17, 18, 19]. Cooling at such low velocities (with electron beams of an energy as low as ~50 eV) poses new problems.

One challenge is the ultra low temperature (in the order of a few milli electron Volt) required. Solutions proposed rely on a "cold" photo cathode [20]. However at present these are capable to deliver only a relatively low current. Magnetisation (to have low effective transverse temperature) is another perhaps additional way. But then, at the low energy, the magnetic field presents a strong perturbation of the ring's optics. Expansion can lower the transverse temperature but it also reduces the current density. In summary: there is a conspiracy of conflicting requirements.

There are other problems specific to the ultra low energy of both the ion and the cooling beam: instabilities, space-charge, intra-beam and gas scattering... to mention only a few. On the positive side: the energy contained in the cooling beam is low so that recovery is probably not required.

Intense research and probably new ideas are required to arrive at a good design of the very low-energy coolers.

STOCHASTIC COOLING

There is advance concerning high energy stochastic cooling in order to extend the luminosity in heavy ion colliders.

Bunched beam "Schottky noise" studies at RHIC are well progressing [21, 22]. The coherent component of the signal at 4–8 GHz ("flag pole on the Schottky hill", Fig. 1) is attributed to the bunch shape plus intra-beam scattering. It is much less violent than the effects observed some time ago in the $\overline{\text{SPPS}}$ and the TEVATRON. Although special measures are necessary, the situation seems manageable for cooling of e.g. gold beams to extend the luminosity lifetime. The power problem can be solved by using an array of high Q (~1000) cavities stagger tuned over the band (4–8 GHz).

Ideas to combine high-energy electron (core) cooling with stochastic (halo) cooling, developed already for low energy at LEAR are being discussed for the High energy Experimental Storage Ring (HESR) planned at GSI Darmstadt [22].

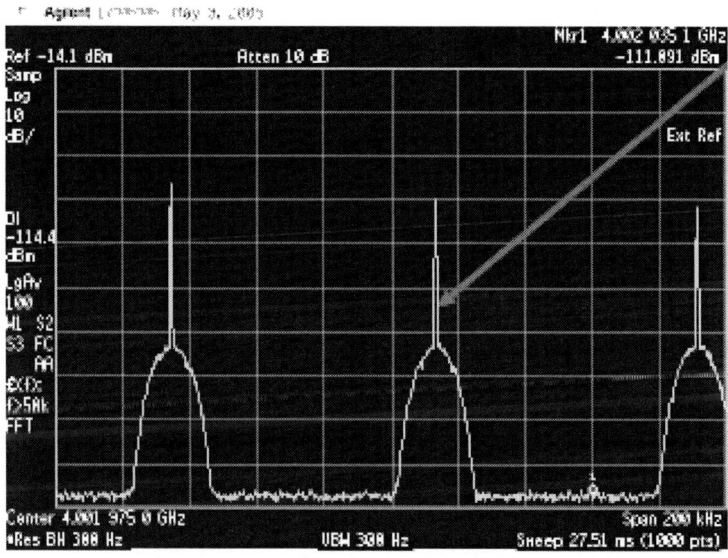

FIGURE 1. Spectrum at 4 GHz of proton bunches in RHIC

STABILITY OF COOLED BEAMS

Electron Cooling

An ion beam in an electron cooler storage ring can suffer from unwanted influences of the cooling beam and from other storage ring coupling impedances. Effects observed include [23]:

- nonlinear lens ("beam-beam") effects of the electron beam leading to ion loss or diffusion (LEAR);
- ion loss at injection (COSY) when the ion beam is larger than the electron beam size (probably similar to the beam-beam effect);
- instability development in a well cooled high intensity ion beam due to interaction with the electron beam "electron heating" (CELSIUS, COSY, HIMAC), probably similar or identical to the "beam-beam" effect;
- "three-body" instability when secondary ions are trapped in the e-beam (LEAR, HIMAC, COSY);
- strong interaction of a well cooled ion beam with parasitic elements in the ring (LEAR, COSY).

A test of the ion beam stability with the novel hollow e-beam coolers will be an important issue. Moreover for the new high-energy coolers, a "transition energy problem" arises:

Longitudinal stability is most critical above transition energy due to the "negative mass effect". Up to now, all electron-cooling rings have operated naturally below transition, whereas the new high-energy coolers have to work above transition. A recent experiment at the ESR (Darmstadt) tuned to $\gamma > \gamma_{tr}$ showed, that e-cooling is possible in this regime but with larger equilibrium spread than below γ_{tr}. In the old Initial Cooling Experiment at CERN, e-cooling above transition was not at all achieved. It will be important, to determine the density limits in the high-energy cooling rings.

Instability Antidotes, Electron Cooling

The beam environment in the cooling ring has to be carefully controlled. Moreover a rather wideband feed back system has proven efficient and necessary to damp coherent instability (LEAR, COSY...). But it does not cure incoherent effects [22]. Therefore (once again): a test of the ion beam stability with the new hollow e-beam coolers will be an important issue!

Stability of Stochastically Cooled Beams

The stability of the stochastically cooled beam in the presence of the usual coupling impedances is basically the same as for any cooling ring that produces dense beams (reduced Landau damping). In addition to this, the stochastic cooling system itself acts as a large "beam coupling impedance".

Beam stability is especially critical in accumulator rings of antiprotons or rare ions, where large stacks (10^{11}–10^{12} particles) have to co-exist with small injected batches ($\sim 10^8$ particles).

Fast cooling and stacking of the injected batch in the presence of the stack requires partial aperture pick-ups and large separation (at the PU-s) of injection and stack orbits by momentum spread and dispersion. Non-dispersive separation by momentum spread (frequency) alone is insufficient if one wants to stack a big number of batches. The classical solution (CERN and FNAL) of a large dispersion ring is expensive and cumbersome. Therefore a revisit of the stacking problem is indicated (see [24]).

THEORY AND SIMULATION

Future projects – RHIC, FAIR, ELENA, FLAIR… – do need efficient tools for numerical simulation of the beam dynamics in the cooler ring. A significant progress was achieved since COOL'03 in the development of the BETACOOL code [6] and its benchmarking ([7]). A lot of work was done at Erlangen [25] and Novosibirsk [2] to establish equations describing the cooling force that can go into BETACOOL. The code can be used to simulate a great variety of different processes in a beam circulating in a storage ring with a given lattice: electron cooling with an intense electron beam (space charge effects), intra-beam and residual gas scattering, the influence of an internal target etc.. Recently the code was extended for the simulation of stochastic cooling [22]. Alternative codes are also being developed at other laboratories. In this context, international collaboration proves extremely fruitful.

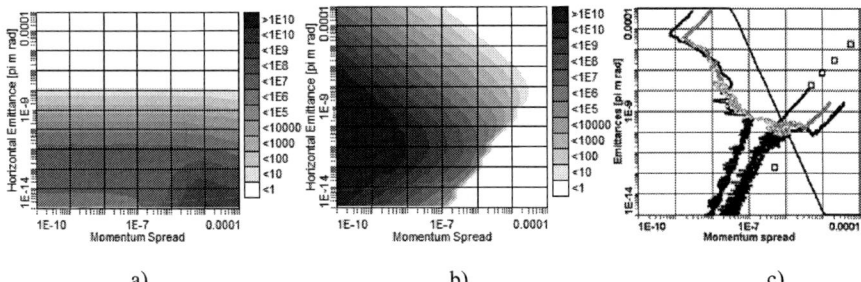

FIGURE 2. An example of results from BETACOOL: evolution of an ion beam of $_{238}U^{92+}$ at 400 MeV/u in ESR undergone by IBS (theoretical dependence by generalized Piwinski model) and electron cooling, (a, b) – the IBS growth rates vs beam emittance and momentum spread: a) – horizontal component, b) – longitudinal one; c) – the results of Molecular Dynamics calculations: evolution of ion beam parameters during electron cooling, solid black line corresponds to cooling rate $4 \cdot 10^4$ s^{-1}, grey circles – 10^4 s^{-1}, square points – ESR experiment. One can see a dependence of "the beam trajectory" on initial parameters ("The Smirnov's effect"[6])

MUON COOLING

The International Scoping Study

A lot of development towards a neutrino factory and the muon beam cooling required for it has been going on recently [5, 26, 27]

An international scoping study of a Neutrino Factory and Super-Beam Facility was launched in spring 2005. In the "Executive summary" [28] it is stated that] "……. The principal objective [of the scoping study]   … will be to lay …foundations for a …conceptual design study of the facility. The … study has been prepared … by the international community …: the ECFA/BENE network in Europe, the Japanese NuFact-J collaboration, the US Muon Collider and Neutrino Factory Collaboration and the UK Neutrino Factory Collaboration. …Rutherford Appleton Laboratory will be the "host laboratory" for the study…Highlights of this programme include the international Muon Ionisation Cooling Experiment (MICE)… which has been approved at the Rutherford Appleton Laboratory (RAL) …It will begin taking data in 2007 with beam from ISIS (RAL)".

The MICE Experiment

The MICE Collaboration [5, 29] includes more than 40 institutions from Belgium, Italy, Japan, Netherlands, Russia, Switzerland, UK, US (spanning 17 hours in time zones): (Louvaine, Bari, Frascati, Genoa, Legnaro, Milano, Napoli, Padova, Roma, Trieste, KEK, Osaka, NIKHEF, BINP, CERN, Geneva University, PSI, Brunel, Daresbury, Edinburgh, Glasgow, Imperial, Liverpool, Oxford, RAL, Sheffield, ANL, BNL, Chicago, Fairfield, Fermilab, IIT, Iowa, Jlab, NIU, UCLA, LBNL, Mississippi, Riverside, UIUC).

The experiment
- aims to show that it's possible to design, engineer and build a section of the ionization cooling channel capable of giving the desired performance for a Neutrino Factory;
- plans to place this section in a muon beam investigating the limits and practicality of ionization cooling.

MICE Status and What We Can Learn from It

MICE is
- accepted as an official (UK) project at RAL;
- funding for the beamline/infrastructure and the tracker come from RAL;
- important hardware and study contributions come from US (MUCOOL collaboration);
- …..
- first beam is expected April 1, 2007

The MICE collaboration has been very successful in getting contributions from many different funding agencies! Great progress has also been made in modeling ionization cooling and emittance measurement. All this can serve as example for the internationalization ("globalization") of other cooling projects.

Muon Cooling Rings

Basically the μ-cooling channel consists of linear accelerator sections interlaced with liquid hydrogen absorbers. This channel has to be fairly long and is expensive. Hence the interest in "ring coolers" [30], where cooling is done over many revolutions We conclude that muon cooling RINGS should/will attract more study!

BEAM ORDERING

The experimental observation in the 1970s of Schottky noise suppression in a cooled proton beam by V. Parkhomchuk et al. inspired a lot of enthusiasm on "crystal beams".

The excitement continues but was somewhat damped in the 1990s when it became clear (due to the work of A. Sessler, G. Wei, H. Okamoto, A. Ruggiero and many others) that 3D crystallisation is subject to a set of tough conditions that cannot be met in existing storage rings.

The observation of 1D ordering by M. Steck and co-workers at GSI in 1996 and its theoretical explanation by the "two-particle model" of R. Hasse – has led to a new boom of interest in beam crystallization in storage rings .

The proposal to use a 1D chain in an ion-electron collider presents a first attractive particle physics application showing the potential of ordered beams.

Concerning 1D ion ordering observed at GSI and (later) also at CRYRING one should mention the problem of proton beam ordering. Recent experiments at COSY demonstrate a saturation of the Schottky noise signal at a level of $\Delta p/p \sim 2 \cdot 10^{-6}$, but a "phase transition jump" (like at NAP-M in Novosibirsk) was not observed [6]. This is to be confronted with molecular dynamics simulation programs [6] which explain the COSY but not the NAP-M results.

The understanding of the optics and the cooling required for beam ordering is progressing due to the work of an "international network of enthusiasts" (including A. Sessler, J. Wei, H. Okamoto, T. Katayama, A. Noda, R. Hasse, A. Sidorin, A. Smirnov and the present authors).

The idea of a dispersionless ring [4, 31, 32] (to avoid "shear" and, related to it "tapered cooling") removes a big tumbling stone from the road to 3D ordered beams.

Beam Ordering Experiments

The RF quadrupole ring PALLAS [33] has shown possibilities and limits of 3D crystal beams at very low energy ($v/c \approx 10^{-5}$). "Shearing forces" in the bends were identified as one major obstacle to 2D and 3D crystals at higher beam energy.

A new ring S–LSR (conceived by a collaboration of the ICR, Kyoto University and the Japanese Institute of Radiological Sciences) with electrostatic and magnetic bends is under commissioning [4, 31, 32]. It can run in a "dispersion free" mode where shear forces are (to first order) absent. For this the bending fields (Fig. 3) have to satisfy the condition:

$$\left(1+1/\gamma^2\right)\cdot E_r = -v\cdot B_z$$

S–LSR can work dispersion free with Mg^+ ions up to $v/c \approx 2\cdot 10^{-3}$ (E $\approx$ 1.5 KeV/nucleon). At higher energy the bending voltage (~ twice the voltage for pure electrostatic bending, see equation above) gets excessive.

S-LSR type rings with mixed electric and magnetic bending, when tuned to have zero linear dispersion have large transition energy, $\gamma_{tr} \Rightarrow \infty$. For high energy one can think of a purely magnetic lattice with $\gamma_{tr} = \infty$ where the dispersion is negative in part of the magnets and $\gamma_{tr}^{-2} = (1/C)\cdot \int D/\rho\, ds = 0$. Is such an "on average shear-less ring" well suited for crystallisation?

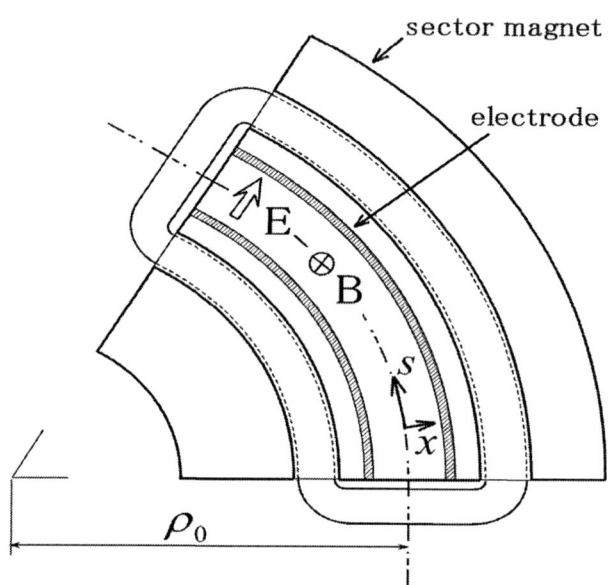

FIGURE 3. S-LSR magnet with electric and magnetic bending, from [32]

One concludes that a lot of fascinating experimental and theoretical work is going on towards crystal beams in general and at higher energy in special.

CONCLUSION

There is a surprising diversity of new and very exiting developments in the – by now mature – field of beam cooling. You will learn much more about them during this workshop. This talk was meant to be an "appetizer". The selection is unavoidably incomplete and biased. We apologize for that.

ACKNOWLEGEMENT

We thank Sergei Nagaitsev and all our other friends and colleagues from Fermilab, for the invitation and the support.

REFERENCES

1. S. Nagaitsev, Antiproton Cooling in the Fermilab Recycler, These Proceedings
2. V. Parkhomchuk, Development of a New Generation of Coolers with a Hollow Electron Beam and Electrostatic Bending; Commissioning of Electron Cooler EC-300; Commissioning of Electron Cooler EC-300, ibid.
3. I. Seleznev, Status of LEPTA Project, ibid.
4. A. Noda, Laser Cooling for 3-D Crystalline State at S-LSR, ibid.
5. D. Kaplan, MICE: the International Ionization Cooling Experiment, ibid.
6. A. Smirnov, Simulation of Beam Dynamics in Cooler Rings, ibid.
7. A. Fedotov, Detailed Studies of Electron Cooling Friction Force; Experimental Benchmarking of the Magnetized Friction Force, ibid.
8. V. Vostrikov, Electron Cooling for Cold Beam Synchrotron for Cancer Therapy, ibid.
9. I. Ben-Zvi, High-Current ERL-Based Electron Cooling System for RHIC, ibid.
10. V. Reva, Budker INP Proposals for HESR and COSY Electron Cooling System, ibid.
11. J. Dietrich, COSY 2-MeV cooling system proposal, ibid.
12. X. Yang, Commissioning of HIREL-CSR and its Electron Coolers, ibid.
13. G. Traquille, LEIR Cooler Status, ibid.
14. D. Prasuhn, Recent and Future Cooling Experiments at COSY, ibid.
15. C. Welsch, Cooling Rates of the USR as Calculated with BETACOOL, ibid.
16. M. Steck, The FAIR Project, ibid.
17. T. Azuma, An LN2-Cooled Electrostatic Ring, ibid.
18. H. Danared, DESIREE – A Double Electrostatic Storage Ring for Merged-Beam Experiments, ibid.
19. A. Wolf, The Heidelberg CSR: Low Energy Ion Beams in a Cryogenic Electrostatic Storage Ring, ibid.
20. D. Orlov, Ultra-cold Electron Target for the Heidelberg TSR, ibid.
21. M. Brennan and M. Blaskiewicz, www.c-ad.bnl.gov/RHIC/retreat2005/presentations/Jun17_AM/Stochstic%20Cooling.pdf; Bunched-Beam Stochastic Cooling for RHIC; Bunched Beam Stochastic Cooling and Coherent Lines, These Proceedings.
22. H. Stockhorst, Cooling Scenario for the HESR Complex, ibid.
23. I. Meshkov, Electron Cooling of Intensive Ion Beam, ibid.
24. T. Katayama, Stacking of 3 GeV pbar with Moving Barrier Bucket Method at GSY-RSER, ibid.
25. G. Zwicknagel, Electron Cooling of Highly Charged Ions in Traps, ibid.
26. R. Palmer, Review of Muon Cooling Development, ibid.
27. R. Johnson, Recent Innovations in Muon Beam Cooling, ibid.
28. http://hepunx.rl.ac.uk/uknf/wp4/scoping/
29. http://mice.iit.edu
30. A. Garren, 6D Cooling of a Circulating Muon Beam, These Proceedings.
31. T. Shirai, S-LSR Cooler Ring Development at Kyoto University, ibid.
32. A. Noda et al., Phys. Rev. Spec. Top. Acc and Beams 7, 120101, (2004)
33. U. Schramm, Laser Cooling for 3-D Crystalline State at S-LSR, These Proceedings.

REPORTS ON OPERATING AND NEW PROJECTS

Status of the FAIR Facility

M. Steck

GSI, Planckstr. 1, D-64291 Darmstadt, Germany

Abstract.
   A new accelerator Facility for Research with Antiproton and Ion Beams (FAIR) has been proposed by GSI. This contribution will give a short overview of the present status of the accelerator complex design with particular emphasis on the installations which employ cooling devices. Stochastic and electron cooling are foreseen in the different storage rings over a large energy range. Stochastic cooling will be applied as first cooling stage to secondary beams of rare isotopes which are produced by projectile fragmentation of heavy ions and to antiprotons which are produced from a high intensity proton beam. In the storage rings which are predominantly used for experiments the highest beam quality of the stored beam will be achieved by electron cooling.

Keywords: heavy ions, antiprotons, beam cooling, intrabeam scattering
PACS: 29.20.Dh, 29.27.Bd, 41.75.Ak

INTRODUCTION

The FAIR (Facility for Antiproton and Ion Research) project is an accelerator project which was originally proposed by GSI [1]. It is based on the use of the existing GSI accelerator complex [2] as injector. After positive evaluation by the 'Wissenschaftsrat', the science advisory committee to the German federal government, it entered into a phase of planning in collaboration with European partners. The German government approved the construction of the facility, requesting the fulfillment of two conditions, first the presentation of a technical plan for a staged construction of the project, and second a contribution of 25 % of the total cost from international partners. GSI is presently elaborating the definition of a cost effective modification of the original proposal together with international partners in order to fulfill the preconditions defined by the German government.

BASIC ACCELERATOR GOALS

The new facility has two major goals: the production of high intensity primary beams of ions ranging from protons to uranium, the manipulation of the secondary beams to achieve the highest beam quality. The secondary beams are antiprotons produced by bombardment of a solid or liquid target with a high intensity proton beam and rare isotope beams produced by projectile fragmentation of heavy ion beams. The secondary beam production comprises highly efficient separation and transport of the secondary beams from the target to dedicated cooling rings. For rare isotope beams, in addition to the injection into storage rings, also the use of a fixed target experimental area is foreseen with a low and a high energy branch. The high intensity heavy ion beams can be used directly for experiments with increased intensities (up to three orders of magnitude) and energies considerably exceeding the values provided by the existing GSI facility. The primary heavy ion beam can be directed to fixed target set-ups or can be transported directly to a storage ring for experiments with stored cooled ions.

SCIENTIFIC GOALS

The new FAIR accelerator facility will provide beam parameters which were not accessible at existing facilities and which open new experimental possibilities for various fields of physics. Among them the following fields have been identified to form the foundations of the research program at FAIR [1]. These are studies of nuclear matter far from stability with rare isotope beams, studies of hadronic matter with antiproton beams, studies of compressed dense hadronic matter in nucleus-nucleus collisions at high energy, the study of high density plasma, and the Quantum Electrodynamics in extremely strong fields.

The FAIR facility offers much higher primary beam intensities and optimum primary beam energy for the generation of rare isotopes. With the new isotope separator SuperFRS optimum use of the primary beams can be made. High intensity proton beams are employed in the efficient conversion to antiprotons. Both secondary beam species benefit from powerful stochastic pre-cooling and the availability of electron cooling for experiments with tremendous reduction of the six-dimensional phase space volume and consequently excellent resolution and accuracy for experiments. Fast cooling will make larger quantities of secondary particles available for experiments.

A high degree of parallel operation is foreseen in the FAIR accelerator complex. The beams from different ion sources can be accelerated and transferred to different experiments on a pulse to pulse basis. Every acceleration cycle can be programmed individually. A high efficiency in the operation of the facility is achieved this way, e.g. in parallel to antiproton production, fixed target experiments with heavy ions can be performed.

THE EXISTING GSI FACILITY AS INJECTOR CHAIN FOR FAIR

The UNILAC linear accelerator for heavy ions [3] and the heavy ion synchrotron SIS18 [4] will serve as a versatile pre-accelerator chain for the new FAIR accelerator complex. The UNILAC can accelerate all ions to an energy of 11.4 MeV/u, which is the injection energy of SIS18. In parallel low energy heavy ion beams will be available in the existing low energy target area for fixed target experiments. An upgrade program has been launched aiming at an increase of beam intensities, which will allow filling of SIS18 up to the low energy space charge limit [5]. Various modifications of the low energy and linac section have already been started.

An additional increase of average intensity will be achieved by faster ramping of SIS18. The installation of a new 110 kV connection to the main power grid will result in an increase of the cycling rate from 0.5 Hz presently to about 4 Hz with an equivalent increase of average beam current for fixed target experiments. The higher ramp rate of

SIS18 will allow the acceleration of up to 5 pulses per second, depending on the final energy, for injection and further acceleration in the new synchrotron SIS100, which is the first acceleration stage of the new FAIR facility.

The process of beam induced desorption due to beam loss after ionization of incompletely stripped ions has been identified as a serious limit to the beam intensity in SIS18, but is also expected to be relevant at higher energies in the new subsequent synchrotrons. The reduction of ionization losses to a level which allows the acceleration of heavy ion beams at the space charge limit is under investigation. Besides general vacuum system upgrades, the installation of dedicated collimation systems for ionized beam particles is considered, first prototype set-ups are in preparation for SIS18.

Proton beams for the antiproton physics program at FAIR will be provided by a new 70 MeV linac. The new linac which is presently under design [6] will be installed parallel to the existing UNILAC. It will merge into the existing injection beam line of SIS18 and inject through the standard injection elements.

THE NEW SYNCHROTRONS OF THE FAIR FACILITY

SIS100

The new synchrotron SIS100 is the central accelerator of the planned intensity and energy increase. The maximum bending power of 100 Tm is achieved with super-ferric dipole magnets and a circumference of 1083 m. The higher bending power, compared to the old synchrotron SIS18, offers two options. First, the acceleration of low charge states without intermediate stripping avoids the inevitable particle losses after the stripping stage. The acceleration of e.g. U^{28+} will result in an intensity of up to 5×10^{11} ions per pulse with an energy of 1.7 GeV/u, which is well suited for efficient production of projectile fragments in a low Z target. Second, the energy increase is achieved by acceleration of high charge states, which e.g. for uranium provides a maximum energy of 8.3 GeV/u, if U^{73+} is accelerated in SIS18 and in SIS100. Pulses of up to 10^{11} U^{73+} ions can be expected. In a similar manner the protons injected from SIS18 can be accelerated in SIS100 from 2 to 29 GeV.

The circumference of SIS100 is five times the SIS18 circumference. SIS100 will be filled with 4 pulses from SIS18, each consisting of two bunches. The 8 bunches with two empty rf buckets in between will be accelerated with the rf system operating on harmonic $h = 10$. A maximum rf voltage of 400 kV at frequencies between 1.1 and 2.7 MHz is required for the proposed fast acceleration. A bunch compression system will provide a single short bunch for the production of secondary beams and the injection into the collector ring CR. The rf system makes up a large fraction of the costs of SIS100.

The main feature of SIS100 is the use of superconducting magnet technology for the achievement of a maximum ramp rate of 4 T/s. The use of super-ferric magnets limits the bending field to 2 T, but allows fast cycling with ramp times below 0.4 s. A dedicated R&D program has resulted in a significant reduction of AC losses [7].

SIS300

The second new synchrotron SIS300 is primarily conceived for acceleration of heavy ions to highest energies. It is designed with the same circumference as SIS100 and will be located in the same tunnel as SIS100. The use of superconducting magnets with a maximum bending field of 6 T and a ramp rate of 1 T/s will allow a maximum energy for bare uranium ions of 34 GeV/u. As the higher energies are achieved with longer cycles times, the average intensity at highest energies is reduced to a few times 10^9 ions per second. The high energy ion beams can be slowly extracted to the high energy fixed target area.

SIS300 can also be used as a stretcher ring below 100 Tm for ions after acceleration in SIS100. The ions are accelerated at maximum ramp rate in SIS100 and transferred to SIS300, which is operated at constant bending power. This mode allows e.g. the production of rare isotope beams with an extraction rate of up to 2.5×10^{11} primary ions per second for experiments in the low and high energy branch after the fragment separator SuperFRS.

SECONDARY BEAM PRODUCTION AND BEAM TRANSPORT

Two beam lines for the production and transport of the secondary beams are foreseen. A 100 Tm beam line will transport protons to the antiproton production target. The target area and all its remote handling facilities still have to be designed in detail, the design will largely follow the antiproton production concept developed at CERN. A magnetic separator and a large acceptance beam line for an antiproton beam of 13 Tm maximum rigidity, corresponding to antiprotons of 3 GeV kinetic energy, will connect the production target to the subsequent collector ring CR.

For the production of rare isotope beams a new large acceptance fragment separator, the SuperFRS, is under design [8]. With a maximum magnetic rigidity of 20 Tm employing superconducting large acceptance magnets, the SuperFRS will allow the transport of rare isotope beams to a fixed target experimental area with a low and high energy branch. A separate ring branch connects it via a 13 Tm beam line to the collector ring CR, which is the first stage of the subsequent storage ring complex. Antiproton and rare isotope beam line are merged in front of the CR and use the same injection beam line and injection elements.

THE 13 TM STORAGE RINGS OF THE FAIR PROJECT

A complex of three new storage rings with a magnetic bending power of 13 Tm serves the preparation of high quality and high intensity beams, both stable primary heavy ion beams or secondary beams [9]. These can be used for experiments in the storage rings or, after deceleration in the last ring, in a subsequent low energy experimental area which is jointly used by antiproton and heavy ion users communities. The polarity of the ring magnets has to be changed between ion and antiproton operation. Presently no fast polarity changes are foreseen, but electronic switches could be installed to execute the polarity change fully computer controlled. The time scale for polarity changes is determined by the time constant for ramping the superconducting CR dipole magnets which have a large coil inductance.

The main technique applied in the storage rings is beam cooling. It provides the basis for the accumulation of high intensity secondary beams and for efficient deceleration. Deceleration is the key technique to improve the resolution in a variety of atomic physics experiments which suffer from Doppler effects. The high beam energy is needed to produce highly charged ions, rare isotope beams or antiprotons, subsequent deceleration allows an improved resolution, as adverse Doppler effects are reduced. The power of this method has been demonstrated at the existing ESR storage ring at GSI [10].

Finally, the beams from SIS100 and SIS300 can be transfered to various fixed target areas. dedicated areas for high energy nucleus-nucleus collisions, for atomic physics and for plasma physics experiment are planned.

The Collector Ring CR

The collector ring CR is a storage ring with a circumference of 212 m designed as a dedicated pre-cooling ring for hot secondary beams of antiprotons or rare isotopes. Due to the large emittances and momentum spread of these beams a stochastic cooling system will be integral part of the CR, which strongly effects the ion optical design of the storage ring. To reduce the longitudinal heating in the production target, the primary beams are compressed to short bunch length (25 ns for protons, 60 ns for ions). A bunch rotation and debunching system in the CR will increase the bunch length in order to reduce the momentum spread for fast cooling with the stochastic cooling system.

The cooling time needs to be minimized for antiprotons as well as for rare isotope beams. For antiprotons the cooling time determines the production rate and finally the luminosity for antiproton experiments, for rare isotope beams the cooling time limits the ability to study short-lived nuclei in experiments with stored cooled beams. Thus a stochastic cooling system for minimum cooling time for both beam species is under design [11]. The difficulty in designing a common system arises from a significant difference in velocity, if the these beams are injected at the maximum bending power of the CR. First prototype electrodes have been studied which are equally efficient for antiprotons at velocity $\beta = 0.97$ and rare isotope at $\beta = 0.83$. These stochastic cooling electrodes can be used without any modification for both secondary beam species. The stochastic cooling is supported by a lattice which is matched to the requirements of stochastic cooling. The frequency slip factor η, which depends on the beam energy and the transition energy of the lattice, is optimized for each beam velocity individually by a different focussing strength of the ring quadrupoles.

A small value of η reduces the required rf voltage for fast bunch rotation. An rf voltage of 400 kV at harmonic $h = 1$ provides optimum conditions for bunch rotation, the maximum voltage is determined by the requirement for rare isotope beams. The stochastic cooling system employs two cooling systems with the frequency range 1-2 GHz and 2-4 GHz for all three phase space planes. By combination of fast bunch rotation and debunching with stochastic cooling total cooling times below 5 s for antiprotons and below

1 s for rare isotope beams are expected. For cost reduction reasons, it is presently considered to reduce the rf voltage to 200 kV and the cooling system bandwidth to 1-2 GHz in a first stage of the project, this would approximately double the total cooling time and consequently reduce the antiproton production rate by a factor of two.

The CR also offers the option to operate it in an isochronous mode, which is beneficial for mass measurement of very short-lived isotopes. This technique is based on tuning the focussing elements for isochronous circulation of particles independent of their momentum [12]. For the given ion optical structure of the CR the momentum acceptance in the isochronous mode covers a range of ± 0.7 %. This mode of operation does not require beam cooling. In the isochronous mode the mass of different isotopes can be measured from the revolution time, which is measured with a special time of flight detector. The ions pass through a thin foil and the emitted secondary electrons are measured to determine the time of passage. The energy loss in the foil is negligible for at least some hundred passages which allows a measurement of the revolution time and consequently the mass of the isotope with an accuracy of better than 10^{-5}.

The Accumulator Ring RESR

After stochastic pre-cooling in the CR the secondary beams will be transferred to the RESR, which has a circumference of 245 m and is located in the same building nearly concentric with the CR. The main task of the the RESR is the accumulation of antiprotons. The pre-cooled antiprotons will be injected into RESR every 5-10 s according to the cooling time in the CR. The cooling time associated with antiproton accumulation in RESR will be short compared to the pre-cooling time in the CR. Therefore the production rate and luminosity for antiproton experiments should not be effected by the accumulation in RESR. The lattice design of the RESR provides a large momentum acceptance allowing the classical accumulation scheme with stochastic cooling of stack tail and stack core, which was developed for the AA at CERN [13]. As an alternative method a barrier bucket technique is under consideration [14]. This method requires the compression of the accumulated antiproton stack to about half the ring circumference into a time interval of less than 0.5 μs. The major challenges of this novel scenario are the generation of less than 100 ns short barriers with voltages of several kV, a very reliable and reproducible timing of the injection components and the design of the stochastic cooling system, which must avoid over-cooling and beam loss due to diffusion and instabilities. The present ion optical lattice seems appropriate for both accumulation methods, a decision about the method will be made based on expected performance and costs of the required technical systems.

Several components for the RESR will be reused from the existing ESR storage ring at GSI [15], such as quadrupole magnets, rf systems, beam diagnostics. The dipole magnets will be identical to the ones used in the NESR in order to minimize design efforts and reduce production costs. As a consequence the dipoles have a large acceptance and can be ramped with a rate 1 T/s. This allows fast deceleration of rare isotope beams in the RESR, which are injected at 740 MeV/u and can be decelerated to a minimum energy of 100 MeV/u in less than 2 s. This range of energies is matched to the requirements of experiments in the NESR which need operation of the NESR at constant magnetic field, particularly in the proposed collider mode of the NESR.

The Storage Ring NESR

The storage ring NESR provides conditions for a large variety of atomic and nuclear physics experiments with stored ion beams. It has a fourfold symmetry with 18 m long straight sections and a circumference of 222 m. High beam quality is achieved with an electron cooling system which for ions covers the full energy range. The electron cooling system occupies one straight section. Another straight section will be equipped with an internal gas jet target providing gaseous targets ranging from hydrogen to xenon. The third straight section is reserved for an electron target, which is conceived as an electron cooler like device with an electron beam of lowest temperature confined in a longitudinal magnetic field. Electron energies of some 10 keV will be sufficient to study ion-electron interactions over a large range of relative velocities. Lowest beam temperatures will be achieved by longitudinal and transverse adiabatic expansion. The forth straight section will be the collision section with an electron ring. Electron bunches circulating in this ring at variable energy between 100 and 500 MeV will collide with short cooled ion bunches for nuclear physics electron-nucleus scattering experiments to determine the radius and charge distribution of nuclei. It is even proposed to fill the small ring with antiprotons and study antiproton-ion collisions.

Another feature of the NESR will be fast ramping of the NESR from the maximum rigidity of 13 Tm to a magnetic rigidity below 1 Tm. This allows the deceleration of highly charged ions and rare isotopes to 4 MeV/u. Application of the deceleration technique to antiprotons will provide antiprotons of minimum energy 30 MeV. The decelerated beams can either be slowly extracted to a fixed target or transferred by fast extraction to the FLAIR facility of the FAIR project [16]. The FLAIR facility offers deceleration of the extracted beams of ions or antiprotons to rest by a system of storage rings, a linac and finally storage in a trap.

The deceleration in the NESR will be supported by electron cooling, for ions over the full energy range, for antiprotons starting from an upper energy of 0.8 GeV, which will allow an intermediate cooling after injection at 3 GeV. Although it is still open, whether normalconducting or super-ferric magnets will be used a maximum ramp rate of 1 T/s is compulsory. As the revolution frequency changes strongly, a frequency swing of the rf system of at least a factor of five is needed in order to use only one change of the harmonic number during deceleration. Diagnostics systems must be designed for the change of bunching frequency and the reduction of electric current during deceleration.

The NESR cooling system is specified for fast cooling of rare isotope beams after stochastic pre-cooling at 740 MeV/u in the CR. A cooling time constant of the order of 100 ms is aimed at for the typical beam parameters after stochastic pre-cooling ($\varepsilon_{x,y} = 0.5 \times 10^{-6}$ m, $\delta p/p = 1 \times 10^{-3}$). This high cooling rate will be achieved by a cooling section length of 5 m and a small electron beam diameter which results in high electron density and which is matched to the ion beam size. For experiments with stored rare isotope beams a longitudinal accumulation scheme employing electron cooling in the NESR after stochastic pre-cooling in the CR is foreseen. This could be applied to beams with sufficiently long lifetime (of the order 10 s). If fast deceleration of ions is required, the acceleration voltage of the electron beam must be ramped within 2 s from the maximum voltage at injection energy to the minimum voltage of 2 kV. Electron cooling will be applied at injection energy, at an intermediate energy and at the final energy. In between the cooling periods the voltage is ramped together with the ring magnets, while the electron current is switched off. The design of the high voltage system for electron acceleration must pay special attention to the necessity of fast ramping, as typical highly stabilized high voltage generators have rather long time constants for voltage stabilization reasons.

Fast cooling of bunched ion beams is required in the collider mode, when bunches of about 15 cm length and transverse size of less than 0.1 mm are required to achieve the luminosity goal of the order of $10^{28} \text{cm}^{-2}\text{s}^{-1}$ in collisions with electron bunches of similar parameters [17]. Electron cooling has to compensate heating by a target, either the internal gas target or the colliding beam, by intrabeam scattering and by impedances. For precision experiments, like e.g. mass measurements of unstable nuclei by Schottky Mass Spectrometry [18], [19], very cold ion beams are required with momentum spreads in the 10^{-7} range or even lower.

For the proposed deceleration of antiprotons electron cooling will be applied after deceleration of the antiprotons in the NESR from their production energy of 3 GeV to an energy below 800 MeV and before extraction at the final energy which can be as low as 30 MeV. The intermediate cooling will allow deceleration with high efficiency. The cooling time for antiprotons after stochastic pre-cooling are expected to stay below 1 min, maybe down to 30 s. This cooling time will determine the total cycle time for the deceleration of antiprotons.

The NESR will be equipped with slow and fast extraction systems designed for beams with a magnetic rigidity below 4 Tm. The extraction is foreseen at both polarities for use with ions or antiprotons which requires a polarity change for all extraction elements.

THE HIGH ENERGY STORAGE RING HESR

As a first step towards an international project the HESR storage ring is designed in a consortium between the research center Jülich, which is the consortium leader, the The Svedberg Laboratory in Uppsala and GSI. Experimental groups from these laboratories also form the core of the future users of this facility which is entirely devoted to a physics program with stored antiproton beams interacting with an internal hydrogen target.

The HESR storage with a circumference of 574 m and a magnetic bending power of 50 Tm allows storage of antiproton up to an energy of 14.1 GeV. The antiprotons will be injected from the RESR at 3 GeV and accelerated or decelerated to an energy in the range 0.8 to 14.1 GeV, according to the requirement of the experiment. Because of the use of superconducting dipole magnets of the RHIC type the ramp rate is limited to 0.025 T/s. Systems for stochastic and electron cooling of the antiproton beam stored at variable energy offer high resolution and luminosity. Electron cooling up to 9 GeV should result in excellent resolution with beams of momentum spreads down to 10^{-5} in a so-called high resolution mode. A first concept for the HESR electron cooling system was worked out by BINP, Novosibirsk [20], work on the system design within the HESR consortium is continued by TSL, Uppsala [21]. At energies above 5 GeV a stochastic cooling system is expected to provide beams of reduced quality, but a higher luminosity of up to 2×10^{32} cm^{-2}s^{-1} for 10^{11} stored antiprotons should be achieved [22].

The outstanding feature of the HESR will be the electron cooling system. To achieve an energy resolution of 100 keV for energies below 9 GeV, as required by high resolution experiments, a strong cooling force must be provided. Only operation of the electron cooling system in the magnetized cooling regime promises sufficient cooling power. Most experiment require coasting antiproton beams, but electron cooling of a bunched antiproton beam is not ruled out. The electron cooling concept is based on a dc electron current which is accelerated in an electrostatic system and confined in a strong longitudinal magnetic guiding field. Such a system resembles low energy electron coolers, but requires a 5 MV acceleration system and a magnetic guiding field in the range 0.1-0.5 T over a cooling section length of 30 m. The high magnetic field strength favors superconducting solenoid magnet technology. This cooling system is expected to provide sufficient cooling rate to counteract the heating rate by the internal target and by intrabeam scattering.

The achievable luminosity and beam quality are strongly interconnected due to the influence of target heating, intrabeam scattering and beam cooling. Recent simulations indicate a luminosity limitation by energy loss straggling of the relativistic antiproton beam in the dense hydrogen target with a maximum thickness of 4×10^{15} atoms/cm^2 [23]. The ultimate luminosity limitation comes from the antiproton production rate. Assuming optimum conditions for primary proton intensity, antiproton production and separation, cooling and accumulation a maximum production rate of 2×10^7 s^{-1} has been estimated, which matches the consumption rate in the high luminosity mode. A presently considered staging of the project might result in lower production rates in the initial phase of the project, and the option of a subsequent upgrade to full performance. As a future extension the possibility to employ beams of polarized antiprotons, produced in a small polarizer ring, has been proposed, which will require the installation of a Siberian snake.

OUTLOOK

The basic conceptual work for the FAIR project is finished. Some modifications of the described concept might be considered in order to make the project more cost effective. The decision about the scope of the project is expected for the end of year, which will allow a final decision by funding agencies. The construction of the facility in collaboration with international partners is expected within a decade after approval by funding agencies.

REFERENCES

1. An International Accelerator Facility for Beams of Ions and Antiprotons, Conceptual Design Report, GSI, 2001.
2. N. Angert, Proc. of the 5th Europ. Part. Acc. Conf., Sitges, World Scientific, Singapore (1996) 125-129.
3. W. Barth, L. Dahl, J. Glatz, L.G. Groening, S.G. Richter, S. Yaramishev, Proc. of the 9th Europ. Part. Acc. Conf., Lucerne, Switzerland (2004) 1171-1173.
4. K. Blasche, B. Franzke, Proc. of the 4th Europ. Acc. Conf., London, England (1994), 133-137.
5. P. Spiller, Proc. of the 2005 Part. Acc. Conf., Knoxville, Tennessee (2005)
6. L. Groening, W. Barth, L. Dahl, R. Hollinger, P. Spädtke, W. Vinzenz, S. Yaramishev, B. Hofmann, Z. Li, U. Ratzinger, A. Schempp, R. Tiede, Proc. of the LINAC 2004, Lübeck, Germany (2004) 42-44.
7. G. Moritz, Proc. of the 9th Europ. Part Acc. Conf., Lucerne, Switzerland (2004) 132-136.
8. H. Geissel, H. Weick, M. Winkler, et al., Nucl. Instr. Meth. B204 (2003) 71-85.
9. P. Beller, K. Beckert, A. Dolinskii, B. Franzke, F. Nolden, C. Peschke, M. Steck, Schriften des Forschungszentrums Jülich, Matter and Materials, Volume 30, ISBN 3-89336-406-8 (2005).
10. M. Steck, K. Beckert, P. Beller, B. Franczak, B. Franzke, F. Nolden, Proc. of the 7th Europ. Part. Acc. Conf., Vienna (2000) 587-589.
11. F. Nolden, K. Beckert, P. Beller, A. Dolinskii, B. Franzke, U. Jandewerth, I. Nesmiyan, C. Peschke, P. Petri, M. Steck, F. Caspers, D. Möhl, L. Thorndahl, contribution to these proceedings.
12. M. Hausmann, K. Beckert, H. Eickhoff, B. Franczak, B. Franzke, H. Geissel, G. Münzenberg, F. Nolden, C. Scheidenberger, M. Steck, T. Winkler, Proc. of the 6th Europ. Part. Acc. Conf., Stockholm (1998) 511-513.
13. H. Koziol, CERN Int. Note PS/AA/Note 84-2, 1984.
14. T. Katayama et al., contribution to these proceedings.
15. B. Franzke, Nucl. Instr. Meth. B24/25 (1987) 18.
16. C. Welsch et al., contribution to these proceedings.
17. I.A. Koop, P.V. Logatchev, I.N. Nesterenko, A.V. Otboev, V.V. Parkhomchuk, V.M. Pavlov, E.A. Perevedentsev, D.N. Shatilov, P.Yu. Shatunov, Yu.M. Shatunov, S.V. Shiyankov, A.N. Skrinsky, A.A. Valishev, P. Beller, B. Franzke, M. Steck, Proc. of the 8th Europ. Part. Acc. Conf., Paris (2002) 620-622.
18. B. Franzke, K. Beckert, H. Eickhoff, F. Nolden, H. Reich, A. Schwinn, M. Steck, T. Winkler, Proc. of the 6th Europ. Part. Acc. Conf., Stockholm, Institute of Physics Publishing, Bristol (1998) 256-258.
19. F. Nolden, K. Beckert, P. Beller, B. Franzke, V. Gostishchev, C. Kozhuharov, Yu. Litvinov, A. Schwinn, M. Steck, contribution to these proceedings.
20. M. Steck, K. Beckert, P. Beller, A. Dolinskii, B. Franzke, F. Nolden, V.V. Parkhomchuk, V.B. Reva, A.N. Skirnsky, V.A. Vostrikov, Proc. of the 9th Europ. Part. Acc. Conf, Lucerne, Switzerland (2004) 1969-1971.
21. D. Reistad, contribution to these proceedings.
22. H. Stockhorst et al., contribution to these proceedings.
23. O. Boine-Frankenheim, to be published.

Antiproton Cooling in the Fermilab Recycler Ring[1]

S. Nagaitsev[a2], A. Bolshakov[b], D. Broemmelsiek[a], A. Burov[a],
K. Carlson[a], C. Gattuso[a], M. Hu[a], G. Kazakevich[c], B. Kramper[a],
T. Kroc[a], J. Leibfritz[a], L. Prost[a], S. Pruss[a], G. Saewert[a],
C.W. Schmidt[a], S. Seletskiy[d], A. Shemyakin[a], M. Sutherland[a],
V. Tupikov[a], A. Warner[a], and P. Zenkevich[b]

[a] FNAL, Batavia, IL 60510, U.S.A.;
[b] ITEP, Moscow, 117259, Russia;
[c] Budker INP, Novosibirsk, 630090, Russia;
[d] University of Rochester, Rochester, NY 14627, U.S.A.

Abstract. The 8.9-GeV/c Recycler antiproton storage ring is equipped with both stochastic and electron cooling systems. These cooling systems are designed to assist accumulation of antiprotons for the Tevatron collider operations. In this paper we report on an experimental demonstration of electron cooling of high-energy antiprotons. At the time of writing this report, the Recycler electron cooling system is routinely used in collider operations. It has helped to set recent peak luminosity records.

Keywords: beam cooling, antiprotons, storage ring
PACS: 29.27.Eg, 29.27.Fh, 29.20.Dh, 41.75.-i, 41.85.Ew

INTRODUCTION

The Run II Luminosity Upgrade Plan at Fermilab requires the Recycler [1] to play a key role as the repository of large stacks of antiprotons (6×10^{12}) with the appropriate phase space characteristics to be used in the Tevatron collider stores. In order to maximize the stacking efficiency of the Fermilab antiproton Accumulator, small stacks of antiprotons will be frequently (every 0.5-1 hour) transferred to the Recycler. In the Recycler, the stacks are initially cooled by stochastic cooling [2] and then stored and cooled by electron cooling until the antiprotons are ready to be used in the Tevatron.

The Run II Luminosity Upgrade Plan foresees the Recycler fully integrated into collider operations in two major steps. First, the Recycler is commissioned to bring its performance to the level that it is ready to begin the implementation of electron cooling. This milestone was achieved on June 1, 2004. In the second phase, the installation of electron cooling and its commissioning takes place. Electron cooling was demonstrated on July 15, 2005. At the time of writing this report the electron

---

1 Work supported by the URA, Inc., under contract DE-AC02-76CH03000 with the U.S. Dept. of Energy.

2 E-mail: nsergei@fnal.gov

cooling system is used for each Tevatron store. However, frequent antiproton transfers have not been implemented yet.

This paper outlines the design parameters of the Recycler Electron Cooling System and then describes the steps that led to the demonstration of electron cooling.

COOLING SYSTEM PARAMETERS

System Layout

Electron cooling of 8.9-GeV/c antiprotons in the Recycler requires an electron beam with kinetic energy of 4.3 MeV [3]. Figure 1 shows the schematic layout of the Recycler electron cooling system.

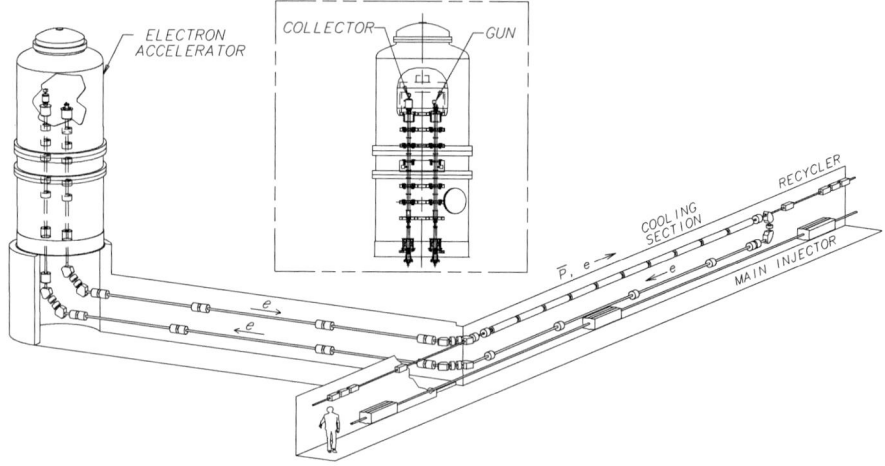

FIGURE 1. Schematic layout of the Recycler electron cooling system and accelerator cross-section (inset).

The dc electron beam is generated by a thermionic-cathode gun, located in the high-voltage (HV) terminal of the electrostatic (Van-de-Graaff type) accelerator. This accelerator is incapable of sustaining dc beam currents to ground in excess of about 100 µA. To attain the electron dc current of 500 mA, a recirculation scheme is employed. The electron beam is first delivered to the cooling section and then returned back to the HV terminal for charge recovery. A typical inefficiency of such a process is 20 ppm for beam currents of up to 500 mA.

The electron cooling system at Fermilab employs a unique beam transport scheme [4]. The electron gun is immersed in a solenoidal magnetic field, which creates a beam with large angular momentum. After the beam is extracted from the magnetic field and accelerated to 4.3 MeV, it is transported to the 20-m long cooling section solenoid using conventional focusing elements. At the entrance to the cooling section solenoid the beam is made round and parallel such that the beam radius, r_b, produces the same magnetic flux, Br_r^2, as at the cathode.

Table 1 presents basic parameters of the Recycler ring and its electron cooling system.

TABLE 1. Electron cooling system and Recycler ring design parameters.

Parameter	Symbol	Value	Units
Electron Accelerator			
Terminal Voltage	U_0	4.34	MV
Beam Current	I_b	0.5	A
Terminal Voltage Ripple, rms	δU	200	V
Cooling Section			
Length	L	20	m
Solenoid Field	B	100	G
Beam Radius	r_b	3.5	mm
Electron Angular Spread, rms	θ_e	≤ 0.2	mrad
Recycler design parameters			
Circumference	C	3.3	km
Momentum	$\beta\gamma Mc$	8.9	GeV/c
Transition γ	γ_t	20.7	
Ave. beta functions	β_{ave}	30	m
Typical emittance (n, 95%)	ε	5-7	µm-rad
Number of antiprotons	N_a	≤ 600	10^{10}
Average pressure	P_{av}	0.5	nTorr

Cooling Scenario

The electron cooling scenario has been reviewed in Ref. [5]. Under the current scenario the electron cooling system is required to decrease the longitudinal 95% emittance of a stored antiproton beam from 100 eV-s to 50 eV-s in 30 minutes for stacks of up to 6×10^{12} particles. This would correspond to providing 36 equally populated bunches with a 1.5-eV-s longitudinal emittance per bunch to the Tevatron collider.

Figure 2 presents a MOCAC code [6] simulation of the electron cooling process in the Recycler with 6×10^{12} antiprotons in a barrier rf bucket. The process of cooling is optimized by keeping the rms antiproton momentum spread constant at 3.5 MeV/c, while continuously reducing the bunch length by moving the rf barriers. The initial transverse emittance in this simulation was 5 µm-rad. The three curves correspond to different values of the electron beam rms angular spread in the cooling section. The design value of 0.2 mrad for the rms angular spread is presented by the bottom curve. Other electron beam parameters are presented in Table 1.

The choice of the rms momentum spread being 3.5 MeV/c is determined by a compromise between the IBS-related longitudinal diffusion and the beam lifetime due to the dynamic momentum aperture of the Recycler ring. Figure 3 shows the calculated longitudinal and transverse IBS growth rates as a function of the rms momentum spread. One can see that for the momentum spread of 3.5 MeV/c the longitudinal diffusion is practically independent of the transverse emittance.

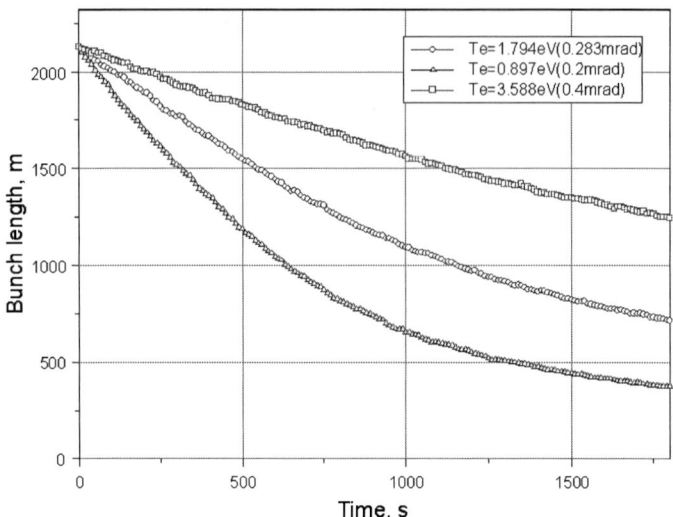

FIGURE 2. The evolution (simulation) of the antiproton bunch length as a function of time (30 minutes full scale) for various electron beam angular spreads. The initial bunch length corresponds to a 100 eV-s longitudinal emittance. The design curve (bottom) indicates that there is a factor of two safety margin in cooling rates.

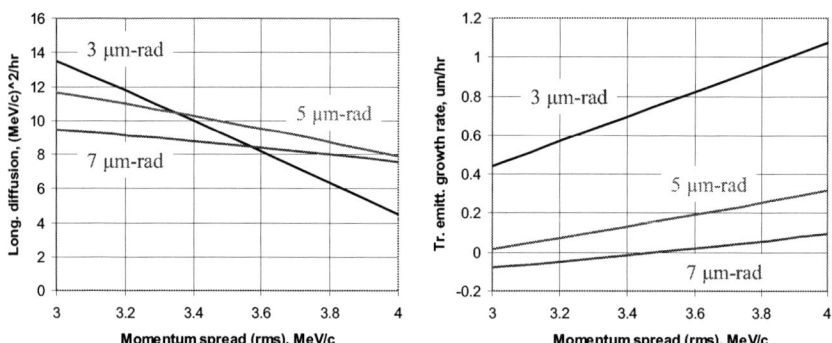

FIGURE 3. The simulated IBS heating rates (Bjorken-Mtingwa model) for 250×10^{10} antiprotons with a constant 54 eV-s longitudinal emittance as a function of the rms momentum spread for various transverse emittances.

COOLING DEMONSTRATION

Energy Alignment

It has been anticipated that in the early stages of commissioning, the electron cooling system might not be able to meet three major design parameters concurrently: (1) the beam current of 0.5 A, (2) the rms angular spread of 0.2 mrad and (3) the perfect energy matching between electron and antiproton beams. We have also anticipated that for the electron beam current and angular spread optimization it would be highly desirable to see the "effect" of cooling first. Since the cooling process is quite slow and the electron energy was not known to the required accuracy, we estimated that the energy matching could be challenging. Our plan for energy matching was based on two assumptions: (1) the Recycler absolute energy is known to 0.1% and (2) the Recycler momentum acceptance is greater than 0.3%. Measuring and adjusting the electron absolute energy to 0.3% [7] would allow us to "land" the electron beam energy somewhere within the Recycler momentum acceptance and to observe the cooling effect even for lower beam currents and higher angular spreads. To observe the cooling process we implemented the following procedure. A small ($<10\times10^{10}$) antiproton beam current was debunched and cooled transversely (by the stochastic cooling system) to a small transverse beam emittance. Using an rf noise source the momentum spread of this beam was increased to create a uniform momentum distribution 0.3-0.4% wide. The antiproton-electron beam interaction was observed with the help of a longitudinal beam Schottky-noise monitor, which measures the momentum distribution function. Figure 4 presents a simulation of this process.

The simulation in Fig. 4 was performed with a coasting antiproton beam perfectly matched in energy with the electron beam of 0.1 A and with 0.5 mrad of rms angular spread. The spike in the distribution, formed by the electron cooling process, increases the distribution function by a factor of 2 – a value easily detectable by a Schottky-noise spectrum analyzer.

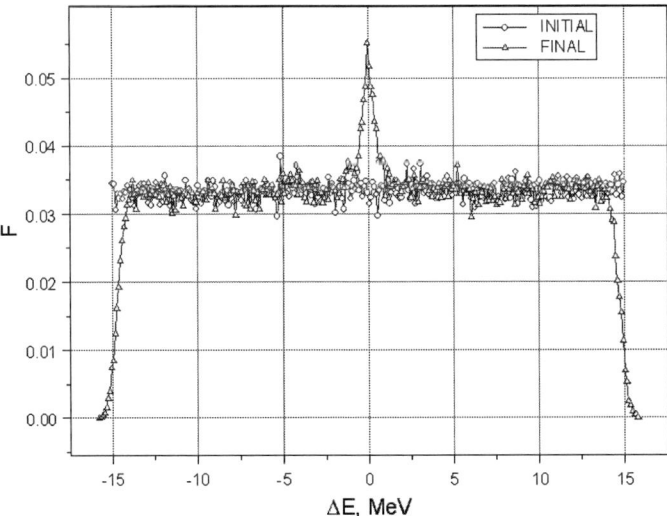

FIGURE 4. The momentum distribution (arb. units) as a function of antiproton energy deviation (simulation by MOCAC code [5]). The initial distribution is uniform in energy. The final distribution is plotted after 30 minutes.

Using this method, the initial observations of cooling in the Recycler demonstrated that the electron energy was within 3 keV of its optimal value. Figure 5 shows the experimental implementation of this procedure in the Recycler after the electron energy adjustment to its optimal value.

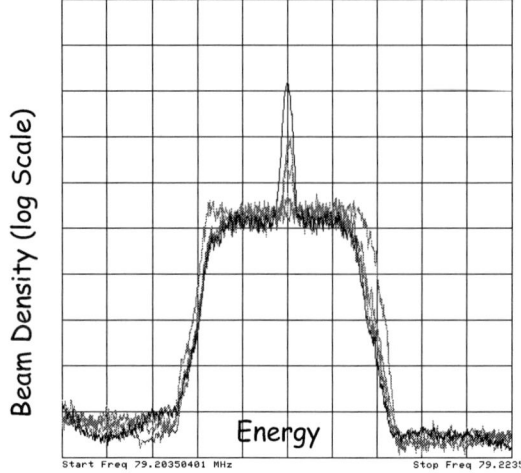

FIGURE 5. The evolution of the antiproton distribution function with electron cooling. The horizontal scale is 25 MeV/div. The antiproton beam intensity was 5×10^{10} and its emittance was 2 μm-rad. Traces were taken 15 minutes apart.

Cooling Observation

The first electron cooling demonstration was performed on July 15, 2005 with a bunched beam of 63×10^{10} antiprotons. Figure 6 shows two antiproton momentum distribution functions, taken 15 minutes apart, while the beam was electron cooled.

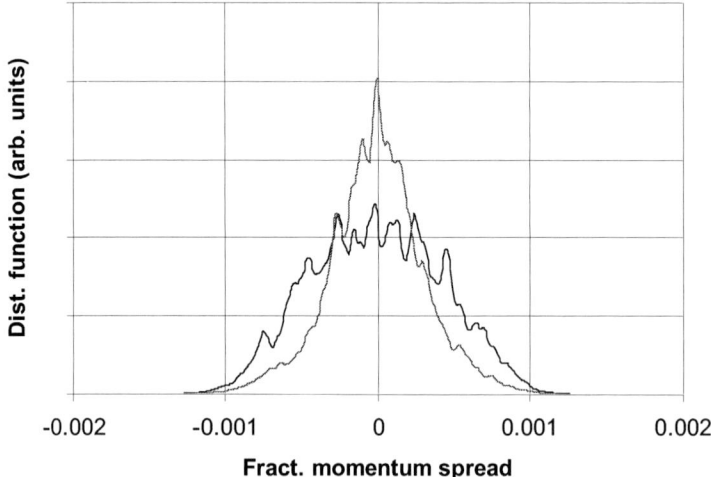

FIGURE 6. The first cooling demonstration of relativistic antiprotons in the Recycler. The antiproton current was 63×10^{10}, the emittance was 4 μm-rad, and the bunch length was 1.7 μs. The electron beam current was 200 mA. Traces were taken 15 minutes apart.

To measure the cooling force we employed the so-called voltage jump method [8]. In this method, the coasting antiproton beam is initially cooled down to a small equilibrium momentum spread. The electron beam energy is then changed instantaneously by several keV. The electron cooling force then drags the antiproton distribution to a new equilibrium momentum, which is M/m times the voltage jump away from the initial equilibrium. Figure 7 presents the evolution of the antiproton momentum distribution function as the antiprotons are being dragged by the electron beam after its energy was jumped by 2 keV. The initial average cooling force, determined from this plot, is about 20 MeV/c per hour – this agrees well with our theoretical predictions.

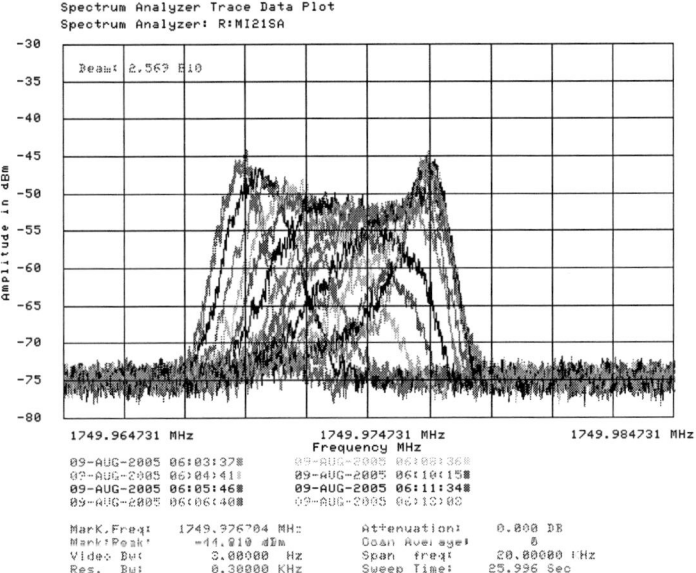

FIGURE 7. The evolution of the antiproton beam momentum distribution function as the antiprotons are being dragged by a 200-mA electron beam to a new equilibrium after the energy jump. Left curve – the initial distribution and right – new equilibrium after the energy jump. Traces are taken approximately 1 minute apart. The number of antiprotons was 4×10^{10}, the transverse emittance was 1.5 µm-rad (n, 95%). The horizontal scale is 1.2 MeV/c per division.

Finally, Figure 8 presents the evolution of both the transverse and the longitudinal emittances of the antiproton beam, being cooled by the electron cooling system. Our initial estimates show that the electron cooling rates are high enough to meet the final requirements of the Run II luminosity upgrades as outlined above.

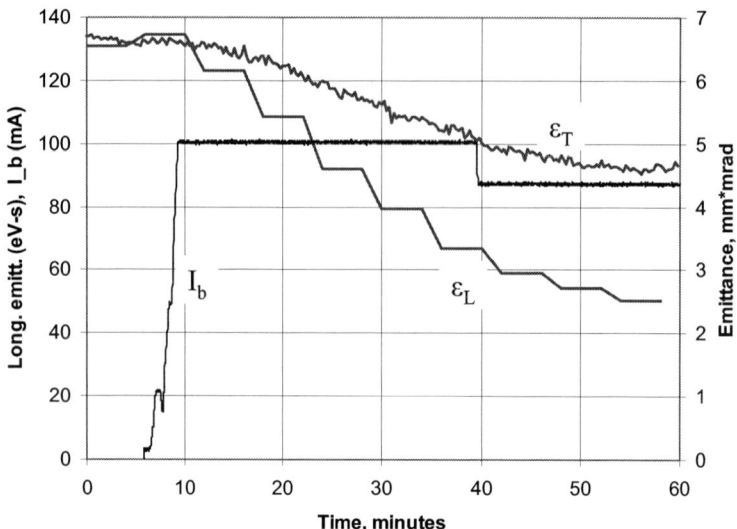

FIGURE 8. Electron cooling of a bunched antiproton beam (78×10^{10}). The electron beam was turned on to 100 mA at 10 minutes. The step-like variation of the longitudinal emittance is due to the extended interval between consecutive data points.

SUMMARY

Electron cooling of 8-GeV antiprotons has been demonstrated and is presently in routine operations. The measured cooling force is in agreement with theoretical predictions. The electron cooling system has been used in Tevatron collider operations since August, 2005. Since then, it has been primarily responsible for the recent advances in the Tevatron peak luminosity.

ACKNOWLEDGMENTS

The authors are grateful to Fermilab's technical staff for installing and maintaining the cooling system and to A.C. Crawford, D. Prasuhn, V. Lebedev, I.N. Meshkov, V.V. Parkhomchuk, and V. Reva for their help and for many useful discussions. We acknowledge contributions made by R. Goodwin, J. Crisp, and L. Carmichael to the protection system; by L. Nobrega and S. Wesseln to the vacuum system design; and by B. Chase and P. Joireman to the design of the BPM system and to beam oscillation measurements.

REFERENCES

1. G. Jackson, FERMILAB-TM-1991 (1996).
2. D. Broemmelsiek, M. Hu and S. Nagaitsev, "Stochastic cooling in barrier buckets at the Fermilab Recycler", EPAC'04, Lucerne, July 2004.

3. "Prospectus for an electron cooling system for the Recycler," edited by J.A. MacLachlan, FERMILAB-TM-2061, Oct 1998.
4. A. Burov, et al., Phys. Rev. Sp. Top. - AB 3, 094002 (2000).
5. A. Burov, "Electron-cooling scenarios at Fermilab", NIM A 532 (2004) 291-297.
6. P. R. Zenkevich et al., "Modeling of Electron Cooling by Monte Carlo Method", Report at the International Workshop on Cooling and Related Topics, Bad Honnef (Germany), 2001.
7. S. M. Seletskiy and A. Shemyakin, Beam-based calibration of the electron energy in the Fermilab electron cooler, to be published in Proc. of PAC'05, Knoxville, USA, May 16-20, 2005
8. H. Danared, et al., Phys. Rev. Lett. 72, 3775 (1994).

Report on Operation of Antiproton Decelerator

Pavel Belochitskii on behalf of AD team

AB Department CERN CH-1211 Geneva 23 Switzerland

Abstract. The Antiproton Decelerator (AD) at CERN operates for physics since 1999. The 3.5 GeV/c antiprotons produced in the target by a 26 GeV/c proton beam coming from CERN PS. Since the experiments need a low energy antiprotons, beam is decelerated in the AD down to an extraction momentum of 100 MeV/c. Due to significant emittance blow up during deceleration, as well as tight requirements from experiments on extracted beam sizes, efficient compression of beam phase space is indispensable. Two cooling systems, stochastic and electron are used in AD. The progress in machine performance is reviewed, along with plans for the future. Special emphasis is given to the proposed new extra low energy antiproton ring (ELENA) for deceleration of antiproton beam further down to an energy of 100 keV (momentum 13.7 MeV/c), which would allow much higher antiproton capture rate with significantly higher beam density.

Keywords:. AD, antiproton, ELENA.
PACS: 29.20 Lq

INTRODUCTION

The Antiproton Decelerator (AD) operates routinely for physics program since 1999. The 100 MeV/c antiprotons are delivered in single bunch or multi-bunch mode to one of the three experiments: ASACUSA, ATHENA, ATRAP. From 2003 the ACE experiment receives 300 MeV/c antiprotons in single bunch ejection.

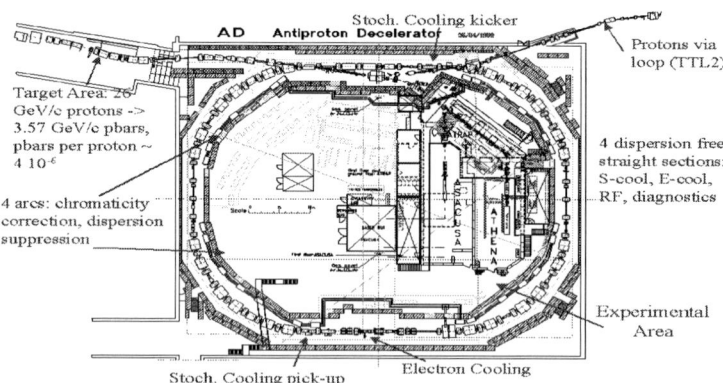

FIGURE 1. AD Hall layout.

The machine performance was gradually improved after the start of operation. The essential contribution to the progress in performance, along with experience comes from the use of new diagnostics. The important parts of routine machine operation are machine development (MD) sessions, where new hardware and software can be tested, or particular aspects of beam physics in AD (like electron cooling) is a subject of study. In addition, MD's are used for machine performance maintenance, which could suffer due to various reasons.

BASIC AD CYCLE

Antiprotons are produced in a target with 26 GeV/c proton beam from CERN PS, then collected and transferred to the AD ring. After injection at 3.57 GeV/c, antiprotons are rotated 90 degrees in the longitudinal phase space, taking advantage of the large AD momentum acceptance and short bunch length of about 25 ns. Then beam is stochastically cooled and decelerated down to 2 GeV/c. Stochastic cooling is repeated again, mainly to reduce a momentum spread to fit requirements of the deceleration RF cavity.

The beam is then decelerated down to 300 MeV/c and cooled down by the electron beam from the electron cooler. After cooling, the beam is decelerated to the ejection momentum of 100 MeV/c (kinetic energy 5.3 MeV). Then the antiprotons are cooled again by the electron beam, rotated 90 degrees in the longitudinal phase space (if experiments demand shorter beam, which is typically the case) and ejected.

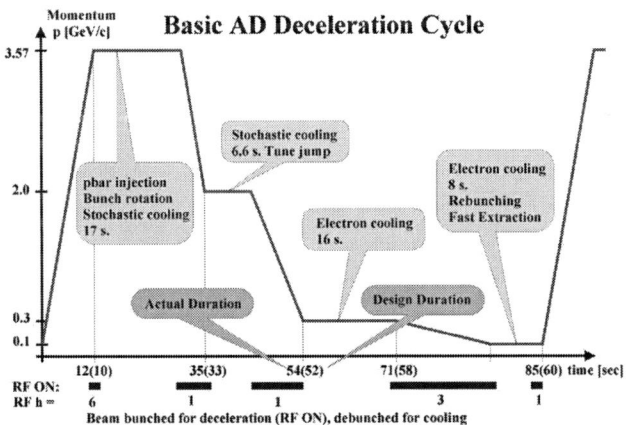

FIGURE 2. Basic AD cycle.

OPERATION

In 2005 the CERN Accelerator Complex was not running, except for the ISOLDE Facility. During the previous year 2004 the total time for physics was substantially increased due to continuation of machine operation on weekends in autumn. The MD time was reduced due to limited manpower available, which contributed as well to increased time for physics. Unfortunately, of 3090 hours scheduled for physics the real time with beam was only 2194 hours (71%).

Two major problems were responsible for this reduction. The PS ejection septum leak happened twice, taking 1 week for replacement with spare unit, and 3 weeks next time to repair spare and install.

Excessive outgassing in the collector region of AD electron cooler was a reason for disassembly, inspection, replacement of all suspect equipment, bakeout, hence stops of machine for 2 weeks.

TABLE 1. Operational statistics.

Run time (h)	2000	2001	2002	2003	2004
Total	3600	3050	2800	2800	3400
Physics	1550	2250	2100	2300	3090
MD	2050	800	700	500	310
Uptime	86%	89%	90%	90%	71%

PROGRESS IN MACHINE PERFORMANCE

Beam Intensity Improvement

The beam intensity is gradually improving with years, with peak numbers shown in Fig.3. This progress is based mainly on higher intensity of a production beam. The optimization of machine acceptances at injection energy also had a positive effect.

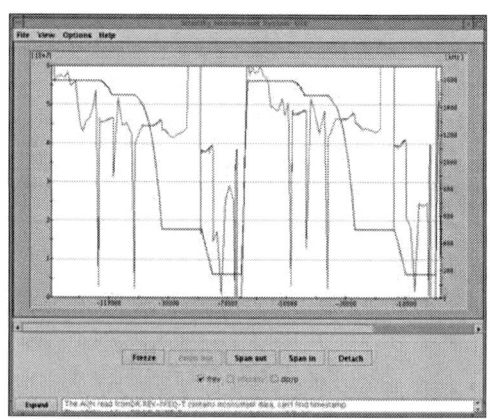

FIGURE 3. Schottky based intensity monitoring during AD cycle.

Ejected Beam Emittances Improvement

Beam bunch rotation before ejection makes it shorter (about 90 ns compared with 220 ns design value), which is appreciated by experiments. During beam bunching the longitudinal emittance grows significantly due to noise of the low level RF system.

This degradation was overcome by extending electron cooling for the time of beam bunching (see Fig.4, left). Unfortunately, this caused beam profile degradation: filamentation into 2 (sometimes even 3) parts and creation of extended tails (Fig. 5, left).

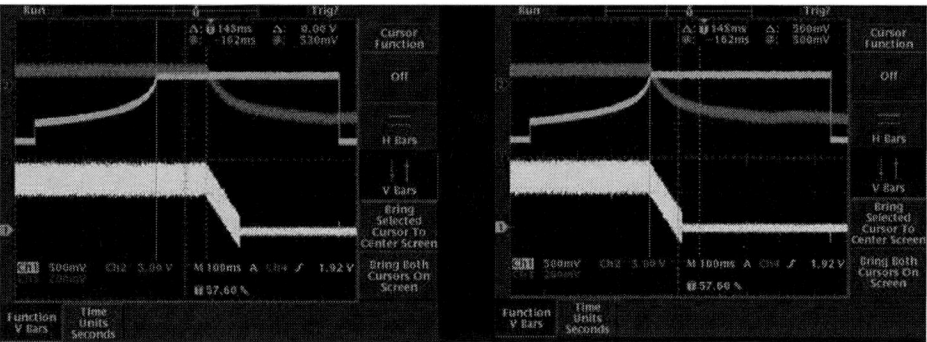

FIGURE 4. Voltage programming during the end of electron cooling and beam bunching. Left figure shows the curves before adjustment, and right figure after adjustment. Two curves on the top of each figure show voltage on the cooler cathode (constant, then decreasing) and voltage in RF cavity (increasing, then constant).

Careful adjustment of two voltage programs (RF cavity and cathode of cooler, Fig.4 on the right) in time with respect each other provided significantly improved transverse emittances (Fig.5, right) while keeping short bunch length of about 110 ns.

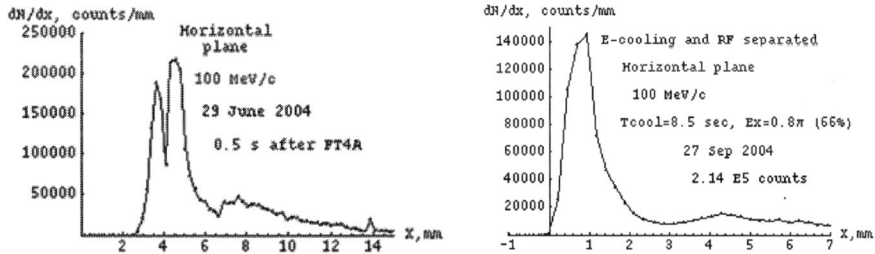

FIGURE 5. Horizontal beam profile measured with scraper before (left) and after (right) adjustment of the voltage programming for cooler cathode and RF cavity.

Reduction of Longitudinal Emittance

Implementation of "tomoscope"-like diagnostics allowed to trace evolution of the longitudinal beam emittance[1]. Particularly, it was found that with reduced voltage from 3kV to 0.5kV on ramp 300MeV/c->100 MeV/c the longitudinal beam emittance at the end of deceleration is reduced by a factor 2. The smaller bucket height causes a larger synchrotron frequency spread which follows by damping of emittance blow up.

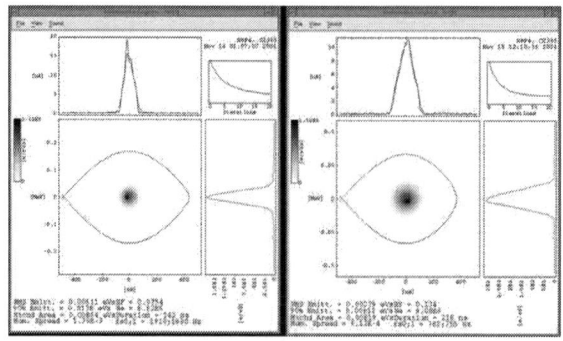

FIGURE 6. Beam inside of RF bucket for different voltage used for beam bunching and deceleration.

MACHINE PERFORMANCE LIMITATIONS

Intensity Limitations

The injection beam intensity is limited by a production beam and by machine acceptance (mainly longitudinal). The longitudinal space charge effect in CERN PS during complicated RF bunch gymnastics puts constraints both on intensity and bunch length of a production beam. Transverse acceptance of AD is optimized, and the longitudinal acceptance of stochastic cooling system is slightly smaller compared with the longitudinal beam emittance after bunch rotation.

The deceleration efficiency which is now about 80% could be increased a bit in the case of implementation of beam diagnostics (tunes and orbit measurements) on ramp. Certain progress might also be expected with more sophisticated correction of the eddy current effects.

Another way to increase intensity (by about 25%) could be achieved with 5 bunch production beam (now 4), which is acceptable for AD with bunch rotation cavities operating at 6th harmonic, but needs certain modifications in CERN PS.

Stacking in the longitudinal phase space could provide about 50% antiprotons/sec gain. This scheme requires modifications in PS RF system and set up in AD.

Cycle Length Limitations

Ramp speed is limited due to eddy current effects. Fast eddy currents in a vacuum chamber of bending magnets (time constant about 2.8 msec) provoke the orbit excursion up to 11mm at the end of the ramp 2 GeV/c->300 MeV/c (most critical point). This can't be compensated. Slow eddy currents in the end plates of a bending magnets cause the orbit excursion up to 45 mm at the same point, but they are partly compensated by special programming of the magnetic field cycle. Yet more sophisticated compensation taking into account differences between the wide and the narrow magnets used in AD is possible and could contribute to machine performance.

Duration of plateaus is defined by the time needed for cooling. Stochastic cooling is well optimized. Electron cooling is slower than expected, probably due to drifting orbit. This drift is caused by slow decay of eddy currents in end plates; it is partly compensated, yet not perfectly.

The cooling performances in AD are summarized in Table below.

TABLE 2. AD cooling performances.

Momentum	Parameters	design	2004
3.57 GeV/c	Stochastic cooling time, sec	20	17
	h/v emittances (2σ), π mm mrad	5	3
	momentum spread (4σ), 10^{-3}	1	1
2.0 GeV/c	Stochastic cooling time, sec	15	6.6
	h/v emittances (2σ), π mm mrad	5	3
	momentum spread (4σ), 10^{-3}	0.3	0.15
0.3 GeV/c	Electron cooling time, sec	6	13.8
	h/v emittances (2σ), π mm mrad	2 / 2	2 / 4
	momentum spread (4σ), 10^{-3}	1	0.1
0.1 GeV/c	Electron cooling time, sec	1	8.4
	h/v emittances (2σ), π mm mrad	1	1 (core)
	momentum spread (4σ), 10^{-3}	0.1	0.1

EXTRA LOW ENERGY ANTIPROTON RING (ELENA)

Motivation

The ejected antiprotons from AD have a kinetic energy of 5.3 MeV, while a few keV energy is required to trap them. Two experiments (ATHENA and ATRAP) use the degrading foil to slow antiprotons down. The drawback of this procedure is significant transverse and longitudinal emittance blow up and beam losses in the foil. The efficiency of capture in this case falls down to 10^{-4}.

The ASACUSA experiment uses Radio Frequency Quadrupole Decelerator for post deceleration of antiprotons down to a few tens of kV. Beam transmission is about 25% in this case and emittance blow up is not far from expected $\sim 1/\beta \gamma$ is observed. After the RFQD thin foil still has to be used to slow beam down.

A significant improvement could be achieved with extra decelerating ring with electron cooling.

Layout

A compact machine with circumference of about 25m is proposed for deceleration of the 5.3 MeV beam delivered by AD down to a kinetic energy of 100 keV. The lattice consists of 4 bending magnets and 8 quadrupoles, grouped in 4 families. One of the long straight sections is used for the electron cooler, and the other for the injection and ejection of the beam. (Fig.7). Special attention has to be paid to a beam diagnostics which is challenging at such a low intensity and energy.

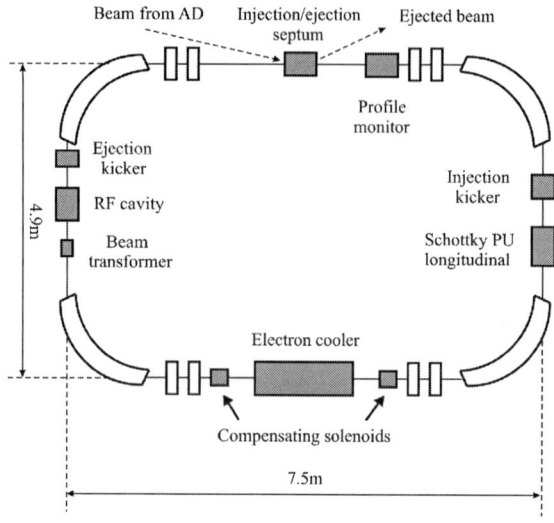

FIGURE 7. ELENA layout.

Electron Cooler and Its Effect on Machine Optics

The cooler is the key element in the machine. Cooling must be fast enough to maintain small beam emittances and counteract IBS and gas scattering at low energies. Careful cooler design has to be done to provide low transverse temperatures of the electron beam at very low energies. With 1m cooling length, integrated correctors, and 90° bend to minimize space it fits properly in one of the straight sections. The vacuum chamber in cooler is assumed to be coated with NEG's.

The simulations done with BETACOOL code [2] showed that about 150 Gs to 200 Gs magnetic field in cooler solenoid needed to make the cooling working fast. At low energy of 100 keV this field (together with fields of compensating solenoids) produces huge tune shifts, and special efforts have to be done to keep optics reasonable. This tune shift scales inversely with energy and is small at injection. During deceleration to low energies it goes up, hence optics has to be flexible to make the necessary adjustments. The tune shift due to the electron beam is noticeable as well (about 0.016) and has to be taken into account also. The parameters of the cooler are given in Table 3.

TABLE 3. Parameters of electron cooler for ELENA.

Cooling length, m	1
Voltage, V	490 / 54
Electron beam current, mA	55 / 2
Beam temperature, eV (transverse / longitudinal)	0.1 / 0.001
Beam radius, cm	2.5
Perveance, μP	5
Magnetic field, Gs	200
Full cooling time at 100 keV, s	2

Other Challenges in ELENA

The beam intensity limitation comes from incoherent tune shift caused by space charge. Based on experience in AD, CERN PS and Booster, a conservative estimate of 0.010 was chosen. This value puts strong limitation to the beam parameters at extraction energy of 100 keV. To relax them, longer extracted bunches of 300 ns have been accepted. Another option is to extract beam from ELENA in several batches. Unfortunately, experiments need about 20s between consecutive batches, and this is difficult to fit in due to lifetime limitation at low energy.

To provide reasonable lifetime ultra high vacuum of $3 \cdot 10^{-12}$ Torr is assumed. This will limit beam emittance blow up due to residual gas scattering to 0.5π mm mrad/s.

Another lifetime limitation comes from intra beam scattering. Due to its dependence on bunch length it is much more severe for bunched beam (deceleration and extraction). With beam parameters at 100 keV after cooling ($N_b=1.5 \cdot 10^7$, $\varepsilon_{x,y}=1\pi$ mm mrad and $\Delta p/p=10^{-4}$, bunch length 1.3m) emittances go up as high as $\varepsilon_{x,y}=2.4 / 0.96 \pi$ mm mrad and $\Delta p/p=6.4 \cdot 10^{-4}$ during 0.5 sec approximately needed for beam bunching and extraction (calculations made with BETACOOL code). Taking this into account, RF programming for beam bunching and before extraction has to have significant margins.

Main parameters of ELENA are summarized in Table 4. To have $1.3 \cdot 10^7$ antiprotons in extracted beam, at least 40% efficiency for beam transfer from AD to ELENA and its further deceleration is needed.

TABLE 4. ELENA main parameters.

Energy, MeV	5.3 – 0.1
Circumference, m	22.6
Working point at 100 keV	1.45 / 1.43
Emittances at 100 keV, π mm mrad	5 / 5
Intensity limitation by space charge for 1 bunch	$1.3 \cdot 10^7$
Average antiproton flux (one bunch), 1/sec	$1.5 \cdot 10^5$
Maximal incoherent tune shift	0.10
Bunch length at 100 keV, m / ns	1.3 / 300
Required vacuum for $\Delta\varepsilon=0.5\pi$ mm mrad/s, Torr	$3 \cdot 10^{-12}$
Beam emittances after 0.5s blow up by IBS ($\varepsilon^i_{x,y}=1\pi$ mm mrad, $\Delta p/p=1 \cdot 10^{-4}$), s	$2.4 / 0.96 / 6.4 \cdot 10^{-4}$

ELENA Allocation in AD Hall

Only small rearrangements have to be done to locate the ELENA ring in the AD Hall (Fig.8). Certain rearrangement of shielding and barracks is needed. ELENA injection line has to be designed and prepared as well. The beam delivery from ELENA to experiments requires the use of electrostatic elements in the transfer lines, which is different to the present case in AD.

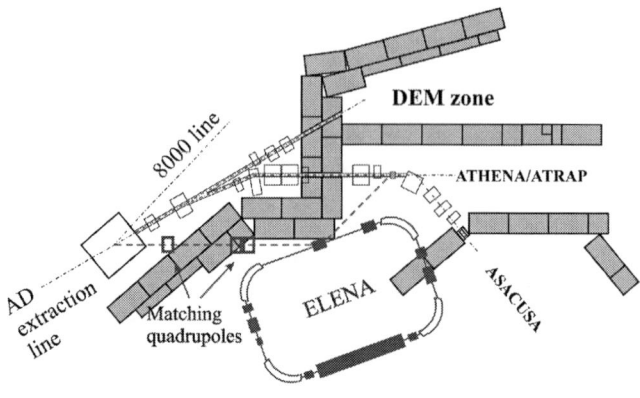

FIGURE 8. ELENA layout in AD Hall.

CONCLUSIONS

Since 2000 AD performance is gradually improving and now $3 \cdot 10^7$ antiprotons in shot are delivered to experiments every 85 seconds. The deceleration efficiency now is as high as 80% due to well working stochastic and electron cooling systems. The beam emittances at ejection are about 1 π mm mrad in both planes and momentum spread after bunch rotation is about 10^{-3}.

To make antiproton capture in trap more efficient, small ring for decelerating antiprotons further down to 100 keV is proposed. This would increase the efficiency of trapping at least one order of magnitude or more.

REFERENCES

1. S. Hancock, e.a. . Tomographic Measurements of Longitudinal Phase Space Density, CERN-PS-99-002-OP.
2. I.N.Meshkov e.a. Simulation of electron cooling process in storage rings using BETACOOL program, in: Proceedings of Beam Cooling and Related Topics, Bad Honnef, Germany, 2001.
3. Charlton e.a. Extra Low ENergy Antiproton ring between the present Antiproton Decelerator and experiments in AD/CERN. CERN-SPSC-2005-029/I-233.

LEIR Cooler Status

Gerard Tranquille

AB Department
CERN
CH 1211 Geneva
Switzerland

Abstract. The LHC program foresees lead-lead collisions in the spring of 2008 with luminosity up to 10^{27} cm^{-2}s^{-1}. To achieve this, the Low Energy Ion Ring (LEIR) has undergone a major upgrade in order to transform lead ion pulses from the LINAC3 into short high-brightness bunches using multi-turn injection, cooling and accumulation. The electron cooler plays an essential role in producing the required beam brightness by rapidly cooling down the newly injected beam and then dragging it to the stack. For the LEIR project, a new cooler has been constructed in collaboration with BINP Novosibirsk. It will deliver up to 600 mA of electron current at an energy of 2.3 keV (corresponding to the lead ion energy of 4.2 MeV/u) and has the advantage that the radius and density profile can be varied to match the characteristics of the injected and circulating lead ion beam. The device was delivered at the end of 2004 and is presently being commissioned. First results of lead ion cooling and stacking with a variable density electron beam are expected in September.

Keywords: LEIR, electron cooling, ion beams.
PACS: 01.30.Cc

INTRODUCTION

For the preparation of dense bunches of lead ions for the Large Hadron Collider (LHC), electron cooling will be essential for accumulation in a storage ring at 4.2 MeV/u. After the completion of the antiproton physics programme at the end of 1996, the Low Energy Antiproton Ring (renamed LEIR for Low Energy Ion Ring) was modified for a final series of experiments to test the lead ion accumulation scheme that is foreseen for the LHC [1]. Tests were carried out in order to determine the optimum parameters for a future state-of-the-art electron cooling device which would be able to cool Linac pulses of lead ions in less than 100 ms. The experiments focused on the generation of a stable high intensity electron beam that is needed to free space in both longitudinal and transverse phase space for incoming pulses.

A feasibility study was made soon after the tests, drawing up the main parameters needed for such a cooler and its integration into the future LEIR machine. Then followed a detailed design study made by BINP Novosibirsk [2] and by May 2003 an agreement had been signed between CERN and BINP for the construction of the new device. After acceptance tests made in Novosibirsk in September 2004, the cooler was delivered to CERN in December. Before its installation in the LEIR ring, many modifications were made to adapt the various elements to CERN standards and an additional series of magnetic measurements were made. All the vacuum elements were then sent for cleaning and preparation for ultra high vacuum (dynamic pressure less than 4×10^{-12} torr) operation in LEIR followed by the bakeout of the ensemble. First cooling tests with O4+ ions were made in November 2005 soon after the successful commissioning of the cooler.

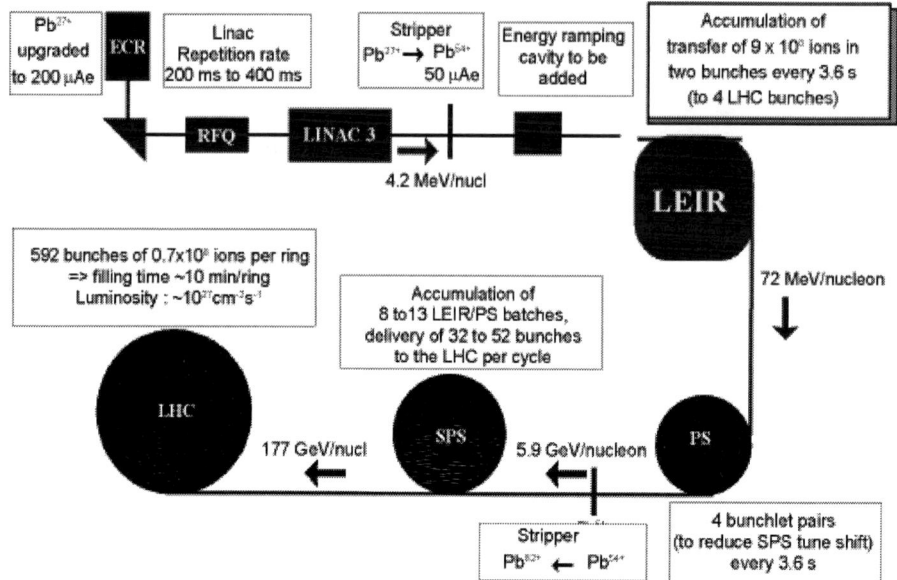

FIGURE 1. LHC ion filling scheme for nominal operation.

THE ELECTRON COOLING DEVICE

The new electron cooler is based on a design previously used for the construction by BINP of the two electron cooling devices for IMP Lanzhou in China. Taking into account recent improvements tested on various electron coolers during the last decade, it uses a high-perveance, variable-density gun followed by an adiabatic expansion provided by an additional solenoid.

The high perveance aims at providing an electron beam with a high density in order to decrease the cooling rate. However, increasing the electron density induces first an increase of the recombination rate (ions may capture an electron from the cooler and, finally be lost hitting the vacuum chamber), which is detrimental to the ion beam lifetime, and secondly increases the electron azimuthal drift velocity and thus increases the cooling time. For these reasons the electron gun has a "control electrode" used to vary the density distribution of the electron beam. In this manner the lifetime of the cooled ion beam will be increased by a reduction of the recombination rate of the ions with the electron beam. The electron beam profile is adjusted in such a way that the density at the centre where the stack sits is smaller and thus the recombination rate is reduced. At larger radii, the density is large and allows efficient cooling of the injected beam executing large betatron oscillations.

The main parameters of the electron cooler are given in table 1. Two operational regimes will be used depending on the momentum of the ions to be cooled. If the small normalized emittances required cannot be reached at injection energy e.g. due to direct space charge detuning, operation of the electron cooler at higher energy allows to cool again at higher energy. In this scenario (unlikely for Pb ion operation, but a possible option for an eventual later upgrade to lighter ions), the LEIR magnetic cycle must contain an additional plateau at a suitable higher energy.

TABLE 1. Main characteristics of the electron cooling device.

Electron beam energy	2.3 keV to 6 keV	6 keV to 40 keV
Electron beam intensity	0.05 A to 0.6 A	up to 3 A
Ion beam momentum·c	88.6 MeV/u to 143 MeV/u	143 MeV/u to 378 MeV/u
Adiabatic factor k	1 to 3	
Electron beam radius	14 mm to 25 mm	
Maximum B field at the gun	0.235 T	
Relative current losses	$<10^{-4}$	
Collector perveance	$25 \cdot 10^{-6}$ A·V$^{-3/2}$ or 25 µP	
Maximum energy spreads	$\Delta_{e\sigma} = 1$ meV, $\Delta_{e\perp} = 100$ meV	
Drift space length	2.5 m	
LEIR machine Twiss parameters at the cooler	$\beta_h = \beta_v = 5$ m, $D = 0$ m	

CONSTRUCTION AND COMMISSIONING

The main components of the electron cooler are shown in figure 2. The principal parts of the device are:

- A gun providing a dense quasi-monoenergetic electron beam. By monoenergetic we suppose that the longitudinal ($\Delta e\sigma$) and transversal ($\Delta e\perp$) electron energy spreads at the gun exit are much smaller than the corresponding ion energy spreads.
- The gun has an additional "control" (or "forming") electrode that enables the density distribution in the electron beam to be varied.
- The beam is expanded by a dedicated solenoid and is bent by a toroid in order to merge with the ion beam to be cooled in the drift space.
- At the end of the drift space the electron beam is bent away from the ion beam and finally collected.
- To improve the collection efficiency, electrostatic plates are used to completely compensate the centrifugal force experienced by the electrons in the toroids.
- The overall system is embedded in a longitudinal magnetic field to counteract the electron beam space charge forces and to "magnetise" the electrons (the electrons execute Larmor circles around the magnetic field lines, and this in turn has significant impact on the cooling forces experienced by an ion in the cooler).

In order to meet the stringent vacuum requirements of LEIR [3], the raw materials used for the cooler construction (316LN stainless steel, CF flanges, bellows, high-voltage feedthroughs) were shipped from CERN to Novosibirsk so that construction of the device could begin in March 2004.

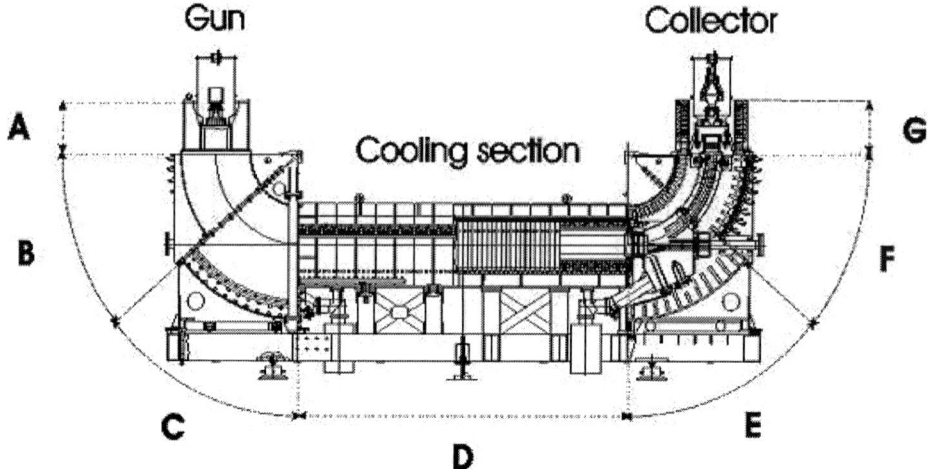

FIGURE 2. LEIR electron cooler general layout.

The "pancake" coils needed to form the longitudinal guiding field were built in the BINP workshop and were ready for testing during the summer of 2004. The magnetic system consists of three main parts: the cooling section (figure 2, section D), the 40° and 50° toroids (figure 2, sections B,C,E & F), and the gun and collector solenoids (figure 2, sections A & G). The series connection of the gun and collector solenoids provides a stable beam size at entrance to the collector. This field value is kept relatively high (0.235 T) also because of the high-perveance gun which requires a strong confinement of the dense electron beam as it is accelerated to the desired energy. However, the vertical component of the magnetic field in the toroids induces a horizontal kick on the ion beam and therefore the maximum operational field in the toroids and the cooling section is 0.075 T. The ratio of the currents in the cooling section solenoid and the gun solenoid is used to control the electron beam size in the cooling zone. The magnetic field in the toroids provides the matching of fields in the gun region and the cooling section.

The magnetic field must be uniform in order to ensure a good cooling efficiency. Accurate measurements were made in Novosibirsk [4] in order to fulfill the required parameters. Independent of the gun requirements, a high magnetic field is welcome for improving the cooling process but the transverse component $B_\perp$ must be kept very small so that the ratio $B_\perp/B$ never exceeds 10^{-4} all along the electron trajectory. To achieve this field quality, careful adjustment of the "pancake" structure of the solenoids was made before final installation in the LEIR ring (figures 3 & 4).

The bakeout procedure aims to obtain a static vacuum in the cooler in the low 10^{-12} torr range. All vacuum elements are heated to temperatures between 100°C and 300°C over a period of a few days and the NEG elements (vacuum chamber coating and cartridges placed in the toroids and gun and collector entrances) are activated towards the end of cycle by heating to 700°C for 2 hours. Unfortunately after the first bakeout a leak was detected on one of the gun ceramics and the measured pressure was an order of magnitude greater than what is needed. After the installation of a differential pumping system around the gun, the pressure level decreased to a more acceptable level of 6×10^{-12} torr.

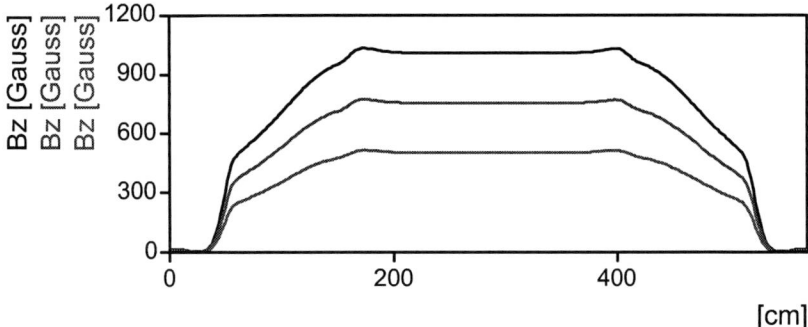

FIGURE 3. Longitudinal field measurements along the cooling axis for a field of 500 G (cyan), 750 G (red) and 1000 G (blue) in the cooling section.

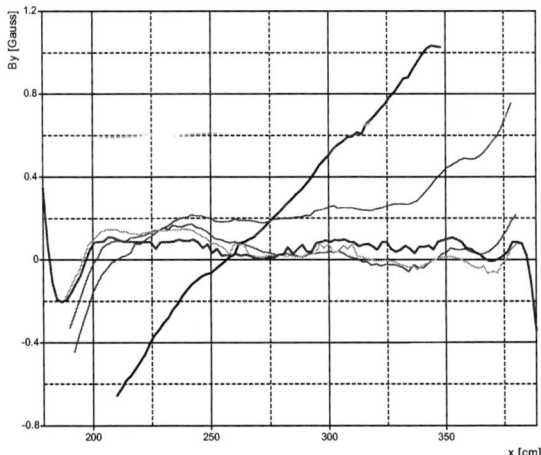

FIGURE 4. Optimisation of the vertical field component in the cooling section by adjustment of the individual "pancakes" for a cooling section field of 750 G. In 5 iterations (blue trace to magenta trace) the ratio of the vertical field component to the longitudinal component was reduced to 10^{-4}.

Electron beam parameter measurements were made at BINP and CERN prior to the first cooling tests. The results show that the performance of the cooler is close to the design specification with stable high intensity beams readily available for ion beam cooling. Due to the lack of time, a complete series of measurements exploring the ultimate limits of the electron gun was not possible, but nevertheless we were able to obtain more the 400 mA of electron current with a nearly "hollow" density distribution at the required energy for ion beam cooling of 4.2 MeV/u in LEIR. Figure 5 shows the beam perveance as a function of the ratio of the voltage on the control electrode with respect to the grid electrode potential at an electron energy of 2.5 keV.

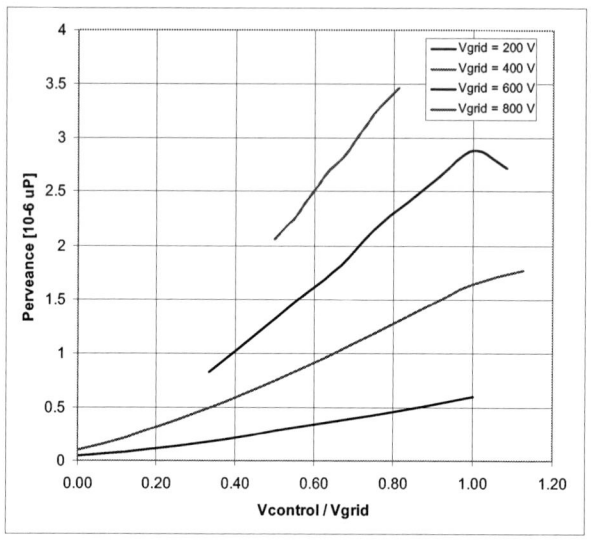

FIGURE 5. Beam perveance as a function of $V_{control}/V_{grid}$ for $E_e = 2.5$ keV.

FIRST COOLING OF O^{4+} IONS

Multiturn injection with simultaneous stacking in momentum and in both tranverse phase spaces is one of the key ingredients for high stacking rates in LEIR :
- This method allows to inject a 200 µs long linac pulse (corresponding to about 70 revolutions in LEIR) with good efficiency (of about 70 % according to simulations) while keeping the stack in the machine.
- The transverse emittances of the injected beam will be relatively small ($\varepsilon_H = 60$ µm and $\varepsilon_V = 40$ µm), but the momentum spread will be large. In fact, the newly injected beam will "sit above" the stack in terms of momentum and then be decelerated and merged with the stack. Such an arrangement is suitable for fast electron cooling, because, typically, longitudinal cooling forces, experienced by an ion of the circulating beam in the cooler, are larger than the transverse ones.

This special multiturn injection is a straightforward extension of simultaneous stacking in horizontal phase space and momentum, a method successfully demonstrated in the framework of the accumulation test [1] in LEAR. For LEIR, stacking in the vertical phase space is also used in order to further increase the length of the Linac pulse that can be injected with good efficiency. A key ingredient for the multiturn injection with stacking in momentum and in transverse phase spaces is a lattice with a relatively large normalized dispersion at the location of the injection septum.

For the LEIR commissioning it was decided to use O^{4+} ions instead of Pb^{54+} ions as the beam vacuum lifetime is expected to be longer. The first injections into the machine were made at the beginning of October 2005 and soon after, the multiturn injection system was put into operation thus increasing the number of injected ions to well above 2×10^{10} particles. Successful cooling was made during November and with the limited diagnostics available we were able to demonstrate the effective cooling of the injected beam. However, it is clear from our observations that much work needs to be done in the coming months to optimise the cooling process. Beam cooling proceeds very quickly (figure 6) after injection, but some 350 ms into cooling instabilities become apparent and the beam intensity decreases exponentially. A transverse damper system will be commissioned at the startup in February next year and it is hoped that we will be able to better control the beam and advance in our cooling studies.

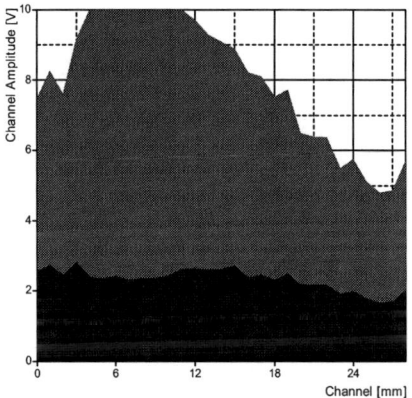

FIGURE 6. Horizontal beam profile measured with an ionisation profile monitor. The blue trace shows the profile at injection and the red profile is after 300 ms of cooling. Saturation occurs due to the large counts on some of the readout channels.

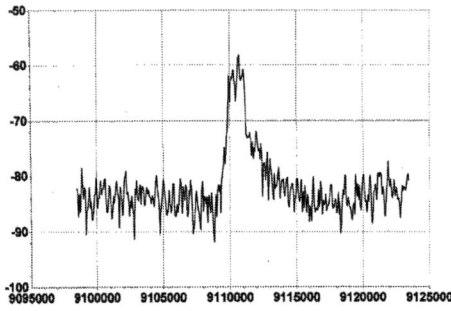

FIGURE 7. Longitudinal Schottky profile of the cooled ion beam. The measured momentum spread is less than 9×10^{-5}.

ACKNOWLEDGMENTS

We would like to thank all those (too numerous to all be named in this paragraph) who have been involved in making the LEIR electron cooler a reality in such a short space of time.

REFERENCES

1. J. Bosser, C. Carli, M. Chanel, R. Maccaferri, D. Mohl, G. Molinari, S. Maury, G. Tranquille , "The Production of Dense Lead-ion Beams for the CERN LHC", Proceedings of ECOOL99, Uppsala, Sweden, 1999, and CERN/PS 99-042 (BD).
2. V. Parkhomchuk, "Feasibility Study for the Design, Construction and Testing of an Electron Cooler for LEIR", BINP internal report (2002).
3. E. Mahner, "The Vacuum System of the Low Energy Ion Ring at CERN: Requirements, Design and Challenges", CERN AT-2005-013.
4. A. Bubley, V. Bocharov, V, Panasyuk, V. Parkhomchuk, "Precise Measurements of a Magnetic Field at the Solenoids for low Energy Coolers", This workshop.

Commissioning of HIRFL-CSR and its Electron Coolers

Xiaodong Yang[*], Vasily Parkhomchuk[†], Wenlong Zhan[*], Jiawen Xia[*], Hongwei Zhao[*], Youjin Yuan[*], Mingtao Song[*], Jie Li[*], Lijun Mao[*], Wang Lu[*], Zhixue Wang[*], BINP Electron Cooler Group[†]

Institute of Modern Physics, Chinese Academy of Sciences, Lanzhou, 730000, P. R. China
†Budker Institute of Nuclear Physics, Laverentyeva 11, Novosibirsk, Russia

Abstract: The brief achievements of HIRFL-CSR commissioning and the achieved parameters of its coolers were presented. With the help of electron cooling code, the cooling time of ion beam were extensive simulated in various parameters of the ion beam in the HIRFL-CSR electron cooling storage rings respectively, such as ion beam energy, initial transverse emittance, and momentum spread. The influence of the machine lattice parameters-betatron function, and dispersion function on the cooling time was investigated. The parameters of electron beam and cooling devices were taken into account, such as effective cooling length, magnetic field strength and its parallelism in cooling section, electron beam size and density. As a result, the lattice parameters of HIRFL-CSR were optimal for electron cooling, and the parameters of electron beam can be optimized according to the parameters of heavy ion beam.

INSTRUCTION

HIRFL-CSR(Heavy Ion Research Facility at Lanzhou---Cooling Storage Ring) is multi-purpose accelerator complex[1], it is consisted two storage ring, the heavy ion beam with energy range 8-50 Mev/u from HIRFL—composed two existing cyclotron SFC(K=69) and SSC (K=450) is used as injector, will be accumulated, cooled and accelerated to the high energy range of 100—400 Mev/u in the main ring(CSRm), then extracted fast to produce RIB or highly charged heavy ions. The secondary beams will be accepted and stored by experimental ring(CSRe) for many internal target experiments or high precision spectroscopy with beam cooling, On the other hand, the beam with energy range of 100-900 Mev/u will also be extracted from CSRm with slow and fast extraction mode for many external target experiments.

Each ring was equipped a electron cooling device, the electron energy in the main ring is 35 keV and 300 keV for the experimental ring. The maximum magnetic field in the electron gun region is 0.45 Tesla,

With the help of electron cooling code, the cooling time of ion beam were extensive simulated in various parameters of the ion beam in the HIRFL-CSR electron cooling storage rings respectively[2,3].

SIMULATION OF ELECTRON COOLING

The cooling rate depends on not only the storage ring lattice parameters, the Betatron function, dispersion of the cooling section, initial emittance and momentum spread of ion, energy and charge state of ion beam, but also on the construction of electron cooling device, the strength of magnetic field, the parallelism of magnetic field in the cooling section, the effective cooling length, and the parameters of electron beam, such as radius, density and transverse temperature of electron beam. These parameters are determined by the storage ring and the technology limitation, on the other hand, they are influenced and restricted each other.

In the case of limit magnetic field, the cooling force is:

$$\vec{F} = -\frac{4Z^2 e^4 n_e}{A m_p m_e} \frac{\vec{v}}{\sqrt{(v^2 + v_{eff}^2)^3}} \ln(\frac{\rho_{max} + \rho_{min} + \rho_L}{\rho_{min} + \rho_L}) \quad (1)$$

n_e ---electron density, e ---electron charge, m_e ---mass of electron, m_p ---mass of proton, Z ---charge state of ion, A ---mass number of ion, $\vec{v}$ ---ion velocity in the moving frame. $v_{effe} = \sqrt{v_{//e}^2 + (\Delta v_{\perp e})^2}$ is not velocity of free electron v_e, is synthesis of longitudinal velocity of electron $v_{//e}$ and transverse drift velocity $\Delta v_{\perp e}$ of electron caused by transverse magnetic field and electrical field. $\rho_{max} = v/(\omega_e + 1/\tau)$ ---maximum collision parameter, $\omega_e = \sqrt{4\pi e^2 n_e/m}$ ---plasma frequency of electron beam, τ ---time of ion single pass the cooling section, $\rho_{min} = e^2/mv^2$ ---minimum collision parameter, $\rho_L = m v_e/eB$ ---Lamor radius of electron with v_e in the cooling section with magnetic field B.

Another experiential formula can be used to understand the relation between cooling time and parameters of storage ring, cooling device, ion beam and electron beam.

$$\tau \approx 4\times 10^{12} \frac{R}{L_c} \cdot \frac{A}{Z^2} \cdot \frac{\beta^4 \gamma^5}{j_e} \cdot \theta^3 \qquad (2)$$

R ---average radius of storage ring, L_c ---length of cooling section, $\beta\gamma$ ---relativistic factor, j_e ---density of electron, θ ---angle between ion and electron in the cooling section.

Table 1 presents the main parameters of HIRFL-CSR.

TABLE 1. Main parameters of HIRFL-CSR.

		CSRm	CSRe
Ion energy (MeV/u)		8-50	25-400
Electron energy (keV)		4.39-27.43	13.7-219.44
Ring circumference (M)		161.00	128.80
Length of cooling section (M)		4.0	4.0
Betatron function (M)	β_H	10.0	7.5
	β_V	17.0	14.9
Dispersion		0	0
Magnetic field in cooling section (T)		<0.15	<0.15
Max electron current (A)		3	3
Initial emittance	ε_H	150	30
$(\pi mm \cdot mrad)$	ε_V	20	30
Initial momentum spread		$\pm 1.5\times 10^{-3}$	$<\pm 5\times 10^{-3}$
Final emittance	ε_H	30	1
$(\pi mm \cdot mrad)$	ε_V	30	1
Final momentum spread		$\pm 5\times 10^{-4}$	$<\pm 1\times 10^{-6}$

For the ion beam given initial emittance and momentum spread, the Betatron function and dispersion of the storage ring determine the size and angle spread of ion beam in the cooling section. When the longitudinal velocity of ion and electron match well, the angle spread θ in the formula (2) is given by angle spread of electron and ion $\theta = \sqrt{\theta_i^2 + \theta_e^2}$. The angle spread of electron is about $\theta_e = \sqrt{T_e/(mc^2\beta^2\gamma^2)} \approx 4mrad$ when the temperature of electron is about 0.1 eV,

The angle spread of ion is determined $\theta_i = \sqrt{\varepsilon/\beta_T}/2$. According to the formula (2), when the angle spread of ion is bigger than electron, the ion will be cooled faster if the Betatron function is bigger, at the same time, the size of ion beam will approach the size of electron beam, the cooling action will be weakened due to the space charge effect. On the other hand, Betatron function decide the size of ion beam $a_i^2 = \varepsilon \cdot \beta_T$, dispersion determines the displacement of ion beam. In order to cool ion beam down fast, one should compromise among these parameters.

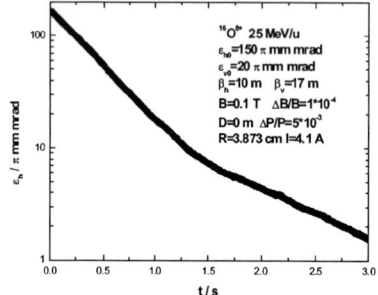 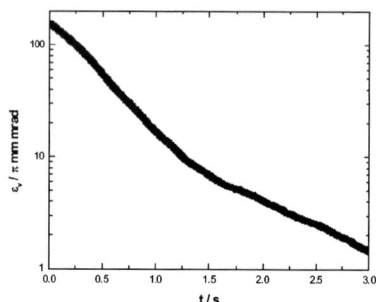

FIGURE 1a. The horizontal emittance vary as the time.

FIGURE 1b. The vertical emittance vary as the time.

Figure 1 shows the transverse emittance decrease with time, it comes from Parkhomchuk code. Figure 2 gives the dependence of cooling time of the transverse direction on the transverse Betatron function, the cooling time denotes the time that the emittances change from initial to the 30 *(πmm · mrad)*, other parameters appear in the figure. Only the x-coordinate value changes, the other was fixed. Intermediate Betatron function yields shorter cooling time than large ones. The optimal Betatron function is about 10—20 M.

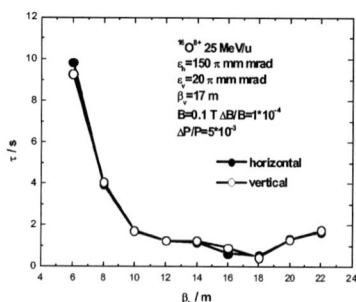

FIGURE 2. Cooling time as a function of horizontal Betatron function.

In order to get fast cooling, the electron cooling device generally was located in the straight section of storage ring where the dispersion is zero, but if the dispersion is not zero in the cooler location, an additional angle was introduced, it will be helpful for cooling, it was dispersive cooling[4]. Figure 3 presents the cooling time as a function of dispersion. If the dispersion in the cooler position is positive, the cooling time becomes shorter than zero dispersion. Figure 4 indicates the cooling time depends on initial momentum spread, fast cooling has been achieved in the case of smaller initial momentum spread.

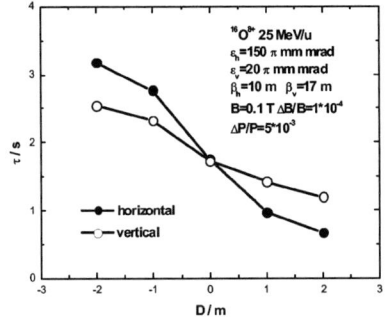 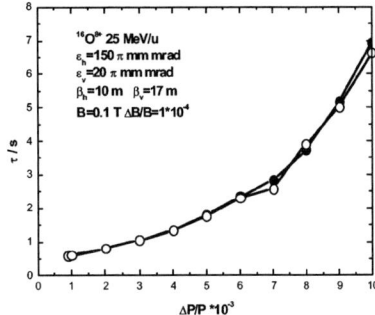

FIGURE 3. Cooling time as a function of dispersion.

FIGURE 4. Cooling time as a function of initial momentum spread.

Figure 5 shows the cooling time varies as initial transverse emittances, the influence of initial emittance is not more strong in the case of optimal situation of electron beam parameters.

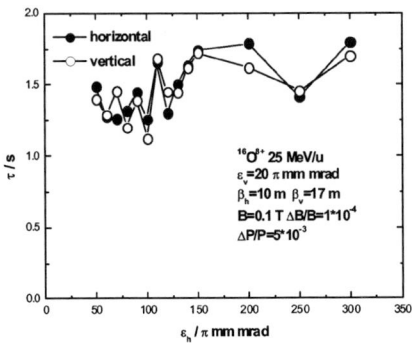

FIGURE 5. Cooling time as a function of initial horizontal emittance.

The cooling time depends on the average longitudinal velocity $\langle v \rangle = \beta c$ of ion and electron, it will be longer in the case of higher energy for the same ion. From the view of electron density $n_e = I/(\pi R_e^2 \beta c)$, this velocity decides the density of electron and time that ion passes through the cooling device in the case of fixed cooling length, the probability of interaction between ion and electron will increase if the velocity is small.

The cooling time is directly proportional to the ion mass, and inversely proportional to the square of ion charge state from the formula (2). Figure 6 presents the cooling time as a function of ion energy, and figure 7 shows the cooling time as a function of ion charge state, the lower energy and higher charge state is helpful to cooling.

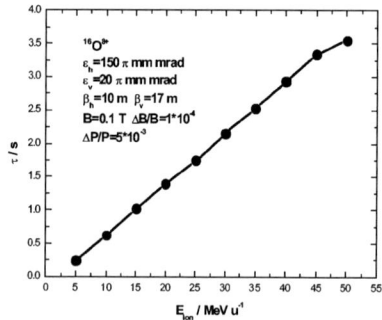
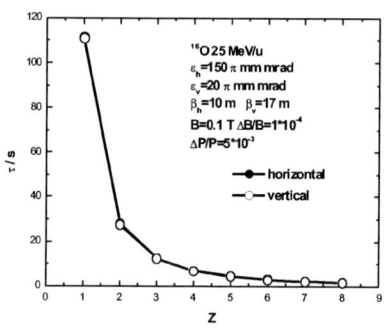

FIGURE 6. Cooling time as a function of ion energy.

FIGURE 7. Cooling time as a function of ion charge state.

The initial ion current is not sensitive to the cooling time as shown in Figure 8, but the length of cooling section strongly influence the cooling time, the result was demonstrated in figure 9.

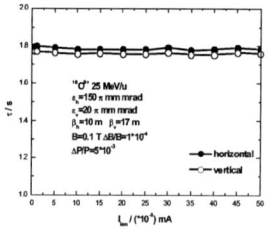
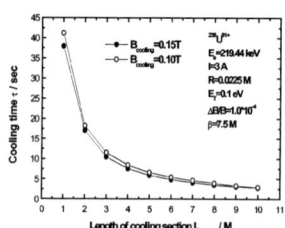

FIGURE 8. Cooling time as a function of ion beam current

FIGURE 9. Cooling time as a function of the length cooling section

The effective transverse velocity v_{eff} is only determined by the magnetic field in the cooler and space charge of electron beam if there is not other factor. The drift velocity in the magnetic field is $v_{drift} = 2\pi n_e R_b c/B$, here R_b is the radius of electron beam, if $v_{eff} = v$ in formula (2), the cooling force approach maximum, and the optimal electron density is $n_e = \beta\gamma^2 B/(2\pi e\beta_T)$, in the practical unit it becomes $j_e \approx 1.6(\beta\gamma)^2 B/\beta_T$, this electron density is not same for different energy. Figure 10 presents the cooling time as a function of magnetic inductive strength in cooling section. The influence of magnetic field parallelism on cooling time is demonstrated in FIGURE 11. The cooling time becomes shorter when the magnetic field strength and its parallelism is higher in the cooling section.

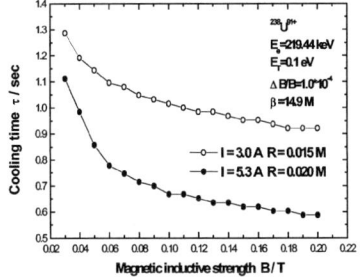

FIGURE 10. Cooling time as a function of the magnetic inductive strength in cooling section

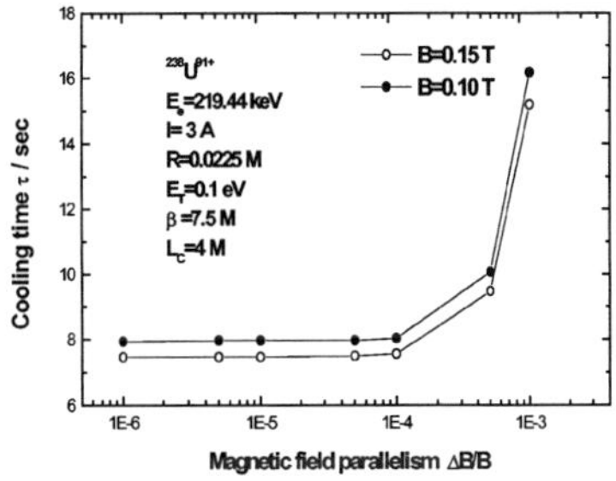

FIGURE 11. Cooling time as a function of the magnetic field parallelism in cooling section

The magnetic field strength and its parallelism in the cooling section influence the electron transverse temperature, specially in case of high energy, the effective velocity $v_{eff} = \gamma v_0 \Delta B/B$ caused by transverse components of magnetic field will be comparable with initial electron temperature, if the ion energy is $400\,MeV/u$, and $\Delta B/B = 10^{-4}$, then $v_{eff} = 3\times 10^4\, m/s$, corresponding temperature is near $53 K$.

From Figure 12, one can see the magnetic field strength slightly influence cooling time. The effect of electron density on cooling time is shown Figure 12, and Figure 13 illustrates the cooling time as a function of electron current for the fixed electron beam radius.

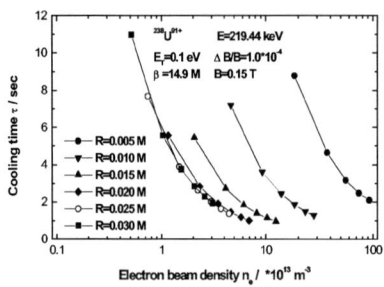

FIGURE 12. Cooling time as a function of the electron beam density

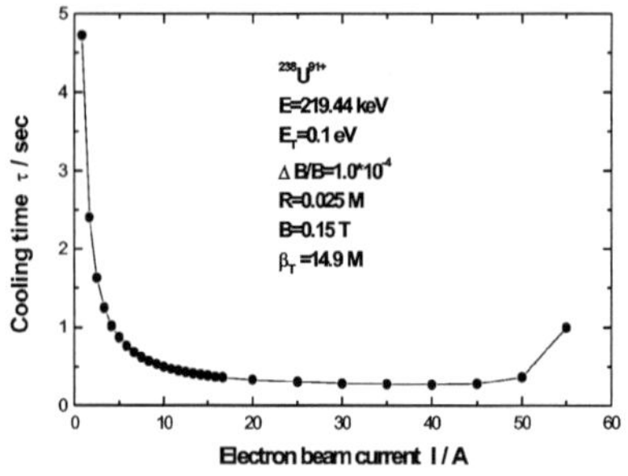

FIGURE 13. Cooling time as a function of the electron beam current

For fixed electron density, the radius of electron decides the region of interaction between ion and electron, if the radius of electron beam is smaller, the outside ion will not be cooled well by electron, the cooling time will become longer, as shown in Figure 14. The electron temperature before the cooling section is determined by the electron gun construction, expansion factor and additional energy caused by through the toroid region, Figure 15 presents the influence of electron transverse temperature.

By the way, the motion of electron beam were studied in the different region of cooler, such as, gun, collector, adiabatic expansion, and toroid, the influences of magnetic field imperfection on the electron beam temperature were presented in references[6,7,8,9].

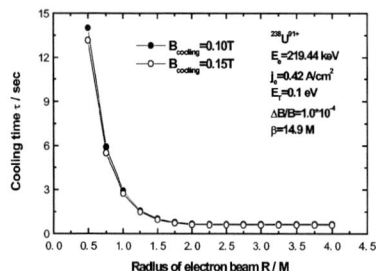
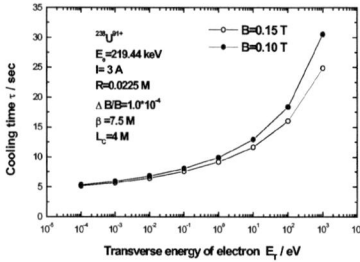

FIGURE 14. Cooling time as a function of the radius of electron beam

FIGURE 15. Cooling time as a function of the transverse temperature of electron

CONCLUSION

From the simulation of electron cooling, it turns out that the lattice parameters of HIRFL---CSR lie in the optimal range for electron cooling. In the CSRm, the ion with higher charge state and lower energy should be chose as injection so that the ion beam will be cooled to required emittance in the shorter time, Electron cooling is more powerful when the injected ion beam has smaller initial emittance and momentum spread . If introduce positive dispersion in the cooling section, the cooling time will become shorter than zero dispersion. For electron beam, its density should approach optimal value in the case of low energy, and as big as possible in the case of high energy. The magnetic field should be strong enough and parallelism is better than 1×10^{-4}.

PREVIOUS COMMISSIONING

Under the cooperation between BINP and IMP, the electron cooling device for HIRFL-CSR main ring was completed in the March 2003, and the cooling device for experimental ring was completed in the beginning of May 2004. Previous commissioning results were reported in references[10,11,12,13]. The control system for coolers was under development right now.

From September 2004, the injection line was tuned with 7 Mev/u C^{4+} beam, the ion beam intensity is about 3 microamperes. The transmission efficiency is about 70%. Three cycles beam were observed in the main ring during first injection. The close orbit has formed on 15 September. The signals from 8 BPM are five times than injection at the same time. Accumulation will be tried next step, The ion beam will be cooled third step, finally acceleration will be tried.

ACKNOWLEDGMENTS

The author would like to acknowledge our Russian colleagues. Sincere honor and respect to the members of international advisory committee.

REFERENCES

1. J. W. Xia, W. L. Zhan, B. W. Wei et al. *Nuclear Instruments and Methods in Physics Research A488* (2002) 11-25
2. X. D. Yang, V. V. Parkhomchuk *High Power Laser and Particle Beam* (in Chinese) Vol.12, No.6, (2000)771-775
3. X. D.Yang, V. V. Parkhomchuk, H. W. Zhao et al. *High Energy Physics and Nuclear Physics* (in Chinese)Vol.27 No.9, (2003) 824-827
4. M. Beutelspacher, M. Grieser, D. Schwalm et al. *Proceedings of EPAC 2002* Vienna, Austria 530-532
5. J. Bosser, C. Carli, M. Chanel et al. *Nuclear Instruments and Methods in Physics Research A532* (2004) 422-426
6. X. D. Yang, M. T. Song, J. W. Xia et al. *High Power Laser and Particle Beam* (in Chinese) Vol.12, No.2, (2000) 245-248
7. X. D. Yang, M. T. Song, J. W. Xia et al. *High Energy Physics and Nuclear Physics* (in Chinese) Vol.24 No.12, (2000) 1179-1184
8. X. D. Yang, M. T. Song, J. W. Xia et al. *High Power Laser and Particle Beam* (in Chinese) Vol.13, No.2, (2001) 223-227
9. X. D. Yang, H. W. Zhao, J. W. Xia et al. *Cyclotrons and their Applications 2001*, Sixteenth International Conference, East Lansing, Michigan 2001, 186-188
10. E. Behtenev, V. Bocharov, V. Bubley et al *Proceedings of EPAC 2004* Lucerne, Switzerland 1419-1421
11. V. Bocharov, V. Bubley, Yu. Boimelstein et al *Nuclear Instruments and Methods in Physics Research A532* (2004) 144-149
12. X. D.Yang, V. V. Parkhomchuk, H. W. Zhao et al. *High Energy Physics and Nuclear Physics* (in Chinese)Vol.27 No.8, (2003) 726-730
13. W. Lu, X. D. Yang, J. Li et al. *Nuclear Physics Review* (in Chinese) Vol.22, No.2, (2005) 186-189

High-Current ERL-Based Electron Cooling System for RHIC

Ilan Ben-Zvi

Brookhaven National Laboratory, Upton, NY 11973, USA

Abstract. The design of an electron cooler must take into account both electron beam dynamics issues as well as the electron cooling physics. Research towards high-energy electron cooling of RHIC is in its 3rd year at Brookhaven National Laboratory. The luminosity upgrade of RHIC calls for electron cooling of various stored ion beams, such as 100 GeV/A gold ions at collision energies. The necessary electron energy of 54 MeV is clearly out of reach for DC accelerator system of any kind. The high energy also necessitates a bunched beam, with a high electron bunch charge, low emittance and small energy spread. The Collider-Accelerator Department adopted the Energy Recovery Linac (ERL) for generating the high-current, high-energy and high-quality electron beam. The RHIC electron cooler ERL will use four Superconducting RF (SRF) 5-cell cavities, designed to operate at ampere-class average currents with high bunch charges. The electron source will be a superconducting, 705.75 MHz laser-photocathode RF gun, followed up by a superconducting Energy Recovery Linac (ERL). An R&D ERL is under construction to demonstrate the ERL at the unprecedented average current of 0.5 amperes. Beam dynamics performance and luminosity enhancement are described for the case of magnetized and non-magnetized electron cooling of RHIC.

Keywords: Electron cooling. High energy. Energy Recovery Linac Magnetized beam. Non-magnetized beam. Superconducting RF gun.
PACS: 29.17.+w, 29.27.-a

INTRODUCTION

The main goal of cooling in a collider is to increase the luminosity, which depends on the details of ion beam's energy and distribution, the properties of the cooler's electron beam and the design of the cooling section. For electron cooling of gold ions in RHIC, the electron energy has to be about 55 MeV and electrostatic acceleration of the electron beam is impossible. RF acceleration of a high-charge bunched electron beam results in an electron transverse velocity spread which is orders of magnitude larger than in conventional coolers. This large temperature of the electron beam has to be carefully minimized by a careful design of the accelerator, and possibly compensated by a strong magnetic field in the cooling solenoid, leading to strong magnetized cooling[1].

The cooling at electron energy of 55 MeV is obviously quite challenging considering that the cooling time is proportional to the energy to the power of 7/2, and that our energies are at $\gamma\sim100$, an order of magnitude higher than even the FNAL Recycler electron cooler, which is at $\gamma\leq10$. Getting the necessary integrated luminosity also brings in various other complications, such as recombination and beam disintegration loss mechanisms.

The R&D towards electron cooling of RHIC[2] proceeds along two directions. The first direction is cooling theory, comprising simulation and experiments in IBS and beam cooling. The other is electron accelerator design and experiments. The cooling theory proceeds to obtain an accurate, benchmarked estimate of the luminosity increase, accounting for the electron beam properties, IBS, cooling friction force, recombination, beam disintegration, instabilities, magnetic field errors, and the properties of the RHIC collider. The electron accelerator design proceeds to develop accelerator components and systems that deliver the best performance for cooling RHIC, carry out the beam dynamics calculations and build components and accelerator systems to benchmark the performance of the challenging accelerator elements.

The performance of the electron beam is fed into the cooling calculations and vice versa. Thus the two research directions have been evolving for a while, getting more consistent and realistic. One of the items under investigation is the comparison of magnetized and non-magnetized electron cooling. Some of the results of such a comparison are discussed below. It should be noted that progress in the generation of high brightness electron beam makes non-magnetized electron beam cooling feasible for RHIC. The issue of recombination in cooling highly charged gold ions seems to be problematic at first glance. Upon careful examination, it is observed that the required luminosity increase can be obtained with a low charge (as low as 2.5 nC) bunches and large beam in the cooling section, thus reducing the recombination rate to well below the beam disintegration rate with no magnetic fields at all in the cooling section. In addition, the use of a low field helical undulator further suppresses recombination[3] to the extent that it is negligible compared to beam disintegration even with a higher electron beam charge.

In this paper, we review the electron cooling of RHIC using magnetized and non-magnetized electrons. We discuss some of the accelerator issues and observe that with the current performance estimated for the cooler's electron beam, non-magnetized cooling is feasible and advantageous for RHIC.

MAGNETIZED ELECTRON COOLING

Luminosity of RHIC with Magnetized Electron Cooling

Using results from the electron beam dynamics calculations on the electron beam parameters and the cooling solenoid, we use the code BetaCool[4] to carry out cooling dynamics simulations.

The simulations show that the design goal for the electron cooler of RHIC, an order of magnitude increase in the integrated luminosity over about a 4 hour (at least) run, can be achieved[5,6], provided that the electron beam can be stretched sufficiently without spoiling its emittance too much.

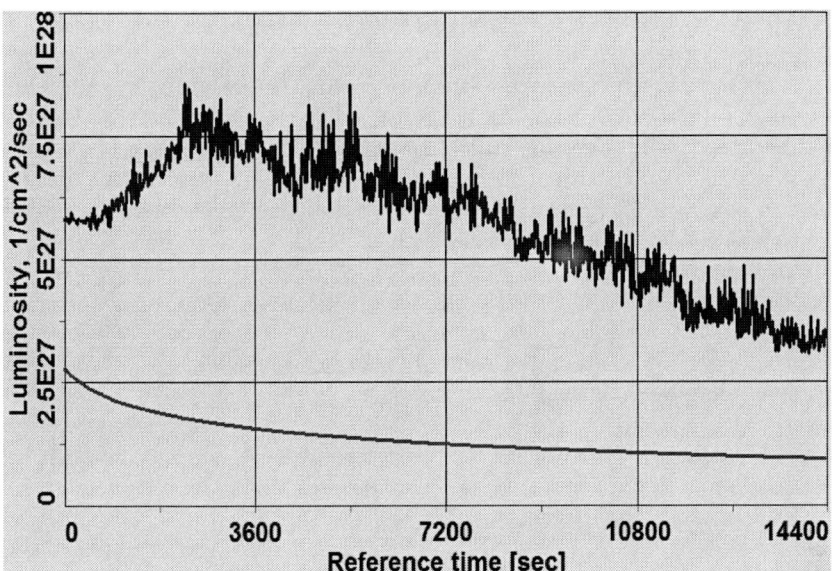

FIGURE 1. Luminosity of RHIC for gold-gold collisions at 100 GeV/A per beam as a function of time in collision with magnetized cooling. The red line represents the luminosity without electron cooling, which decays due to IBS. The blue curve represents the luminosity of the cooled collider. The decline in luminosity here is due to ion beam loss by disintegration in the interaction points.

The order of magnitude luminosity increase (from 7×10^{26} to about 7×10^{27} for gold at 100 GeV/A) can be also achieved for some other ion species and at various energies.

The integrated luminosity under cooling in Figure 1 is calculated from the percentage of the beam burned during 4 hours, for 3 IPs, 112 bunches of 10^9 gold ions, with interaction point beta*=0.5 meters.

The decay in the luminosity curves is either from IBS (for no cooling) or beam disintegration due to the high luminosity (under cooling).

The challenge for the electron beam comes mainly from the requirement for a large charge at a low emittance. The required charge can be estimated from the critical number of electrons, defined as the number of electrons required to achieve equilibrium between electron cooling and IBS heating. The critical number can be written as:

$$N_{ec} \approx \frac{r_i}{r_e} \frac{N_i}{\eta} \frac{\Lambda_{ibs}}{\Lambda_c} \frac{1}{g_f}, \quad \text{where} \quad g_f = \left(\frac{v_{longitud.}}{v_{tranverse}}\right)^2 = \frac{\sigma_p^2}{\gamma^2 (\varepsilon/\beta_a)} \quad (1)$$

For cooling of the whole beam to take place at all, the number of electrons has to be higher than the critical number. Below the critical number, one may observe cooling of the core of the ion beam.

Taking the following parameters of RHIC: Number of ions per bunch $N_i=10^9$, fraction of cooling solenoid filling the RHIC circumference $\eta=0.0078$, IBS Coulomb log $\Lambda_{ibs}=20$, ion velocity form factor $g_f=0.2$, and assuming that the cooler will have magnetized cooling logarithm $\Lambda_c=2$ one gets critical number of electrons about $N_{ec}=1$-$3\ 10^{11}$, depending on the IBS model.

In simulations we see clearly the effect of cooling below and above the critical number. First, a gold bunch is cooled in a BetaCool simulation by 1.2×10^{11} electrons, which is estimated to be below the critical number. We observe cooling of the core of the ions, forming a bi-Gaussian distribution. While cooling with 3×10^{11} electrons, estimated to be above the critical number, we observe that good cooling is provided for the complete beam. The RMS emittance of the beam does not decrease as a function of time under cooling (or may even grow) below the critical number. Thus the simple approximate formula works reasonably well. Below the critical number the luminosity still increases, but the luminosity gain is smaller and the beam-beam parameter is larger. In addition, the centrally peaked "double Gaussian" distribution may lead to beam-beam instabilities for a much smaller value of the beam-beam parameter[7].

Therefore one can see that for RHIC, the charge of the electron beam must be at least 20 nC in order to provide magnetized cooling. In addition, we have requirements for the beam emittance and magnetization and the solenoid length, to achieve the values of the cooling logarithm and fill factor that enter the critical number and affect the cooling speed. The numbers used to achieve the target luminosity growth of Figure 1 are two solenoids at 40m long each, solenoid field of B=5T, electron, emittance 50μm, energy spread 3×10^{-4}.

In the following section we will see what it takes to generate this electron beam.

Generating the Magnetized Electrons

The 20 nC, 54 MeV, 9.4 MHz beam has a beam power of about 10 MW, This is a high enough beam power that calls for Energy Recovery Linac (ERL) for electrical power savings and beam-dump radiation mitigation. The ERL configuration is shown in Figure 2. The critical items in the system are the RF gun, which has to produce a CW high-charge, low emittance beam, the ERL which must be capable of about 200 mA current with no beam breakup, the debuncher which has to stretch the beam and reduce its energy spread without emittance degradation and the solenoids, which are long, high-field and very precise.

In order to produce the high-charge, low emittance beam, we are developing a superconducting laser-photocathode RF gun[8]. A 3-D graphic of the gun's prototype is shown in Figure 3. The advantage of using an RF gun is the ability to provide a high electric field gradient and large energy gain as close to the cathode as possible. This is essential in order to avoid a large emittance growth due to linear and non-linear space-charge forces due to the large charge. The RF superconductivity's small surface resistance allows us to operate the gun in a continuous (CW) mode with affordable electric power. This prototype gun has a single cavity accelerating section, so it will be capable only of 2 to 3 MeV energy. The RHIC electron cooler will use a twin-cavity, or as is better known as a "1 ½ cell" gun[9], since the cavity containing the cathode is shorter due to the lower average speed of the electrons at that point. The 1 ½ cell gun will provide a 4 to5 MeV beam with improved emittance. At the beam current of 200 mA, the input RF power of the gun will be close to 1 MW, all of which going into beam power.

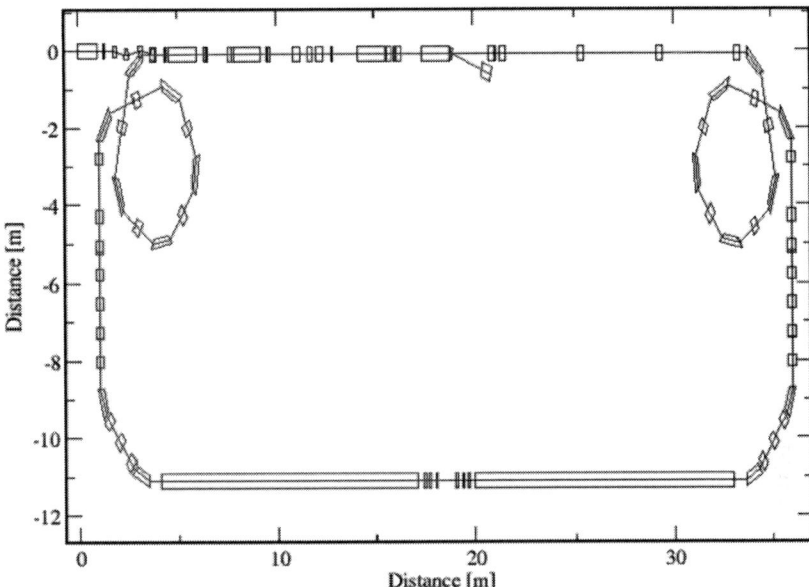

FIGURE 2. Layout of the ERL and cooling section for the RHIC magnetized beam electron cooler. The beam starts at the top left corner, which is the location of the electron gun. The beam is next merged with the returning high-energy beam. Both beams enter the linac, comprising most of the top horizontal line. The bottom contains the long solenoids, in which the electrons move together with the RHIC ion beam. The transport system between the linac and the solenoid contains a beam debunching optics and RF cavity, and the inverse operation is done on the return leg.

In an ERL the emittance of the gun by itself is something that is difficult to measure and not very relevant. One has to consider the properties of the electron bunch following the beam merging system (which may introduce huge emittance growth unless properly designed) and the linac. All of these elements must perform together to recover part of the linear emittance growth, in what is known as emittance compensation[10].

The beam merger which we will use has been developed by our group[11]. It is called the "Z-bend" merging system, and it is capable of reducing the emittance growth in the bending plane to a negligible level. The linac, like the gun, is superconducting at 703.75 MHz (a harmonic of the RHIC revolution frequency) and has been designed[12] to accelerate a high-current electron beam with negligible emittance growth and high-stability.

The performance of the system from the gun to the end of the linac (including the Z-bend merging system) is given in Table 1.

The notable feature of the system is that the low emittance, at about 30 microns for a high bunch charge of 20 nC and a very high magnetization emittance of 380 microns. This is not necessarily the best possible result, since a different bunch shape (elliptical bunch instead of the uniform-Gaussian distribution used for the computation in Table 1) can improve the results even further. However, the main issue in getting this beam to the cooling solenoid is the debunching operation. It is necessary to debunch the beam for two reasons: First, to reduce the longitudinal energy spread, second, to reduce the electron charge density in the cooler and thus avoid a short Debye length which would otherwise result is a lower Coulomb logarithm.

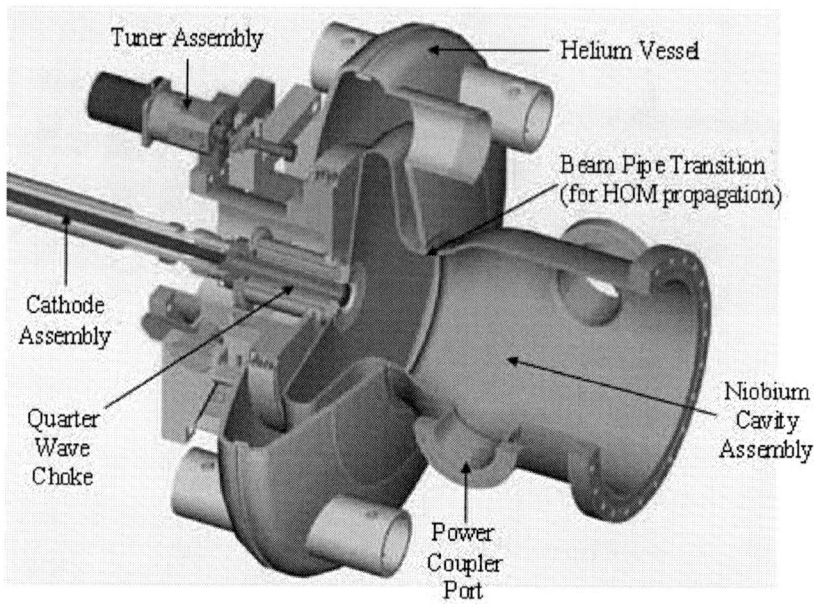

FIGURE 3. Layout of the superconducting laser photocathode RF gun. Details that are not shown to avoid complicating the figure are the cryostat, fundamental power coupler and details of the photocathode insertion mechanism (load-lock). The solid niobium cavity is shown in blue. The photocathode assembly is retractable into the load-lock chamber for cathode replacement under ultra-high vacuum conditions. The large beam pipe is designed to allow High-Order Mode (HOM) power to escape the cavity towards a ferrite microwave damping section located in the beam pipe outside the cryostat (not shown).

Table 1. Beam dynamics parameters from gun to end of linac.

Charge	20nC
Radius (Transverse uniform distribution)	12mm
Magnetization	380mm.mr
Longitudinal Gaussian distribution	4degrees, 16ps
Maximum field on axis of gun cavity	30MV/m
Initial phase	30deg.
Energy at gun exit	4.7MeV
Energy spread at gun exit	rms 1.87%
Bend angle	10degrees
Energy at linac exit	55MeV
Final emittance (normalized rms)	30mm.mr
Final longitudinal emittance	100deg.keV

The difficulty with the debunching transport system is the large magnetization and high space-charge. However, a careful design should keep the final emittance at about 50 microns (thermal in the solenoid), a value that has been used in the cooling simulations.

NON-MAGNETIZED ELECTRON COOLING

Due to the high energy of the RHIC ion beam, the effective longitudinal velocity introduced by the magnetic field error limits the cooling rate from magnetized cooling to a degree that non-magnetized cooling can successfully compete. Since the superconducting magnet experts feel reluctant to promise a field error much smaller than 10^{-5} for the long, multi-Tesla field superconducting solenoid, the benefit of fast magnetized cooling which peaks for ions moving at about the effective electron velocity, is greatly diminished. Using a non-magnetized beam with a lower charge (under 5 nC) enables us to reduce the transverse electron emittance to at or below the ions' normalized emittance. This provides for good cooling rates and has also the following other benefits: First, the elimination of the very long, high-field high-precision solenoids, which save a considerable amount of money and complication. Second, the reduction of the electron current by about a factor of 5 or more, simplifying the system and making it easier. Third, elimination of the bunch stretcher (debuncher), leading to simplification of the beam transport system and additional cost savings. Fourth, uniform cooling of the ions, avoiding fast cooling of the core and thus the generation of a peaked distribution which is problematic for reasons the beam-beam interaction. Fifth, since one does not have to deal with solenoid errors, it is possible to increase the ion beam size in the cooling section and thus lower the ion velocities and get better cooling speed. Sixth, the ability to use the analytic Budker formulae which provide a high level of confidence in the cooling rate. One may also add the recent demonstration of high-energy cooling with non-magnetized beam at Fermilab[13].

Luminosity Performance and Recombination Issues

For a non-magnetized beam, analytic and precise expressions were developed to calculate the friction force. From the friction force, one can arrive at the cooling rate, luminosity growth etc., taking into account the various relevant accelerator physics issues such as exact electron and ion velocity distributions, the competition between IBS and cooling, the effects of beam disintegration and recombination. Naturally, tracking a realistic ion distribution and allowing for the betatron and synchrotron motions is essential. At the RHIC cooling R&D project are using the BetaCool code, working in collaboration with the JINR Dubna group. In this particular case, the code integrates the friction force (2):

$$\vec{F} = -\frac{4\pi n_e e^4 Z^2}{m} \int L \frac{\vec{V}_i - \vec{v}_e}{\left|\vec{V}_i - \vec{v}_e\right|^3} f(v_e) d^3 v_e \qquad L = \ln \frac{\rho_{max}}{r_0} \qquad (2)$$

The result from BetaCool for the luminosity is given in Figure 4.

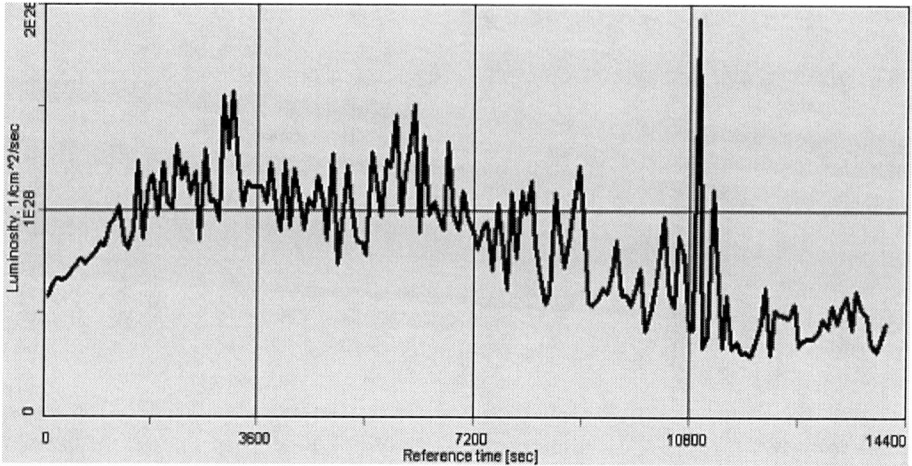

FIGURE 4. Luminosity of RHIC for gold-gold collisions at 100 GeV/A per beam as a function of time in collision with non-magnetized cooling. The decline in luminosity here is due to ion beam loss by disintegration in the interaction points in combination with recombination losses. The spikes are artifacts of the simulation.

The various cooler and beam parameters used in this simulation are given in Table 2. The discussion of the recombination (which is included in this simulation) and the electron beam parameters follow below.

Clearly, as one compares Figure 1 and Figure 4, the luminosity gain through electron cooling is just as good or better for the non-magnetized case. Also one observes a uniform cooling of the ion bunch, with no buildup of density spikes.

Table 2. Non-magnetized electron beam cooling parameters.

Rms momentum spread of electrons	0.1%
Rms normalized emittance	2.5 microns
Rms radius of electron beam in cooling section	0.3 mm
Rms bunch length	5 cm
Charge per bunch	5nC
Cooling sections	2x30 m
Ion beta-function in cooling section	200 m
IBS – Martini's model	exact RHIC lattice

The great surprise one may have from this result is the small role that recombination has to play here, since now we are dealing with highly charged gold ions and very cold electrons. It may help to remember that the decrease in electron bunch charge and the expansion of the beam size (from a beta function of about 60 meters to about 200 meters, possibly more) reduce considerably the electron charge density in the cooler and thus the recombination rate. Furthermore, the beam disintegration in the interaction points dominates the ion loss, and once enough ions are lost the recombination rate also declines. In addition, we can also use a helical undulator to rotate the electrons, producing a coherent velocity that reduces recombination with just a small sacrifice in cooling speed. Using an undulator with a period of λ_w=0.05 m, a field of B=0.002 T wound on a radius of R=0.05 m, we require a current of I=70 Amp to generate the field. Then we get (see eq. 3) the rotation radius as r_0=0.7μm, and the focusing results a beta function of β_w=180 m. The loss in Coulomb log due to the electrons helical motion with a radius of 0.7 microns is only 0.8, something that can be neglected. The coherent rotation is equivalent to better than an electron temperature of 22 eV, enough to make recombination negligible in the disintegration dominated system.

$$\theta = \frac{K}{\gamma} \approx \frac{93.4B\lambda}{\gamma} \qquad r_0 = \frac{\theta\lambda}{2\pi} \qquad (3)$$

We carried out simulations[14] showing that the luminosity increase of an order of magnitude in RHIC can be achieved with no undulator and with an electron bunch charge of 2.5nC and with an emittance of 2 microns.

Generating the Non-Magnetized Electrons

Assuming the same gun as described above for the magnetized beam, and pretty much the same accelerator layout, we[9] use the parameters in Table 3.

Table 3. Beam parameters for gun for non-magnetized electrons.

Bunch length	16degrees (63ps) from head to tail.
Laser pulse shape	Ellipsoid
Lunch phase	about 35deg.
Maximum field on axis	30MV/m
Energy out of gun	4.7 MeV

The main difference is the use of an elliptical bunch shape[15] for the electron beam. This is significant, and leads to a large decrease in emittance, as can be seen in Table 4.

Table 4. RMS normalized emittance vs. bunch charge.

Charge/bunch (nC)	Emittance after linac (μm)
2.5	1.7
3.2	2.0
5	2.9

The longitudinal emittance is under 300 degree*keV, and following 3rd harmonic correction is reduced it to under 100 degree*keV, allowing electron energy spread of under 10^{-4}.

ACKNOWLEDGMENTS

I would like to thank and acknowledge the work done on this research by the many members of the Collider-Accelerator Department's electron-cooling group, accelerator physics and engineering groups as well as Superconducting Magnet Division and Instrumentation Division. Likewise I would like to thank our collaborators in industry (AES and Tech-X), National Laboratories (JLab, FNAL), universities (Indiana) and international institutions (BINP, JINR, Celsius, GSI).

This manuscript has been authored by Brookhaven Science Associates, LLC under Contract No. DE-AC02-98CH10886 with the U.S. Department of Energy. The United States Government retains, and the publisher, by accepting the article for publication, acknowledges, a world-wide license to publish or reproduce the published form of this manuscript, or allow others to do so, for the United States Government purposes. Support was provided by DOE Office of Nuclear Physics. Partial support by the U.S. Department of Defense High Energy Laser Joint Technology Office and Office of Naval Research is acknowledged.

REFERENCES

[1] I. Ben-Zvi and V.V. Parkhomchuk, Collider-Accelerator Department Accelerator Physics Notes, C-AD/AP/47 Brookhaven National Laboratory, Upton NY USA (2001).
[2] RHIC E-cooler Design Report (ZDR-2004), http://www.agsrhichome.bnl.gov/eCool
[3] Private communications, Ya. Derbenev and V. Litvinenko.
[4] The BETACOOL program, http://lepta.jinr.ru
[5] Cooling Dynamics Studies And Scenarios For The RHIC Cooler, A. Fedotov et al, Proceedings, 2005 Particle Accelerator Conference, Knoxville, Tennessee, USA, May 16-20, 2005.
[6] Simulations Of High-Energy Electron Cooling, A. Fedotov et al, Proceedings, 2005 Particle Accelerator Conference, Knoxville, Tennessee, USA, May 16-20, 2005.
[7] C. Montag, N. Malitsky, I. Ben-Zvi, V. Litvinenko, Beam-Beam Simulations for Double-Gaussian Beams, Proceedings, 2005 Particle Accelerator Conference, Knoxville, Tennessee, USA, May 16-20, 2005.
[8] R. Calaga, I. Ben-Zvi, X. Chang, D. Kayran, V. Litvinenko, High Current Superconducting Gun at 703.75 MHz, proceedings of the 2005 International SRF Workshop, Cornell University, USA, July 10-15, 2005.
[9] X.-Y. Chang, Ph.D. dissertation, Stony Brook University, Stony Brook NY, 2005.
[10] L. Serafini and J.B. Rosenzweig, Envelope analysis of intense relativistic quasilaminar beams in rf photoinjectors: A theory of emittance compensation Phys. Rev. E **55**, 7565–7590 (1997).
[11] D. Kayran and V. Litvinenko, A Method of Emittance Preservation in ERL Merging System, Proceedings 2005 International FEL Conference, Stanford CA, USA August 22-26, 2005.
[12] R. Calaga, I. Ben-Zvi, Y. Zhao and J. Sekutowicz, High Current Superconducting Cavities at RHIC, Proc. EPAC'04, 5-9 July 2004, Lucerne, Switzerland
[13] S. Nagaitsev, Antiproton cooling in the Fermilab Recycler, Proceedings of the COOl'05 International Beam Cooling Workshop, Galena, IL,USA September 18-23, 2005.
[14] A. Fedotov, private communication.
[15] C. Limborg-Deprey, Maximizing Brightness in Photoinjectors, Proceedings 2005 International FEL Conference, Stanford CA, USA August 22-26, 2005.

FLAIR Project at GSI

C.P. Welsch*, M. Grieser, J. Ullrich, A. Wolf
for the FLAIR collaboration

Max-Planck Institute for Nuclear Physics, Heidelberg, Germany
*Present address: CERN, Geneva, Switzerland

Abstract. The future Facility for Antiproton and Ion Research (FAIR) at Darmstadt will produce the highest flux of antiprotons in the world. Within the planned complex of storage rings, it will also be feasible to decelerate the antiprotons to about 30 MeV kinetic energy, opening up the unique possibility to create low energy antiprotons and thus, establish low-energy antiproton physics at GSI. In the Facility for Low-energy Antiproton and Ion Research (FLAIR) the antiprotons shall be slowed down by means of two cooler storage rings. In the second one, the Ultra-low energy electrostatic Storage Ring (USR), energies ranging from 300 keV to 20 keV will be available for various in-ring experiments as well as for efficient injection of antiprotons into traps. In the limit of such small beam energies, the realization of efficient electron cooling, employing electron energies of only a few eV is one of the new challenges. In this contribution, a review of the FLAIR facility is given and its deceleration and cooling scheme is elucidated in comparison to the present AD operation scheme. Special emphasis is placed on the problems related to electron cooling at ultra-low energies.

Keywords: Electrostatic storage ring, electron cooling, antiproton physics, LSR, USR.
PACS: 29.17.+w , 29.20.Lq , 34.10.+x, 34.50.Fa

INTRODUCTION

Low-energy antiproton physics is currently being done at the Antiproton Decelerator (AD) at CERN, Geneva [1, 2]. Since the start of the physics program in the year 2000 100 MeV/c antiprotons, ejected in single or multiple batches, have been routinely produced and delivered to the three experiments ASACUSA, ATHENA and ATRAP. Since 2003, a 4th experiment, ACE, uses single batch antiprotons ejected at 300 MeV/c. An overview of the deceleration cycle and subsequent cooling steps is given in Fig. 1.

Due to the low intensity of about 10^5 $\bar{p}$/s and the availability of only pulsed extraction, the physics program is limited to the spectroscopy of antiprotonic atoms and antihydrogen formed in charged particle traps or by stopping antiprotons in low-density gas targets. Furthermore, the output energy of the AD (5 MeV kinetic energy) is still significantly higher than the <100 keV energy best suited for these experiments.

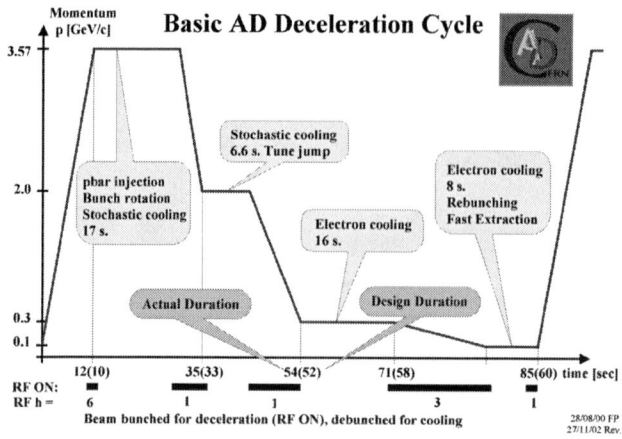

FIGURE 1. Overview of the basic AD deceleration cycle [1].

A next-generation low-energy antiproton facility must overcome these limitations by providing cooled beams at higher intensities and at least a factor 10 lower energy. In addition, it should have the possibility of slow (i. e. continuous) extraction, which will allow nuclear/particle physics type experiments requiring coincidence measurements to be performed.

FLAIR will allow overcoming these limitations. Its accelerator structure to decelerate the antiprotons consists of two storage rings, a magnetic (LSR) and an electrostatic (USR) one, and a universal trap facility (HITRAP). These components of the facility can provide stored as well as fast and slow extracted cooled beams at energies between 30 MeV and 300 keV (LSR), between 300 keV and 20 keV (USR), and cooled particles at rest or at ultra-low eV energies (HITRAP). This will allow a large variety of new experiments to be performed, as can be found in the FLAIR letter of intent [3].

Among the unique experiments only possible at such a facility are nuclear physics studies using antiprotons as a hadronic probe to investigate the structure of nuclei, including radioactive isotopes produced at the future facility, and many atomic-collision type experiments with internal targets in both storage rings with effective intensities as large as 10^{12} $\overline{p}$/s. Moreover it becomes possible with longer storage lifetimes and in combination with a low-energy positron storage ring [4] to produce a continuous flux of $\overline{H}$ atoms at energies close to 1 atomic unit, which can be suitable for $\overline{H}$ collision experiments [3].

PBAR PRODUCTION AND DECELERATION SCHEME

The antiproton generation and deceleration cycle at FAIR follows in principle the CERN layout during the time of the AA/AC rings. It is foreseen to produce the antiprotons by an intense 29 GeV proton beam (delivered by the SIS100) shot on a special production target. Up to 2.8×10^{13} protons per cycle will thus produce 1×10^{8}

antiprotons every 5 seconds. The antiprotons are then precooled at 3 GeV in the Collector Ring (CR) by a stochastic cooling system. Afterwards, the antiprotons are transferred to the accumulator/decelerator ring RESR, where the accumulation of up to 7×10^{10} antiprotons is foreseen. The antiprotons are either reinjected to the SIS100 and accelerated for experiments with high energy antiprotons in the High Energy Storage Ring (HESR) or directly transferred to the New Experimental Storage Ring (NESR), which is used to decelerate the beam to 30 MeV that is then transferred to the FLAIR hall. [5]

A preliminary layout of the FLAIR hall is given in the following Fig. 2. The "working horse" of the facility is the Low-energy Storage Ring (LSR) in area F3, which shall deliver both low-energy highly charged ions at energies between 15 MeV/u and 300 keV/u to the experimental areas F1 and F2 and antiprotons in the energy range between 30 MeV and 300 keV to the areas F2 and F4-F10. In this article, only the operating scheme of the two storage rings with an antiproton beam is discussed.

It is foreseen to use CRYRING from MSL, Stockholm, as the LSR since it is operating in the energy range of interest, has already been operated with acceleration and deceleration, is equipped with the necessary electron cooler, matches the requirements for vacuum and will become available during the next few years.

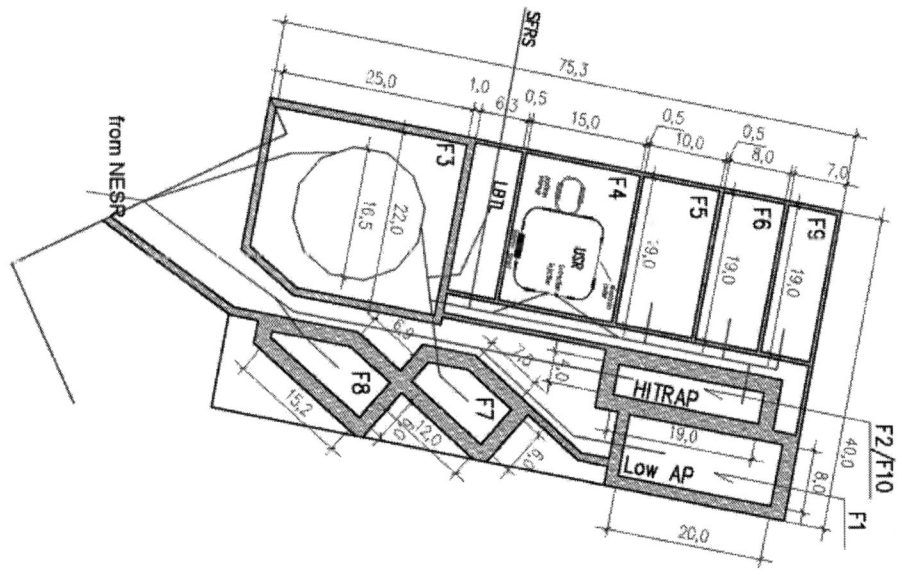

FIGURE 2. Overview of the FLAIR facility [6].

Before being extracted from the NESR, the antiproton beam will be cooled and can therefore be immediately decelerated after injection into the LSR to an intermediate energy of around 4 MeV. At that energy the beam will be electron-cooled for one or a few seconds, then decelerated to the extraction energy of 300 keV/u, where it will be electron-cooled again before actually being extracted towards the second storage ring, the USR. The optimum sequence for deceleration and cooling will be investigated at MSL.

THE ULTRA-LOW ENERGY STORAGE RING

The last synchrotron used for decelerating the antiprotons to lowest energies is the ultra-low energy storage ring. It uses electrostatic elements only to avoid problems with remanence and hysteresis effects, offer a compact and simple design and might also be built up as a cryogenic machine to provide longer life times. An overview of the machine is shown in Fig. 3 and its main parameters are given in Table 1. A detailed description of the ring lattice can be found in [7].

TABLE 1. Summary of the USR design parameters.

General Parameters	
Energy range	20 keV – 300 keV
Circumference	22.28 m
Base pressure	$< 5.10^{-11}$ mbar
Betatron tunes (h/v)	2.29 / 1.08
Chromaticity	-2 / -1.5
# of pbars at 20 keV	10^7
Initial momentum spread	10^{-3}
90° Deflectors	
Height	160 mm
Radii	970 mm and 1030 mm
Shield Distance	15 mm
Voltage |U|	< 20 kV
Quadrupoles	
Length	200 mm
Distance between lenses	150 mm
Aperture Radius	50 mm
Shield Distance	10 mm
Voltage	+/- 10 kV
Steerer Length	100 mm
Steerer Plate Distance	120 mm

The USR shall serve as a true multi-purpose machine providing an electron-cooled antiproton beam in the complete energy range between 20 keV and 300 keV for both, in-ring experiments and effective injection into traps. The luminosity for in-ring experiments will be increased by at least five orders of magnitude as compared to a single pass situation and cross sections as small as 10^{-22} cm^2 will become accessible under single collision conditions.

The design foresees an integrated in-ring reaction microscope [8] such that total, as well as any differential cross sections up to full differential cross sections (FDCS) including ionization-excitation reactions will become measurable serving as benchmark data for theory.

Although some experience with electrostatic storage rings is available [9-11], none of the existing machines is energy-variable, provides electron cooling at different energies, houses an integrated reaction microscope or realizes ultra-short pulses in the order of a few nanoseconds as they will be required for collision experiments.

For this reason, a prototype machine, the cryogenic storage ring (CSR), is at present being built up at MPI-K, Heidelberg and will serve as a testing facility to address the above mentioned open questions [12, 13].

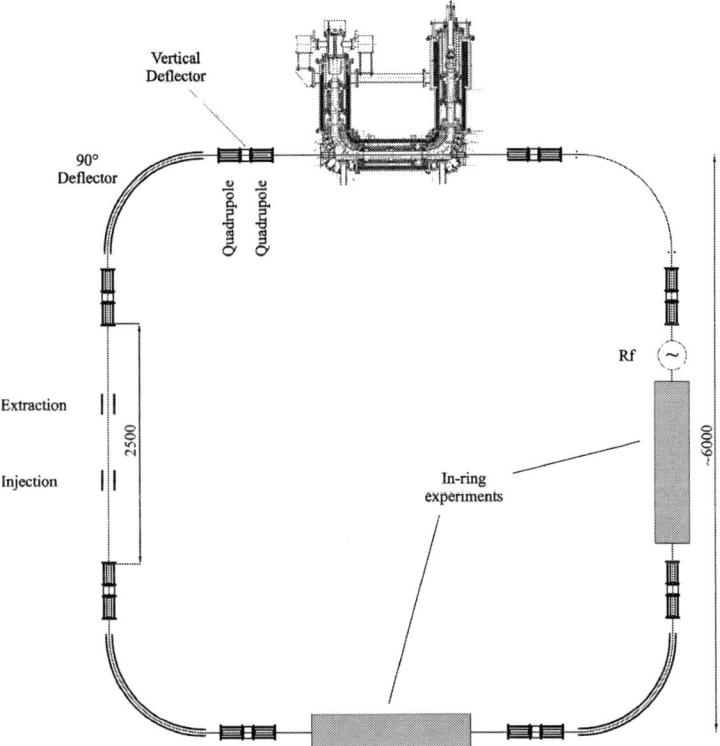

FIGURE 3. Layout of the ultra-low energy storage ring (USR).

A clear advantage of such a compact storage ring in comparison with alternative structures like decelerating RFQs is the availability of a <u>cooled</u> beam over the whole energy range which is in addition available to the experiments via both slow and fast extraction.

Besides the benefits for external experiments, the performance of the storage ring itself can also be improved by a beam cooling system. An electron cooler will counteract small angle scattering of the circulating antiprotons and limit beam losses to single scattering. As a result, the lifetime of the beam is clearly enlarged and rises to several seconds even at lowest energies, Fig. 4. In this calculation, a vacuum pressure of 5×10^{-11} mbar at room temperature is assumed and a maximum tolerable scattering angle of $\Theta_{max}=2$ mrad.

Furthermore, assuming the maximum number of particles N of a non-bunched beam that can be stored at a given beam energy is directly proportional to the beam emittance ε

$$N = \frac{2\pi}{r_p} \cdot B \cdot \beta^2 \cdot \gamma^3 \cdot \varepsilon \cdot (-\Delta Q) \tag{1}$$

where $r_p=1.53 \times 10^{-18}$ m, ΔQ is the tolerable (Laslett) tune shift in the machine and B the so called bunching factor defined as the ratio of bunch length and machine circumference. For the following calculations, a non-bunched beam is assumed.

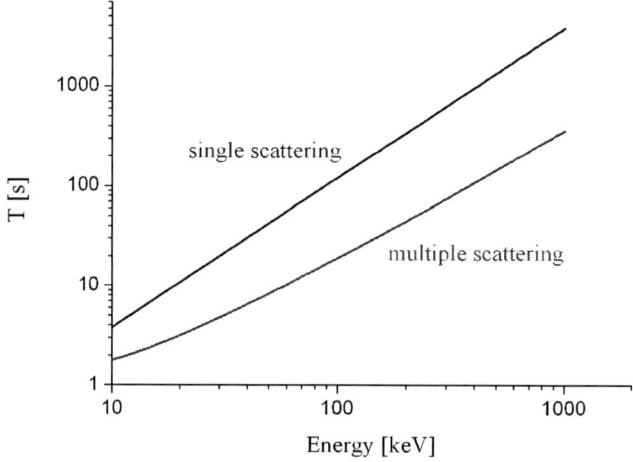

FIGURE 4. Lifetime of antiprotons as function of energy at a base pressure of 5×10^{-11} mbar. Shown are the calculated graphs for multiple and single scattering.

Experience at CERN shows that a tune shift of $-\Delta Q=0.24$ is still acceptable at the proton synchrotron PS [14] for stable operation. For a conservative estimation, a tune shift of $-\Delta Q=0.1$ was assumed for the calculation of the plot shown in Fig. 5. In the heavy ion storage ring TSR at MPI-K, a 23 MeV electron-cooled proton beam could be stably stored with an intensity $I_{C6+}= 1$mA [15]. This corresponds to a proton number $N=5.3 \times 10^9$, which is also shown in Fig. 5. With an expected emittance of 1 mm mrad after the cooling process even at lowest energies, a minimum number of 10^7 antiprotons is available in the USR at all energies.

Using these results in combination with the envisaged antiproton production rates, one gets an overall picture of the antiproton intensities that will be available at FLAIR, Fig. 6. The USR values assume a cooled beam with an emittance of $\varepsilon=1$ mm mrad and 90% losses during the cooling and deceleration cycle.

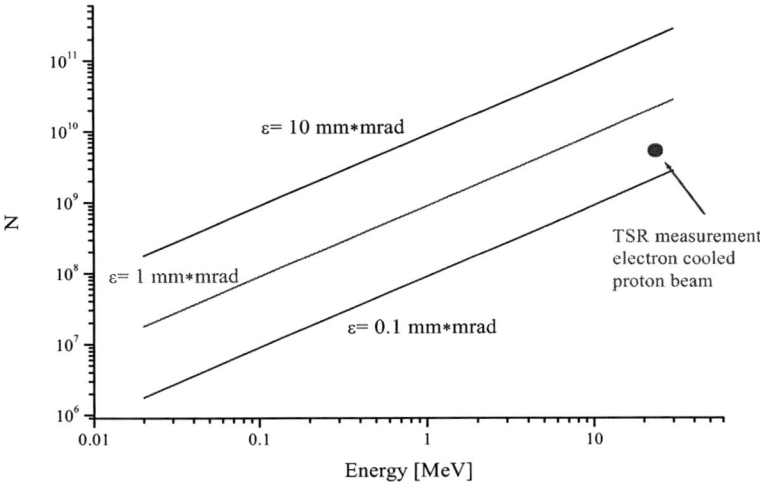

FIGURE 5. Maximum number of antiprotons that can be stored in the USR at a given emittance. A tune shift of $\Delta Q=0.1$ was assumed for the calculations. The point indicates a measurement at the TSR of a C^{6+} beam at 1 mA.

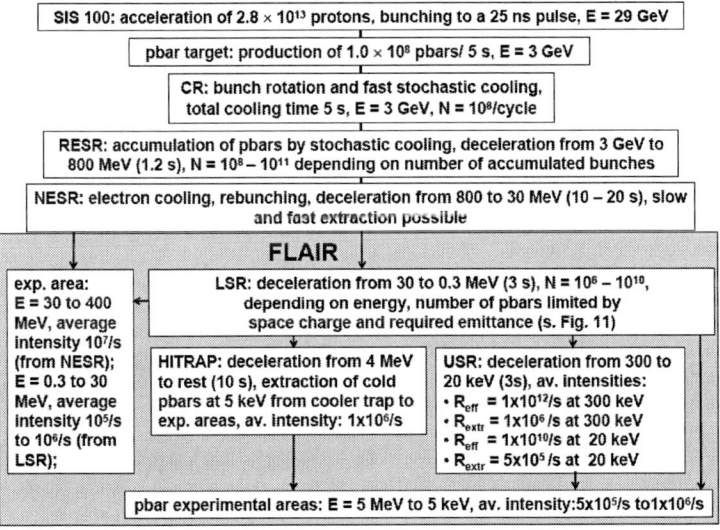

FIGURE 6. Estimated antiproton intensities available at FLAIR. R_{eff}: effective antiproton rates in the ring, R_{extr}: average rates of extracted antiprotons.

The electron energy for cooling of the antiproton beams in the envisaged USR energy range is above 10 eV. Cooling times have been estimated and were confirmed by simulations with the BETACOOL code [16]. A summary is shown in table 2. A precooled antiproton beam from the LSR with an emittance of 1 mm mrad at 300 keV is assumed and the calculated Spitzer cooling time τ during cool-down given.

TABLE 2. Prelimary design parameters of the USR electron cooler and analytically estimated cooling rates.

Length	0.8 m
Magnetic field	0.1 kG
Beta function (h/v)	7.5 m / 2 m
Horizontal dispersion	0.77 m
Electron beam radius	0.5 cm
Electron beam current	0.05 mA
T_{e-} (transv./long.)	4 meV / 0.5 meV
Perveance	2 μP
Cathode diameter	2 mm
$\tau_{20\,keV}$	< 1s

The cooling performance can be improved by using a GaAs photocathode [17], [18] which should yield "cold beam" cooling times shorter by about a factor of 30.

Before a detailed layout of the electron cooler can be designed, the influence of the internal target on the circulating beam and thus the required cooling power need to be determined as well as the expected values of the longitudinal electron temperature and its influence of the cooling performance. For the final decision whether the electron cooler should be equipped by a photocathode or a thermal cathode it is also important to investigate the cool-down times for a hot beam in particular with regard to the larger electron density achievable with a photocathode [19].

At lowest energies, other critical points are the influence of the electron cooler on the stored ion beam and the homogeneity of the magnetic guiding field in the solenoid. First, the magnetic field of the cooler has to be kept as low as possible to minimize the disturbance of the motion of the stored particles. Compensating coils at the entrance and exit of the cooler and additional quadrupoles for matching the beam have to be installed and the changes to the beam lattice taken into account. Second, the differences of the magnetic field inside the guiding solenoids need to be kept below 10^{-3} putting high demands on both the voltage supplies and the cable quality.

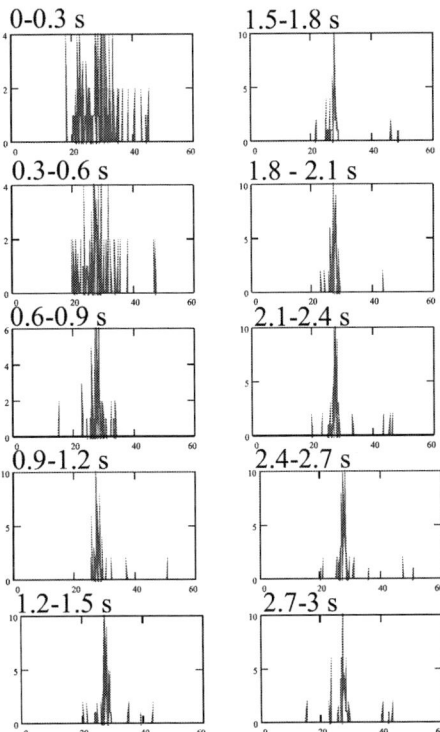

FIGURE 7. Measured transverse beam profiles of a 480 keV proton beam in the TSR during the cooling process. Shown is the count rate (arbitrary units) as a function of transverse position [mm].

In the limit of small beam energies of around 20 keV, the realization of efficient electron cooling, employing electron energies of only a few eV is a new challenge and shall be addressed with the CSR machine as a prototype for the future FLAIR installations. In particular, cryogenic GaAs photocathodes have already been shown to provide much lower initial electron temperature at the electron source (~10 meV instead of 100 meV for thermoemission cathodes) as demonstrated by the operation of a cold electron beam target from GaAs photocathodes at the TSR (MPI-K).

It should be noted that the scaling of electron cooling times from previous experimental results to the low-energy conditions of interest here is not straightforward, as many details of the theoretical description are not fully understood. Systematic measurements of electron cooling times with protons at low energies started recently at the TSR. Fig. 7 shows the measured transverse beam profile of a 480 keV proton beam during the cooling process, averaged over 20 injections. The cooling time deduced from the measurements shown in Fig. 10 of $\tau = 1$ s confirms our estimated values. Comparable measurements have also been carried out at CRYRING and confirm the above given values [3].

CONCLUSION

In this article an overview of the storage ring installations of the facility for low-energy antiproton and ion research at GSI is presented and compared to the present AD installation at CERN.

After a short review of the proposed antiproton production and deceleration scheme at FAIR, the final energy ramping between 30 MeV and 20 keV in the LSR and the USR is described. The necessary ultra-low-energy electron cooling in a variable energy electrostatic accelerator has never been realized before and needs considerable R&D. This challenge among with the problem how to realize nanosecond pulse lengths as required for in-ring experiments, will be faced at the cryogenic storage ring at MPI-K which will serve as a prototype facility for the USR.

In 2004 the FLAIR letter of intend was approved and green light towards a technical proposal was given. In the meantime, this proposal received excellent judgements by both the physics advisory committee (PAC) and the working group in scientific and technical issues (STI) and FLAIR became an integral part of the core facility.

REFERENCES

1. S. Maury, "The Antiproton Decelerator (AD)", CERN/PS 99-50 (HP)
2. T. Eriksson, "The CERN Antiproton Decelerator AD: Performance, Developments and Future Possibilities" to be published in AIP Conf. Proc. (LEAP 2005)
3. http://www.flair.eu.tt
4. G. Trubnikov et al., "The Low Energy Positron Storage Ring for Positronium Generation: Status and Developments", Proc. European Part. Acc. Conf., Paris, France (2002)
5. P. Beller et al, "Layout of the Storage Ring Complex of the International Accelerator Facility for Research with Ions and Antiprotons at GSI", Proc. European Part. Acc. Conf., Lucerne, Switzerland (2004)
6. FLAIR technical proposal, see http://www.flair.eu.tt
7. C.P. Welsch et al, "An ultra-low-energy storage ring at FLAIR", NiM A 546 (2005) 405–417
8. J. Ullrich, et al., Rep. Prog. Phys. 66 (2003) 1463
9. S.P. Møller, "ELISA – an Electrostatic Storage Ring for Atomic Physics", Proc. European Part. Acc. Conf., Stockholm, Schweden (1998)
10. T. Tanabe et al, "An Electrostatic Storage Ring for Atomic and Molecular Science", Nucl. Instr. and Meth. A 482 (2002) 595
11. C.P. Welsch, et al., "Electrostatic Ring as the Central Machine of the Frankfurt Ion Storage Experiments", PRST-AB, 7, 080101 (2004)
12. R. von Hahn, et al.,"CSR - a cryogenic storage ring at MPI-K", Proc. European Part. Acc. Conf., Lucerne, Switzerland (2004)
13. D. Zajfman et al, J. Phys.: Conf. Ser. 4 (2005) 296-299
14. M. Beutelspacher, "Systematische Untersuchungen zur Elektronenkühlung am Heidelberger Schwerionenspeicherring TSR", Doktorarbeit, Heidelberg (2000)
15. R. Rathmann, et al.,"New Method to Polarize Protons in a Storage Ring and Implications to Polarize Antiprotons.", Phys. Rev. Lett. 71, 1379 (1993)
16. C.P. Welsch, A. Smirnov, " Cooling Rates of the USR as Calculated with BETACOOL", these Proceedings
17. S. Pastuszka et al., J. Appl. Phys. 88, 6788 (2000)
18. D. Orlov et al., Appl. Phys. Lett. 78, 2721 (2001)
19. D. Orlov, these proceedings

Status of the LEPTA Project

A. Kobets, Y. Korotaev, V. Malakhov, I. Meshkov, V. Pavlov, R. Pivin,
I. Seleznev, A. Sidorin, A. Smirnov, G. Trubnikov S. Yakovenko

JINR, Dubna, Russia

Abstract. The Low Energy Particle Toroidal Accumulator (LEPTA) was commissioned in September 2004 at JINR. The facility is dedicated to studies of particle beam dynamics in a storage ring with longitudinal magnetic field focusing (so called "stellatron"), application of circulating electron beam to electron cooling of antiprotons and ions in adjoining storage ring, electron cooling of positrons, and positronium in-flight generation. The last modes of the ring operation enables setting of numerous experiments with positronium in-flight and generation of directed and "monoenergetic" flux of antihydrogen. The positronium (Ps) atoms appear in recombination of positrons with cooling electrons inside the cooling section of the ring. An assembling of the storage ring LEPTA was completed during year 2004.

Peculiarity of the storage ring is focusing of circulating particles with longitudinal magnetic field which covers whole orbit. As result, the particle motion in the ring is coupled in transverse plane. First results of the experimental study of the particle dynamics in the ring performed with circulating electron beam are presented. The beam life time was achieved above 20 ms at electron energy of 4 keV and vacuum pressure of 30 nTorr. The limitations of the beam life time and the possibility of its enhancement are discussed in the report.

This work is supported by RFBR grant #05-02-16320 and INTAS grant #03-54-5584.

Keywords: Electron Cooling, Positron, Storage Ring, Positronium.
PACS: 29.20.Dh; 29.27.Bd

INTRODUCTION

The Low Energy Particle Toroidal Accumulator (LEPTA) is designed for studies of particle beam dynamics in a storage ring with longitudinal magnetic field focusing (so called "stellatron"), application of circulating electron beam to electron cooling of antiprotons and ions in adjoining storage electron cooling of positrons and positronium in-flight generation. The last mode of the ring operation enables setting of numerous experiments with positronium in-flight and generation of directed and "monoenergetic" flux of antihydrogen [1÷5]. The LEPTA facility (Fig.1) includes small positron storage ring of the circumference of 17.2 m equipped with electron cooling system and positron injector consisting of a low energy positron source based on β^+-active ^{22}Na isotope and Penning-type trap for preliminary storing of positrons [6]. The energy of positron beam circulating in the ring is in the range 8÷10 keV, the value of focusing magnetic field is equal to 400 G.

The peculiarity of the LEPTA ring is the longitudinal focusing magnetic field for both circulating positron beam and cooling electron beams. The longitudinal magnetic field provides the positron magnetisation and, as a consequence, long lifetime of the circulating positrons. However, to form closed orbit of circulating beam one needs to use additional helical quadrupole coil. In the presence of longitudinal magnetic field the beam superposition and separation requires especial design of injection system.

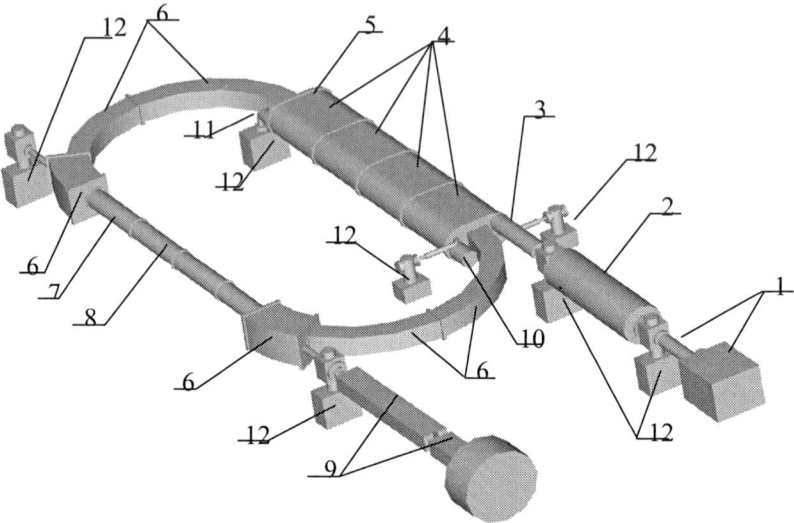

FIGURE 1. Design of the LEPTA. 1 – positron source, 2 – positron trap, 3 – positron transfer section, 4 – septum solenoids, 5 – kicker (inside septum solenoid), 6 – toroidal solenoids, 7 – solenoid and helical quadrupole inside it, 8 – electron cooling section, straight solenoid, 9 – channel for experiments with positronium in-flight, 10 – electron gun of cooling electron beam, 11 – collector of the electrons coming in to it after complete turn around the ring, 12 – vacuum pumps.

PARTICLE DYNAMICS IN LEPTA RING

The long-term stability of the circulating beam is provided by additional helical quadrupole, which forms a quadrupole magnetic field.

FIGURE 2. Helical quadrupole.

The helical quadrupole provides longtime stability of the circulation beam owing to the beam rotation as a whole around its axis. At the first stage of the experiment the dependence of the beam rotation angle on the quadrupole winding current was measured at special test bench (Fig. 3) using "pencil" electron beam at 3 keV energy.

The dependence is in a good agreement with theoretical estimation for any value of the pencil beam radial position inside the designed aperture:

$$\varphi = \frac{G^2}{kB^2} s, \quad (1)$$

where $k = \frac{2\pi}{h}$, φ is beam rotation angle, G is magnetic field gradient, which is proportional to the helical quadrupole current, B is longitudinal magnetic field, s is the quadrupole length, h is the quadrupole helix step. The required gradient of the quadrupole field was calculated using especially developed computer code BETATRON [6].

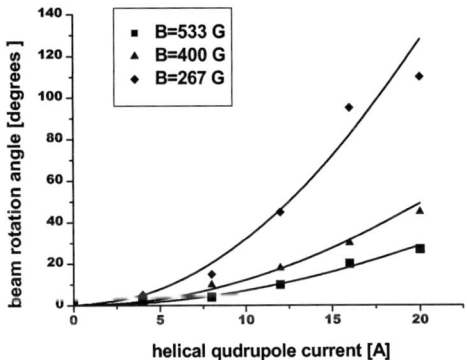

FIGURE 3. Dependence of the beam rotation angle on the helical quadrupole current at different longitudinal magnetic fields.

CIRCULATING ELECTRON BEAM IN THE LEPTA RING

Turning of the closed orbit is provided correctors, which are located at the entrance and at the exit of the toroidal sections. Circulating electron beam was observed with two PU stations placed at the entrance and exit of the straight section 8 (Fig. 1). When quarupole was OFF a few first turns of the pulsed injected beam were observed after the back slope of the kicker pulse (Fig. 4).

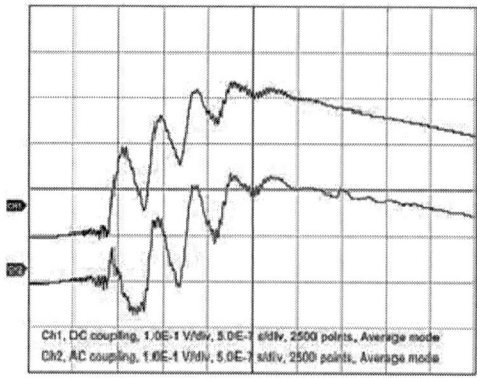

FIGURE 4. A few first turns of the beam in the ring. Signals from vertical PU station. Quadrupole is OFF.

Switching ON the helical quadrupole allows to achieve a stable particle motion in the ring (Fig. 5).

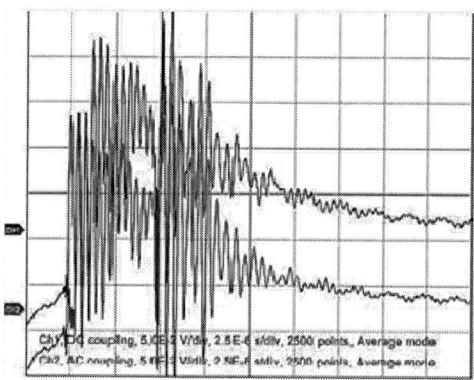

FIGURE 5. Circulating beam. Signals from two electrodes of the vertical PU station. Quadrupole is ON.

At the beginning signal of the revolution frequency can be observed, further signal of the slow frequency exists. The signals of slow frequency from different plates of PU station have the opposite phase (Fig. 5) that confirms the rotation of the beam as a whole around equilibrium orbit.

The measurement of this slow (betatron) frequency dependence on the helical quadrupole current was done (Fig. 6). Then one could find the experimental value of the betatron tune as the following:

$$Q_{slow} = \frac{f_{slow}}{f_{revolution}} \qquad (2)$$

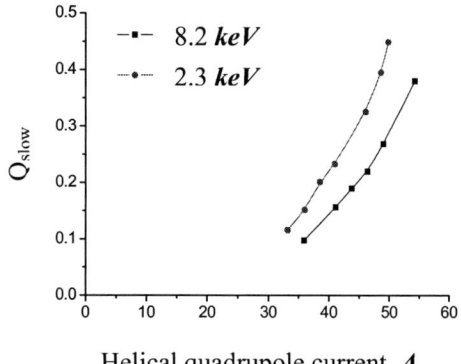

FIGURE 6. The results of the slow betatron tunes measurements for two different energies 2.3 and 8.2 keV.

This parameter can be calculated also in the drift approximation:

$$Q_{slow} = arccos\left(\frac{1}{2}Sp\right)\frac{1}{2\pi}, \qquad (3)$$

where Sp is the spur of the ring transfer matrix. The experimental results are in a good agreement with analytical dependence (3) (Fig. 7).

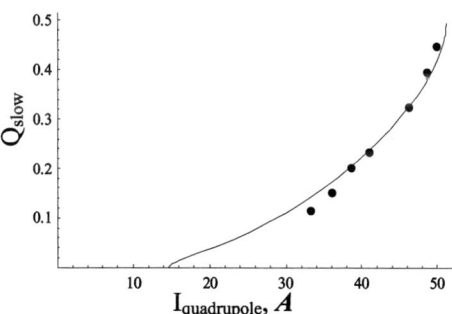

FIGURE 7. The fitting of measurements of the slow betatron tunes (beam energy is 2.3 keV) with theoretical dependence on the helical quadrupole current.

Betatron tune was measured also with excitation of the slow mode resonance using an external sin signal applied to PU electrodes (Fig. 8).

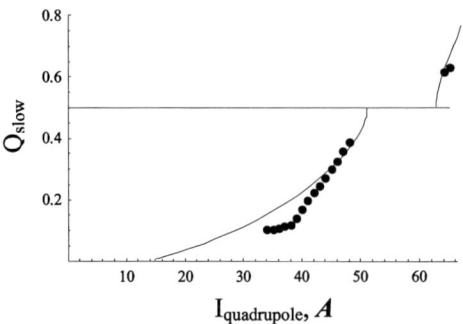

FIGURE 8. Results of the measurements of Q_{slow} by excitation of the beam with an external sin signal (points) and fitting with analytical function (3) (solid curve).

Dependence of the slow betatrone tunes on the quadrupole current shows an existence of an instability region. Some width of the region appears if the quadrupole effective length is not equal to an integer number of the helix steps (Fig. 9).

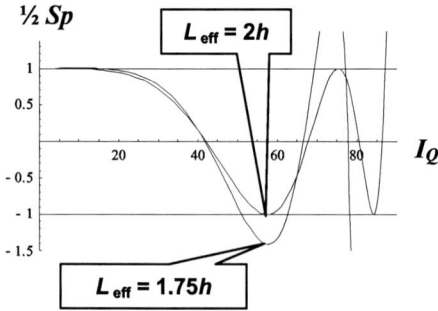

FIGURE 9. The dependence of the ½ Spur on helical quadrupole current for two different quadrupole effective lengths: $L_{eff} = 2h$ and $L_{eff} = 1.75h$ where h – the helix step.

A difference of the real helical quadrupole length with the design one originates from variation of the winds steps at the ends of this quadrupole. It has been done to provide an adiabatic variation of the magnetic field gradient at the entrance and the exit of the quadrupole.

The beam lifetime was measured in the energy range from 1 up to 10 keV (Fig. 10).

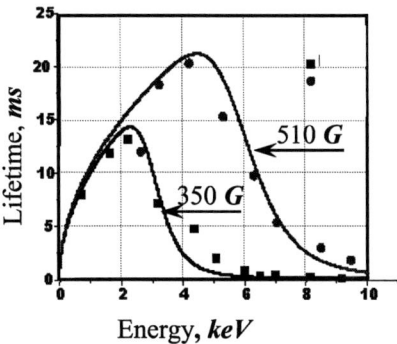

FIGURE 10. Experimental results of the beam life time measurements and fitting them with analytical dependence.

Fig. 10 shows the existence of two regions. In the first one the beam lifetime growths with energy, in the second one it decreases with energy. This behavior of the beam life time can be explained by peculiarity of the focusing system of the LEPTA ring. Electrons circulating in the ring are magnetized, therefore they travel in the ring along magnetic field lines. In the LEPTA ring there are a few regions with distortions of the magnetic field at the connections of the solenoids of different cross sections. If the lenght of the distortion is less than Larmor radius electron crossing this region can get a kick in transverse direction and some displacement. Thus, in these regions a violation of an adiabatic motion of electrons can take place. Values of the kick and displacement defined with the distortion of magnetic field amplitude and length. In this case at fixed value of the magnetic field the energy growth leads to increase of Larmor radius and, as result, to growth of the displacement value. This mechanism has diffusional character and life time defined by this process can be described as the following:

$$\tau_B = \frac{1}{N_D} \cdot \frac{C}{c} \cdot \left(\frac{b}{a}\right)^2 \cdot \left(\frac{mc^2}{eBD}\right)^2 \cdot \sqrt{\frac{2\varepsilon}{mc^2}} \cdot \left(\frac{\Delta B}{B}\right)^{-2} \cdot \exp\left\{\frac{2eBD}{mc^2} \cdot \sqrt{\frac{mc^2}{\varepsilon}}\right\}, \quad (4)$$

where N_D – number of regions with perturbed magnetic field, C – ring circumference, c – the speed of light, b – aperture, a – beam radius, B – magnetic field, D – the length of the region with perturbed magnetic field, $\Delta B/B$ – the amplitude of the magnetic field perturbation, ε – electron energy.

The life time defined by electron scattering on residual gas atoms ("vacuum life time") can be estimated with the Formula

$$\tau_{vacuum} = \frac{K}{P} \cdot \sqrt{\frac{\varepsilon}{mc^2}} \cdot \left(\frac{eBb}{mc^2}\right)^2, \quad (5)$$

where P – residual gas pressure, K – numerical coefficient. And total life time is

$$\tau_{total} = \left(\frac{1}{\tau_{vacuum}} + \frac{1}{\tau_B} \right)^{-1}, \quad (6)$$

The experimental dependence of beam lifetime on electron energy is in a good agreement with theoretical one (Fig. 10).
Results achieved during commissioning of the LEPTA ring with electron beam are presented to table 1.

TABLE 1. Parameters of the circulating electron beam.

Parameters	Value
Longitudinal magnetic field, G	300 – 510
Energy of the circulating beam, keV	1 – 10
Current of the quadrupole, A	32 – 57
"Slow" betatron tune	0.1 – 0.43
Injection current, mA	10.0
Injection efficiency	0.5
Number of circulating particles	$3 \cdot 10^{10}$
Life time at 4 keV, ms	22
Residual gas pressure, $Torr$	$3 - 7 \cdot 10^{-8}$

SUMMARY

New storage ring LEPTA was commissioned at JINR. First experiments with the pulsed electron beam show the validity of the injection scheme. The electron beam life time has been measured and its limitations were defined.

REFERENCES

1. I.N. Meshkov, A.N. Skrinsky, Antihydrogen beam generation using storage ring, NIM A 379 (1996) 41, preprint JINR E9-95-130, Dubna, 1995.
2. I.N.Meshkov, A.O.Sidorin, Conceptual design of the low energy positron storage ring, NIM A 391 (1997), 216.
3. Yu.V.Korotaev, I.N.Meshkov, S.V.Mironov, A.O.Sidorin, E. Syresin, The low energy positron storage ring for positronium generation, Proc. of EPAC 1998, p. 853
4. I.Meshkov, A.Sidorin, A.Smirnov, E.Syresin The particle dynamics in the low energy storage rings with longitudinal magnetic field, Proc. of EPAC 1998, p. 1067.
5. V.Antropov, E.Boltushkin, V.Bykovsky, et al. Particle dynamics in the Low Energy Positron Toroidal Accumulator, first experiments and results, Proc. of EPAC 2004.
6. M.Amoretti, C.Amsler, G.Bonomi, A.Bouchta, P.D.Bowe, C.Carraro, M.Charlton, M.J.T.Collier, M.Doser, V.Filippini, K.S.Fine, A.Fontana, M.C.Fujiwara, R. Funakoshi, P.Genova, A.Glauser, D.Grögler, J.Hangst, R.S.Hayano, H.Higaki, M.H.Holzscheiter, W.Joffrain, L.V.Jørgensen, V.Lagomarsino, R.Landua, C.Lenz Cesar, D. Lindelöf, E.Lodi-Rizzini, M.Macri, N.Madsen, D.Manuzio, G.Manuzio, M.Marchesotti, P.Montagna, H.Pruys, C.Regenfus, P.Riedler, J.Rochet, A.Rotondi, G.Rouleau, G.Testera, D. P. van der Werf, A.Variola, T.L.Watson, T.Yamazaki, Y.Yamazaki, The ATHENA antihydrogen apparatus, NIM A 518 (2004) 679.
7. I.Meshkov, A.Sidorin, A.Smirnov, E.Syresin, G.Trubnikov, The computer simulation of the particle dynamics in the storage ring with strong coupling of transverses modes, Proc. of EPAC 2000.

S-LSR, Cooler Ring Development at Kyoto University

Toshiyuki Shirai[1], Shinji Fujimoto[1], Masahiro Ikegami[1], Akira Noda[1], Hikaru Souda[1], Mikio Tanabe[1], Hiromu Tongu[1], Koji Noda[2], Shinji Shibuya[3], Takeshi Takeuchi[3], Takeshi Fujimoto[3], Soma Iwata[3], Atsushi Takubo[3], Hiromi Okamoto[4], Yosuke Yuri[4], Manfred Grieser[5], Evgeny M. Syresin[6]

[1] *ICR, Kyoto University, Uji, Kyoto, 611-0011, Japan*
[2] *NIRS, Inage, Chiba 263-8555, Japan*
[3] *AEC, Ltd, Inage, Chiba 263-8555, Japan*
[4] *Hiroshima University, Higashi-Hiroshima, Hiroshima 739-8530, Japan*
[5] *MPI fur Kernphysik, Heidelberg 69029, Germany*
[6] *JINR, 141980, Dubna, Moscow Region, Russia*

Abstract. A compact ion cooler ring, S-LSR is under construction in Kyoto University. One of the subjects of S-LSR is a realization of the crystalline beams using the electron beam and the laser cooling. The ring is designed to be satisfied several required conditions for the beam ordering, such as a small betatron phase advance, a small magnetic error and a precise magnet alignment. The design phase advance per a period is less than 127 degree. The calculated closed orbit distortion and the stopband is less than 1 mm and 0.001 without correction, respectively.

Keywords: Ion Cooler Ring, Electron Cooling, Crystalline Beam.
PACS: 29.27.Bd

INTRODUCTION

In Kyoto University, a new ion cooler ring (S-LSR) is now under development. Figure 1 shows the layout of the ring and the table 1 shows the main parameters of the ring. The circumference of the ring is 22.557 m and the maximum magnetic rigidity is 1 Tm. S-LSR has an electron beam cooling system and a laser cooling system [1]. The maximum electron energy is 5 kV and the maximum electron beam current is 400 mA. The laser cooling system consisted Dye lasers excited by a solid state green laser, and sub-harmonic generators. The system covers from the visible light to ultraviolet light.

The peculiarity of the ring is that it is optimized for the beam cooling, especially the realization of the ultra-cold beam. One of the goals of S-LSR project is a realization of the crystalline beams. Many analytical and numerical studies predict the possibility of the crystalline beam and the required condition of the formation. S-LSR is designed and the technical developments have to been carried out to be satisfied these conditions.

The strategy to achieve the crystalline beam is step by step. The first step is to achieve the 1-D crystal (string) of the 7 MeV proton beam using the electron beam cooling. The second step is to achieve the 1-D crystal and 2-D crystal (zigzag) of the low energy Mg^+ beam using the 3 dimensional laser cooling [2]. Even with the shearing force, the zigzag structure can be stable in the molecular dynamics (MD) simulation. The third step is that the formation of the simple 3-D crystal (1 shell) using the laser cooling and an electrostatic deflector in the vacuum chamber of the bending magnet, which has a role to cancel the shearing force [3].

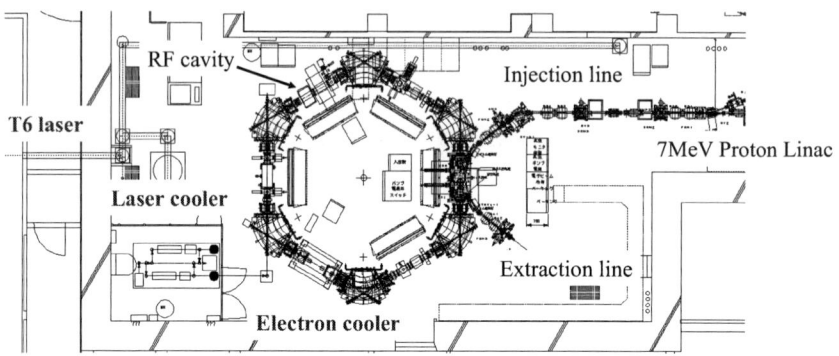

FIGURE 1. Layout of S-LSR.

TABLE 1. Main specification of S-LSR.

Ring		Bending magnet	
Circumference	22.557 m	Maximum field	0.95 T
Average radius	3.59 m	Curvature radius	1.05 m
Length of straight section	2.66 m	Gap height	70 mm
Number of periods	6	Quadrupole magnet	
Max. magnetic rigidity	1 Tm	Length	0.20 m
		Bore radius	70 mm

Development of the Cooler Ring

The key issues of the design of S-LSR is reduce the beam heating as small as possible in order to realize the ultra cold beam and the crystalline beam. Even if the electron beam cooling and the laser cooling are used, it is difficult to overcome the strong beam heating from the ring it self. The following items are considered in the design stage and the development has been carried out.
1. High symmetry of the ring lattice. This is important to reduce the phase advance per a superperiod. In S-LSR, the number of the superperiod is 6 to keep the reasonable length of the straight section for the cooling.
2. Small phase advance per a superperiod. It is less than 127 degree in the normal operation mode, even in the zero-dispersion mode using the electrostatic deflector in the bending magnet chamber [4]. The minimum phase advance is less than 90 degree in the special mode. This condition is necessary to avoid the beam heating due to the linear resonance.

3. Small individual differences among the magnets. The differences of the magnets break symmetry of the ring and increase the stop band of the resonance, which is the source of the beam heating. The all magnets are made from the block irons and fabricated precisely.
4. Precise alignment of the magnets. The alignment errors also create the magnetic field error and have the same effect as above.
5. Compensation system of the shearing force. Even if the beam crystal is created, the complicated crystal is destroyed by the shearing force in the bending magnet. In order to cancel the effect, an electrostatic deflector is placed in the vacuum chamber of the bending magnet [5].

Lattice Design

Figure 2 shows the twiss parameters of one period for the crystalline modes. The betatron tune is (1.45, 1.44) at the left figure. The phase advance in this mode is 86 degree, which is less than the important criteria of 90 degree. It is important to avoid the envelope instability. The electron cooling is possible but we do not find the suitable method of the three dimensional laser cooling in this mode.

The right one in the figure 2 shows the twiss parameters when the betatron tune is (2.08, 1.07). In this mode, if we set the synchrotron tune of 0.07, we obtain the strong transverse-longitudinal resonance coupling and the three dimensional laser cooling becomes possible. The phase advance per a period is still lower than the 127 degrees at this tune value.

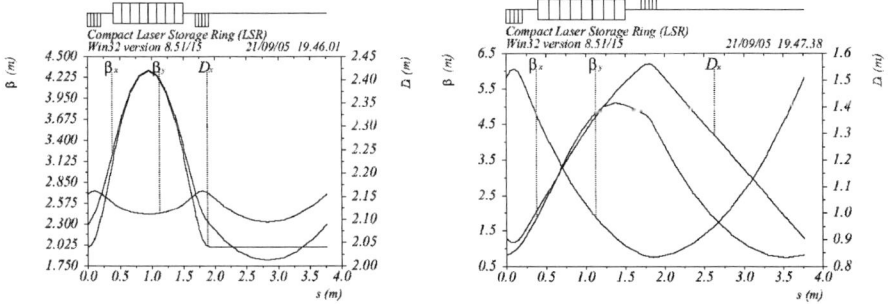

FIGURE 2. Twiss parameters in one period of S-LSR. The operating tune is (1.45, 1.44) in the left, (2.08, 1.07) in the right figure.

Magnetic Field Measurement

The figure 3 shows the bending magnet and the quadrupole magnet for S-LSR. They are made from the block ion and fabricated carefully to reduce the individual differences. The differences enhance the closed orbit distortion and the stop band of the resonance. One more important point is to reduce the nonlinear magnetic field. It induces the nonlinear resonance and it leads to the beam heating with the high tune depression. The design of the magnets were done by the three dimensional magnetic simulation code (TOSCA) and optimized to minimize the high order component [6][7].

We measure the magnetic field of the bending magnet using the three Hall probes (Group3 TP141) and the field gradient of the quadrupole magnets was measured by a shift coil method. The difference of the BL products among 6 bending magnets is shown in the figure 4 (left). It is within $+/-2 \times 10^{-4}$ without corrections. The difference of the GL products among 12 quadrupole magnets is shown in the figure 4 (right). It is within $+/-2.5 \times 10^{-3}$.

The resultant closed orbit distortion is 1 mm and the stop band is 0.001 due to the above differences. It will corrected by the correction current or correction coils in the operation.

 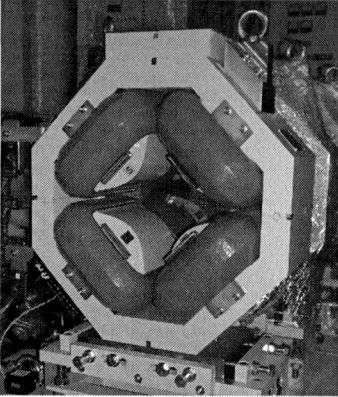

FIGURE 3. View of the bending magnet (left) and the quadrupole magnet (right).

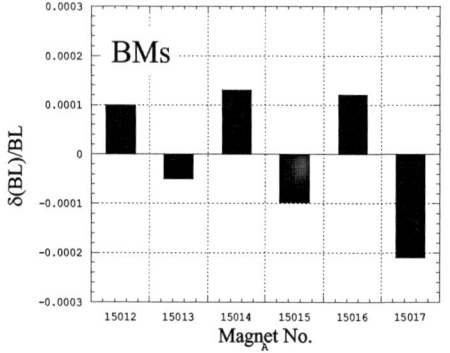 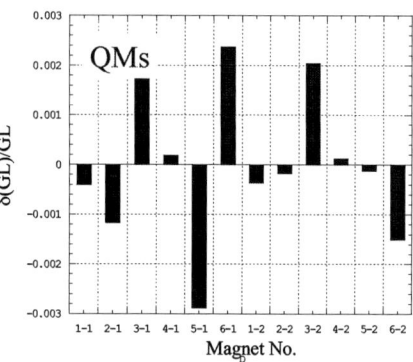

FIGURE 4. Layout of the S-LSR.

Magnet Alignment

The precise alignment of the dipole and quadrupole magnets is also important to suppress the closed obit distortion and the stop band. It was carried out using the laser tracker (LEICA, LTD800) and the level (LEICA, N3). The measurement accuracy of the laser tracker is 40 mm in this situation. After the fine alignment, we measured the position of the alignment targets on the magnets again by the laser tracker. The results are shown in figure 5. The left one shows the results of the bending magnet and the right one is quadrupole magnets. In the both cases, 65 % of the alignment targets on the magnets are placed within the error of +/-50 μm and all targets are placed within +/- 100 mm.

The effect of these alignment errors on the closed orbit distortion and the stopband is smaller than that from the differences of the magnets which is discussed in the previous section.

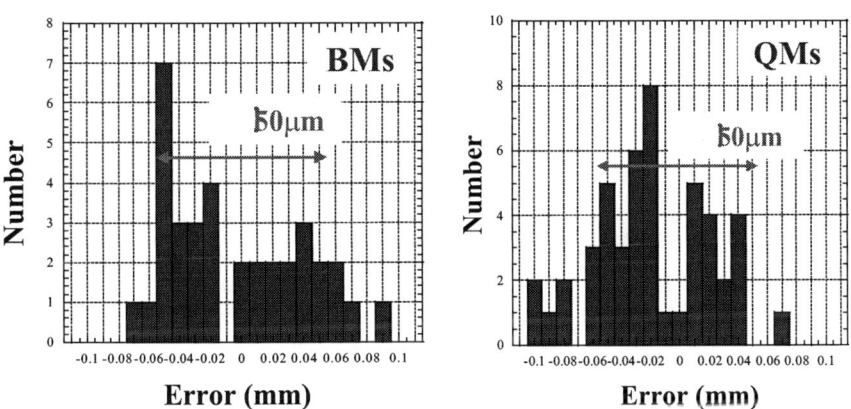

FIGURE 5. Alignment error of the bending magnets (left) and the quadrupole magnets (right).

ACKNOWLEDGMENTS

This project is performed as a part of "Advanced Compact Accelerator Research Project" with the collaboration groups.

REFERENCES

1. A. Noda et al., Proc. of Symposium on Accel. Sci. and Tech. (2001) 125.
2. A. Noda et al., Proc. of EPAC 2002, Paris, France (2002) 2748.
3. J.P. Schiffer and P. Kienle, Z. Phys. A321, 181 (1985).
4. J. Wei, X.-P. Li, A. M. Sessler, Phys. Rev. Lett. 73 (1994) 3089.
5. L. Tecchio, Nucl. Instr, and Meth, in Phys. Res. A 391 (1997) 147.
6. T. Kihara et al., Phys. Rev. E, 59, 3594-3604 (1999)
7. J. Wei, X-P Li, A. M. Sessler, Reprt BNL-52381 (1993).M. P. Brown and K. Austin, *The New Physique*, Publisher City: Publisher Name, 2005, pp. 25-30.

Antiproton – Ion Collider for FAIR Project

P.Beller[1], B.Franzke[1], P.Kienle[2,4], R.Kruecken[2], I.Koop[3],
V.Parkhomchuk[3], Y.Shatunov[3], A.Skrinsky[3], V.Vostrikov[3], E.Widmann[4]

[1]GSI, Darmstadt, Germany; [2]TUM, Munich, Germany; [3]BINP, Novosibirsk, Russia;
[4]SMI, Vienna, Austria.

Abstract. An antiproton-ion collider (AIC), with extensive using of electron cooling, is proposed to determine rms radii for protons and neutrons in unstable and short lived nuclei by means of antiproton absorption at medium energies. The experiment makes use of the electron-ion collider complex with appropriate modifications of the electron ring to store, cool and collide antiprotons of 30 MeV energy with 740 MeV/unit ions in the NESR. Antiprotons are collected, cooled, decelerated up to 30 MeV and transferred to the electron storage ring. The radioactive nuclei beams are transferred to the CR and cooled at 740A MeV and transported via the RESR to NESR, in which especially short lived nuclei are accumulated continuously to increase the luminosity. Luminosities of about 10^{23} cm^{-2}s^{-1} may be reached with 10^6 ions accumulated in the NESR in coasting mode of operation, used for Schottky spectroscopy of the fragments.

Keywords: FAIR, Electron cooling, Luminosity.

INTRODUCTION

An antiproton-ion collider [1] is proposed to independently determine rms radii for protons and neutrons instable and short lived nuclei by means of antiproton absorption at medium energies [2]. The experiment makes use of the electron ion collider complex [3] with appropriate modifications of the electron ring to store, cool and collide antiprotons of 30 MeV energy with 740A MeV ions in the NESR (Fig.1). Antiprotons are collected, cooled and slowed to 30 MeV. Hereafter the antiprotons are transferred to the electron storage ring using a new transfer line. Radioactive nuclei are produced by projectile fragmentation and projectile fission of 1.5A GeV primary beams and separated in the Super FRS. The separated beams are transferred to the collector ring (CR) and cooled at 740A MeV and transported via the RESR to NESR, in which especially short lived nuclei are accumulated continuously to increase the luminosity. In Tabs. 1, 2 the parameters of NESR and antiproton ring are presented.

TABLE 1. Main Parameters of NESR.

Parameter	Units	Value
Circumference	m	222.11
Reference nucleus		^{132}Sn^{50}
Maximum energy for reference nucleus	MeV/ unit	740
Momentum acceptance		±1.75%
Acceptance (hor/vert)	π mm mrad	160/100
Betatron tune (hor/vert)		3.4/3.2

TABLE 2. Main Parameters of Antiproton Storage Ring.

Parameter	Units	Value
Circumference	m	45.215
Energy	MeV	20 – 125
Revolution frequency	MHz	0.68 – 3.1
Betatron tunes (hor/vert)		3.8/2.8

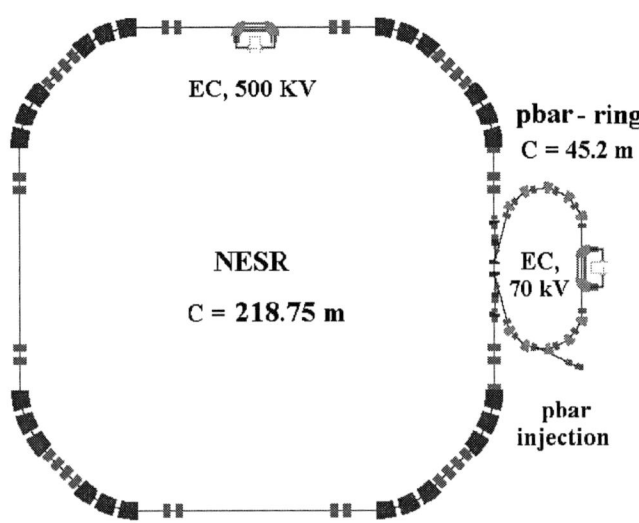

FIGURE 1. Schematic layout of the antiproton – ion collider.

ELECTRON COOLING

Here, we estimate the electron cooling force using the following phenomenological expression which has been proposed in Ref. [4]

$$\Delta \vec{p} = \vec{F} \cdot \tau = -\frac{4e^4 n_e \vec{V} \tau}{m_e (\sqrt{V^2 + V_{eff}^2})^3} \ln\left(1 + \frac{\rho_{max}}{\rho_L + \rho_{min}}\right). \quad (1)$$

For simulation of the time behavior of the normalized emittances $\varepsilon_{nx}, \varepsilon_{ny}$ and the momentum spread σp (1σ-values) we used the following equation system

$$\frac{\partial \varepsilon_{nx}}{\partial t} = \lambda_{IBS}^x (\varepsilon_{nx}, \varepsilon_{ny}, \sigma p) + \lambda_{heat}^x (\varepsilon_{nx}, \varepsilon_{ny}, \sigma p) - \lambda_{cool}^x (\varepsilon_{nx}, \varepsilon_{ny}, \sigma p),$$

$$\frac{\partial \varepsilon_{ny}}{\partial t} = \lambda_{IBS}^y (\varepsilon_{nx}, \varepsilon_{ny}, \sigma p) + \lambda_{heat}^y (\varepsilon_{nx}, \varepsilon_{ny}, \sigma p) - \lambda_{cool}^y (\varepsilon_{nx}, \varepsilon_{ny}, \sigma p), \quad (2)$$

$$\frac{\partial \sigma p}{\partial t} = \lambda_{IBS}^{\sigma p} (\varepsilon_{nx}, \varepsilon_{ny}, \sigma p) + \lambda_{heat}^{\sigma p} (\varepsilon_{nx}, \varepsilon_{ny}, \sigma p) - \lambda_{cool}^{\sigma p} (\varepsilon_{nx}, \varepsilon_{ny}, \sigma p).$$

In our calculation we consider the cooling drag force λ_{cool}, the heating induced by the thermal motion of the electron λ_{heat}, the IBS process λ_{IBS}. The increments and decrements of the rms values of ε_{nx}, ε_{ny}, σp were calculated for a single turn. If the tune-shift becomes more than 0.1 then the cooling force is supposed to be zero. So, the space charge phenomena are simulated.

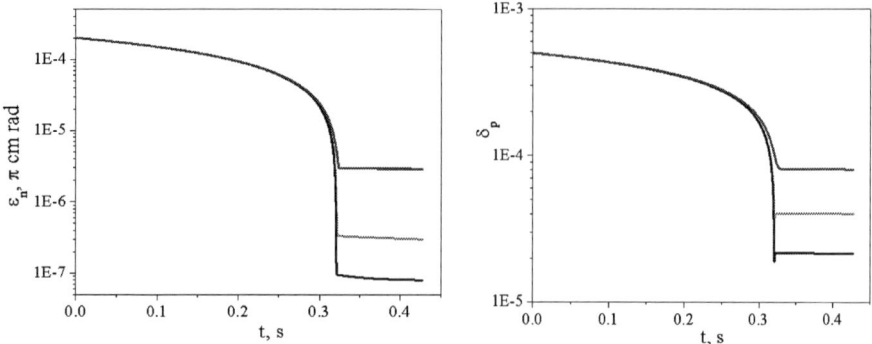

FIGURE 2. Time evolution of normalized transverse emittance (left) and momentum spread (right), for different antiproton beam intensity, from top to bottom $N_i = 10^{10}$, 10^9, 10^8 accordingly, $E_{\bar{p}} = 30$ MeV.

In Fig. 2 the evolution of normalized transverse emittance and longitudinal momentum spread of antiproton beam are presented. The simulation was made for follow parameters: the beta-function in cooling section is 4.5/6.5 m (h/v), the antiproton energy is $E_{\bar{p}} = 30$ MeV, the electron current is $I_e = 1$ A. The initial antiproton beam normalized transverse emittance is 2 π mm mrad, momentum spread is $\pm 5\cdot 10^{-4}$. As it is seen, the ion beam emittance and momentum spread are effectively decreased during 330 ms. The further cooling is limited due to achieving of the beam tune shift value $\Delta v = 0.1$, if beam intensity more than 10^9. At lower beam intensity the further cooling limited by IBS.

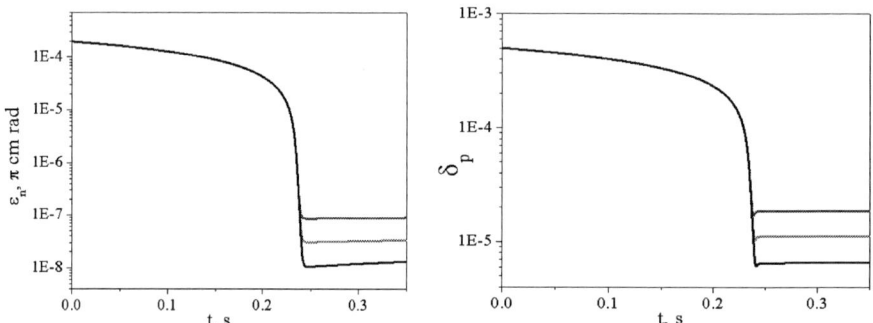

FIGURE 3. Time evolution of normalized transverse emittance (left) and momentum spread (right), for different ion beam intensity, from top to bottom $N_i = 10^7$, 10^6, 10^5 accordingly, $E_i = 740$ MeV/u.

In Fig. 3 the evolution of normalized transverse emittance and longitudinal momentum spread of ion beam in the NESR are presented. The simulation was made for follow parameters: the beta-function in cooling section is 30 m, the $^{132}Sn^{50}$ ion energy is $E_i = 740$ MeV/u, the electron current is $I_e = 1$ A. The initial ion beam transverse emittance is 1.4 mm mrad, momentum spread is ±5·10⁻⁴. As it is seen, the ion beam emittance and momentum spread are effectively decreased during 250 ms. The further cooling is limited due to IBS.

The Fig. 4 shows the tune shift as function of time for different number of ions (left) and antiprotons (right). Simulations shows that the Laslett tune shift of ion beam is a far from threshold value. On the contrary, the transverse emittance of pbar beam is limited by space charge phenomena.

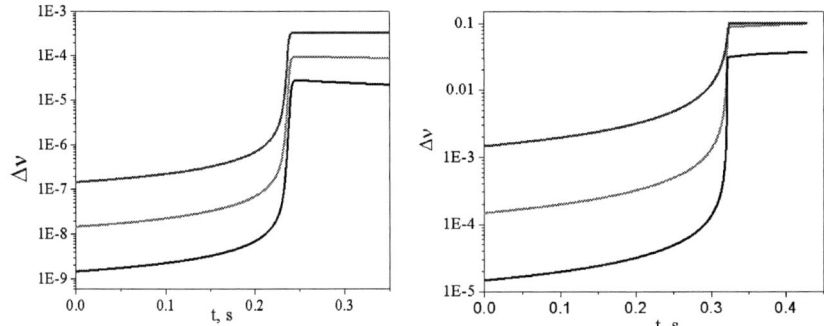

FIGURE 4. Time evolution of the tune – shift value for $^{132}Sn^{50}$ ion with energy E_i = 740 MeV/u (left), from top to bottom N_i = 10^7, 10^6, 10^5, and for antiprotons with energy 30 MeV (right), from top to bottom N_i = 10^{10}, 10^9, 10^8 accordingly.

The motion of a particle passing the IP is perturbed by the fields of the counter – moving beam. Such perturbations result in resonant phenomena. Their strengths and relevant increases in the particle oscillations amplitude depend on the values of the particle oscillation tunes. It may result in the instability of both coherent and incoherent oscillations of colliding beams, limiting the value of the collider luminosity. For the case of pbar-RI collider, the beam-beam parameters for pbar and ion beams of the round cross-sections read

$$\xi_{\bar{p}} = Z_i \frac{N_i r_p L_{IP}(1+\beta_i \beta_{\bar{p}})}{4\pi \Pi_i \gamma_{\bar{p}} \beta_{\bar{p}}^2 \varepsilon_i}, \quad \xi_i = \frac{Z_i}{A_i} \frac{N_{\bar{p}} r_p L_{IP}(1+\beta_i \beta_{\bar{p}})}{4\pi \Pi_{\bar{p}} \gamma_i \beta_i^2 \varepsilon_{\bar{p}}}. \quad (3)$$

Beam-beam parameters are $\xi_{\bar{p}} = 4.3 \cdot 10^{-3}$, $\xi_i = 2.3 \cdot 10^{-4}$ for the follow parameters: E_i=740 MeV/u, $E_{\bar{p}}$ = 30 MeV, N_i=10^6, $N_{\bar{p}}$ = 10^9, ε_i=2·10⁻⁸ cm·rad, $\varepsilon_{\bar{p}}$ = 1.8·10⁻⁶ cm·rad. So, clear those values of beam-beam parameters are safe.

The design of electron cooler device for NESR is presented in details in Ref.[5]. The design of electron cooler device for antiproton storage ring is based on electron cooler manufactured by BINP for LEIR storage ring [6]. List of main parameters is presented in Tab.3.

TABLE 3. Parameters of EC for Antiproton Storage Ring.

Parameter	Units	Value
Maximum electron energy	keV	70
Maximum electron current	A	2
Electron beam diameter	mm	5 – 20
Magnet field in cooling section	T	0.2
Length of cooling section	m	3.5

LUMINOSITY

The luminosity for head-on collisions of two coasting beams, with interaction length L_{IP}, is given by

$$L_0 = \frac{n_i n_{\bar{p}}}{\pi} c(\beta_i + \beta_{\bar{p}}) \int_0^{\frac{L_{IP}}{2}} \frac{ds}{\sqrt{\sigma_{xi}^2(s) + \sigma_{x\bar{p}}^2(s)}\sqrt{\sigma_{yi}^2(s) + \sigma_{y\bar{p}}^2(s)}}, \quad (4)$$

where suffixes i and $\bar{p}$ mark the values related to ion and antiproton beams, A is the atomic number of the ion, n_i and $n_{\bar{p}}$ are the line density of particles, $c\beta_i$ and $c\beta_{\bar{p}}$ are the velocity of particles, $\sigma^2(s) = \varepsilon\beta_0\left(1 + \frac{s^2}{\beta_0^2}\right)$ is the beam size.

For modeling parameters: E_i=740 MeV/u, $E_{\bar{p}} = 30$ MeV, N_i=10^6, $N_{\bar{p}} = 10^9$, ε_i=$2 \cdot 10^{-8}$ cm·rad, $\varepsilon_{\bar{p}} = 1.8 \cdot 10^{-6}$ cm·rad, $\beta_{\bar{p}} = 30\, cm, \beta_i = 40/25\, cm$ the estimated luminosity value is $L_0 = 9.5 \cdot 10^{22}\, cm^{-2} s^{-1}$. In Fig. 5 the luminosity dependence on antiproton beam energy is shown.

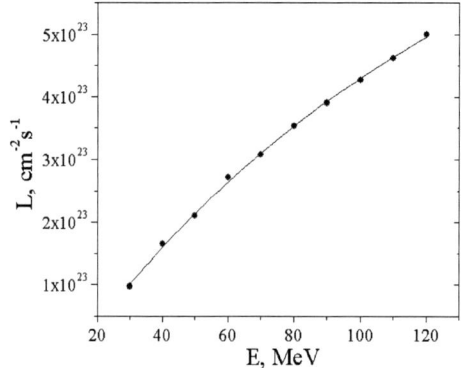

FIGURE 5. Luminosity versus antiproton beam energy.

REFERENCES

1. AIC Technical Proposal, 2004.
2. P. Kienle, Nucl. Instr. Meth. B 213 (2004) 191.
3. ELISE Technical Proposal.
4. V.V. Parkhomchuk. *New insights in the theory of electron cooling*. Nucl. Instr. Meth. Phys. Res., A 441 (2000).
5. Electron cooling for NESR Technical Proposal (2004).
6. V.V. Parkhomchuk et al. First test of the LEIR- cooler at BINP. Proc. of this Workshop.

GENERAL TOPICS

Transverse-Longitudinal Phase-Space Manipulations and Correlations[*]

Kwang-Je Kim[†] and Andrew Sessler[‡]

[†]*Argonne National Laboratory, 9700 S. Cass Avenue, Argonne, IL 60439 USA*
[‡]*Lawrence Berkeley National Laboratory, Berkeley, CA 94720 USA*

Abstract. Manipulations on transverse and longitudinal phase-space distribution of an electron beam are discussed within the constraints imposed by symplectic conditions. A few examples are presented: transverse-longitudinal emittance exchange to improve performance of a high-gain free-electron laser (FEL) for hard x-rays, and the flat beam technique and its application to compact Terahertz devices and ultrashort-pulse generation. It is shown that emittance transfer to some degree would be advantageous for FELs and that introducing correlations would allow just such transfers. Also, it is shown that transverse-longitudinal correlations would be distinctly advantageous for FELs. Conventional and exotic methods of producing such correlations are described. Practical difficulties associated with each of the conventional methods are described, although the nonconventional methods appear to hold promise.

Keywords: Phase space, beam correlations, symplectic, transverse-longitudinal emittance transfer and exchange, beam conditioning, free electron lasers, Smith-Purcell, ultrashort pulse

PACS: 29.27.-a, 41.85.Ct, 41.85.Lc, 41.75.Ak, 41.75.Lx, 41.75.Fr, 41.75.Ht, 41.50+h

1. INTRODUCTION

Particle beams need often to be manipulated in 6-D phase space to optimize the beam applications. Beam cooling is an excellent example of an advanced phase-space manipulation in a non-Hamiltonian system. There are also numerous important examples of Hamiltonian phase-space manipulation operating within transverse or in longitudinal subspace separately or across the different subspaces. In this paper we explore some examples of phase-space manipulations in Hamiltonian systems and how to achieve them.

Section 2 presents the limits imposed by the symplectic condition of a Hamiltonian system on transverse and longitudinal phase-space manipulation. With these limits in mind, we consider phase-space manipulation. In the absence of correlation, we establish a theorem referred to as the emittance exchange theorem stating that the emittance of one subspace cannot be transferred partially to another subspace, that is, it can only be exchanged as a whole with that of another subspace. An optical system producing an exact exchange is presented.

---

[*]Work supported by the U.S. Department of Energy, Office of Basic Energy Sciences, under Contract Nos. W-31-109-ENG-38 and DE-AC02-05CH11231.

In Section 3 examples are presented of transverse-longitudinal *exchange* and transverse-longitudinal *transfer* for high-gain FEL. It is shown that the exchange would be advantageous for short bunches while the transfer would be for long bunches. Due to the emittance exchange theorem, correlations are necessary to effect transfers.

The production of a flat beam is discussed in Section 4 as an example of non-symplectic beam manipulation. The application of flat beam techniques to ultrashort-pulse generation and to a compact Terahertz generator based on Smith-Purcell radiation are also presented in Section 4.

In Section 5 the concept of "conditioning" is introduced, and it is shown in Section 6 that transverse-longitudinal correlations would be distinctly advantageous for FELs. Conventional and exotic methods of producing such correlations are described in Section 7. Practical difficulties associated with each of the conventional methods are described. However, the exotic methods are all seen to have promise. Some final thoughts are presented in Section 8.

2. SYMPLECTIC CONDITIONS

2.A. Beam Manipulation and Symplectic Conditions

Although phase space is six dimensional, we will consider in this paper 4-D subspaces, either in the transverse phase space (x,x',y,y') or in the transverse-longitudinal phase space (x,x',z,δ). A point in transverse-longitudinal phase space is represented by a column vector X:

$$X = \begin{pmatrix} x \\ x' \\ z \\ \delta \end{pmatrix}. \tag{2.1}$$

Here δ is the relative energy deviation. A beam manipulation is a linear transformation of the phase space

$$X \to MX. \tag{2.2}$$

The 4×4 matrix M is referred to as the transfer matrix.

A realizable transfer matrix in a Hamiltonian system must satisfy the symplecticity condition

$$\tilde{M} J M = M J \tilde{M} = J, \tag{2.3}$$

where the tilde denotes the transpose operation and

$$J = \begin{bmatrix} J_{2D} & 0 \\ 0 & J_{2D} \end{bmatrix}. \tag{2.4}$$

Here J_{2D} is the unit symplectic matrix in 2-D:

$$J_{2D} = \begin{bmatrix} 0 & 1 \\ -1 & 0 \end{bmatrix}. \tag{2.5}$$

It follows from Eq. (2.3) that

$$Det(M) = 1. \tag{2.6}$$

Here Det is the determinant. Equation (2.6) is a statement on conservation of the phase-space volume, which is equivalent to Liouville's theorem.

The symplecticity requirement, Eq. (2.3), places more conditions on the transfer matrix. The 4×4 transfer matrix M consists of four 2×2 submatrices A, B, C, D as follows:

$$M = \begin{bmatrix} A & B \\ C & D \end{bmatrix}. \tag{2.7}$$

Including Eq. (2.6), there are altogether six constraints on the sixteen elements of the matrices $A, B, C,$ and D imposed by Eq. (2.3) [1].

2.B. Beam Matrices, Invariants, and Emittances

A global characterization of the phase-space distribution is the set of second-order moments conveniently represented by the symmetric beam matrix Σ:

$$\Sigma = \langle X\widetilde{X} \rangle = \begin{bmatrix} \langle x^2 \rangle & \langle x'x \rangle & \langle \delta x \rangle \\ \langle xx' \rangle & & \\ \vdots & & \vdots \\ \langle x\delta \rangle & \cdots & \langle \delta^2 \rangle \end{bmatrix}. \tag{2.8}$$

Corresponding to Eq. (2.2), the beam matrix transforms as follows:

$$\Sigma \rightarrow M \Sigma \widetilde{M}. \tag{2.9}$$

Two invariants of the beam matrix under an arbitrary symplectic transformation M are [2]

$$\varepsilon_{4D} = Det(\Sigma) \quad \text{and} \tag{2.10}$$

$$I^{(2)} = -\frac{1}{2}Tr(\Sigma J \Sigma J), \qquad (2.11)$$

where Tr is the trace, and the quantity ε_{4D} is the 4-D phase-space volume occupied by the beam.

To see the meaning of $I^{(2)}$, we introduce the 2×2 submatrices of the beam matrix as follows:

$$\Sigma = \begin{bmatrix} \Sigma_x & \Sigma_c \\ \tilde{\Sigma}_c & \Sigma_z \end{bmatrix}. \qquad (2.12)$$

The matrix Σ_c represents the correlation. The *projected* emittances ε_x and ε_z in their respective subspaces are defined as follows:

$$\varepsilon_x = \sqrt{Det(\Sigma_x)} = \sqrt{\langle x^2 \rangle \langle x'^2 \rangle - \langle xx' \rangle^2}, \qquad (2.13)$$

$$\varepsilon_z = \sqrt{Det(\Sigma_z)} = \sqrt{\langle z^2 \rangle \langle \delta^2 \rangle - \langle z\delta \rangle^2}. \qquad (2.14)$$

One can show

$$I^{(2)} = \varepsilon_x^2 + \varepsilon_z^2 + 2Det(\Sigma_c). \qquad (2.15)$$

In general the projected emittances are not conserved. However, when the correlation vanishes

$$\Sigma_c = 0, \qquad (2.16)$$

we have

$$\varepsilon_{4D} = \varepsilon_x^2 \varepsilon_z^2, \qquad (2.17)$$

$$I^{(2)} = \varepsilon_x^2 + \varepsilon_z^2. \qquad (2.18)$$

The projected emittances, Eqs. (2.13) and (2.14), are simply referred to as the emittances when and only when the correlation vanishes. Then $I^{(2)}$ is the sum of the emittances.

2.C. Emittance Exchange Theorem

Suppose we start from a system 1 in which the beam matrix is uncoupled to another uncoupled system 2. From the invariance of ε_{4D} and $I^{(2)}$ it follows:

$$\varepsilon_{x1}^2 \varepsilon_{z1}^2 = \varepsilon_{x2}^2 \varepsilon_{z2}^2 \quad \text{and} \qquad (2.19)$$

$$\varepsilon_{x1}^2 + \varepsilon_{z1}^2 = \varepsilon_{x2}^2 + \varepsilon_{z2}^2, \qquad (2.20)$$

where the quantities in systems 1 and 2 are indicated by subscripts 1 and 2, respectively. The solutions of Eqs. (2.19) and (2.20) are either

$$\varepsilon_{x1} = \varepsilon_{x2}, \quad \varepsilon_{z1} = \varepsilon_{z2} \qquad (2.21)$$

or

$$\varepsilon_{x1} = \varepsilon_{z2}, \quad \varepsilon_{z1} = \varepsilon_{x2}. \qquad (2.22)$$

That is, the emittances of the subspaces are either conserved or completely exchanged. In other words, emittances cannot be partially transferred from x- to z-subspace by going from an uncoupled system to another uncoupled system. This fact, also valid in 6-D and higher dimensions [2], will be referred to in the following as the emittance exchange theorem. The theorem appears to have been first pointed out in beam dynamics context by E. Courant [3].

Projected emittance can be partially transferred by introducing a suitable correlation. The emittance exchange theorem does not apply for a nonsymplectic transformation. An example of a nonsymplectic transformation giving rise to an arbitrary emittance ratio keeping the product constant, i.e., the flat beam technique, is discussed in Section 4.

2.D. Optical System for Transverse-to-Longitudinal Emittance Exchange

An optical system for transverse-to-longitudinal emittance exchange was studied in ref. [4]. There it is shown that the necessary and sufficient condition for exact exchange is that the submatrix A of the transfer matrix Eq. (2.7) vanishes identically:

$$A = 0. \qquad (2.23)$$

A beam transport system producing the transverse-to-longitudinal emittance exchange can be constructed from doglegs and a dipole-mode cavity producing a transverse kick. A dogleg consists of a pair of bending magnets of opposite polarities separated by a free space of length L. The corresponding transfer matrix is

$$M_D(\eta,\xi,L) = \begin{bmatrix} 1 & L & 0 & \eta \\ 0 & 1 & 0 & 0 \\ 0 & \eta & 1 & \xi \\ 0 & 0 & 0 & 1 \end{bmatrix}, \qquad (2.24)$$

where η and ξ are the dispersion and momentum compaction, respectively. The transfer matrix for a dipole-mode cavity of kick strength k is of the form [4]

$$M_C(k) = \begin{bmatrix} 1 & 0 & 0 & 0 \\ 0 & 1 & k & 0 \\ 0 & 0 & 1 & 0 \\ k & 0 & 0 & 1 \end{bmatrix}. \tag{2.25}$$

An approximate scheme for emittance exchange consisting of a dipole mode cavity in the middle of two doglegs of opposite kicks was discussed in ref. [4]. The scheme is adequate when the ratio of the emittances involved is not too large. A better scheme is necessary when the emittance ratio is large. An exact exchange optics was found recently by K.-J. Kim that is similar to the one in ref. [4], but with the second dogleg in the same direction as the first one [5]. The configuration is shown in Fig. 2.1. The corresponding transfer matrix is

$$M_{EX} = M_D(\eta, \xi, L) M_C(k) M_D(\eta, \xi, L). \tag{2.26}$$

If the dispersion η and the kick strength k are chosen so that

$$1 + k\eta = 0, \tag{2.27}$$

the transfer matrix becomes

$$M_{EX} = \begin{bmatrix} 0 & 0 & kL & \eta + kL\xi \\ 0 & 0 & k & k\xi \\ k\xi & \eta + kL\xi & 0 & 0 \\ k & kL & 0 & 0 \end{bmatrix}. \tag{2.28}$$

This matrix clearly satisfies the condition for exact exchange, Eq. (2.23).

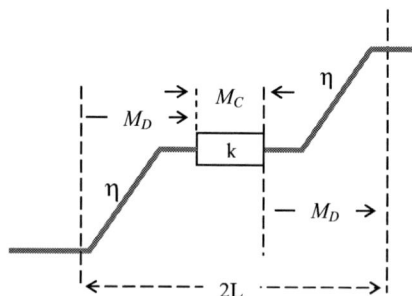

FIGURE 2.1. Schematic diagram for exact longitudinal-to-transverse emittance exchange, consisting of two identical doglegs (M_D) and a transverse cavity (M_C). The matrix for M_D and M_C are given by Eqs. (2.24) and (2.25), respectively.

3. TRANSVERSE-LONGITUDINAL EXCHANGE AND TRANSFER FOR HIGH-GAIN FELs

3.A. An Example of Emittance Exchange

For operation of high-gain x-ray free-electron lasers (FELs), such as the LCLS under construction [6] or the "Greenfield FEL" [7], it is critical to have small transverse emittance. The normalized emittance specified for these projects are

$$\gamma\varepsilon_x \text{ and } \gamma\varepsilon_y \sim 1 \times 10^{-6} \text{ m-rad.} \quad (3.1)$$

If the transverse emittance can be reduced by a factor of 10,

$$\gamma\varepsilon_x \text{ and } \gamma\varepsilon_y \sim 1 \times 10^{-7} \text{ m-rad,} \quad (3.2)$$

then the length of the undulator can be reduced significantly, thereby reducing the construction cost as well as the operational complexity. With the present gun technology, however, the emittance value of the 1-nC electron bunches is already challenging at the level given by Eq. (3.1).

On the other hand, the uncorrelated energy spread σ_E is quite small, less than 2 keV [8]. Assuming $\sigma_E \approx 1.5$ keV, the spread in γ is

$$\sigma_\gamma = \gamma\sigma_\delta = 3 \times 10^{-3}. \quad (3.3)$$

The LCLS operates at electron energy 15 GeV. Thus the rms spread of relative energy deviation is

$$\sigma_\delta = 1 \times 10^{-7}. \quad (3.4)$$

This is three orders of magnitude smaller than necessary for avoiding gain reduction.

Is there a way to trade the excessively small energy spread (Eq. (3.4)) with the transverse emittance so that the transverse emittance can be reduced to Eq. (3.2)? We find that the answer is affirmative by manipulating the beam as follows. First, the length of the bunch is chosen so that the longitudinal emittance becomes

$$\gamma\varepsilon_z = \gamma\sigma_z\,\sigma_\delta = 1 \times 10^{-7} \text{ m-rad.} \quad (3.5)$$

In view of Eq. (3.3), we find

$$\sigma_z = 33 \text{ μm.} \quad (3.6)$$

Next, we use the flat beam technique discussed in the next section to produce

$$\gamma\varepsilon_x = 1 \times 10^{-5} \text{ mrad}, \gamma\varepsilon_y = 1 \times 10^{-7} \text{ m-rad.} \quad (3.7)$$

Note that the product of the two emittances in Eq. (3.7) is the same as the product of those in Eq. (3.1). We then exchange the x- and z-emittances to obtain

$$(\gamma\varepsilon_x, \gamma\varepsilon_y, \gamma\varepsilon_z) = (1 \times 10^{-7}, 1 \times 10^{-7}, 1 \times 10^{-5}) \text{ m-rad} \qquad (3.8)$$

The longitudinal phase-space distribution is adjusted so that

$$\sigma_z = 3.3 \times 10^{-6} \text{ m}, \sigma_\gamma = \gamma \times 10^{-4} = 3. \qquad (3.9)$$

If we assume that the current at the cathode is I = 100 A, then the total charge per 33-μm bunch is

$$Q = \sqrt{2\pi} \frac{33 \times 10^{-6}}{3 \times 10^{8}} \times 100 = 28 \text{ pC}. \qquad (3.10)$$

The final bunch length, Eq. (3.9), is ten times smaller than the initial value in Eq. (3.6). Therefore, the final current is

$$I_f = 1 \text{ kA}. \qquad (3.11)$$

As a numerical example, we consider a "Greenfield" FEL operated at $\lambda_r = 0.4$ Å [7]. A permanent magnet undulator with a period of 3 cm is assumed to be used. We also assume that the beta function in the undulator is optimized to produce the shortest gain length. Two possible beam configurations are considered. The electron beam in case (a) is similar to the LCLS: $\gamma\varepsilon_x = 1 \times 10^{-6}$ m-r and a peak current $I = 3.5$ kA after bunch compression; while the electron beam in case (b) is the manipulated one as discussed in the above: $\gamma\varepsilon_x = 1 \times 10^{-7}$ m-r and a peak current $I = 1$ kA. Both beams have the same uncorrelated relative energy spread at 1×10^{-4}. Figure 3.1 shows the power gain length computed from Xie's fitting formula [9] versus the undulator parameter K for both cases. The gain length for case (b) is smaller by a large factor— about four for K=2. The length of the undulator will be reduced by a similar factor, reducing the construction cost as well as the operational complexity.

Is the small longitudinal emittance, Eq. (3.5), consistent with the space-charge force? The answer is affirmative as can be seen as follows: the longitudinal emittance arising from the space-charge effect in the rf photocathode gun can be written as [10]

$$\gamma\varepsilon_z^{SC} = \frac{\pi}{4\sin\phi_0} \frac{1}{E_0} \frac{2mc^2}{I_A} \mu_z(A). \qquad (3.12)$$

Here ϕ_0 is the emission phase, E_0 is the peak electric field, $I_A \approx 17$ kA is the Alfvén current, and $\mu_z(A)$ is a form factor depending on the aspect ratio $A = \sigma_x/\sigma_z$. The approximate formulae of the form factor for a Gaussian beam is

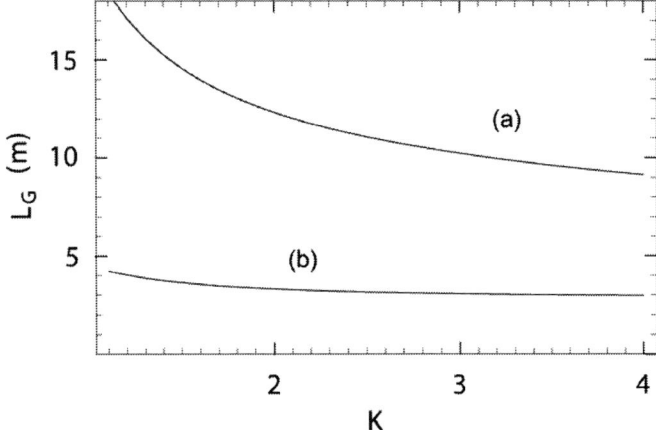

FIGURE 3.1. Power gain length L_G of an x-ray FEL at 0.4 Å versus the undulator parameter K for (a) a beam with a normalized transverse emittance 1×10^{-6} m-r and a peak current 3.5 kA and (b) a beam with a normalized transverse emittance 1×10^{-7} m-r and a peak current 1 kA. The relative rms energy spread in both cases is 1×10^{-4} (courtesy of Z. Huang).

$$\mu_z(A) = \frac{1.1}{1 + 4.5A + 2.9A^2}. \quad (3.13)$$

Therefore the space-charge emittance will be small for a beam of large aspect ratio, that is, a very short bunch length. If we use $\sigma_x = 0.6$ mm and $\sigma_z = 33$ μm, we obtain $A \approx 18$ and $\mu_z \approx 0.001$. Using the LCLS values $E_0 = 120$ MV/m, $I = 100$ A, and assuming $\pi/(4 \sin \phi_0) = 1$, we obtain the longitudinal emittance given by Eq. (3.5).

3.B. Emittance Transfer and Example

In the last section we considered emittance *exchange*. We now turn our attention to emittance *transfer*; that is, to the process that is less than a complete exchange. Due to the emittance exchange theorem, the transfer will require an appropriate correlation in the beam matrix. We shall see that there are many circumstances in which it is not necessary, or even advantageous, to have a complete exchange, but quite advantageous to transfer emittance to some degree. We present one such case here.

In the last section, we started from a very small longitudinal emittance for a short electron bunch. Usually, however, the bunches are long, with $\sigma_z \approx 1$ mm. We then have $A \approx 1$ and $\mu_z \approx 0.1$, and the longitudinal emittance becomes $\gamma \varepsilon_z = \gamma \sigma_z \sigma_\delta = 1 \times 10^{-5}$ m-rad. Thus, the longitudinal emittance is larger than the transverse emittance. Can one transfer emittance from a dimension where the emittance is smaller to one where it is larger? We shall see how this can be done slowly over many turns in a storage ring.

In the usual coupling resonance used to effect transfers or interchanges (complete transfers), the transfer takes place in the direction of large to small. However, after a while (the "half-way" point) emittance is being transferred from the larger to the smaller. How can that be? Clearly there must be correlations that have been established amongst the variables spanning the four-dimensional phase space (two degrees of freedom).

A careful study was made [11], numerically, of the nature of the distribution in phase space in a simple coupling resonance as emittance is transferred from one degree of freedom to the other. It was observed that at the half-way point there were strong correlations between p_x and z and between p_z and x. Of course all this can be (and has been) done analytically, but the first observations were numerical and followed the process described here.

An ideal case, with emittances 1 (x-direction) and 100 (z-direction), $v_x = 5.1$ and $v_z = 0.1$, and coupling potential 0.01 xz was studied. The transfer of emittance is shown in Fig. 3.2. In order to accomplish this transfer the initial distribution was taken with correlations as shown in Fig. 3.3. It is seen that there is no difficulty in producing a transfer that would be most advantageous for FELs.

How can such correlations be established? At this writing there has been no work done on this subject, although there will be (ref. [11]). However, considerable thought has gone into producing the correlations necessary for beam conditioning (see Section 7). It appears that the correlations needed for emittance transfer will be even easier to produce (linear in amplitude of oscillation rather than nonlinear) and, therefore, the work on conditioners will readily produce a practical system for introducing the correlations necessary for emittance transfer.

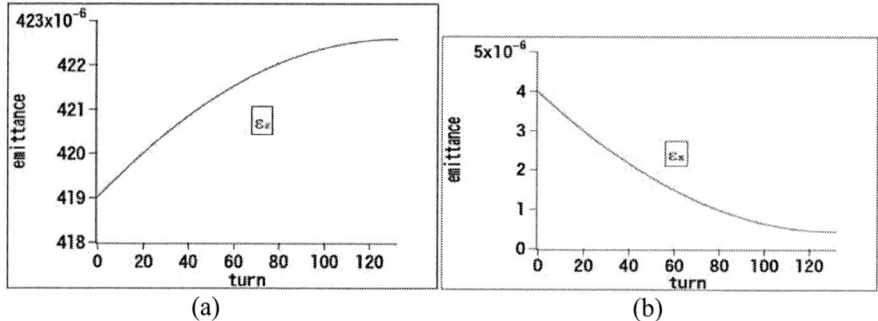

FIGURE 3.2. Evolution of the longitudinal and transverse emittance as a function of the turn number.

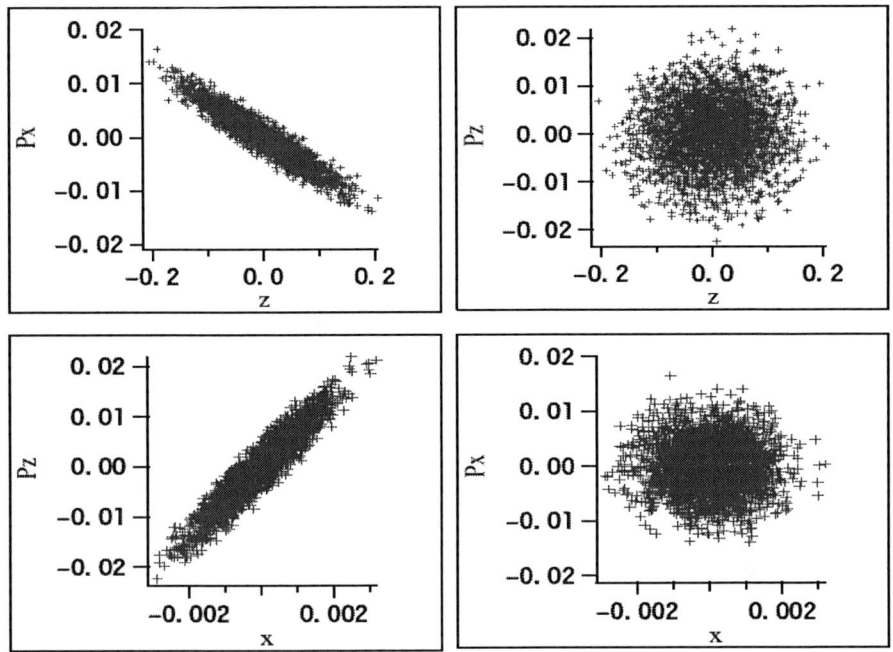

FIGURE 3.3. The initial phase-space distribution that is needed for the emittance transfer shown in Figure 3.2.

4. FLAT-BEAM GENERATION AND APPLICATION

In this section we consider beam manipulation in transverse 4-D phase space spanned by (x, x', y, y'). In particular we discuss generation and applications of flat beams, that is, uncoupled beams (no $(x\text{-}y)$ correlation) with large emittance ratio

$$\varepsilon_x \gg \varepsilon_y. \tag{4.1}$$

The flat beam technique was invented by Brinkmann, Derbenev, and Flöttmann [12] by inverting an earlier work by Derbenev [13] on flat-to-round beam transformation. The technique was further studied theoretically in references [14] and [15]. Here we follow the derivation in reference [15].

4.A. Flat-Beam Technique

A flat beam is generated from a cylindrically symmetric, correlated beam by immersing a laser-driven photocathode in an axial magnetic field B [12]. Particle motion in an axial magnetic field is most conveniently described in a frame rotating around the z-axis with the rotation rate per unit distance s in the z-direction [16] given by

$$\frac{d\theta}{ds} = \kappa = \frac{qB(s)}{2P_s}. \tag{4.2}$$

Here q is the particle charge and P_s is the momentum in the z-direction. In the rotating frame the phase-space coordinate just outside the cathode is [16]

$$X_0 = X_{th} + X_{rot},$$

where

$$X_{th} = \begin{pmatrix} x_{th} \\ x'_{th} \\ y_{th} \\ y'_{th} \end{pmatrix}, \quad X_{rot} = \kappa \begin{pmatrix} 0 \\ -y_{th} \\ 0 \\ x_{th} \end{pmatrix}. \tag{4.3}$$

In the above, X_{th} represents the thermal motion within the cathode while X_{rot} is due to the rotation of the coordinate system. The beam matrix corresponding to the thermal motion is

$$\Sigma_{th} = \langle X_{th} \hat{X}_{th} \rangle$$

$$= \begin{bmatrix} \sigma_c^2 & 0 & 0 & 0 \\ 0 & \sigma_{c'}^2 & 0 & 0 \\ 0 & 0 & \sigma_c^2 & 0 \\ 0 & 0 & 0 & \sigma_{c'}^2 \end{bmatrix}. \tag{4.4}$$

The beam matrix corresponding to X_0 is

$$\Sigma_0 = \langle X_0 \tilde{X}_0 \rangle = \begin{bmatrix} \varepsilon_{\text{eff}} T_0, & \mathcal{L} J_{2D} \\ -\mathcal{L} J_{2D} & \varepsilon_{\text{eff}} T_0 \end{bmatrix}. \tag{4.5}$$

Here

$$T_0 = \begin{bmatrix} \beta_0 & 0 \\ 0 & 1/\beta_0 \end{bmatrix}, \tag{4.6}$$

$$\beta_0 = \frac{\sigma_c}{\sqrt{\sigma_{c'}^2 + \kappa^2 \sigma_c^2}}, \tag{4.7}$$

$$\mathcal{L} = \frac{1}{2}\langle x_0 y_0' - y_0 x_0' \rangle = \kappa^2 \sigma_c^2, \text{ and} \tag{4.8}$$

$$\varepsilon_{\it eff} = \sqrt{\varepsilon_{th}^2 + \mathcal{L}^2}. \tag{4.9}$$

Note the rotation rate κ diverges as the electron longitudinal momentum vanishes. However, the normalized beam moments obtained by replacing the slopes x' and y' by $P_x = \beta\gamma x'$ and $P_y = \beta\gamma y'$, respectively, are finite.

The quantity $\varepsilon_{\it eff}$ is the projected emittance since the (x-y) correlation is nonvanishing, given by the angular momentum $\mathcal{L}$. Under a cylindrically symmetric beam transport $\mathcal{L}$ does not change. However, the correlation can be removed by a triplet of skew quadrupoles, which are not cylindrically symmetric [17]. The resulting beam matrix is

$$\Sigma_f = \begin{bmatrix} \varepsilon_x T_f & 0 \\ 0 & \varepsilon_y T_f \end{bmatrix}, \tag{4.10}$$

where T_f is a matrix similar to T_0 in Eq. (4.6) but with β_0 replaced by β_f. Since the transformation from the initial state (Eq. (4.3)) to the final state corresponding to the beam matrix (Eq. (4.10)) is symplectic, the two invariants ε_{4D} and $I^{(2)}$ must be conserved. Computing these quantities for Σ_0 and Σ_f, we find the condition

$$\varepsilon_{th}^2 = \varepsilon_{\it eff}^2 - \mathcal{L}^2 = \varepsilon_x \varepsilon_y \tag{4.11}$$

and

$$2\left(\varepsilon_{\it eff}^2 + \mathcal{L}^2\right) = \varepsilon_x^2 + \varepsilon_y^2. \tag{4.12}$$

The solution of Eqs. (4.11) and (4.12), are, for $\varepsilon_x > \varepsilon_y$:

$$\varepsilon_x = \varepsilon_{\it eff} + \mathcal{L}, \quad \varepsilon_y = \varepsilon_{\it eff} - \mathcal{L}. \tag{4.13}$$

When $\mathcal{L} \gg \varepsilon_{th}$ the beam is extremely flat:

$$\frac{\varepsilon_x}{\varepsilon_y} \approx \left(\frac{2\mathcal{L}}{\varepsilon_{th}}\right)^2 \gg 1. \tag{4.14}$$

The flat-beam technique transforms the initial thermal emittances, which are identical in the x- and y-directions, to two different emittances whose ratio is arbitrary. Since correlations are absent in both the initial (thermal) as well as the final states, does the technique violate the emittance exchange theorem? The answer is no because the process of electron emission from inside the cathode to the outside of the cathode, $X_{th} \to X_0$, is not symplectic. However, the determinant of the corresponding transformation matrix is unity. Thus the 4-D emittance before the cathode is the same as the final 4-D emittance, as shown by Eq. (4.11).

The flat-beam technique was experimentally demonstrated at the A0 facility at Fermilab [18-20]. An emittance ratio of about 100 was achieved recently in this facility [21]. Figure 4.1 shows the schematics of this experiment together with measured and computed beam evolution.

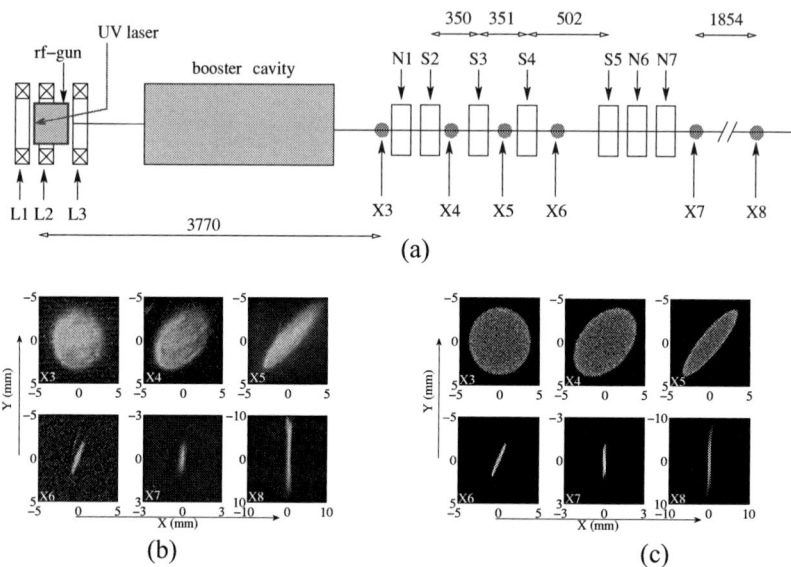

FIGURE 4.1. Demonstration of the flat-beam technique at Fermilab (courtesy of Yin-e Sun). (a) Schematics of the accelerator and beam manipulation layout. (b) Evolution of the measured beam distribution. (c) Evolution of the computed beam distribution.

4.B. Application in Order to Generate Short Pulses of X-rays

There is strong scientific demand for x-rays with sub-picosecond duration. High-gain x-ray free-electron lasers based on linacs can provide such ultrafast x-rays since electron beams are tightly bunched to increase the gain. However, spontaneous emission x-ray sources from third-generation synchrotron radiation facilities based on high-brightness electron storage rings are rather long, of the order of 100 ps. Recently, A. Zholents proposed using rf orbit deflection to generate sub-picosecond x-ray pulses [22] in order to compress 100-ps x-ray beams from a storage ring to 1-ps or 10-ps x-ray beams from an electron recirculation linac (ERL) to 100 fs.

In this method, an electron beam receives z-dependent transverse deflection in an rf cavity operating in the TM_{110} mode. The beam then passes through an undulator emitting an x-ray beam with similar z-x′ correlation. After the x-ray beam has drifted to a sufficient distance, the z-x′ correlation will be transformed to the z-x correlation. A short x-ray pulse can then be obtained by using a vertical slit, thereby "slicing out" a small temporal portion of the pulse, or by using asymmetrically cut crystals for pulse compression. This is schematically shown in Fig. 4.2. Application of this method for the APS ring has been studied taking into account various degrading effects [23-25].

Under ideal conditions, the minimum rms pulse length achievable by pulse compression in this scheme is given by [22]

$$\sigma_t = \frac{E}{\hbar \omega_{RF} V} \sqrt{\sigma_{y',e}^2 + \sigma_{y',rad}^2}. \qquad (4.15)$$

Here E is the electron energy, V is the deflecting voltage, ω_{RF} is the ring rf frequency, and $\sigma_{y',e}$ and $\sigma_{y',rad}$ are the angular divergence due to the electron beam and to the undulator radiation, respectively. Equation (4.15) shows that the vertical divergence, and therefore the vertical emittance, of the electron beam gives the ultimate limit of the compression technique. An ERL-based x-ray light source project at Lawrence Berkeley National Laboratory dedicated to ultrafast x-ray science [26] uses the flat-beam technique together with the pulse compression to produce 100-fs pulses of x-ray beams.

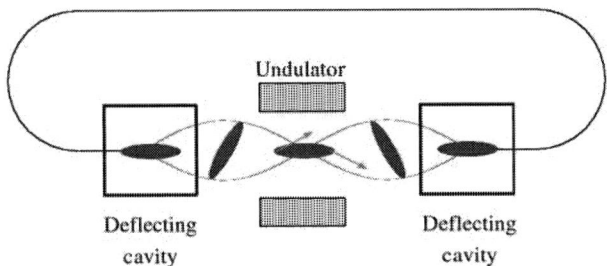

FIGURE 4.2. Pulse compression via transverse-longitudinal correlation [22]. [Figure reprinted courtesy of K. Harkay et al., Proc. of the 2005 Particle Accelerator Conference, 668 (2005). © 2005 IEEE]

4.C. Application to Smith-Purcell FEL in Order to Generate THz Radiation

Interest in Smith-Purcell radiation recently increased after an experiment at Dartmouth, using 30-35 kV electron beams from a scanning microscope, observed nonlinear behavior in the Smith-Purcell radiation power as a function of the beam current [27]. If the nonlinearity is due to the FEL gain mechanism, a Smith-Purcell FEL is promising as a miniature device producing intense THz radiation. However, another experiment at the University of Chicago using the same parameters as at Dartmouth failed to produce any evidence of an FEL gain [28]. Figure 4.3 shows a schematic of these experiments. Nonlinearity was indeed seen initially; however, it disappeared with the cooling of the grating surface. Thus it is suspected that the nonlinearity observed at Dartmouth could have been due to the blackbody radiation emitted when grating was heated from electron impact. To resolve these issues, we performed an extensive theoretical investigation [29] and found that the electron beams used for the experiments at Dartmouth and the University of Chicago were not suitable, both in beam current and in the shape of the transverse phase-space distribution [30].

The analysis was done for the case where the electron beam is thin and uniform in the direction perpendicular to the beam direction. The beam moves close to the grating surface, as shown in Fig. 4.4, and interacts with a surface mode of the grating. The surface mode is a freely propagating mode, arising mathematically as a singularity of the reflection matrix off the grating. It turns out that the group velocity of the surface mode is in the direction opposite to the beam velocity [31]. The system is therefore a backward wave oscillator (BWO), in which the optical intensity grows exponentially to saturation due to the inherent feedback mechanism in a BWO. The condition under which the oscillation occurs is

$$\frac{dI}{dy} > \frac{dI_s}{dy} \equiv 7.71_A \frac{(\beta\gamma)^4 \lambda}{(2\pi)^2 \chi L^3} e^{2\Gamma_0 h}. \quad (4.16)$$

Here, y is the coordinate in the direction parallel to the grating groove; I_s is known as the start current; $I_A = 17$ kA is the Alfvén current; β is the electron velocity divided by the velocity of light; γ is the relativistic factor, which is close to unity in our case; λ is the free-space wavelength of the surface mode, about 690 microns; χ is the coupling factor representing the efficiency with which an external evanescent wave couples to the surface mode; $\Gamma_0 = 2\pi/\lambda\gamma\beta$; and h is the distance from the grating surface to the beam axis.

From the last factor in Eq. (4.16) we see that the grating-electron beam distance should be small:

$$h < \frac{1}{2\Gamma_0} = \frac{\lambda\beta\gamma}{4\pi}. \quad (4.17)$$

In the present case the RHS is about 20 microns. Therefore the beam size in the vertical direction should be about the same, 20 microns. Horizontally, however, the beam size is only constrained by diffraction of spatial wavelength $\beta\lambda$:

$$\Delta y \approx \sqrt{\frac{\lambda\beta}{4\pi} L}. \quad (4.18)$$

Here L is the length of the grating. In the present case, Δy is about 500 microns. Thus the beam is indeed flat. The start current $I_s = \Delta y \, dI_s/dy$ computed from Eq. (4.16) is about 30 mA. Such a beam may be constructed from a line source or by the flat-beam technique.

Figure 4.5 shows the growth of the optical intensity in a Smith-Purcell BWO above and below the threshold condition.

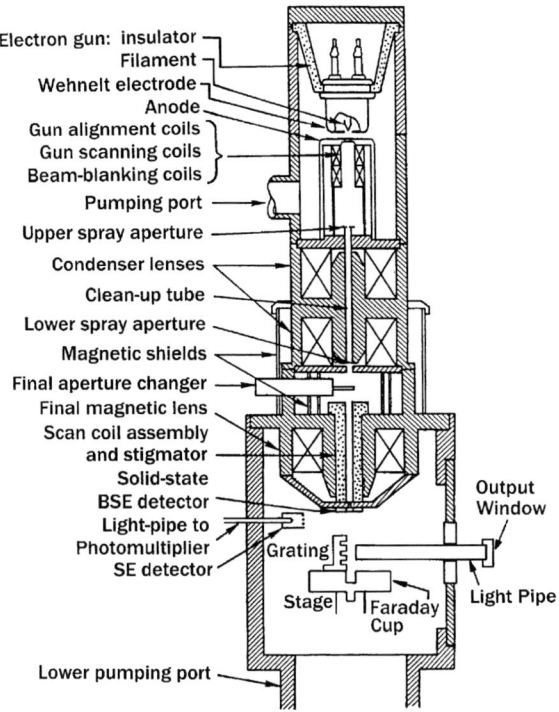

FIGURE 4.3. A schematic of a Smith-Purcell experiment based on a scanning electron microscope.

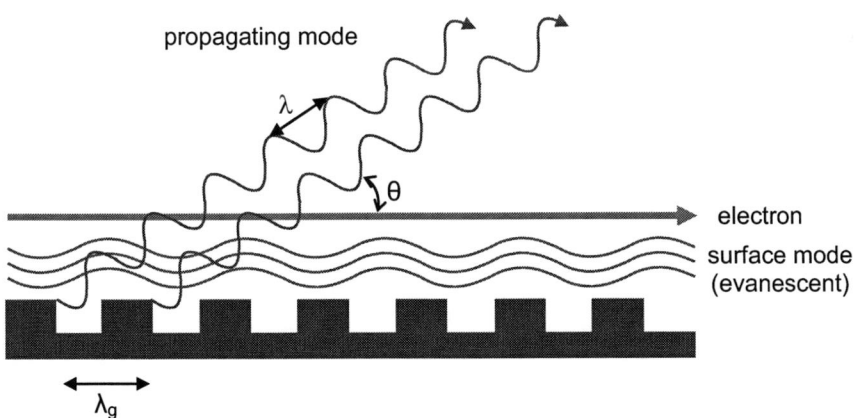

FIGURE 4.4. A schematic of a Smith-Purcell free-electron laser in which the electron beam interacts strongly with the evanescent surface mode.

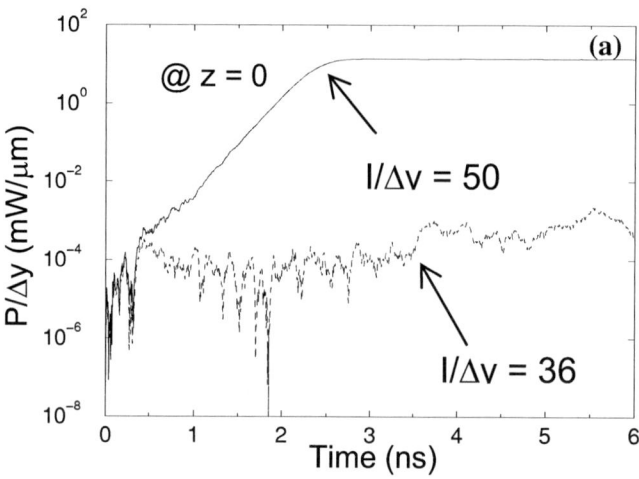

FIGURE 4.5. Intensity build-up for a Smith-Purcell BWO (courtesy of K.-J. Kim and V. Kumar). The solid and dotted curves are for above and below the threshold, respectively.

5. BEAM CONDITIONING CRITERIA

The concept of "beam conditioning" was introduced many years ago [32-33]. Briefly, the resonance condition for an FEL requires a specific average velocity: after each undulator period, electrons fall behind the laser field by exactly one wavelength. The usual resonance condition assumes zero emittance. Adding correlations of transverse amplitude with energy brings more particles into resonance, which is simply the idea of "conditioning."

For a zero-amplitude particle, the typical angle is K/γ, where K is the normalized strength of the undulator. Then the slippage after one undulator period should be

$$\lambda = \Delta z = \left(1 - \frac{v_z}{c}\right)\lambda_w \simeq \frac{1}{2}\left(1 - \frac{v_z^2}{c^2}\right)\lambda_w = \frac{1+K^2}{2\gamma^2}\lambda_w. \tag{5.1}$$

This is the basic resonance condition. For large γ, the angle and $v_\perp/c$ are roughly the same.

For nonzero emittance, the average angle in terms of the normalized emittances is

$$\langle \theta_\varepsilon^2 \rangle \simeq \frac{2\pi(\varepsilon_{Nx} + \varepsilon_{Ny})}{\gamma \lambda_\beta}. \tag{5.2}$$

Here $\lambda_\beta = 2\pi\beta_x = 2\pi\beta_y$. However, the angles from the emittances and undulator are uncorrelated, and add in quadrature. Thus we have a modified equation for v_z:

$$\langle \theta_\varepsilon^2 \rangle + \frac{v_z^2}{c^2} \simeq 1 - \frac{1+K^2}{\gamma^2}. \tag{5.3}$$

To have uniform v_z requires an energy shift $\Delta\gamma$ from the zero emittance case to balance out the emittance term:

$$\frac{2\pi(\varepsilon_{Nx}+\varepsilon_{Ny})}{\gamma \lambda_\beta} \simeq \langle \theta_\varepsilon^2 \rangle = \frac{1+K^2}{\gamma^2}\frac{2\Delta\gamma}{\gamma}. \tag{5.4}$$

Note $\Delta\gamma/\gamma \ll 1 \Rightarrow \lambda_\beta \gg \pi\gamma(\varepsilon_{Nx}+\varepsilon_{Ny})/(1+K^2)$.

Using the resonance condition and taking $\varepsilon_{Nx}=\varepsilon_{Ny}=\varepsilon_N$, we arrive at

$$\Delta\gamma = \pi \frac{\lambda_w}{\lambda}\frac{\varepsilon_N}{\lambda_\beta}. \tag{5.5}$$

This is the basic conditioning criterion. Before we turn to the subject of producing the conditioning criterion by various beam devices in Section 7, we show, in Section 6, that FEL performance will be greatly improved provided a beam is conditioned.

6. EXAMPLES OF FEL PERFORMANCE WITH AND WITHOUT CONDITIONING

The subject of beam conditioning and how it would greatly improve FEL performance was explored in [33]. The example presented here is taken from a more recent paper [34] containing several examples which show that in virtually all cases conditioning greatly improves FEL performance. Here we only give a single example, namely a possible scheme to achieve highly energetic (30 keV) photons, radiation wavelength 0.4 Å. There are two options: a low-energy case and high-energy case. In both cases the peak current is taken to be 3.5 kA, with a nominal emittance of 1.2 µm, but we shall also consider emittances as low as 0.1 µm. The nominal betatron wavelength is $\lambda_\beta \approx 110$ m in both cases.

For the high-energy case we consider beam parameters 27.8 GeV and $\Delta\gamma/\gamma = 1 \times 10^{-4}$, and undulator parameters $\lambda_w = 3$ cm and $K = 2.62$. For the low-energy case we consider beam parameters 12.1 GeV and $\Delta\gamma/\gamma = 1.2 \times 10^{-4}$, and undulator parameters $\lambda_w = 3$ cm and $K = 0.71$. The results are displayed in Figs. 6.1 and 6.2.

Conditioning is undoubtedly advantageous, even offering the possibility of entering FEL ranges (very short x-rays) not otherwise attainable.

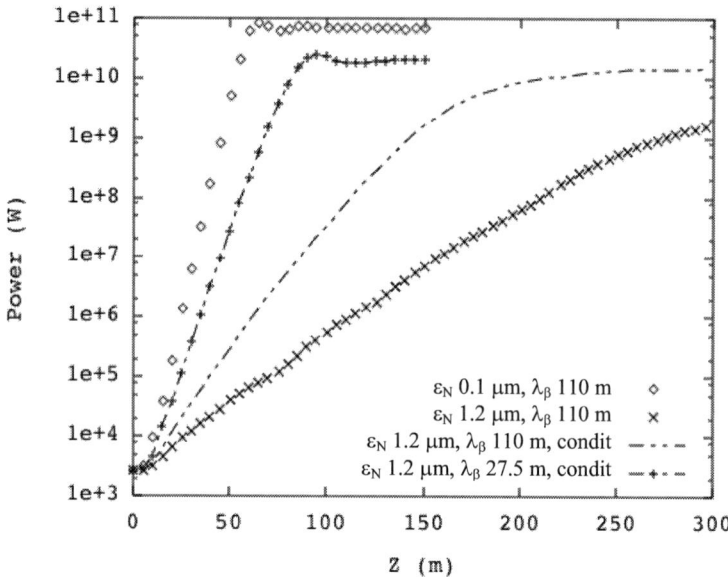

FIGURE 6.1. Performance of an x-ray FEL with conditioned electron beams. [Figure reprinted with permission from A. Wolski et al., PRST-AB 7, 080701 (2004). Copyright 2004 by the American Physical Society.]

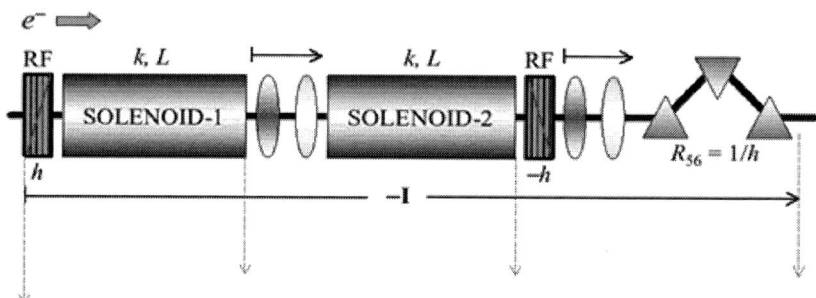

FIGURE 6.2. Schematic of 'Two Phase' FEL conditioning (courtesy of P. Emma and G. Stupakov).

7. METHODS OF CONDITIONING

7.A. Conventional

Conditioning as first developed [33] was restricted to very short beams. The authors of that paper were aware, and commented upon, the Panofsky-Wentzel theorem and how it would deflect anything but a delta function bunch in z and cause an increase in "effective emittance." The authors suggested—incorrectly—that it would be simple to correct the deleterious effect and hence condition long beams.

The analysis was extended by N. A. Vinokurov [35], who presented a different method of conditioning than that presented in [33]. His method actually preserved emittance in long bunches, although he did not explicitly state that fact (but probably was well aware of it).

A particular type of conditioner was considered by P. Emma and G. Stupakov [36], and they found that the conditioner produced very large bunch deflection as a function of longitudinal position. The deflection created a projected emittance that was far larger than would be acceptable. They speculated that *all* conditioners would inevitably lead to the same phenomena.

This was shown not to be the case by the group of Wolski et al. [34] who independently developed the same method as had been previously proposed by Vinokurov. Subsequent work, built on this approach was made by P. Emma and G. Stupakov [37]. The layout is simple: an rf cavity, followed by a nonlinear channel with the beta at the ends independent of energy, and then another rf cavity. The first cavity provides energy that sweeps across the bunch (in z). The nonlinear channel gives a delay to particles proportional to the square of their amplitude. The final rf cavity removes the sweep in energy, but leaves the variation of energy with amplitude; just what is needed for conditioning. Because the betas are independent of energy (to first order) all energies can be matched into a wiggler, and there is no emittance increase.

The only difficulty—not one of principle—is that with practical rf gradients and typical magnets the amount of conditioning that can be given to a beam is far less than that desired.

7.B. Laser-Compton

Conditioning can be achieved by Compton backscattering of photons as was realized by the team of [34] and, independently, by Carl Schroeder [38], because photons scattering off moving electrons cause the electron to lose energy. The upshift in energy is $4\gamma^2$. It is easy to tailor the laser pulse so that the electrons near the axis lose more energy than those at larger radii. It is necessary to choose the conditioning energy $\gamma m_e c^2$ to be sufficiently low such that there are a good number, N, of photons, of energy $h\nu$ required to give the necessary shift $\Delta\gamma_c m_e c^2$ in the conditioning energy. We have:

$$N = \frac{\Delta\gamma_c m_e c^2}{h\nu 4\gamma^2}. \tag{7.1}$$

This condition can be easily achieved by modern lasers, as shown in [38], where many other aspects are also considered.

7.C. Laser-Wiggler

Sasha Zholents realized [39] that it would be advantageous to replace the rf cavities in the scheme of Wolski et al. with an inverse free-electron laser, i.e., a wiggler and a laser beam. In this way he has been able to achieve the desired conditioning, with realistic parameters.

He has presented parameters for the LCLS, which he conditions at an energy of 1.5 GeV, where conditioning requires a variation of energy from zero amplitude to a maximum amplitude of 12 MeV. He accomplishes this with a laser of 6 mJ, wavelength of 800 nm, and pulse length of 100 fs; a wiggler of 10 periods, each 10 cm long; and a nonlinear channel of 16 FODO cells, 90° advance per period, using 25-cm-long quads with gradient of 4.3 kG/cm, and total length of 18 m.

7.D. Plasma Channel

In a work to be published by Jonathan Wurtele, Gregg Penn, and Andrew Sessler, consideration has been given to replacing the nonlinear channel with a plasma channel. Because the focusing is very strong in a plasma channel, this device can result in a practical conditioner. The idea is to send a laser through a gas in a tube, blow out all the electrons, and make an ion channel; then send the high-energy beam just behind the laser before the slow electrons return.

In the plasma channel $\beta = (2\gamma)^{1/2} c/\omega_p$, where $\omega_p = 6 \times 10^{12} (n \text{ (cm}^{-3})/10^{16})^{1/2}$ and $\lambda_p = 2\pi c/\omega_p$. For example, at $n = 10^{17}$ cm^{-3} and 1 GeV, $\omega_p = 2 \times 10^{13}$ s^{-1}, $1/\omega_p = 50$ fs, $\lambda_p = 100$ μm, and $\beta = 0.1$ cm.

They have produced an example of conditioning for the LCLS at a beam energy of 100 MeV, where the beam length is 1 mm and has a radius of 2.4 microns. The performance of this plasma conditioner is shown in Fig. 6.3. In a plasma channel of length 2 m, and $n_p = 1.12 \times 10^{16}$ cm^{-3}, the beta is 0.1 cm, $\lambda_p = 315$ μm. Conditioning is achieved, but only over a length that is small compared to λ_p. However, producing very short bunches of x-rays can be quite advantageous for many experiments.

beam energy 1 GeV
beam length 20 micron
beam emittance 1.2 micron
beam radius 1.1 micron

Plasma channel:
length 5 m, $n_p = 2.8 \times 10^{16}$ cm^{-3}
beta = 0.2 cm, $\lambda_p = 199$ μm

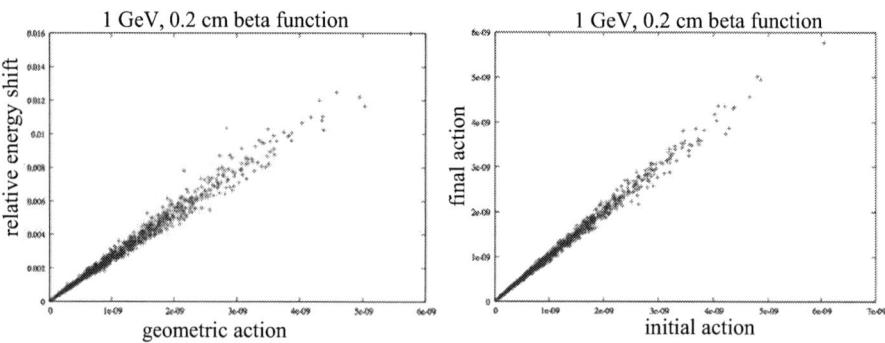

FIGURE 6.3. A particle simulation study of the effect of a plasma channel conditioner. Perfect conditioning (energy vs action) would correspond to all the particles lying on the line (of proper slope) in the first picture and not have their emittance blown up (final action the same as initial action in the second picture). It can be seen that the particles lie close to the lines in both pictures; i.e., that the plasma channel conditioner works very well.

8. CONCLUSIONS

In this paper we have explored some of the consequences of symplectic conditions on emittance exchange and emittance transfer and their applications. We have not considered the shrinking of phase-space volume—"cooling"—but rather the sophisticated process of manipulating phase space. We have seen in the examples we have provided—and there are many more—the distinct advantages of what we would call "phase-space manipulation."

We have discussed the emittance exchange theorem stating that with no correlations in a symplectic system, emittance cannot be partially transferred. We have presented a new optical system producing exact emittance exchange. An example of the exchange was discussed that could drastically improve the performance of future high-gain FELs for hard x-rays.

An example of phase-space manipulation involving a non-symplectic process is the flat beam technique with which the initial emittances in two transverse dimensions can be partitioned arbitrarily. A beam with large emittance ratio can be produced with applications to ultrashort x-ray pulse generation and Smith-Purcell BWO.

We have also looked at the consequence for FELs of correlation of energy with transverse amplitude ("conditioning"). Some correlations have already become an often-used tool of the beam physicist: correlations of energy with longitudinal position is commonly used for beam bunching. Angular momentum beams are used for making flat beams. Here, we have suggested some ways in which even more complicated correlations can be put on a beam. It is our belief that the future will see the development of these methods and surely different—and better—methods for phase-space manipulations. It is our opinion that phase-space manipulation will become an often-used and very important tool of the beam physicist.

REFERENCES

1. For a clear derivation, see D.A. Edwards and M.J. Syphers, *An Introduction to the Physics of High Energy Accelerators* (New York: Wiley-Interscience, 1993).
2. A.J. Dragt, F. Neri, and G. Rangarajan, Phys. Rev. A **45**, 2572 (1992).
3. E. Courant, in "Perspectives in Modern Physics, Essays in Honor of Hans A. Bethe," R.E. Marshak, ed., Interscience Publishers, 1966.
4. M. Cornacchia and P. Emma, Phys. Rev. ST Accel. Beams **5**, 084001 (2002).
5. P. Emma, Z. Huang, K.-J. Kim, and Ph. Piot, in preparation.
6. Linac Coherent Light Source, Conceptual Design Report, SLAC-R-593, 2002.
7. J. Galayda, K.-J. Kim, and J. Murphy, " Greenfield FELs," presentation to the BESAC Subcommittee on BES 20 Year Facility Road Map, Washington DC, Feb. 22-24, 2003.
8. M. Hüning and H. Schlarb, Proceedings of the 2003 Particle Accelerator Conference, 2074 (2003), http://www.jacow.org.
9. M. Xie, Proceedings of the 1995 Particle Accelerator Conference, 183 (1995).
10. K.-J. Kim, Nucl. Instrum. Methods A **275**, 201 (1989).
11. H. Okamoto, A.M. Sessler, and J. Wei, private communication of work in progress.
12. R. Brinkmann, Y. Derbenev, and K. Flöttmann, Phys. Rev. ST Accel. Beams **4**, 053501 (2001).
13. Y. Derbenev, University of Michigan Report No. UM-HE-98-14, 1998.
14. A. Burov, S. Nagaitsev, and Y. Derbenev, Phys. Rev. E **66**, 016503 (2002).
15. K.-J. Kim, Phys. Rev. ST Accel. Beams **6**, 104002 (2003).
16. K.-J. Kim and C.-x. Wang, Phys. Rev. Lett. **85**, 760 (2000).
17. A. Burov and V. Danilov, FERMILAB-TM-2043 (1998).

18. D. Edwards et al., Proceedings of the XX International Linac Conference, 122 (2000); D. Edwards et al., Proceedings of the 2001 Particle Accelerator Conference, 73 (2001), http://www.jacow.org.
19. E. Thrane et al., Proceedings of the XXI International Linac Conference, 308 (2002).
20. Y.-e Sun, "Angular Momentum-Dominated Beam," U. of Chicago Ph.D thesis, 2005.
21. Ph. Piot, Y.-e Sun, and K.-J. Kim, "Photoinjector-generation of a flat electron beam with transverse emittance ratio of 100," submitted for publication.
22. A. Zholents, P. Heimann, M. Zolotorev, and J. Byrd, Nucl. Instrum. Methods A **425**, 385 (1999).
23. M. Borland, Phys. Rev. ST Accel. Beams **8**, 074001 (2005).
24. M. Borland and V. Sajaev, Proceedings of the 2005 Particle Accelerator Conference, 3886 (2005), http://www.jacow.org.
25. K. Harkay et al., Proceedings of the 2005 Particle Accelerator Conference, 668 (2005), http://www.jacow.org.
26. J. Corlett et al., Proceedings of the 2001 Particle Accelerator Conference, 2635 (2001), http://www.jacow.org.
27. J. Urata, M. Goldstein, M.F. Kimmit, A. Naumov, C. Platt, J.E. Walsh., "Supperradiant Smith-Purcell Emission," Phys. Rev. Lett. **80**, 516 (1998).
28. O.H. Kapp, Y.-e Sun, K.-J. Kim and A.V. Crewe, "Modification of a scanning electron microscope to produce Smith-Purcell radiation," Rev. Sci. Instrum. **75**(11), 4732 (2004).
29. V. Kumar and K.-J. Kim, submitted for publication to Physical Review E.
30. K.-J. Kim and V. Kumar, to be published.
31. H.L. Andrews and C.A. Brau, "Gain of a Smith-Purcell free-electron laser," Phys. Rev. ST Accel. Beams **7**, 070701 (2004).
32. A.M. Sessler, D. H. Whittum, L.-H. Yu, "RF Beam Conditioning for the FEL," February 1991 (unpublished).
33. A.M. Sessler, D.W. Whittum, and L.-H. Yu, "Radio-frequency beam conditioner for fast wave free electron generators of coherent radiation," Phys. Rev. Lett. **66**, 309 (1992).
34. A. Wolski et al., "Beam Conditioning for Free Electron Lasers: Consequences and Methods," Phys. Rev. ST Accel. Beams **7**, 080701 (2004).
35. N.A.Vinokurov, Nucl. Instrum. Methods A **375**, 264 (1996).
36. P. Emma and G. Stupakov, "Limitations of Electron Beam Conditioning for Free-Electron Lasers," Phys. Rev. ST Accel. Beams **6**, 030701 (2003).
37. P. Emma and G. Stupakov, "Controlling Emittance Growth in an FEL Beam Conditioner," Proceedings of the 2004 European Particle Accelerator Conference, 503 (2004), http://www.jacow.org.
38. C.B. Schroeder et al., "Electron Beam Conditioning by Compton Backscattering," Phys. Rev. Lett. **93**, 194801 (2004).
39. A.A. Zholents, "Laser Assisted Electron Beam Conditioning for Free Electron Lasers," Phys. Rev. ST Accel. Beams **8**, 050701 (2005).

Optics of Electron Beam in the Recycler

A. Burov[1], G. Kazakevich[2], T. Kroc[1], V. Lebedev[1], S. Nagaitsev[1],
L. Prost[1], S. Pruss[1], A. Shemyakin[1], M. Sutherland[1], M. Tiunov[2], A.Warner[1]

[1]*Fermi National Accelerator Laboratory, P.O. Box 500, Batavia IL 60543*
[2]*Budker Institute of Nuclear Physics, 630090 Novosibirsk, Russia*

Abstract. Electron cooling of 8.9 GeV/c antiprotons in the Recycler ring (Fermilab) requires high current and good quality of the DC electron beam. Electron trajectories of ~0.2 A or higher DC electron beam have to be parallel in the cooling section, within ~ 0.2 mrad, making the beam envelope cylindrical. These requirements yielded a specific scheme of the electron transport from a gun to the cooling section, with electrostatic acceleration and deceleration in the Pelletron. Recuperation of the DC beam limits beam losses at as tiny level as ~0.001%, setting strict requirements on the return electron line to the Pelletron and a collector. To smooth the beam envelope in the cooling section, it has to be linear and known at the transport start. Also, strength of the relevant optic elements has to be measured with good accuracy. Beam-based optic measurements are being carried out and analysed to get this information. They include beam simulations in the Pelletron, differential optic (beam response) measurements and simulation, beam profile measurements with optical transition radiation, envelope measurements and analysis with orifice scrapers. Current results for the first half-year of commissioning are presented. Although electron cooling is already routinely used for pbar stacking, its efficiency is expected to be improved.

Keywords: Envelope match, angular momentum dominated beam, electron cooling.
PACS: 29.27.Bd, 29.27.Eg, 29.27.Fh

INTRODUCTION

The Recycler ring (RR) is used for stochastic and electron cooling of 8.9 GeV/c antiprotons coming from the Accumulator [1]. A layout of the electron cooling line is presented in Fig. 1. The designed beam envelope is shown in Fig. 2, generated by the OptiM code [2, 3]. Main features of the electron cooling line are.

- Electrons are emitted from a thermo-cathode, accelerated and decelerated in an electrostatic accelerator (Pelletron);
- Ideally, electron trajectories in the cooler are straight lines parallel to the axis;
- For focusing purposes, there is ~ 100 G magnetic field in the 20 m long cooler;
- Magnetic flux at the cathode is equal to the flux at the cooler;
- Matrices Pelletron-Cooler and Cooler-Pelletron are rotation-invariant;
- Possibility for zero dispersion in the return line is foreseen.

ELECTRON ANGLES

Cooling efficiency strongly depends on the effective angle between the pbars and electrons. To have maximal cooling, the electron rms angle should not exceed the proton angle, at least for the tail protons.

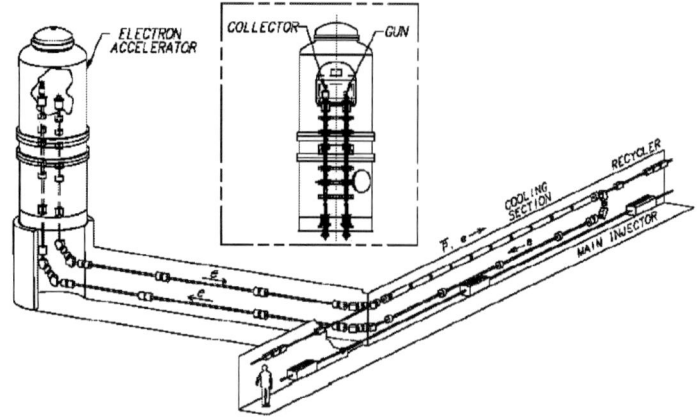

FIGURE 1. Layout of 100 m long electron cooling line

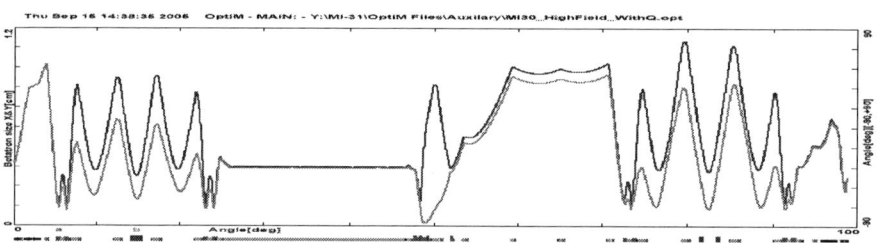

FIGURE 2. Envelope half-axes (design) of the electron cooling line (OptiM simulations). At the bottom, black color is for acceleration and deceleration sections, gold is for solenoids, blue for dipoles, red is for quads.

Assuming 95% normalized pbar emittance of 5 mm mrad, the 1D r. m. s. proton angle in the cooler with 30 m beta-function is 50 μrad. Electron angles in the cooling section are contributed by the following sources:
- Electron thermal angles; for 0.1 eV of the cathode temperature and equal sizes of the beam at the cathode and the cooler, the 1D r. m. s. thermal angle of electrons is 50 μrad.
- Imperfections of the magnetic field in the cooler (static); the last measurements yielded the field r. m. s. angle of 50 μrad [4].
- Perturbations from the Main Injector ramps contribute ~ 40 μrad.
- Non-linear aberrations of the optical elements in the supply line, mainly doublet solenoids, are estimated to give rise to ~ 20 μrad at the envelope surface; they scale as the offset cubed.
- Non-linearity of the beam angles distribution due to a deviation of the gun optics from the ideal Pierce regime. As a result, the beam has some halo, which optics is far from being similar to the almost linear beam core. Because of that, present halo-matching procedure based on the scraper

measurements in the cooler leads to the core envelope oscillations with ~ 400 μrad, according to simulations and the drag force measurements (see more below). This source of the electron angles is currently dominant; hopefully, that envelope mismatch will be significantly reduced in the near future.

MAIN INJECTOR RAMPS CONTRIBUTION

RR is located in the same tunnel as the Main Injector (MI). AC magnetic fields excited by the MI ramps give rise to drifts of the electron beam. Although these fields are suppressed by a compensation loop and shielding, some remnant effects still exist. To see how significant they are, the electron orbit was measured during 2 s of the MI cycle [5] with the sample frequency 700 Hz.

To distinguish the beam signal from the electronic noise, the raw data of 11+11 cooler BPMs were fitted by the helical trajectory for every time sampling point. For sufficient electron current and its modulation depth, the fit was found to be close to the raw signal, so the noise contribution was small. The AC r. m. s. beam angle was calculated then as 40 μrad, which looks to be small enough for the purposes of cooling.

ENVELOPE MISMATCH

Envelope Quality

To find tolerances for the electron envelope mismatch, let us consider an antiproton with an r. m. s. offset a_p and angle $\theta_p = a_p / \beta_p$, where β_p is RR beta-function in the cooler. Cooling of this antiproton would not be reduced by the electron angles if they are smaller than the antiproton angle: $\theta_e(a_p) < a_p / \beta_p$. Assuming the mismatch being linear with the offset, this requirement is identical to $\theta_e(a_e) < a_e / \beta_p$, where a_e is the electron beam radius. Taking into account that the electron angle relates to its offset variation $\Delta a_e = \theta_e \beta_e$ with $\beta_e = p_e c / eB$ as the Larmor beta-function, the matching requirement for maximal cooling can be expressed as

$$\theta_e(a_e) < a_e / \beta_p \Rightarrow \Delta a_e / a_e < \beta_e / \beta_p. \qquad (1)$$

For $B = 100\,\mathrm{G} \Rightarrow \beta_e = 160\,\mathrm{cm}$, and $\beta_p = 30\,\mathrm{m}$, this gives $\Delta a_e / a_e < 0.05$, and with $a_e = 3.5$ mm, it is equivalent to $\Delta a_e < 0.2$ mm, $\theta_e(a_e) < 120$ μrad. Note that the linear matching condition (1) does not depend on the pbar emittance.

To match the envelope, two issues have to be known well enough. First, for some optical settings, the envelope parameters somewhere in the line have to be known. Second, properties of the related optical elements have to be known with sufficient accuracy. When both problems are solved, any initial envelope can be matched by a proper change of settings of the well-modeled optical elements. The second problem is being solved by measuring differential trajectories, or responses of the beam trajectory on kicks applied by different correctors. The BPM data for the differential trajectories (normally for a set of 4 independent correctors and the energy offset) are fitted by variable optical parameters of the focusing elements. So far, our main approach to the envelope initial condition was based on measurements with 11 orifice scrapers located equidistantly in the cooling section [6]. This sort of measurements is sensitive to the beam halo, not the core. In case of significant non-linearity in the beam angle profiles, the core envelope is not smoothed together with the halo.

Envelope Measurement by Orifice Scrapers

Every scraper of the cooling section is a copper plate with a round orifice; they are located every 2 m. Normally, all the scrapers are moved out of the chamber. For the envelope measurement, one of the scrapers is moved in, with the center of its orifice approximately coinciding with the chamber axis. Then, the beam is shifted in some direction, and in parallel to the axis, until it starts touching the scraper. The BPM data for the beam center are taken at this point. After that, the beam is shifted in other direction, and everything repeats; normally, 8 directions are used. Then, the entire procedure is repeated for other scrapers.

When the data for all or a sufficient number of the scrapers are taken, the envelope parameters are found in a two-step fitting procedure. At the step number one, the beam ellipse and the scraper offset are found for every scraper involved. At the step number two, initial conditions for the beam envelope at the entrance of the cooler are found by fitting all these ellipses, using 4D phase space coupled optics formalism [2].

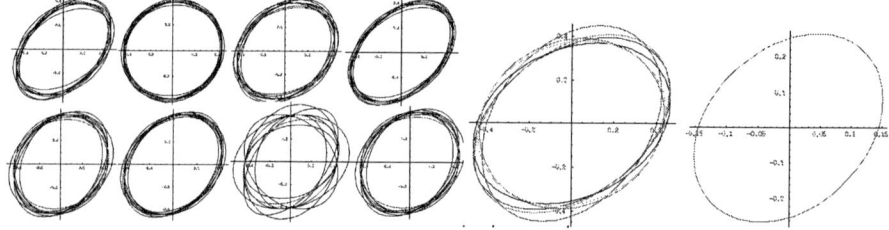

FIGURE 3. Envelope matching the orifice scraper data. On the left, the blue ellipses show the beam cross-section found from the data for every of 8 involved scrapers (in cm). Several blue ellipses for every scraper represent the matching error. Red ellipses on the left show the beam envelope in the cooler's magnetic field of 105 G, matched to the blue cross-sections. On the center, the found beam envelope is shown for the first 6 scrapers. On the right, the beam angles along the envelope are shown (in mrad), as a parametric plot. Average beam radius is 4.3 mm, 2D r. m. s. angle - 220 μrad.

Recent results of this analysis are presented in Fig. 3. It is clear that although the envelope is not quite round (which is not necessary), its scalloping along the cooler is rather small; namely, the 2D r. m. s. angle at the envelope has been calculated as 220 μrad. Remember that due to the beam non-linearity, this number can significantly underestimate mismatch angles of the beam core.

OTR Measurements and Simulations

Other device for the envelope measurements is an optical transition radiation (OTR) image analyzer (see details in Ref [7]). Essentially, this device shows 2D beam density distribution in a transverse to the beam line plane at the OTR location about 2 m downstream the acceleration exit in the vertical direction. Fig. 4 (left) shows beam profiles detected by OTR and calculated by UltraSAM-BEAM code [8, 9] with various settings of a nearest upstream lens. Good agreement between the measurements and calculations convinces us that both of them are essentially correct. Fig. 4 (right) gives the simulation results for beam density, radial and tangential velocity profiles at ~ 1 MeV of the kinetic energy. Clearly, the velocity profiles are significantly non-linear at the halo. This means that if the beam halo is matched in the cooler, its core would have significant angles. Estimations, based on these results, show that the core envelope angles for matched halo are as high as ~ 400 μrad. In other words, for this gun regime, the non-similarity of the beam core and halo is important: when the halo angle is, say, 200 μrad, the core angle at the nominal radius can be anything between 200 and 600 μrad. In the near future, either more linear gun regime will be used, or the data analysis will take into account the beam non-linearity.

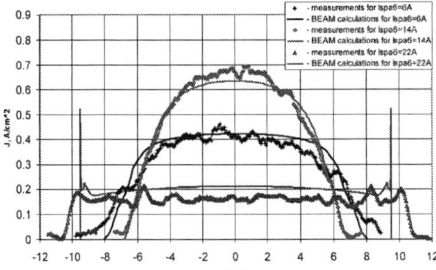 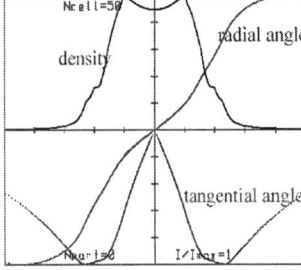

FIGURE 4. On the left: OTR measurements (dots) and UltraSAM-BEAM simulations of the beam transverse density profile for 3 different settings of an upstream lens and ~600 mA of the beam current. On the right: profiles for the beam density (blue), radial angle (green) and tangential angle (red) at ~1 MeV of the beam kinetic energy and ~200 mA of the current.

REFERENCES

1. S. Nagaitsev et al., "Antiproton cooling in the Fermilab Recycler", *this conf.* (*COOL05*, Galena, IL, 2005).
2. V. Lebedev and A. Bogazc, "Betatron Motion with Coupling", e-print JLAB-ACC-99-19 (2001).
3. V. Lebedev, "OptiM", at http://www-bdnew.fnal.gov/pbar/organizationalchart/lebedev/OptiM/optim.htm .
4. V. Tupikov et al., "Magnetic Field Measurement and Compensation at Fermilab Electron Cooler", *this conf.*
5. P. Joireman, measurements of Aug. 2005.
6. T. Kroc et al., "Electron Beam Size Measurements in the Fermilab Electron Cooling System", *this conf.*
7. A. Warner et al., "OTR Measurements and Modeling of the Electron Beam Parameters at the E-cooling Facility", *this conf.*
8. A.V. Ivanov, M.A. Tiunov. "UltraSAM - 2D Code for Simulation of Electron Guns with Ultra High Precision". Proceeding of *EPAC-2002*, Paris, 2002, pp.1634-1636.
9. M. A. Tiunov, "BEAM – 2D-Code Package for Simulation of High Perveance Beam Dynamics in Long Systems", Proc. of *Space Charge Effects in Formation of Intense Low-Energy Beams,* JINR, Dubna, Russia, 1999, pp. 202-208.

Experimental Study of Dispersion Control Utilizing both Magnetic and Electric Fields

Mikio Tanabe[*], Masahiro Ikegami[*], Akira Noda[*], Toshiyuki Shirai[*], Hikaru Souda[*], Hiromu Tongu[*], Shinji Shibuya[†], Koji Noda[†]

[*]*Institute of Chemical Research, Kyoto University, Gokashou, Uji, Kyoto pref., 611-0011, Japan*
[†]*National Institute of Radiological Sciences, NIRS, Anagawa 4-9-1 Inage, Chiba, Chiba pref., 263-8555, Japan*

Abstract. An experiment to control dispersion of beams in one bending section has been carried out. This experiment is based on a theory that the dispersion of accumulated beams can be controlled, if they are bent by a cross field composed of magnetic and an electric fields. Suppression of the dispersion can ease a shear which affects the 3-dimensionally ordered structure of the ultimate-low-temperature beams. In order to realize this scheme experimentally, we have manufactured a set of electrodes to create precise electric fields whose strength is 6.6×10^4 V/m for ^{24}Mg$^+$, 35keV beam. The electrodes have been inserted to the gap of dipole magnet. 3-dimensional field calculation shows that the error of the electric fields is less than 0.1% within $\pm$ 5mm from the reference orbit. We also tested the effect of the electric field using a single set of bending elements. The result showed that the linear dispersion can be controlled or canceled by changing the ratio of magnetic and electric fields.

Keywords: Storage ring, Laser cooling, Beam ordering, Dispersion
PACS: 41.85.-p, 41.85.Ja, 29.20.Dh,

INTRODUCTION

These days, some kinds of laser cooling experiment in a storage ring have been curried out [1-4], and some methods to cool transverse temperature of circulating beam have also been pointed out [5-7]. Furthermore, if the beams are 3-dimensionally cooled to an ultimate-low temperature state, it has been predicted that a phase transition to crystalline state happens [8-9].

A small ion cooling and storage ring "S-LSR", now under construction at Kyoto University, has equipment for an experiment to cool 35keV ^{24}Mg$^+$ beam by laser cooling. The aim of this experiment is to realize ordering or crystallized beam. If one aims at generation of a horizontally extended crystalline beam, one encounters the problem of shearing force. The shearing force affects the stability of the crystalline beam structure. The shearing force is caused by the existence of dispersion of storage rings [10]. If the dispersion can be suppressed, the stability of the crystalline structure is greatly improved. Suppression of the dispersion at a bending section is realized by using an electric field generated by cylindrical electrodes together with a bending magnetic field. The electrostatic potential causes deceleration or acceleration of particles depending on their radial position, and this effect eases the shearing force. If the following relationship between the magnetic and electric field strengths is fulfilled, both the linear dispersion and the shearing force canceled out in the bending section.

$$vB - 2E = 0. \qquad (1)$$

And by changing the ratio of magnetic and electric fields, we can control the dispersion.

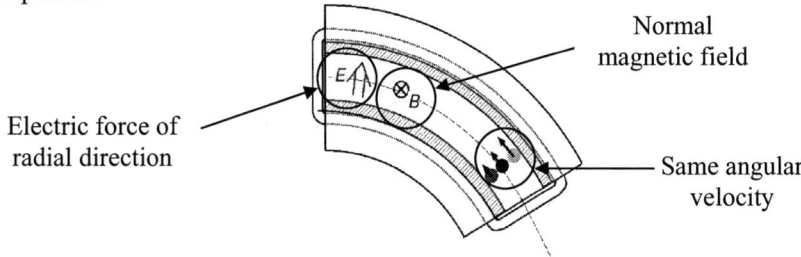

FIGURE 1. A bending section of a storage ring and the magnetic and electric fields to control dispersion are shown. The direction of the magnetic field is perpendicular to the median plane, and the direction of the electric field is radial.

DESIGN OF A SET OF ELECTRODES

To make precise electric fields in bending sections of S-LSR, we have designed a set of electrodes shown in figure 2. The designed electrodes consist of two main electrodes and eight intermediate electrodes. The main electrode is a part of cylindrical stainless steel, which is 26mm in height, 5mm in width, and has shims at the edges. The intermediate electrode is a stainless wire whose radius is 1mm. The positions of the intermediate electrodes and the sizes of the shims are determined from the results of 3-dimensional calculation of the electric fields. The effective length of the electric fields is also adjusted to match that of the magnetic fields. There are also two holes for laser path, and the affection to the electric fields by these holes is compensated by changing the structure of the outer main electrode.

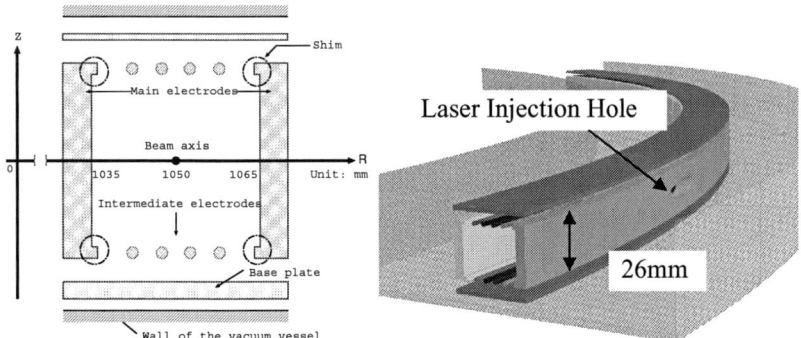

FIGURE 2. The cross sectional view (left) and the 3-dimensional view of the electrodes to control dispersion is shown.

145

As shown in figure 3, the errors of the electric fields are calculated to be less than ±0.1% in the area ±5 mm from the center of the reference orbit.

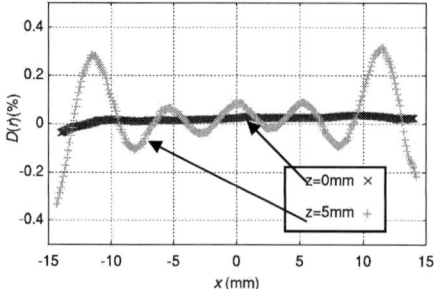

FIGURE 3. The errors of the electric fields at the center of the electrode are shown.

EXPERIMENTS TO CONTROL DISPERSION

In order to test the effects of designed electrodes before the construction of S-LSR, we carried out experiments to bend beam through only one bending section.

Setups and Conditions

The experimental setups are shown in the figure 4. First, ion beam is generated at the ion source. Next, the beam goes through the straight section, and the beam is cut to be 5mm in the horizontal and vertical sizes. Then the beam is bent at the bending section and finally, it is stopped by a fluorescent screen, and the positions of the beam are detected by a CCD camera.

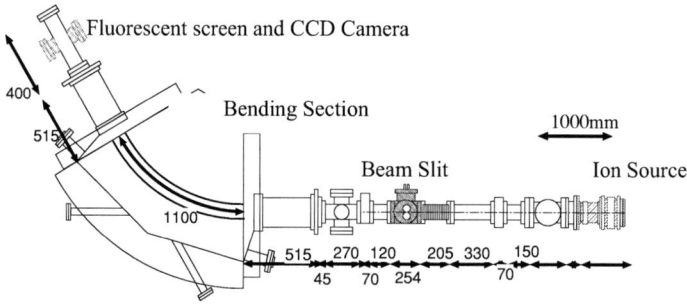

FIGURE 4. The experimental setups are shown in this figure.

The experimental conditions are shown in the table 1 and 2. As shown in table 2, we tested some combinations of magnetic and electric fields, such as bent by magnetic field only, dispersion canceling condition, and magnetic or electric field stronger conditions.

TABLE 1. The Experimental Conditions.

Condition	Value
Beam element	N_2^+
Beam kinetic energy	25keV
Emittance	5π mm mrad.
Vacuum	$\sim 10^{-5}$Pa

TABLE 2. The Combination of Magnetic and Electric Fields.

Magnetic Field Strength [T]	Electric Field Strength[V/m]	Condition
0.115	None	B only
0.252	5.71×10^4	$vB > 2E$
0.230	4.76×10^4	$vB = 2E$
0.205	3.81×10^4	$vB < 2E$

Results and Discussion

The results of bent beam by the magnetic field and under dispersion-canceling condition are shown in the figure 5. As shown in these graph, dispersion is able to be canceled if the equation (1) is fulfilled. When the beam was bent only by the magnetic field, the position deviation was proportional to the momentum deviation almost linearly because the linear dispersion is dominant. Compared with magnetic only bent, the relation between the momentum deviation and the position deviation under the dispersion canceling condition becomes secondary function, since the linear dispersion has been canceled out.

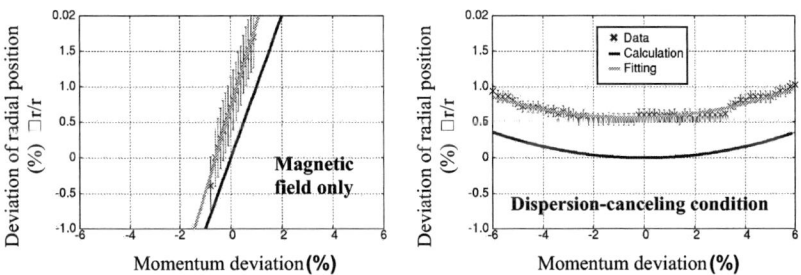

FIGURE 5. The results of the experiment to bend beam by magnetic fields only (left) and under the condition of dispersion control (right) are shown.

The results of bent beam by changing the ratio of magnetic and electric fields are shown in the figure 6. As shown in these graph, dispersion is able to be controlled from positive to negative, when the ration of magnetic and electric field strengths are changed. The parabolic shapes in the figure 6 imply that the second order effect appears, since the linear term has been suppressed.

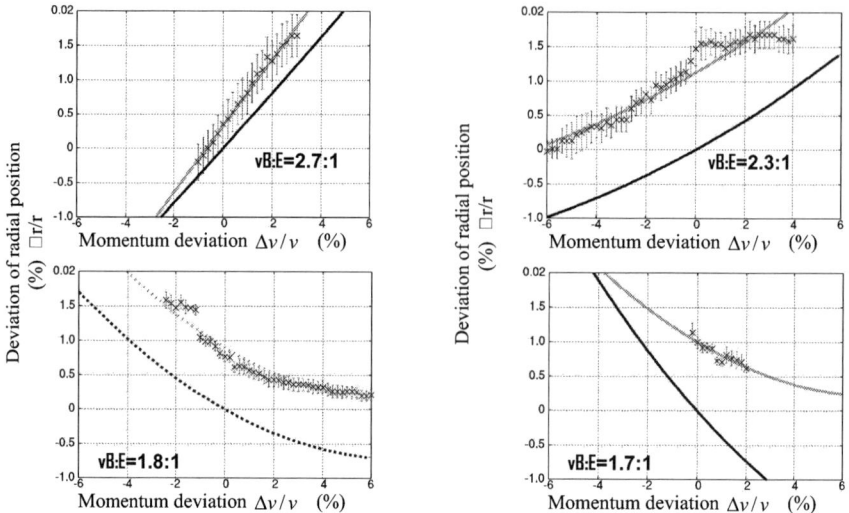

FIGURE 6. These graph show how dispersion changes when the ration of magnetic and electric field changes.

CONCLUSIONS AND FUTURE PLANS

We have designed and tested a set of electrodes to control dispersion at the bending sections of S-LSR. The experiments to bend beam through one bending section showed that dispersion of beam can be controlled by changing the ratio of magnetic and electric fields.

ACKNOWLEDGMENTS

The present work is financially supported with Advanced Compact Accelerator Development by Ministry of Education, Culture, Sports, Science and Technology, Japan and 21COE at Kyoto University.

REFERENCES

1. S. Schröder, et.al, Phys. Rev. Lett. **64**, 2901-2904 (1990)
2. J. S. Hangst, et.al, Phys. Rev. Lett. **67**, 1238-1241 (1991)
3. T. Schätz, U. Schramm, D. Habs, Nature (London) **412**, 717 (2001)
4. U. Schramm, T. Schätz, and D. Habs, Phys. Rev. Lett. **87**, 184801 (2001)
5. H. Okamoto, Phys. Rev. E **50**, 4982-4996 (1994)
6. H. Okamoto, A. M. Sessler, D. Möhl, Phys. Rev. Lett. **72**, 3977-3980 (1994)
7. T. Kihara, H. Okamoto, Y. Iwashita, K. Oide, G. Lamanna, J. Wei, Phys. Rev. E, **59**, 3594 (1999)
8. J. Wei, X. P. Li, and A. M. Sessler, Phys. Rev. Lett. **73**, 3089-3092 (1994).
9. J. Wei, H. Okamoto, and A. M. Sessler, Phys. Rev. Lett. **80**, 2606-2609 (1998)
10. M. Ikegami, A. Noda, M. Tanabe, M. Grieser, H. Okamoto, Phys. Rev. ST. A. B. **7**, 120101 (2004)

Transverse Echo Measurements in RHIC[1]

Wolfram Fischer

Brookhaven National Laboratory, Upton, New York 11973

Abstract. Diffusion counteracts cooling and the knowledge of diffusion rates is important for the calculation of cooling times and equilibrium beam sizes. Echo measurements are a potentially sensitive method to determine diffusion rates, and longitudinal measurements were done in a number of machines. We report on transverse echo measurements in RHIC and the observed dependence of echo amplitudes on a number of parameters for beams of gold and copper ions, and protons. In particular we examine the echo amplitudes of gold and copper ion bunches of varying intensity, which exhibit different diffusion rates from intrabeam scattering.

INTRODUCTION

Beam echoes [1] are a potentially very sensitive method to measure diffusion rates. Longitudinal beam echoes were observed in several machines, and used for diffusion rate measurements [2, 3, 4]. In the SPS a transverse echo response could be observed by applying 2 dipole kicks of different strength [5]. Here we report on transverse measurements in RHIC, in which echoes were created by applying a dipole kick, followed by a quadrupole kick.

We are using the notation in Ref. [1]. After applying a dipole kick a, the beam response decoheres with time $\tau_d = T_0/4\pi\mu$, where T_0 is the revolution time and μ the the betatron tune shift at for particles at one rms beam size σ. If a second kick is applied after time τ, the dipole signal can recohere to a dipole echo η after time $\tau_{echo} = 2\tau$ (see Fig. 1 for a transverse echo in RHIC). The echo response depends on the normalized quadrupole strength $Q = \beta/f$ where β is the lattice function and f the quadrupole focal length. Second order perturbation theory predicts a time dependent dipole response [1, 6]

$$A(t) = \frac{\eta(t)}{a} = F\left(\frac{\tau_0}{\tau_d}, \frac{t-\tau_{echo}}{\tau_d}\right) \quad \text{with} \quad F(x,y) = \frac{x}{[(1+x^2-y^2)^2+4y^2]^{3/4}} \quad (1)$$

where $\tau_d = T_0/4\pi\mu$. The relative echo amplitude $A_{max} = \eta_{max}/a$ is reduced with diffusion, and echo measurements can therefore be used to infer diffusion rates. The time of an echo measurement is considerably shorter than the time needed to observe the expansion of the beam size due to diffusion, and is also much less dependent on a precise emittance measurement. For $\tau_0 \ll \tau_d$ and small dipole kicks a the maximum echo response η_{max} was calculated as a function of a constant diffusion coefficient D_0 in Ref. [7]. This formula was found to be not applicable in the experimental parameter range reported here [8].

---

[1] Work supported by US DOE, contract No DE-AC02-98CH10886.

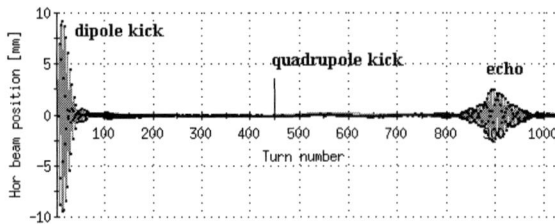

FIGURE 1. Transverse echo of a gold ion beam in RHIC. The beam is injected under a horizontal angle. After 450 turns a quadrupole kick is applied. The transverse echo appears after 900 turns.

MEASUREMENTS

A dipole kick is created by injecting the beam under a horizontal angle, leading to oscillations of about 10 mm, or 4 σ (see Fig. 1). In about 100 turns the dipole signal decoheres due to lattice nonlinearities, created by arc octupoles. After some time τ, a one-turn quadrupole kick is applied, and an echo is observable at time $\tau_{echo} = 2\tau$. Typical parameters relevant to the measurements are shown in Tab. 1 for three different ion species.

The echo amplitude was observed under variation of a number of parameters: the dipole kick amplitude a, the quadrupole kick amplitude Q, the detuning μ, the quadrupole kick time τ, the horizontal tune Q_x, and the bunch intensity N_b. Echoes could only be observed with dipole kicks of a few σ, nonlinear detuning μ an order of magnitude larger than the natural detuning, and quadrupole kick times τ no larger than a few hundred turns. The observed echo amplitudes were not sensitive to small changes in the dipole amplitude, or the horizontal tune, and proportional to the quadrupole amplitude [8]. Large chromaticity or coupling can reduce the echo signal. Operation near a strong resonance leads to particles trapped in island, which created a non-decaying non-zero dipole moment.

We show the echo amplitudes for variations in μ (Fig. 2), τ (Fig. 3), and N_b (Fig. 4)

TABLE 1. Typical parameters for transverse echo measurement in RHIC with beams of gold and copper ions, and protons.

parameter	unit	Au	Cu	p
mass and charge number A, Z	...	197, 79	63, 29	1, 1
relativistic γ	...	10.5	12.1	25.9
revolution time T_0	μs		12.8	
rms emittance, unnorm. ε	mm·mrad	0.16		0.10
detuning μ	...		0.0014	
decoherence time τ_d	turns		57	
dipole kick a	mm / σ		$10 / \approx 4$	
normalized quadrupole kick Q	...		0.025	
time τ_0	turns		10	
quadrupole kick time τ	turns		450	200
synchrotron period T_s	turns	450	540	3900
bunch intensity N_b	10^9	0.1–1.0	0.1–1.3	65–95

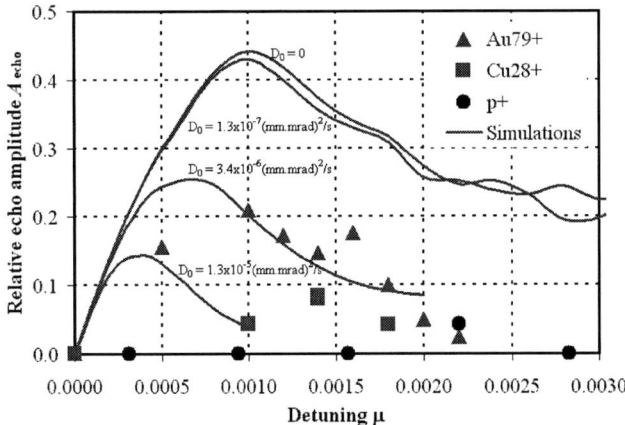

FIGURE 2. Relative echo amplitude A_{echo} as a function of the nonlinear detuning μ for bunches of gold and copper ions, and protons. For all species the quadrupole kick time was $\tau = 450$ turns. The simulations are for different diffusion coefficient for the normalized emittance. Gold data are from Ref. [8].

for gold and copper ions, and protons. The detuning μ was changed with arc octupoles. μ is calculated with a SixTrack [9] model of RHIC, and its value is consistent with the observed decoherence time (see Tab. 1). Without additional octupoles, no echoes can be observed. For gold and copper ion beams, the echo amplitude A_{echo} increases with increasing octupole strength reaches a maximum in the range $\mu = 0.010 - 0.015$. The amplitude falls off with further strength since the phase memory time of the particles in the transverse distribution is reduced. For protons only one weak echo is observed agains a rather large background from particles trapped in islands.

The relative echo amplitude as a function of the quadrupole kick time τ is shown in Fig. 3. For gold and copper beams the echo amplitudes were largest around 500 turns, for proton beams around only 200 turns, indicating a stronger transverse diffusion mechanism. The octupole strength for all these measurements was the same, which implies a 50% larger μ for p bunches.

Fig. 4 shows the echo amplitude as a function of the bunch intensity for gold and copper ions. For both ion species the echo amplitudes are reduced for larger bunch intensities, consistent with intrabeam scattering as the dominant diffusion source. Proton data over a sufficiently large range of N_b are not available for the same μ and τ.

SIMULATIONS

The particle motion was simulated in one dimension only, with a model that consisted of linear transfer maps, and three octupoles to adjust the nonlinear detuning. The octupoles were spaced such as to minimize resonance driving. 10000 particles were placed with an offset to simulate the dipole kick, tracked for the time τ, received a quadrupole kick, and tracked for at least another time τ. The dipole moment of the distribution is calculated turn-by-turn to obtain the relative echo amplitude. Diffusion is introduced through random kicks to the particle momentum after each turn. The random kicks

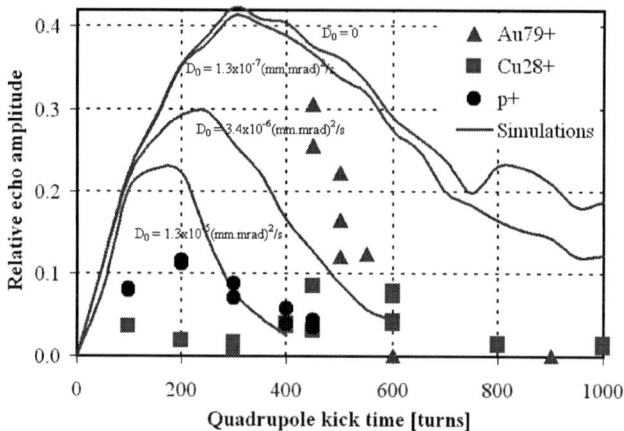

FIGURE 3. Relative echo amplitude A_{echo} as a function of the quadrupole kick time τ. For all species the nonlinear detuning was $\mu = 0.0014$. The simulations are for different diffusion coefficient for the normalized emittance. Gold data are from Ref. [8].

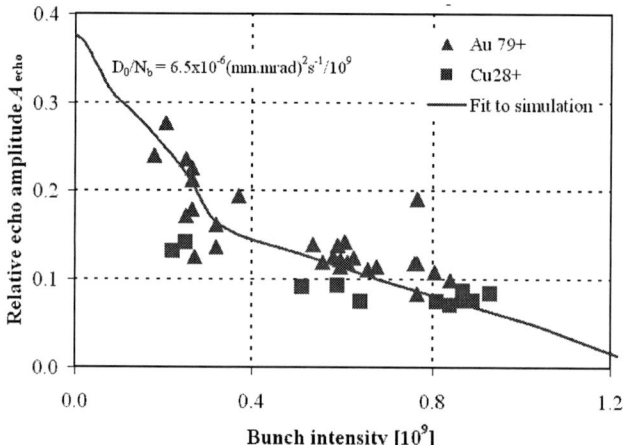

FIGURE 4. Relative echo amplitude A_{echo} as a function of the bunch intensity with $\mu = 0.0014$ and $\tau = 450$ turns. The coefficient D_0/N_b was chosen so that the simulated echo amplitudes fit the observed ones. Gold data are from Ref. [8].

follow a Gaussian distribution with a width that is constant for the whole phase space[2].

In Fig. 2 simulated curves for different diffusion coefficients are shown. For small diffusion coefficients and large detuning a significant number of particles are trapped in islands, leading to echo amplitudes that fall only slowly with increasing detuning

---

[2] Simulations using a width growing exponentially with action did not show significantly different qualitative results.

μ. For the gold data, the shape of the experimental data can be approximately reproduced with a constant diffusion coefficient for the normalized emittance of $D_0 = 3.4 \times 10^{-6}$ (mm mrad)^{2}s^{-1}.

Fig. 3 shows the simulated echo amplitudes as a function of the quadrupole kick time τ for varying diffusion coefficients. The weak copper ion echoes, and the fact that no gold ion echoes were observed for small quadrupole kick times is not reproduced by the simulations. However, the simulations indicate that protons exhibit stronger transverse diffusion than the heavier ions.

In Fig. 4 shows a fit to the simulated data that translates an increasing bunch intensity linearly into an increasing diffusion rate. The diffusion rate of $D_0 = 6.5 \times 10^{-6}$ (mm mrad)^{2}s^{-1} for bunches of 10^9 ions corresponds to an emittance growth time of about 100 h, the same order of magnitude that was measured observing the free expansion of bunches [10]. Note that the density of kicked beams is reduced, and that intrabeam scattering growth rate (Z^2/A) of copper ions is about a factor 2 smaller than the growth rate of gold ions.

SUMMARY

Transverse echoes were observed in RHIC at injection, with beams of gold and copper ions as well as protons. The echo amplitude was recorded as a function of detuning, quadrupole kick time, and the bunch intensity. The measurements were compared with simulated echo amplitudes, allowing the extraction of diffusion rates. The measurements revealed stronger transverse diffusion for protons than for heavier ions, indicating that a diffusion mechanism other than intrabeam scattering is dominant in for protons. For gold and copper ions the measured diffusion rates decrease approximately linearly with increasing bunch intensity, consistent with intrabeam scattering as the dominant diffusion source.

ACKNOWLEDGMENTS

The author would like to thank M. Blaskiewicz and O. Boine-Frankenheim for discussions, and T. Satogata for help with the orbit measurement system.

REFERENCES

1. G. Stupakov, "Echo", in "Handbook of accelerator physics and engineering" edited b A.W. Chao and M. Tigner, World Scientific (1999).
2. L.K. Spentzouris, J.-F. Ostiguy and P.L. Colestock, "Direct measurement of diffusion rates in high energy synchrotrons using longitudinal beam echoes", PRL Vol. 76, No 4, pp. 620 (1996).
3. O. Brüning et al., "Beam echos in the CERN SPS", proceedings PAC'97 (1997).
4. J. Kewisch and M. Brennan, "Bunched beam echos in the AGS", proceedings EPAC'98 (1998).
5. G. Arduini, F Ruggiero, F. Zimmermann, and M.P. Zorzano, "Transverse beam echo measurements on a single proton bunch at the SPS", CERN SL-Note-2000-048 MD (2000).
6. R.W. Gould, T.M. O'Neil, J.H. Malmberg, PRL 19, p 219 (1967).
7. G. Stupakov and A. Chao, "Effect of diffusion on bunched beam echo", proceedings PAC'97 (1997).
8. W. Fischer, R. Tomas, and T. Satogata, "Measurement of transverse echoes in RHIC", proceedings PAC'05 (2005).
9. F. Schmidt, "SixTrack, User's Reference Manual", CERN/SL/94-56 (AP) (Update March 2000).
10. W. Fischer et al., "Measurements of Intra-Beam Scattering Growth Times with Gold Beam below Transition in RHIC", proceedings PAC'01 (2001).

Studies of Beam Dynamics in Cooler Rings

J.Dietrich[1], I.Meshkov[2], A.Sidorin[2], A.Smirnov[2], J.Stein[1]

[1]*FZJ, Juelich, Germany*
[2]*JINR, Dubna, Russia*

Abstract. This report describes the numerical simulation of the crystalline proton beam formation in COSY [1] using BETACOOL code [2]. The study includes the description of experimental results at NAP-M [3] storage ring where the large reduction of the momentum spread was observed for first time. The present simulation shows that this behavior of proton beam can not be explained as ordered state of protons. The numerical simulation of crystalline proton beams was done for COSY parameters. The number of protons when the ordering state can be observed is limited by value 10^6 particles and momentum spread less then 10^{-6}. Experimental results for the attempt to achieve of ordered state of proton beam for COSY is presented. This work is supported by RFBR grant # 05-02-16320 and INTAS grant #03-54-5584.

Keywords: Electron Cooling, Beam Ordering, Crystalline Beam, BETACOOL.
PACS: 29.20.Dh; 29.27.Bd

NAP-M EXPERIMENTS

The dependence of momentum spread on ion number in a cooled beam has very specific character: at certain conditions the momentum spread drops up to very low value and remains constant with the decrease of the ion beam intensity.

For the first time such a "disappearance" of the beam momentum spread was observed in experiments on NAP-M, where the suppression of the Schottky noise of the cooled proton beam up to very low level was registered (Fig.1) [3]. Then the assumption of some orderliness of the cooled beam was declared and soon the idea of the crystalline beam was proposed [4].

TABLE 1. Parameters of storage rings.

	NAP-M	COSY
Circumference, m	47,25	183,5
Proton energy, MeV	65	45,6
Gamma transition, γ_{tr}	1,069	2,4
Betatron tunes, Q_x/Q_y	1,34 / 1,24	3,62 / 3,68
Dipole field stability	~10^{-5}	~2×10^{-5}
Electron cooler		
Cooling section length, m	1,0	1,4
Beam current, A	1,0	0,05 ÷ 1,0
Beam radius, cm	0,5	1,27
Magnetic field, kG	1,0	0,8
Electron energy stability	~10^{-5}	~2×10^{-5}

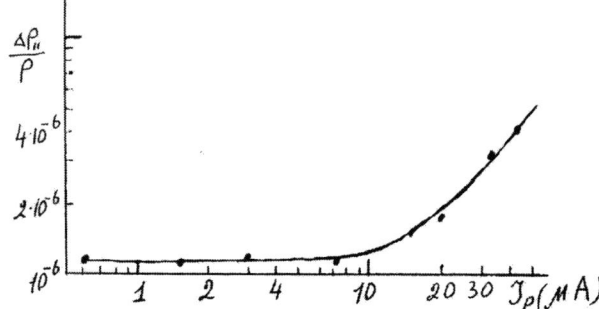

FIGURE 1. NAP-M experiment (1979). The dependence of momentum spread on proton number (1 µA = 2,64·10⁶ protons).

To verify the idea of ordering proton beam on NAP-M the numerical simulation of beam dynamics was done for NAP-M lattice structure (Fig.2) which was reconstructed from original articles. The main parameters of NAP-M are presented in Table 1. The example of input file for MAD program is the following:
```
dr1: drift, l=7.1
sb1: SBEND, L=4.7124, ANGLE=1.5708, E1=0.415, E2=0.415
NAPM: line=(dr1,sb1,dr1,sb1,dr1,sb1,dr1,sb1)
```

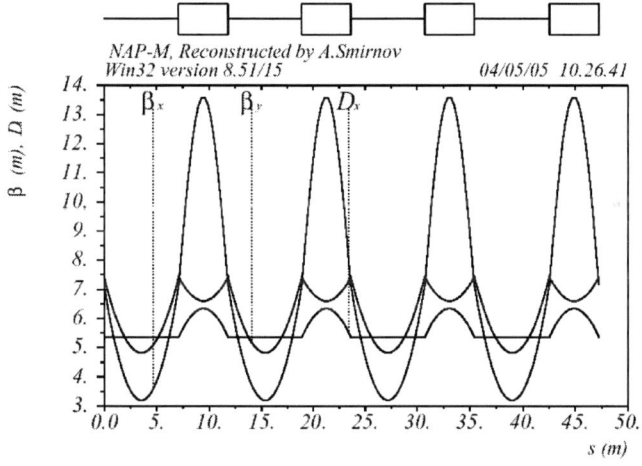

FIGURE 2. Lattice structure of NAP-M.

Fig.3 shows the result of simulation for NAP-M using Molecular Dynamics technique [5] for proton number of 10^6 at constant cooling time 20 µsec. If one overlaps transverse and longitudinal component of IBS (Fig.3,a,b) one can find the particle number when the ordering state can be reached. In the case of particle number of 10^6 the summary picture (Fig.3,c) has the "channel" between of heating growth rates. The simulation results using Molecular Dynamics technique shows that the proton beam can reach the ordered state in NAP-M.

In the case of the proton number of 2×10^6 the "channel" disappears in the summary picture and the ordering state can not be reached. These results show that the ordered state of proton beam on NAP-M can be observed for particle number less then 10^6. Therefore the experimentally observed large reduction of the momentum spread (Fig.1) can not be explained by the ordering process.

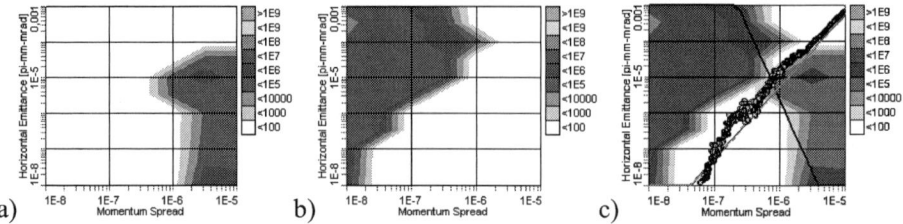

FIGURE 3. Growth rates (1/sec) for NAP-M (Molecular Dynamics). $N_p = 10^6$.
a) horizontal component of IBS, **b)** longitudinal component of IBS, **c)** overlapping of **a** and **b** pictures and beam evolution during cooling process. Gray straight line corresponds to equilibrium between transverse and longitudinal temperature, black straight line – ordering state criteria [6].

COSY SIMULATION

COSY ring has parameters at the injection energy a similar to NAP-M (Table 1). This ring can be used for the study of the ordering proton beams. The main difference from NAP-M is the superperiodicity of the lattice structure. NAP-M has 4 superperiodicity but COSY has only one.

The longitudinal components of IBS growth rates (Fig.4,a) have a large difference from NAP-M lattice structure. The longitudinal component has a very specific island of growth rates in the range of the transition point to the ordered state. The same island was found in simulation for other ring with small superperiodicity. The physics of this island existence is not explained yet. The experimental verification of this behavior of IBS growth rates at low temperature of ion beams is a very interesting and important task.

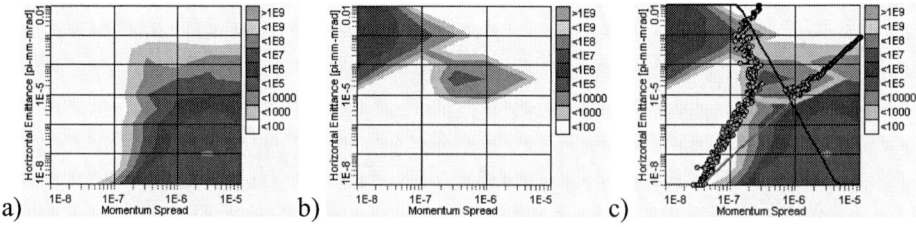

FIGURE 4. Growth rates (1/sec) for COSY (Molecular Dynamics). $N_p = 10^6$.
a) horizontal component of IBS, **b)** longitudinal component of IBS, **c)** overlapping of **a** and **b** pictures and beam evolution during cooling process for different initial conditions.

The numerical simulation with Molecular Dynamics techniques of the cooling process shows that the ordering state for COSY parameters can be reached if the proton beam has specific initial parameters: large transverse emittances and small momentum spread (Fig.4,c).

If initial parameters of the proton beam do fit to the equilibrium temperature (upper points of the gray straight line on Fig.4,c) the proton beam reaches the equilibrium between cooling and heating and can not come onto the ordered state.

COSY EXPERIMENTS

The goal of the proposed experiments is an achievement of the ordered state of the proton beam. Simulation shows that the ordered state can be observed if the proton number less than 10^6 and the momentum spread less than 10^{-6}.

The electron cooling of the proton beam in COSY was done for the different electron beam current values (Fig.5). Momentum spread was measured as FWHM (full width on half maximum) of longitudinal Schottky signal when only one peak is observed. For larger number of protons after the cooling the longitudinal Schottky signal has two peaks (well known plasma waves propagated in the beam) and other method of calculation of momentum spread is needed.

In experiments the COSY ring was operated in a single injection mode. Number of protons at one injection cycle was about $2 \div 5 \times 10^9$. To speed up the process of proton losses the horizontal scraper was used. It decreased the ring aperture and shortened the proton lifetime. After a few minutes of cooling process with the inserted scraper the proton number reaches the value less than 10^8 and longitudinal Schottky signal is transformed to single peak. Then the scraper was returned to initial position and the proton number continues to decrease at constant lifetime.

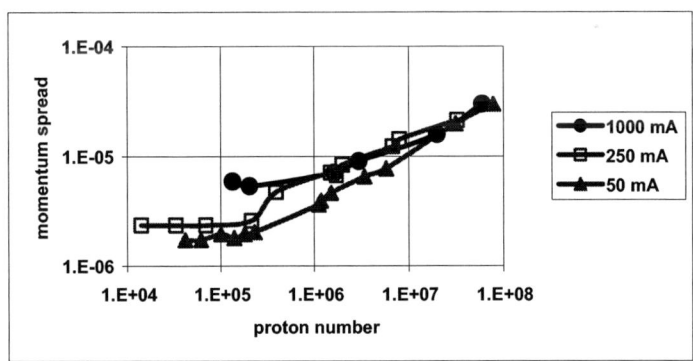

FIGURE 5. The dependence of momentum spread on the proton number for different values of electron beam current.

When the proton number achieves the value less than 10^6 the momentum spread stop to decrease and remains constant. No sudden reduction of momentum spread was observed in this experiment. It means that the proton beam does not achieve an ordered state at present parameters.

Results of COSY experiments are very similar to experiments at NAP-M (Fig.1). They show that the minimum momentum spread depends on the value of the electron beam current (Fig.6) and does not depend on particle number in the range of small number of protons below 10^5. It means that the equilibrium is defined by the parameters of the electron beam when intrabeam scattering disappears.

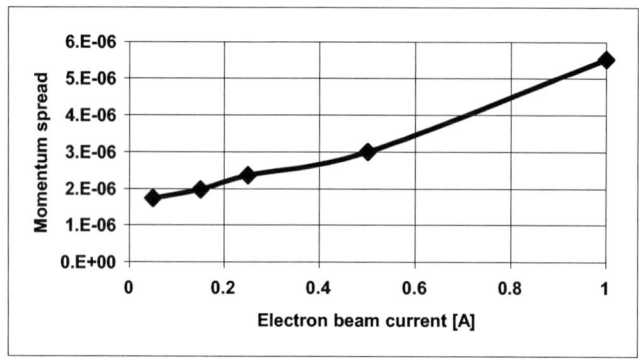

FIGURE 6. The dependence of minimum momentum spread on electron beam current.

To verify the simulation results which predicted the island of the longitudinal component of intrabeam scattering (Fig.4,c) the additional transverse heating with white noise was applied to the pick-up electrodes. In the experiment the additional heating leads to increasing of particle losses or to the excitation of beam selfmodulation in the longitudinal direction. No decreasing of the momentum spread is observed. The special method of transverse heating is needed to verify the break in the longitudinal component of IBS heating rates

ACKNOWLEDGMENTS

We are grateful to K.Henn, I.Mohos and D.Prasuhn for helping of experiments on COSY.

REFERENCES

1. R.Maier, "Cooler Synchrotron COSY – performance and perspectives", NIM A 390 (1997) 1-8.
2. http://lepta.jinr.ru/betacool.htm
3. G.I.Budker, N.S.Dikansky, V.I.Kudelainen et al., Proc. 4th All-Union Conf. on Charged-Particle Accelerators [in Russian], Vol. 2 (Nauka, Moscow, 1975) 309; Part.Accel. 7 (1976) 197; At.Energ. 40 (1976) 49. E.Dementev, N.Dykansky, A Medvedko at al., Prep. CERN/PS/AA 79-41, Geneva (1979).
4. Parkhomchuk V.V. Proc. of Workshop on Electron Cooling and Related Applications, Kernforschungszentrum Karlsruhe (1984) p. 71.
5. T.Katayama, I.Meshkov, D.Mohl, A.Sidorin, A.Smirnov, E.Syresin, H.Tsutsui, "Simulation Study of Ordered Ion Beams", Preprint RIKEN-AF-AC-42 (2003).
6. T.Katayama, I.Meshkov, A.Sidorin, A.Smirnov, E.Syresin, "Ordered State of Ion Beams", Preprint RIKEN-AF-AC-40 (2002).

IBS in a CAM-Dominated Electron Beam

A. Burov[*], I. Gusachenko[§], S. Nagaitsev[*] and A. Shemyakin[*]

[*]*Fermi National Accelerator Laboratory, P.O. Box 500, Batavia IL 60543*
[§]*Novosibirsk State University, Novosibirsk, 630090, Russia*

Abstract. Electron cooling of the 8.9 GeV/c antiprotons in the Recycler ring requires high-quality dc electron beam with the current of several hundred mA and the kinetic energy of 4.3 MeV. That high electron current is attained through beam recirculation (charge recovery). The primary current path is from the magnetized cathode at high voltage terminal to the ground, where the electron beam interacts with the antiproton beam and cooling takes place, and then to the collector in the terminal. The energy distribution function of the electron beam at the collector determines the required collector energy acceptance. Multiple and single intra-beam scattering as well as the dissipation of density micro-fluctuations during the beam transport are studied as factors forming a core and tails of the electron energy distribution. For parameters of the Fermilab electron cooler, the single intra-beam scattering (Touschek effect) is found to be of the most importance.
Keywords: Touschek effect, coupled optics, canonical angular momentum.
PACS: 29.27.Bd, 29.27.Eg, 29.27.Fh

INTRODUCTION

In the Fermilab e-cooler[1], energy distribution of the electrons at the collector affects the current loss and, consequently, possibility of operations in the DC mode. Core of the energy distribution is formed by multiple intra-beam scattering (IBS), as well as the dissipation of density micro-fluctuations. Extended tails of the electron energy distribution, significant for the charge recovery, are formed by single IBS, or Touschek effect. Because the electron beam is CAM-dominated[2]; conventional IBS results (as Bjorken-Mtingwa, Piwinski-Martini) cannot be applied. Both multiple and single IBS phenomena are treated here on a base of the Landau collision integral[3]; the present paper is a more extended and refined version of Ref.[4].

TABLE 1. Fermilab e-cooler parameters.

Parameter	Symbol	Value	Units
Beam current	I_e	0.1-0.5	A
Electron momentum	p	4.8	MeV/c
Cathode radius	r_c	0.38	cm
Electron temperature at the cathode	T_c	0.11	eV
Length of trajectory	l	99	m
Collector potential with respect to the cathode	U_{coll}	2 – 4	kV
Longitudinal magnetic field on the cathode	B_c	90	G

CORE OF THE DISTRIBUTION

An electron with a longitudinal velocity $v_\| \ll c$ in the beam frame has the energy deviation $U = pv_\| = \gamma\, mcv_\|$ in the laboratory frame. This value is not changed by acceleration and deceleration; thus, according to kinematics, the beam longitudinal temperature $T_\| \equiv \overline{mv_\|^2} \cong mT_c^2/p^2$ gets to be extremely small with the acceleration.

IBS transfers high transverse temperature of electrons, which is not changed by the acceleration, to low longitudinal temperature; the evolution of the longitudinal distribution is described by the Landau kinetic equation[3]. When $T_\| \ll T_\perp$, the kinetic equation on the longitudinal distribution function $f_\|(v_\|)$ reduces to a pure diffusion with the diffusion coefficient independent on the longitudinal velocity:

$$\frac{\partial f_\|}{\partial t} = \frac{D}{2}\frac{\partial^2 f_\|}{\partial v_\|^2}, \quad D = 4\pi n_e r_e^2 c^4 L_C \iint d^2v\, d^2v'\, \frac{f_\perp(\mathbf{v})f_\perp(\mathbf{v}')}{|\mathbf{v}-\mathbf{v}'|}. \tag{1}$$

Here $\mathbf{v}$ is 2D transverse velocity vector, and the electron density n_e is supposed to be constant (Pierce regime) over the elliptic cross-section with half-axes a_ξ, a_η. Assuming the transverse distribution $f_\perp(\mathbf{v})$ to be Gaussian with main axes r. m. s. velocities $\overline{v_{\tilde{x}}^2}, \overline{v_{\tilde{y}}^2}$, it yields

$$D = \frac{2\pi^{3/2} n_e r_e^2 c^4 L_C}{\sqrt{\overline{v_{\tilde{x}}^2}\,\overline{v_{\tilde{y}}^2}}} \int_0^{2\pi} \frac{d\varphi}{2\pi} \frac{1}{\sqrt{\dfrac{\cos^2\varphi}{\overline{v_{\tilde{x}}^2}} + \dfrac{\sin^2\varphi}{\overline{v_{\tilde{y}}^2}}}}. \tag{2}$$

At the end of the transfer line, beam acquires energy spread given by an integral over the beam line:

$$\overline{U_{IBS}^2} = \frac{2\sqrt{\pi} I_e r_e^2 L_C}{ec^2 \varepsilon_{4n}} (mc^2)^2 \int_0^l dz \int_0^{2\pi} \frac{d\varphi}{2\pi} \frac{1}{\sqrt{\dfrac{\cos^2\varphi}{\overline{v_{\tilde{x}}^2}} + \dfrac{\sin^2\varphi}{\overline{v_{\tilde{y}}^2}}}}, \tag{3}$$

where $\varepsilon_{4n} = a_\xi a_\eta \sqrt{\overline{v_{\tilde{x}}^2}\,\overline{v_{\tilde{y}}^2}}$ is a normalized 4D emittance, $L_c = \ln(r_{max}/r_{min})$ is the Coulomb logarithm. For the design optics[5], this yields $\sqrt{\overline{U_{IBS}^2}} \approx 75\,\text{eV}$.

Another source of longitudinal temperature growth is dissipation of density fluctuations in the beam. Excessive potential energy (beam frame) $U_{exc} \approx 2e^2 n^{1/3}$ (see Ref[6]) transforms into the r. m. s. energy spread

$$\overline{U_{DF}^2} \cong 2(\gamma\beta)^{5/3} r_e \left(\frac{I_e}{\pi e c a_e^2}\right)^{1/3} (mc^2)^2.$$

For $I_e = 0.2$ A, this yields $\sqrt{U_{DF}^2} \cong 50$ eV, and $\sqrt{U^2} = \sqrt{U_{IBS}^2 + U_{DF}^2} = 90$ eV.

TAILS OF THE DISTRIBUTION

A portion of particles $\Delta(U)$ which energy deviation exceeds a given collector potential U is referred here as losses. Losses as low as $\Delta(U)=10^{-6}$ correspond to $U=4.75\sqrt{U^2} = 430$ V low-energy tail of the Gaussian distribution $f(U)$. However, when the loss level is so small, it is determined rather by single scatterings (Touschek), than by multiple ones. In non-relativistic case, the Rutherford differential cross-section gives a cross-section for events when one of the scattering particles acquires longitudinal velocity larger than $v_\|$, assuming v_0 as a relative velocity:

$$\sigma(v_0,v_\|) = \int_0^{2\pi} d\varphi \cdot 2 \int_0^{\arccos(2v_\|/v_0)} \frac{d\sigma}{do} \sin\theta\, d\theta = \frac{4\pi r_e^2 c^4}{v_0^2 v_\|^2}\left(1 - \frac{4v_\|^2}{v_0^2}\right). \quad (4)$$

The instantaneous loss rate in the beam frame is

$$\frac{1}{n_e}\frac{dn_e}{dt} = \frac{n_e}{2}\int d\mathbf{v}d\mathbf{v}'\sigma(|\mathbf{v}-\mathbf{v}'|,v_\|)|\mathbf{v}-\mathbf{v}'|f_\perp(\mathbf{v})f_\perp(\mathbf{v}').$$

For the Gaussian transverse distribution, the above 4D integral reduces to a single integral

$$\Delta(U) = \frac{\sqrt{\pi}I_e r_e^2 L_C}{ec^2 \varepsilon_{4n}}\left(\frac{mc^2}{U}\right)^2 \int_0^l dz \int_0^{2\pi} \frac{d\varphi}{2\pi}\frac{S(\Xi(\varphi))}{\sqrt{\frac{\cos^2\varphi}{v_{\tilde{x}}^2} + \frac{\sin^2\varphi}{v_{\tilde{y}}^2}}} \quad (5)$$

$$\Xi(\phi) = \frac{U}{p}\sqrt{\frac{\cos^2\phi}{v_{\tilde{x}}^2} + \frac{\sin^2\phi}{v_{\tilde{y}}^2}}, \qquad S(x) = (1+2x^2)\text{erfc}(x) - \frac{2x}{\sqrt{\pi}}\exp(-x^2).$$

For the 99 m long cooling line, the losses vary moderately (<50%) for drastically different beam envelopes. For the short U-bend line[7], the calculated dependence on the envelope is much stronger. Increase of the beam size leads to significant drop of the losses. Fig. 1 shows calculated (line) and measured (dots) dependences of the beam losses versus the collector potential for design optics of the cooling line (left) and the U-bend line (right). For the cooling line, the growing discrepancy at high voltage is thought to be caused by secondary electrons escaped from the collector.

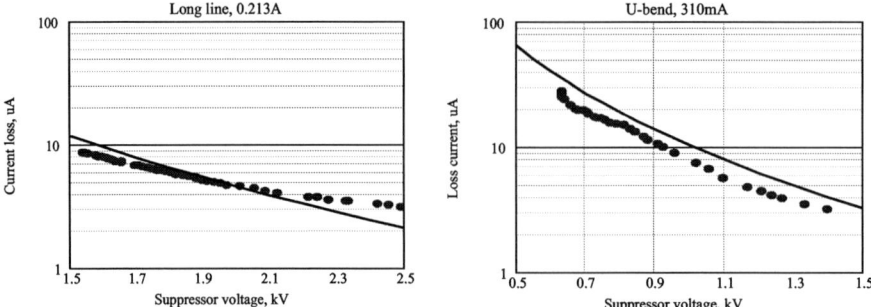

FIGURE 1. Measured and calculated current loss in long line (0.213A) and U-bend (0.310A).

CONCLUSIONS

1. Single IBS places a limitation for the minimum collector voltage U_{coll}. Acceptable level of losses of $\Delta = 2.5 \cdot 10^{-6}$ corresponds to the minimum value of $U_{coll} = 2$ kV.
2. The r.m.s. energy spread of the beam due to the multiple IBS is below the acceptable limit for the electron cooling process.
3. Tails of the multiple IBS Gaussian distribution are insignificant in comparison with the single large-angle IBS tails.
4. Any increase of the beam size is beneficial for both the multiple and single IBS as it leads to a lower transverse temperature and lower beam density.
5. Loss dependence on the envelope seems to be weak in the long line and very strong in the U-bend.
6. Geometrical beam parameters are obtained using 4D formalism for coupled optics.

The authors are thankful to Valeri Lebedev (FNAL) for fruitful discussions and to Denis Artamonov (NSU, Novosibirsk, Russia) for his valuable help in the data measurements.

REFERENCES

1. S. Nagaitsev et al., "Antiproton cooling in the Fermilab Recycler", *these Proc.* (*COOL05*, Galena, 2005).
2. A. Burov, Ya. Derbenev, S. Nagaitsev and A. Shemyakin, "Optical principles of beam transport for relativistic electron cooling", Phys. Rev. ST-AB **3**, p. 094002 (2000).
3. E. M. Lifshitz and L.P. Pitaevskii, "Physical Kinetics", Pergamon, 1981.
4. A. Burov, A. Shemyakin and S. Nagaitsev, FERMILAB-TM-2133 (2000).
5. A. Burov et al., "Optics of Electron Beam in the Recycler", *these Proc.*
6. N. Dikansky et al., Ultimate Possibilities of Electron Cooling, BINP Preprint 88-61 (1988).
7. A. Shemyakin et al., "Attainment of a high-quality electron beam for Fermilab cooler", *these Proc.*
8. V. Lebedev and A. Bogazc, "Betatron Motion with Coupling", e-print JLAB-ACC-99-19 (2001).
9. V. Lebedev, "OptiM", at http://www-bdnew.fnal.gov/pbar/organizationalchart/lebedev/OptiM/optim.htm

APPENDIX: PRINCIPAL AXES OF THE BEAM DISTRIBUTION

Geometrical beam parameters can be calculated via generalized Twiss functions for coupled optics[8]. All functions for the simulation were calculated by OptiM code[9]. Particle coordinate vector can be represented through the linear combination of two complex vectors $\hat{v}_1(s)$ and $\hat{v}_2(s)$ satisfying the equation $\mathbf{M}(s|s')\hat{v}_i(s) = e^{i\mu(s|s')}\hat{v}_i(s')$.

$$\hat{x}(s) = \text{Re}(\sqrt{2I_1}e^{-i\psi_1}\hat{v}_1(s) + \sqrt{2I_2}e^{-i\psi_2}\hat{v}_2(s))$$

$$\hat{x} = \begin{pmatrix} x(s) \\ x'(s) - \dfrac{eB_s}{2pc}y(s) \\ y(s) \\ y'(s) + \dfrac{eB_s}{2pc}x(s) \end{pmatrix}, \quad \hat{v}_1 = \begin{pmatrix} \sqrt{\beta_{1x}(s)} \\ -\dfrac{i(1-u(s))+\alpha_{1x}(s)}{\sqrt{\beta_{1x}(s)}} \\ \sqrt{\beta_{1y}(s)}e^{i\nu_1(s)} \\ -\dfrac{iu(s)+\alpha_{1y}(s)}{\sqrt{\beta_{1y}(s)}}e^{i\nu_1(s)} \end{pmatrix}, \quad \hat{v}_1 = \begin{pmatrix} \sqrt{\beta_{2x}(s)}e^{i\nu_2(s)} \\ -\dfrac{iu(s)+\alpha_{2x}(s)}{\sqrt{\beta_{2x}(s)}}e^{i\nu_2(s)} \\ \sqrt{\beta_{2y}(s)} \\ -\dfrac{i(1-u(s))+\alpha_{2y}(s)}{\sqrt{\beta_{2y}(s)}} \end{pmatrix}$$

B_s is longitudinal magnetic field, I_i, ψ_i are new canonic variables, actions and phases for the two transverse modes. Defining emittances (un-normalized) as $\varepsilon_1 = 2I_{1\max}$, $\varepsilon_2 = \langle I_2 \rangle$, the 4D emittance $\varepsilon_T = r_c\sqrt{mT_c}/p$ is a product of two emittances defined above[8]: $\varepsilon_T^2 = \varepsilon_1\varepsilon_2$ with

$$\varepsilon_1 = \frac{B_c r_c^2}{B\rho}, \quad \varepsilon_2 = \frac{mcT_c}{peB_c}, \quad \frac{\varepsilon_1}{\varepsilon_2} = \frac{B_c^2 r_c^2}{T_c}\frac{e^2}{mc^2} \approx 8\cdot 10^3,$$

$B\rho = pc/e = 16.2$ kG·cm for the 4.35 MeV electrons. Emittance ε_1 corresponds to the total magnetic flux on the cathode and is responsible for the beam hydrodynamic motion and its shape. Emittance ε_2 relates to the thermal velocities, driving intrabeam scattering. For the hydrodynamic mode, $x = \sqrt{2I_1\beta_{1x}}\cos\psi_1$ and $y = \sqrt{2I_1\beta_{1y}}\cos(\nu_1 - \psi_1)$, yielding the half-axes $a_{\xi,\eta}$, the beam cross-section S, and the axes tilt angle φ_1:

$$a_{\xi,\eta}^2 = I_1[(\beta_{1x}+\beta_{1y}) \pm \sqrt{\beta_{1x}^2 + 2\beta_{1x}\beta_{1y}\cos 2\nu_1 + \beta_{1y}^2}] \quad (6)$$

$$S = \pi\varepsilon_1\sqrt{\beta_{1x}\beta_{1y}}|\sin\nu_1|, \quad \text{tg}\,2\varphi_1 = \frac{2\sqrt{\beta_{1x}\beta_{1y}}\cos\nu_1}{\beta_{1y} - \beta_{1x}}.$$

Emittance of the thermal mode ε_2 determines the r.m.s. transverse velocities in the beam frame in φ_2-tilted principal axes

$$\overline{v_{\tilde{x},\tilde{y}}^2} = \frac{B\rho}{B_c\lambda_{\tilde{x},\tilde{y}}}\frac{T_c}{mc^2}, \quad \lambda_{\tilde{x},\tilde{y}} = \frac{1}{2}[(\beta_{2x}+\beta_{2y}) \pm \sqrt{\beta_{2x}^2 + 2\beta_{2x}\beta_{2y}\cos 2\nu_2 + \beta_{2y}^2})$$

$$\text{tg}\,2\varphi_2 = \frac{2\sqrt{\beta_{2x}\beta_{2y}}\cos\nu_2}{\beta_{2y} - \beta_{2x}}.$$

Stability Studies under Dipole Oscillation Model for RHIC E-Cooling

Gang Wang

Physics Dept., State University Of New York at Stony Brook, N.Y.11790

Abstract. In the presence of the electron beam in the cooling solenoid, both longitudinal and transverse instability could take place for the circulating Au beam. A threshold of the electron beam density has been derived for RHIC gold ion beam longitudinal instability to take place. The transversal instability growth rate has been calculated for magnetized cooling scheme, which shows strong dependence on the neutralization factor, η, of the cooling solenoid. The instability thresholds for the non-magnetized electron cooling scheme have been calculated both in the longitudinal direction and the transverse direction. While the longitudinal instability threshold for the designed parameters is quite similar as the magnetized cooling case, the transversal instability threshold is two orders of magnitude bigger than the magnetized cooling scheme and a coherent damping effect is introduced to the ion beam by the two beam transversal interaction in the cooling section.

Keywords: Electron Cooling, Coherent instability, Rhic.

INTRODUCTION

In 1998, a substantial shorten beam life time was observed as soon as the E-Cooler was turned on in Celsius and this phenomenon has been called 'electron heating'. Similar phenomena have also been observed by other facilities such as NAP-M, Fermi lab, Indiana, TARN II and COSY. Although nonlinear electric field is regarded as an important reason for the fast beam loss in Celsius due to the fact that the electron beams have smaller radius than the ion beam, the coherent ion-electron beam interaction may also play a role. For RHIC e-cooler, since the electron beam and the ion beam have essentially the same beam size, the nonlinear electric field effects are greatly reduced and the coherent ion-electron interaction could be important for the ion beam stability. V.V.Parkhamchuk and V.B.Reva to estimate the growth rate due to transversal coherent oscillation induced by electron beam have developed a dipole oscillation model. It is also shown that this coherent effect could be amplified in the presence of the ion clouds ionized from the residue gas. In section 2, the longitudinal-longitudinal coupling of the electron beam and the ion beam in the cooling section has been studied using the Parkhomchuk model for Rhic magnetic and nonmagnetic schemes. In section 3, the transverse–transverse two stream coupling has been studied and the growth rate for Rhic magnetized cooling scheme has been calculated. In section 4, the conclusions have been made and the effects of the ion clouds in the cooling section have been discussed and the dependence of the growth rate on the neutralization factor has been shown.

LONGITUDINAL COUPLING

The evolution of the ion beam longitudinal electrostatic perturbation could be rewritten into the following matrix form under short wavelength approximation[1],

$$\begin{pmatrix} s_i \\ s_i' \end{pmatrix}_{n+1} = M_{ring} \begin{pmatrix} s_i \\ s_i' \end{pmatrix}_n \quad (1)$$

,where M_{ring} is the one turn transfer matrix for the plasma oscillation and s_i is the local longitudinal oscillation displacement of the considered portion of the ion beam. The transfer matrix for the ring is composed of the cooling section part and the rest of the ring, i.e.

$$M_{ring} = M_{cool} M_{out}$$

Within the cooling section, the ion motion is coupled with the cooling electron motion and the transfer matrix can be written as

$$M_{cool} = \begin{pmatrix} \xi(\cos(\frac{\omega_0}{c}l_{cool})-1)+1 & \frac{c}{\omega_0}\left[\xi\sin(\frac{\omega_0}{c}l_{cool})+(1-\xi)\frac{\omega_0}{c}l_{cool}\right] \\ -\xi\frac{\omega_0}{c}\sin(\frac{\omega_0}{c}l_{cool}) & \xi(\cos(\frac{\omega_0}{c}l_{cool})-1)+1 \end{pmatrix} \quad (2)$$

where l_{cool} is the length of the cooling section, $\xi = \frac{\omega_{pi}^2}{\omega_0^2}$ and the plasma frequencies involved are defined in the same way as in reference [1]. Outside the cooling section, the longitudinal oscillation frequency for the perturbation is just the ion plasma frequency, ω_{pi} and the transfer matrix is given by

$$M_{out} = \begin{pmatrix} \cos\left[\frac{\omega_0}{c}(Cir-l_{cool})\right] & \frac{c\cdot\sin\left[\frac{\omega_0}{c}(Cir-l_{cool})\right]}{\omega_0} \\ -\frac{\omega_0}{c}\sin\left[\frac{\omega_0}{c}(Cir-l_{cool})\right] & \cos\left[\frac{\omega_0}{c}(Cir-l_{cool})\right] \end{pmatrix} \quad (3)$$

where Cir represents for the RHIC circumference. The determinant and the eigenvalue increment for one turn transfer matrix M_{ring} have been plotted for the magnetized electron cooling design in Fig.1. The electron density threshold for the determinant to be bigger than 1 is determined by equation,

$$\omega_0 \tau = 2\pi \quad (4)$$

For Rhic non-magnetized electron cooling scheme, the designed length of the cooling section is 100 meters, which correspond to the lab frame electron density threshold of $1.17 \times 10^{17} m^{-3}$ or a minimal bunch length of $1.1 cm$ for $2.5 nC$ electron charge per bunch.

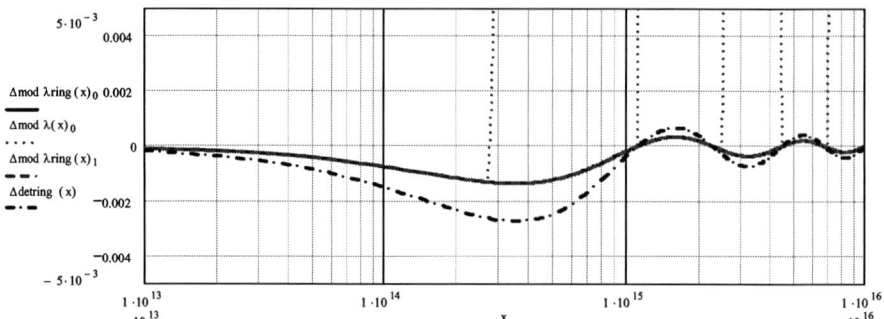

FIGURE 1. The eigenvalues and determinant increments of the longitudinal Longmuir Oscillation for the non-magnetized electron-cooling scheme. The x-axis is the beam frame electron density in units of m^-3. The blue dot-dash curve is the determinant of the one turn transfer matrix and the red solid curves are its the eigenvalues. The dot red curve is the maximal eigenvalue of the cooling-section transfer matrix.

For Rhic magnetized electron cooling scheme, the designed cooling section length is about 30 meters, increasing the electron density threshold to $1.24 \times 10^{18} m^{-3}$. Since the electron total charge per bunch has also been increased to $20nC$, the minimal bunch length is $0.8cm$, similar to the non-magnetized case.

The eigenvalue of the one turn transfer matrix become bigger than one at the same place when the determinant become bigger than one though it is not true for the cooling section transfer matrix M_{cool} as shown in FIG.1.

TRANSVERSAL COUPLING

Consider the two beams interaction inside the electron cooling section which has a transverse wiggler field to suppress the recombination rate. The transverse oscillation of the heavy ion beam centroid can be expressed into a matrix form[2]

$$\begin{pmatrix} R_i \\ R_i' \\ 1 \end{pmatrix}_{n+1} = M_{ring} \begin{pmatrix} R_i \\ R_i' \\ 1 \end{pmatrix}_n$$

, where R_i is the displacement of the ion beam centroid from the golden orbit. The one turn transfer matrix is composed of the electron cooling section transfer matrix and the twiss matrix with two $-\frac{l_{cool}}{2}$ drift matrices to compensate double counting of the cooling section drift, i.e.

$$M_{ring} = R_x L_{drift} M_{cool} L_{drift}$$

, where

$$M_{cool} = \begin{pmatrix} 1+\xi'(\cos(\Omega'_0 l_{cool})-1) & \frac{1}{\Omega'_0}[\Omega'_0 l_{cool}(1-\xi') + \xi'\sin(\Omega'_0 l_{cool})] & a(l_{cool}) \\ -\xi'\Omega'_0 \sin(\Omega'_0 l_{cool}) & 1+\xi'(\cos(\Omega'_0 l_{cool})-1) & a'(l_{cool}) \\ 0 & 0 & 1 \end{pmatrix}$$

and $\xi' = \frac{\omega_{ie}^2}{\omega_0^2}$, $\Omega_0' = \frac{\omega_0}{c}$. The plasma frequencies have been defined in the same way as in reference [2]. The wiggler field introduces a dipole error represented by $a(l_{cool})$ and $a'(l_{cool})$, which is in the order of 10^{-5}. The instability threshold is determined by the maximal eigenvalue as shown in FIG.2. The designed electron density for RHIC non-magnetized ecooling scheme is $6.2 \times 10^{16} m^{-3}$, which is two orders of magnitude smaller than the threshold shown in FIG.2, $6.6 \times 10^{18} m^{-3}$. The determinant decrement of the one turn transfer matrix at the designed electron density is 4.9×10^{-4}, which corresponds to a coherent damping time about 2000 turns.

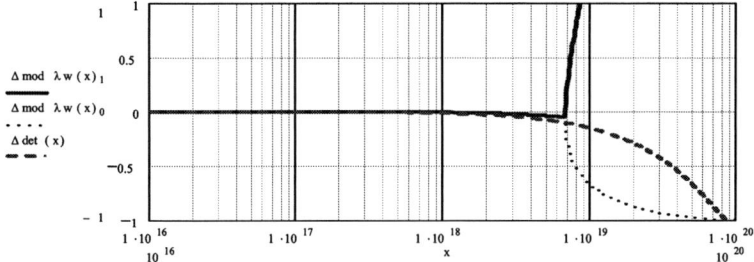

FIGURE 2. The eigenvalues and determinant increments of the transverse Oscillation for the non-magnetized electron-cooling scheme. The x-axis is the beam frame electron density in units of m^-3. The purple dash curve is the determinant of the one turn transfer matrix. The red solid curve and the blue dot curve are its two eigenvalues.

For the magnetized electron-cooling scheme, the ion beam is coupled with the transverse drift motion of the electron beam and one of the eigenvalues of the one turn transfer matrix is always bigger than 1 as shown in FIG.3.[2] For the designed beam frame electron density, $n_e = 3.3 \times 10^{15} m^{-3}$, the growth rate is 2.8×10^{-7}, which correspond to a instability rise time of 46 seconds.

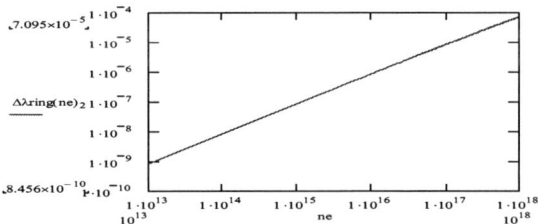

FIGURE 3. The maximal eigenvalues amplitude of the transverse Oscillation for the magnetized electron-cooling scheme. The x-axis is the beam frame electron density in units of m^-3 and the y-axis is the maximal amplitude of the complex eigenvalue minus one.

CONCLUSION AND DISCUSSION

As shown in the previous section, the transverse instability increment for the magnetized electron-cooing scheme can not be avoid by reducing the electron density and the increment rate strongly depends on the neutralization level of the electron beam[3] as shown in FIG.4

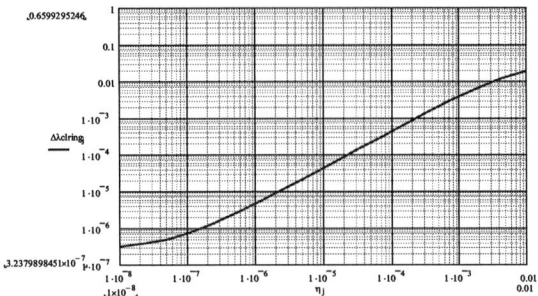

FIGURE 4. The dependence of the instability increment on the neutralization factor for the magnetized electron cooling scheme. The x-axis is the neutralization factor of the electron beam and the y-axis is the eigenvalue increment of the transverse transfer matrix for the magnetized cooling scheme.

In order to reduce the natural neutralization level, clearing electrodes may become necessary since making a gap between two bunches is not sufficient to clear the ion clouds as shown in FIG.5. The current non-magnetized electron cooling design is free from the cooling section dipole instability both longitudinally and transversely. Furthermore, the cooling electron beam introduces a coherent damping effect for the transverse oscillation, which may be used to compensate the coherent kick due to the misalignment of the magnets or the coherent beam-beam interaction.

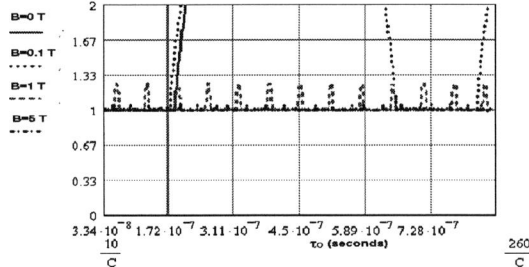

FIGURE 5. The dependence of the instability increment of the ion cloud transverse motion on the spacing of the incoming bunches for different solenoid fields. For strong solenoid field, the ion cloud motion is stable except for the lamour oscillation resonances.

ACKNOWLEDGMENTS

I would like to thank Professor Ilan Ben-Zvi to suggest this subject and I also want to thank Professor V.Parkhomchuk, Dr. V.Reva and Dr.Zenkevich for their helpful discussions.

REFERENCES

[1] V. Parkhomchuk, I. Ben-Zvi , *Electron Cooling for RHIC*, CA/AP Note # 47, April 2001.
[2] V.V.Parkhomchuk, V.B.Reva, J.Exp. Theor.Phys. 91 (5) (2000) 975
[3] P. Zenkevich, A. Dolinskii, I. Hofmann, Nucl. Instr. And Meth. Phys. A 532 (2004) 454-458

Hamiltonian Analysis of the Particle Motion in an Accelerator with the Longitudinal Magnetic Field

V.B.Reva

Budker Institute of Nuclear Physics, Novosibirsk, Russia

Abstract. The particle motion at a presence of a large magnetic field directed along the particle trajectory demands the special description. This article deals with the decomposition of the Hamiltonian on the two parts: fast and slow motion. The first part describes the fast rotation around the magnetic line of longitudinal field. The second part describes the slow drift of rotation center from one magnetic line to another. The supposed method enables to write the simple Hamiltonian to each motion type and to formulate the matrix formalism for any element of an accelerator device (quadruple, skew- quadruple, drift gap, bend with a filed index). The Hamiltonian decomposition has physical clearness when the longitudinal field is larger than another fields but it is correct for the arbitrary parameters. At the small longitudinal field the coupling term in Hamiltonian between two modes is essential. The dispersion property of fast and slow modes is derived easy from Hamiltonian also. This method expands easily for nonlinear motion of such modes. This results may be used at analyzed the electron motion in the cooling device, the muon motion in the muon ionization cooler [1] or another system with strong solenoidal coupling [2].

Keywords: modified betatron, storage ring with longitudinal field.
PACS: 29.27.

HAMILTONIAN METHOD

A non-relativistic Hamiltonian for a single particle in the electromagnetic field can be written as [3]

$$H = \frac{1}{2m}\left[\left(q_x - \frac{e}{c}A_1\right)^2 + \left(q_y - \frac{e}{c}A_2\right)^2 + \left(q_s - \frac{e}{c}A_3\right)^2 (1+hx)^{-2}\right] \quad (1)$$

Here (x,y,s) are coordinates in the reference orbit Frenet coordinate system $\vec{r} = \vec{r}_0 + x\vec{n}(s) + y\vec{b}(s)$, (q_x, q_y, q_z) is matched canonical momenta, $h = 1/\rho$ is the curvature of the reference orbit, ρ is the radius of curvature, ($A_1 = \vec{A}\cdot\vec{n}$, $A_2 = \vec{A}\cdot\vec{b}$, $A_3 = \vec{A}\cdot\vec{\tau}(1+hx)$) is vector potential of the magnetic field. Keeping only the linear and quadratic terms the magnetic field can be read as [4]

$$A_x = -\frac{1}{2}B_s(s)y, \quad A_y = \frac{1}{2}B_s(s)x$$

$$A_s = -b_0(s)x - \frac{1}{2}(b_1(s) - h(s)b_0(s))x^2 + \left(a_1(s) + \frac{1}{2}B_s'(s)\right)xy + \frac{1}{2}b_1(s)y^2 \quad (2)$$

The independent components a_n and b_n are multipole coefficients of the transverse magnetic field on the reference orbit

$$\left.\frac{\partial^n B_y}{\partial x^n}\right|_{x=y=0} = n! \cdot b_n, \quad \left.\frac{\partial^{n-1} B_x}{\partial x^{n-1}}\right|_{x=y=0} = (n-1)! \cdot a_{n-1}, \quad (3)$$

b_s is the longitudinal magnetic field.

The resulting Hamiltonian of the particle motion is

$$H = \frac{p_s^2}{2} + \frac{p_x^2}{2} + \frac{p_y^2}{2} + \frac{x^2}{2}\left(K_x^2 + K + \frac{R^2}{4}\right) - N x \cdot y \\ - K_x p_s x - \frac{R}{2}(p_y x - p_x y) + \frac{y^2}{2}\left(-K + \frac{R^2}{4}\right) \quad (4)$$

Here (B_x, B_y, B_s) is the horizontal, vertical and longitudinal components of the magnetic fields, $K_x = \frac{e B_y}{q_{s0} c}$, $R = \frac{e B_s}{q_{s0} c}$, $K = K_x \frac{1}{B_y} \frac{\partial B_y}{\partial x}$, $N = K_x \frac{1}{B_y} \frac{\partial B_x}{\partial x}$, $N_s = K_x \frac{1}{B_y} \frac{\partial B_s}{\partial s}$. All parameters of motion K, K_x, R, N, N_s are function of longitudinal position and time as $f(\bar{s} + \xi)$. The longitudinal and transverse momenta are normalized on the total particle momentum.

At condition of the strong longitudinal magnetic field the motion can be decompose on the fast Larmour rotation around the magnetic force line and slow drift of Larmour center. The center of the Larmour rotation moves along the magnetic field force line and drifts slowly in the plane (x,y). This drift motion is induced by the small transverse component of the magnetic and electrical forces, non-homogenuity of the longitudinal magnetic field or the centrifugal force. Thus the particle motion can be decomposed to fast and slow motion mode. This physical picture is convenient if the longitudinal magnetic force is strong.

Let us to do the change of variables

$$\begin{bmatrix} P_1 = p_x - \frac{1}{2} R y \\ Q_1 = \frac{p_y}{R} + \frac{1}{2} x \end{bmatrix}, \quad \begin{bmatrix} P_2 = p_y - \frac{1}{2} R x \\ Q_2 = \frac{p_x}{R} + \frac{1}{2} y \end{bmatrix}. \quad (4)$$

Taking into account that p_x and p_y are the canonical momenta $\vec{p} = \vec{\rho} - \frac{e}{c}\vec{A}$ one can obtain the correlation between the new variable and usual momenta (ρ_x, ρ_y) and coordinates (x, y)..

$$\begin{bmatrix} P_1 = \frac{\rho_x}{p_{s0}} - \frac{e B_s}{p_{s0} c} y \\ Q_1 = \frac{p_{s0} c}{e B_s} \frac{\rho_y}{p_{s0}} + \frac{x}{2} \end{bmatrix}, \quad \begin{bmatrix} P_2 = \frac{\rho_y}{p_{s0}} \\ Q_2 = \frac{p_{s0} c}{e B_s} \frac{\rho_x}{p_{s0}} \end{bmatrix}. \quad (5)$$

The mode (P_1,Q_1) describes the coordinates (X,Y) of the center of the Larmour circle. The mode (P_2,Q_2) relates to the rotation amplitudes of the particle around the magnetic force line. The equations describing these motion modes are weak coupled at the limit of the strong longitudinal magnetic field. In the limit of the infinite magnetic field this motion modes may be considered as uncoupling. But at a low value of the longitudinal magnetic field the coupling is strong. Thus, this situation is opposite to the classical case when the initial uncoupling vertical and horizontal motion is coupled by a weak magnetic field.

The generating function for this variables changing is

$$\Psi(x,y,Q_1,Q_2) = R\left(-\frac{1}{2}xy - Q_1 Q_2 + xQ_2 + yQ_1\right) \quad (6)$$

The new Hamiltonian is

$$H = -\frac{K}{R^2}\frac{P_1^2}{2} + (K_x^2 + K)\frac{Q_1^2}{2} + \frac{1}{R}\left(N - \frac{N_s}{2}\right)P_1Q_1 + \\ + \left(1 + \frac{K_x^2}{R^2} + \frac{K}{R^2}\right)\frac{P_2^2}{2} + (R^2 - K)\frac{Q_2^2}{2} + \frac{1}{R}\left(N - \frac{N_s}{2}\right)P_2Q_2 - \\ - \frac{1}{R}(K_x^2 + K)P_2Q_2 + \frac{K}{R}P_1Q_2 - N\left(Q_1Q_2 + \frac{1}{R^2}P_1P_2\right) + \frac{N_2}{2}\left(Q_1Q_2 - \frac{1}{R^2}P_1P_2\right) + \\ + \frac{p_s^2}{2} + \frac{K_x}{R}P_2p_s - K_xQ_1p_s \quad (7)$$

The first string describes the motion of the slow mode (P_1,Q_1) with large "pseudo"-mass $M=R^2/K$. The second string describes the fast oscillation of the mode (P_2,Q_2) with frequency R. The third string is coupling between modes (P_1,Q_1) and (P_2,Q_2). In the case $R \gg K, K_x, N, N_s$ it is small and can be consider with perturbation method. The last string is the longitudinal motion and dispersion terms.

Drift Section

This case the Hamiltonian deals with the particle motion in the simple straight solenoid ($K=0, K_x=0, N=0, N_s=0$)

$$H = \frac{1}{2}p_s^2 + \frac{1}{2}P_2^2 + \frac{R^2}{2}Q_2^2 \quad (8)$$

The Hamiltonian doesn't depends from variable P_1 and Q_1. Thus, P_1 and Q_1 are the motion integral and the center of the Larmour circle is immovable. The fast rotation is described by the standard Hamiltonian of an oscillator.

The matrix for the slow motion is

$$\begin{pmatrix}\tilde{Q}_1 \\ \tilde{P}_1\end{pmatrix} = \begin{bmatrix}1 & 0 \\ 0 & 1\end{bmatrix} \cdot \begin{pmatrix}Q_1 \\ P_1\end{pmatrix} \quad (9)$$

Here (P_1,Q_1) are variables of slow mode before the element and $(\tilde{P}_1,\tilde{Q}_1)$ are variables after element of drift section.

Quadruple Lens with Longitudinal Magnetic Field

The total Hamiltonian with coupling term is

$$H = -\frac{K}{R^2}\frac{P_1^2}{2} + K\frac{Q_1^2}{2} + \left(1+\frac{K}{R^2}\right)\frac{P_2^2}{2} + (R^2 - K)\frac{Q_2^2}{2} - \frac{K}{R}(P_2Q_1 - Q_2P_1) + \frac{p_s^2}{2} \quad (10)$$

and it can be read after simplification as

$$H = -\frac{K}{R^2}\frac{P_1^2}{2} + K\frac{Q_1^2}{2} + \frac{P_2^2}{2} + R^2\frac{Q_2^2}{2} + \frac{p_s^2}{2} \quad (11)$$

The fast rotation doesn't change significantly. The motion of the Larmour center can be described as

$$\begin{aligned} P_1' &= -KQ_1 \\ Q_1' &= -\frac{K}{R^2}P_1 \end{aligned} \quad \Longrightarrow \quad \begin{aligned} \frac{\partial}{\partial \xi}(P_1 + RQ_1) &= -\frac{K}{R}(P_1 + RQ_1) \\ \frac{\partial}{\partial \xi}(P_1 - RQ_1) &= \frac{K}{R}(P_1 - RQ_1) \end{aligned} \quad . \quad (12)$$

The beam is reshaped to an ellipse with axes tilted on angle 45° to (x,y) coordinate system. Along one axis the beam is stretched, along other axis is compressed. The particle motion in skew quadruple lens is analogously.

The matrix of elements for slow mode are

$$\begin{pmatrix} \tilde{Q}_1 \\ \tilde{P}_1 \end{pmatrix} = \begin{bmatrix} \cosh\left(\frac{K}{R}s\right) & -\frac{1}{R}\sinh\left(\frac{K}{R}s\right) \\ -R\sinh\left(\frac{K}{R}s\right) & \cosh\left(\frac{K}{R}s\right) \end{bmatrix} \cdot \begin{pmatrix} Q_1 \\ P_1 \end{pmatrix} \quad \text{(quadrupole),} \quad (13)$$

and

$$\begin{pmatrix} \tilde{Q}_1 \\ \tilde{P}_1 \end{pmatrix} = \begin{bmatrix} \exp\left(-\frac{K}{R}s\right) & 0 \\ 0 & \exp\left(\frac{K}{R}s\right) \end{bmatrix} \cdot \begin{pmatrix} Q_1 \\ P_1 \end{pmatrix} \quad \text{(skew-quadrupole)} . \quad (14)$$

Bending Magnet

The particle moves in the bending magnet with some field index. Along the particle trajectory the longitudinal magnetic field is applied. The Hamiltonian of such motion is

$$H = -\frac{K}{R^2}\frac{P_1^2}{2} + (K_x^2 + K)\frac{Q_1^2}{2} + \frac{P_2^2}{2} + R^2\frac{Q_2^2}{2} + \frac{p_s^2}{2} + \frac{K_x}{R}P_2 p_s - K_x Q_1 p_s \quad (15)$$

At $K < 0$ the dynamic of the mode (P_1, Q_1) is similar to some oscillator. The centers of the Larmour circles move along an ellipse curve. At the field index n=0.5 this curve is circle. The length corresponding one turn along this curve is

$$L = 2\pi \frac{R^2}{\sqrt{|K|(K_x^2 - |K|)}} \quad (16)$$

In the case $K = 0$ the field index $n=0$ and particle motion is

$$P_1' = K_x^2 Q_1 \quad \rightarrow \quad \begin{array}{l} X = const \\ Y' = \dfrac{K_x^2}{R} X \end{array} \quad (17)$$
$$Q_1' = 0$$

Thus, the horizontal position of the particle is constant. At nonzero horizontal shift the centrifugal force doesn't balanced by the bending magnetic field B_y. As result the particle has unlimited shift in the vertical direction.

The matrix of element for slow mode is

$$\begin{pmatrix} \tilde{Q}_1 \\ \tilde{P}_1 \end{pmatrix} = \begin{bmatrix} \cos(\lambda s) & \dfrac{|K|}{\lambda R^2}\sin(\lambda s) \\ -\dfrac{\lambda R^2}{|K|}\sin(\lambda s) & \cos(\lambda s) \end{bmatrix} \cdot \begin{pmatrix} Q_1 \\ P_1 \end{pmatrix}, \quad \lambda = \dfrac{\sqrt{|K|(K_x^2-|K|)}}{R} \quad (18)$$

Taking into account the dispersion terms and ignoring the term $K_x^2/R^2 \ll 1$ the Hamiltonian can be read as

$$H = -\dfrac{K}{R^2}\dfrac{P_1^2}{2} + \dfrac{1}{2}(K_x^2+K)\left(Q_1 - \dfrac{K_x}{K_x^2+K}p_s\right)^2 + \dfrac{1}{2}\left(p_2 + \dfrac{K_x}{R}p_s\right)^2 \\ + R^2\dfrac{Q_2^2}{2} + \dfrac{P_s^2}{2}\dfrac{K}{K_x^2+K} \quad (19)$$

The type of particle motion with some longitudinal momentum spread isn't changed. The center of ellipse painting the motion of the Larmour circle is shifted on value

$$\Delta X = \dfrac{K_x}{K_x^2+K}p_s \quad (20)$$

The fast rotation the particle acquires the additional momentum induced by the bending force

$$\dfrac{\Delta \rho_y}{p_{s0}} = -\dfrac{K_x}{R}\dfrac{\Delta \rho_s}{p_{s0}} \quad (21)$$

The term

$$\dfrac{p_s^2}{2}\dfrac{K}{K_x^2+K} \quad (22)$$

describes the "effective mass" of particle in bending.

REFERENCES

1. R.B. Palmer. *Nucl. Instr. Meth.* **A 532** (2004), p.255-259..
2. I.Meshkov, A.Sidorin, A.Smirnov et al. *Nucl. Instr. Meth.* **A 441** (2000), p.145-149.
3. Kolomenskiî A A, Lebedev A N *Teoriya Tsiklicheskikh Uskoriteleî* (*Theory of Cyclic Accelerators*) (Moscow: Fizmatgiz, 1962) [Translated into English (Amsterdam: North-Holland, 1966)]
4. Lee Teng. *Proceeding of Particle Accelerator Conference*, 1995, vol.5, pp.2814-2816.

STOCHASTIC COOLING

Stochastic Cooling Developments at GSI

F. Nolden*, K. Beckert*, P. Beller*, A. Dolinskii*, B. Franzke*, U. Jandewerth*, I. Nesmiyan*, C. Peschke*, P. Petri*, M. Steck*, F. Caspers[†], D. Möhl[†] and L. Thorndahl[†]

*GSI, Darmstadt, Germany
[†]CERN, Geneva, Switzerland

Abstract. Stochastic Cooling is presently used at the existing storage ring ESR as a first stage of cooling for secondary heavy ion beams. In the frame of the FAIR project at GSI, stochastic cooling is planned to play a major role for the preparation of high quality antiproton and rare isotope beams. The paper describes the existing ESR system, the first stage cooling system at the planned Collector Ring, and will also cover first steps toward the design of an antiproton collection system at the planned RESR ring.

Keywords: Storage rings, Beam characteristics
PACS: 29.20.Dh, 29.27.Fh

ESR STOCHASTIC COOLING

The stochastic cooling system at the GSI storage ring has been applied mainly to the preparation of secondary heavy ion beams for subsequent electron cooling. The successive application of the two cooling methods has made it possible to observe Schottky spectra of rare isotopes already 6 seconds after their production. As a possible application, the measurement of an isomeric state with a lifetime at rest of just 1.33 s has been reported [1].

STOCHASTIC COOLING FOR THE FAIR PROJECT

Overall Requirements

Two of the major pillars of the planned GSI FAIR project are experiments with rare isotopes and with antiprotons [2] - [4].

Rare isotope beams are produced by shooting a primary beam from the planned SIS100 synchrotron on a fragmentation or fission target, separating one or more species in the superconducting fragment separator SuperFRS, precooling them at a fixed specific energy of 740 MeV/u in the CR collector ring, decelerating them in the RESR ring and finally delivering them to the NESR storage ring. The latter ring is equipped with electron cooling and serves for various kinds of in-ring experiments.

The antiproton beams are produced by shooting an intense 29 GeV proton beam on an antiproton production target. Antiprotons at 3 GeV are guided to the CR, where a first stage of stochastic precooling is performed. Then they are extracted and injected into

TABLE 1. Basic parameters of the CR.

	rare isotopes	antiprotons
circumference	210.45 m	
max. magnetic rigidity	13 Tm	
energy	740 MeV/u	3.0 GeV
Lorentz β	0.83	0.97
Lorentz γ	1.80	4.20
max. number of particles	$1 \cdot 10^9$	$1 \cdot 10^8$
charge state	2 up to 92	-1

the RESR ring, which serves as an accumulator. Accumulated antiprotons can either be shot towards the HESR high energy storage ring for in-ring experiments, or towards the NESR, where they can be decelerated and be delivered to a low-energy experimental area, where further deceleration down to rest is foreseen.

Stochastic Cooling in the CR Ring

The CR serves as a first stage cooler both for rare isotopes and for antiprotons. Its purpose is comparable to the former ACOL ring at CERN and to the FNAL Debuncher.

It differs to both rings due to the fact that rare isotope beams at a much lower velocity have to be cooled, as well. In order to optimize the cooling scenario for both purposes, two dedicated beam optical modes have been developed. These modes differ mainly by their transition point γ_T. The optical design of the CR [5] was adapted to the needs of stochastic precooling. While the main effort for the antiproton beams consisted in ensuring enough good mixing, the rare isotope optical mode needed special measures to be taken to decrease undesired mixing between pick-up and kicker.

The injection and extraction parameters are given in table 2. In any case, the injected bunch is so short (< 50 ns) that bunch rotation followed by adiabatic debunching can be used to decrease the momentum width prior to stochastic cooling. The extraction parameters for rare isotopes are challenging. They were chosen such that subsequent electron cooling in the NESR will work in the domain of magnetized cooling. The antiproton parameters are close to the ones at the former ACOL ring at CERN.

For the different energies, a new type of pick-ups and kickers is being developed, which works well for both kinds of particles. There will be no need to exchange hardware components if the beam type is changed.

Besides the difference in energy, the signal to noise ratio of the pick-up signal depends strongly on the charge state. In fact, the signal from a bare uranium beam is by 39 dB larger than the one from an antiproton beam given the same number of particles. The design of the pick-up is therefore mainly dictated by the needs from the antiprotons. Cryogenic pick-ups with a system temperature of about 20 K will probably be installed.

TABLE 2. Initial and final beam for CR stochastic cooling.

	rare isotopes		antiprotons	
	$\delta p/p$ (2σ)	ε_{xy} [mm mrad]	$\delta p/p$ (2σ)	ε_{xy} [mm mrad]
injected	1.5 %	200	3.0 %	240
debunched	0.4 %	200	0.7 %	240
cooled	0.05 %	0.5	0.1 %	5
total phase space reduction	$1.3 \cdot 10^6$		$1.6 \cdot 10^4$	
total cooling time [s]	1		5	

The following discussion requires some basic equations. Let us write the simplified expression for the cooling rate τ in the form

$$\frac{1}{\tau} = \frac{2W}{N} \left[2Bg - g^2(M+U) \right] \quad (1)$$

where W is the system bandwidth and N is the number of particles in the beam. The undesired mixing is described by the factor B which can be approximately written as

$$B = \cos m_c \phi_u \quad (2)$$

where m_c is the central cooling harmonic. The undesired mixing angle

$$\phi_u = \omega T_{p-k} \eta_{p-k} \frac{\delta p}{p} \quad (3)$$

depends on the (angular) revolution frequency ω, the time of flight T_{p-k} between pick-up and kicker, the local frequency slip η_{p-k} between pick-up and kicker, and finally on the off-momentum $\delta p/p$.

The desired mixing parameter M can be written aproximately as

$$M = \left(m_c \eta \frac{\delta p}{p} \right)^{-1} \quad (4)$$

In (1) U denotes the overall noise to signal ratio.

Both the desired and undesired mixing factors depend on frequency slip factors

$$\eta = \gamma^{-2} - \alpha_p \quad (5)$$

with the momentum compaction

$$\alpha_p = \frac{1}{s_2 - s_1} \int_{s_1}^{s_2} ds \frac{D(s)}{\rho(s)} \quad (6)$$

which is related to the transition γ_T by

$$\alpha_p = \gamma_t^{-2} \quad (7)$$

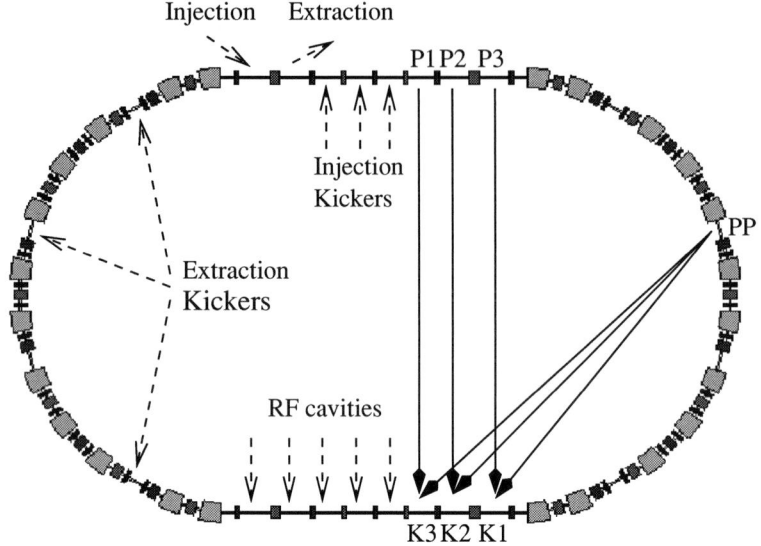

FIGURE 1. Layout of the CR with stochastic cooling lines.

In the case of undesired mixing, s_1 is the location of the pick-up and s_2 is the location of the kicker. In the case of desired mixing, s_1 and s_2 differ exactly by one ring circumference.

Figure 1 displays the layout of CR. The ring has a two-fold mirror symmetry. The left arc and the neighboring straight sections are used for injection and extraction elements and for the bunch rotation cavities which are also used for bunching prior to extraction.

The right arc and its neighboring straight sections are completely dedicated to the various stochastic cooling systems. There are 3 pick-up (P1-P3) and 3 kicker tanks (K1-K3) in the straight section and one pick-up tank in the arc (PP) for rare isotope cooling. There will be two frequency bands (1-2 GHz and 2-4 GHz) for the various cooling systems.

Table 3 shows some main parameters of the varous cooling lines. α_p, γ_T, and η are defined by (6), (7), and (5). For the rare isotopes, the ring is operated below transition. The maximum dispersion in the dipoles is 8 m, leading to large horizontal good field apertures of 400 mm. Therefore a superferric design of these magnets is favored. The local slip factors are still so large, that all cooling lines have to make use of the short path from the Palmer pick-up to the kickers K1-K3. This is obvious from the maximum delay

$$\delta T_{\text{bad}} = T_{\text{p-k}} \eta_{\text{p-k}} \delta p/p \qquad (8)$$

(cmp. (3)). This delay is calculated from the worst case momentum deviation after debunching. It leads to a maximum cooling frequency

$$f_g = \frac{1}{4 \, \delta T_{\text{bad}}} \qquad (9)$$

TABLE 3. Parameters of the CR stochatic cooling paths with the momentum width after debunching.

Item	P1-K3	P2-K2	P3-K1	PP-K1	PP-K2	PP-K3
orbit distance [m]	92.5	85.1	76.7	44.9	49.1	52.8
signal path length [m]	50.0	50.0	50.0	39.0	41.6	44.2
rare isotope parameters						
α_p (total)			0.1318			
γ_T (total)			2.754			
η (total)			0.1787			
purpose	vert.	long.	hor.	hor.	long.	vert.
α_p (local)	0.150	0.163	0.181	0.193	0.176	0.164
γ_T (local)	2.582	2.477	2.352	2.279	2.383	2.471
η (local)	0.161	0.148	0.130	0.118	0.134	0.147
T_{p-k} [ns]	372	342	308	180	197	212
δT_{bad} [ps]	239	202	160	85.1	106	125
f_g [GHz]	1.05	1.24	1.56	2.94	2.36	2.01
t_{el} [ns]	196	166	133	43	51	57
antiproton parameters						
α_p (total)			0.0740			
γ_T (total)			3.676			
η (total)			-0.0172			
purpose	long.	vert.	hor.			
α_p (local)	0.084	0.091	0.101			
γ_T (local)	3.447	3.306	3.139			
η (local)	-0.027	-0.035	-0.045			
T_{p-k} [ns]	318	292	264			
δT_{bad} [ps]	-60.9	-71.0	-82.5			
f_g [GHz]	4.10	3.52	3.03			
t_{el} [ns]	142	117	88			

beyond which the undesired mixing phase ϕ_u exceeds $\pi/2$, i.e. one gets a coherent blowing-up effect instead of cooling. Obviously, the pick-ups P1-P3 cannot be used during the initial phase of rare isotope cooling. Therefore all signals have to be derived from the Palmer type pick-up PP. This pick-up will be equipped only with 1-2 GHz electrodes.

Once the momentum spread of the rare isotope beams has decreased below 0.1 %, it is envisaged to switch over to the pick-up tanks P1-P3 which offer a superior signal quality. The respective parameters are listed in table 3, as well, but for the debunched momentum width in order to avoid confusion.

For the antiproton beams, the ring is operated slightly above transition. The η values are small enough to apply notch filter cooling which is superior to Palmer cooling because of its better thermal noise rejection. The full bandwith is applicable from the beginning.

Table 3 also contains values for the available electric length t_{el} of amplifiers and other electronic components. This length is calculated by comparing the time of flight from pick-up to kicker with the signal propagation time along a straight line from the end of a pick-up tank to the beginning of a kicker tank. It is assumed that the signal travels at $0.95c$. t_{el} is the difference of these times. The shortest value shows up for the horizontal

TABLE 4. Initial betatron amplitudes [mm] at P1-P3 and K1-K3 (for the kickers, entry and exit values have to be interchanged).

	rare isotopes				antiprotons			
	A_x		A_y		A_x		A_y	
	entry	exit	entry	exit	entry	exit	entry	exit
P1→ (←K3)	54	42	37	49	54	34	33	53
P2→ (←K2)	42	58	49	38	34	54	53	34
P3→ (←K1)	58	41	40	60	53	32	34	59

TABLE 5. Initial amplitudes at PP for rare isotopes.

	entry	exit
A_x	55	51
A_y	60	69
$D\delta p/p$	27	24
$D\delta p/p + A_x$	82	75

cooling line PP-K1 used for rare isotopes. For all lines from the Palmer pick-up a strictly straight connection is desirable. The cooling lines from P1-P3 could be installed at some elevated height in order to permit the installation of power supplies in the central part of the CR hall.

Table 4 displays the betatron amplitudes for the initial beam emittances at the pick-ups P1-P3 and the kickers K3-K1 (for the kickers, entry and exit labels must be interchanged). The amplitudes for rare isotopes and for antiprotons are comparably large. Also, the horizontal and vertical amplitudes are always both large, which tends to result in a low signal amplitude. On the other hand, any other optical design would have increased considerably the ring circumference and the total cost. Due to the FODO lattice, there is also a considerable betatron function slope along the straight sections (typical length 2.8 m) where the tanks are located. It is therefore foreseen to adapt the distance between electrodes inside the tank to the envelopes. The electrode arrays will be made plunging.

Table 5 shows the betatron and dispersion amplitudes for rare isotopes at the Palmer pick-up. The dispersion amplitudes are smaller than the betatron amplitudes, and the vertical amplitudes are large, as well. Adiitionally, there will be Schottky band overlap between the longitudinal and horizontal sideband signals. Both effects lead to enhanced diffusion. Simulations indicate, however, that the overall cooling time of 1 s can still be reached for bare uranium beams. Examples of such calculations can be found in [6]

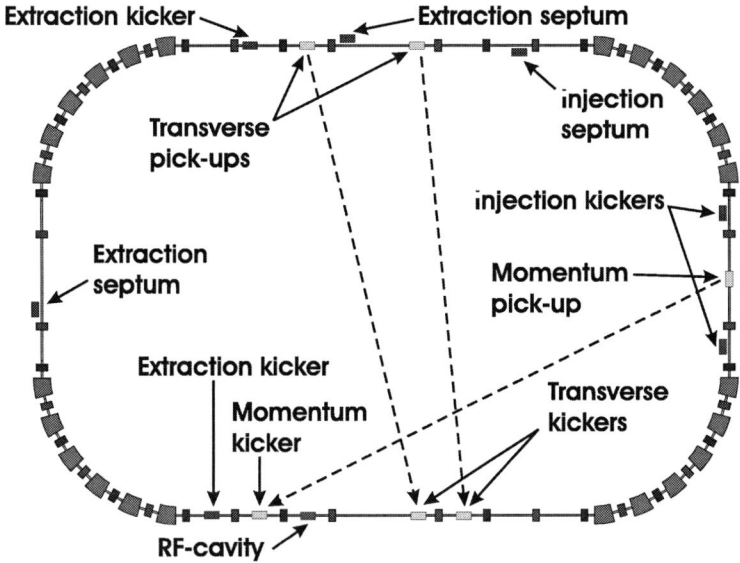

FIGURE 2. Layout of the RESR.

Stochastic Accumulation in the RESR Ring

In the RESR ring, which is situated in the same hall as the CR [3], [4], antiprotons will be accumulated up to $1 \cdot 10^{11}$ particles.

Two alternative methods have been investigated. The classical scenario is analogous to the schemes which were worked out first for the CERN AA ring by van der Meer [7]. The record performance has been attained recently at Fermilab's Accumulator [8], [9]. The essence of the method consists in moving the injected beam by rf to a handover orbit from where a group of stochastic cooling subsystems gradually shifts the beam towards regions of increasing phase space density. Ideally, the phase space density inceases exponentially, whereas the system gain decreases exponentially. The momentum pick-ups will be installed at a location with a dispersion of 8 m. The transition γ_T is 3.67. Figure 2 shows the layout of the RESR with stochastic cooling paths.

A second scheme which was investigated makes use of barrier buckets [10], which are also presently used for accumulation at the FNAL Recycler [11].

Stochastic Cooling in the HESR Ring

The experimental high energy storage ring HESR for antiprotons is the third storage ring of the FAIR project where stochastic cooling plays a role. A dedicated report on this topic has been prepared by the group at FZ Jülich [12], which is the consortium leader for the HESR in the frame of the FAIR project.

REFERENCES

1. F. Nolden et al., Experience and prospects of stochastic cooling of radioactive beams at GSI, *Nucl. Inst. Meth. A*, **532** (2004), 329-334
2. B. Franzke et al., Fast cooling of antiproton and radioactive ion beams in future storage rings at GSI, *Nucl. Inst. Meth. A*, **532** (2004), 97-104
3. P. Beller et. al., Layout of the Storage Ring Complex of the International Accelerator Facility for Research with Ions and Antiprotons at GSI, *Proc. EPAC '04* (2004) 1174-1176
4. P. Beller et al., Storage rings in the FAIR project at GSI, *Proc. STORI 2005, Research Center Jülich (series Matter and Materials)* (2005), to be published
5. A. Dolinskii et. al., Nonlinear Effects Studies for a Large Acceptance Collector Ring, *Proc. EPAC '04* (2004) 1177-1179
6. I. Nesmiyan et. al., Simulation Calculations of Stochastic Cooling for Existing and Planned GSI Facilities, *Proc. EPAC '04* (2004) 2170-2172
7. S. van der Meer, Stochastic Stacking in the Antiproton Accumulator, *CERN internal report CERN/PS/AA/78-22* (1978)
8. D. McGinnis et. al., Antiproton Production Rate Increase, *these proceedings*
9. P. Derwent et. al., Performance and upgrades of the Fermilab Accumulator Stacktail Stochastic Cooling, *these proceedings*
10. T. Katayama et. al., Stacking of 3 GeV pbar with Moving Barrier Bucket Method at GSI-RSER, *these proceedings*
11. S. Nagaitsev et. al., Antiproton cooling in the Fermilab Recycler, *these proceedings*
12. H. Stockhorst et. al., Cooling Scenario for the HESR Complex, *these proceedings*

Bunched Beam Stochastic Cooling Project for RHIC

J.M. Brennan and M. Blaskiewicz

Brookhaven National Laboratory, Upton New York, USA

Abstract. The main performance limitation for RHIC is emittance growth caused by IntraBeam Scattering during the store. We have developed a longitudinal bunched-beam stochastic cooling system in the 5-8 GHz band which will be used to counteract IBS longitudinal emittance growth and prevent de-bunching during the store. Solutions to the technical problems of achieving sufficient kicker voltage and overcoming the electronic saturation effects caused by coherent components within the Schottky spectrum are described. Results from tests with copper ions in RHIC during the FY05 physics run, including the observation of signal suppression, are presented.

Keywords: Stochastic cooling, bunched beam cooling, heavy ion collider.
PACS: 29.27..-a ; 29.27.Eg

INTRODUCTION

Intra-Beam scattering causes the beam to jump the separatrix and escape the 200 MHz buckets that store ions bunches during a RHIC physics store. This has two deleterious effects; one, beam is lost from the useful collision vertex at the experiments, and two, the de-bunched beam drifts into the abort gap and can cause excessive losses when the beam is cleared from the machine at the end of a store. Stochastic cooling of momentum can counteract IBS and prevent the de bunching. Herein we describe the study of the viability of a bunched-beam stochastic cooling system for RHIC and report on results with beam of tests of the key hardware components that have been developed for the special technical problems presented by the 100 GeV/n heavy ion collider.

The main concern as to the viability of such a system stems from the often observed anomalous coherent components of the Schottky spectra of high frequency bunches in colliders[1]. We believe solutions to the technical problem of filtering these components have been developed. A second technical challenge of such a cooling system is the high voltage that the kicker must produce. By synthesizing the kick with a set of harmonically related high-Q cavities the voltage can be generated with modest microwave power supplied by small solid state amplifiers. The availability of commercial fiber optic components enables realizing much of the broadband signal processing in analog optic networks.

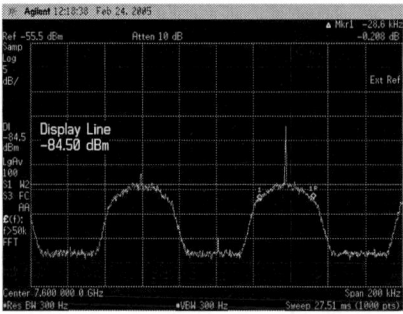

Figure 1. Beam spectrum showing Schottky component (broad) and coherent component (spikes). The coherent component is stronger on some revolution lines than others, see below.

Coherent Components in the Schottky Spectrum

Figure 1 shows a spectrum of a longitudinal pickup at 7.6 GHz from copper ions in RHIC at 100 GeV/n. The two identical broad bands comprise true Schottky signals and the narrow spikes are the coherent components. Although these coherent components are not as severe with ions as they are with protons, they nevertheless can be a serious problem for the signal processing electronics of a cooling system. They can cause saturation in amplifiers that generate non-linearities that can not be removed by further filtering. They are "anomalous" in the sense that one would expect that at 7.6 GHz the bunch Fourier transform would not have strength as great as the Schottky signal. We have learned that this expectation is erroneous. The evidence for this assertion is shown in figure 2. The left and center figures are two spectra with 10 MHz span centered at 15 MHz and 7.6 GHz. The envelopes of the two spectra are identical to one another, and to that shown on the right, which is a calculation of the spectrum in which the bunches are taken as delta functions. The three envelopes then reflect only the bunch filling pattern in the ring, which contains some bunches at harmonic 60 spacing and some at harmonic 120 spacing (9.5 MHz) and three bunches are missing to create an abort gap. The equivalence of these envelopes implies that the current from each bunch contributes coherently to the total current, and that the shape of each bunch is stationary with respect to the rf frequency. One does not need to invoke hot spots or solitions to explain the coherent lines. However, it is clear that the electronics will have to be able to handle the high peak voltages that are generated by these bunch shapes.

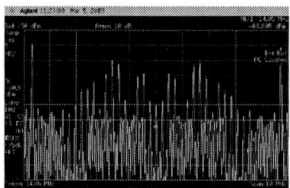 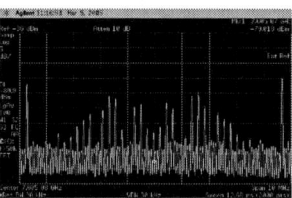 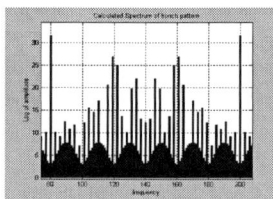

Figure 2. Beam spectra with 10 MHz span, left is spectrum from wall current monitor at 14 MHz, center is S.C. pickup at 7.6 GHz, right is delta function calculation of beam bunch fill pattern including abort gap.

Filtering the Coherent Components

The filter shown in figure 3 is very effective in reducing the peak voltages by a factor of eight while reducing the signal strength in its periodic passbands by only 6 dB. Because it is composed of passive components (power splitters and coax cables) which do not saturate it can be situated very early in the amplifier chain. The fact that its passbands are separated by 200 MHz is compatible with the kicker technology described below.

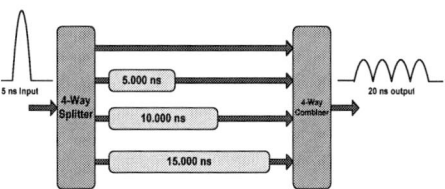

Figure 3. The front-end filter that reduces the peak voltage by a factor of 8. The spectral power at each harmonic of 200 MHz is reduced by only 6 dB.

KICKER VOLTAGE

With an energy spread of $\sigma = 0.3 \times 10^{-3}$, 10^9 particles in 5 ns bunches and 4 GHz bandwidth, a cooling system requires a kicker voltage of approximately 3 kV for Au^{+79}. If this were realized with a conventional 50 Ω broadband kicker the required power would be prohibitive. We have chosen to synthesize the kick with set of high-Q cavities, spaced every 200 MHz between 4 and 8 GHz. The spacing is sufficient because the bunch length is 5 ns.[2] The structures are 4-cell and 2-cell (depending on frequency) TM_{010} cavities with 20 mm beam bore and approximately 100 mm length. The loaded Q is constrained by the filling time and the 100 ns bunch spacing. The multi-cell design yields R/Q in order of 120 Ω, so that each cavity can be driven with a dedicated 40 Watt solid state amplifier. The linearity and dynamic range of solid state amplifiers are superior to traveling wave tube amplifiers. The intrinsic Q of the copper -plated aluminum structures is >6000, so that they can be operated CW without water cooling. Loading resistors, which set the cavity bandwidth to 10 MHz, are external to the vacuum. During beam injection and energy ramping the cavities split and open to provide clear aperture.

Figure 4. End view of cavities in vacuum chamber. Closed aperture is 20 mm diameter. Open horizontal aperture is 50mm.

Low-Level Drive for the Kickers

The pickup signal is transmitted from the pickup to the low-level room via a fiber optic link. Before the signal is converted back to electrical it is stretched to 80 ns duration with the optical network shown in figure 5. The two delays are set via optical trombones to 20 and 40 ns to within1 ps. The optical switch (Semiconductor Optical Amplifier) admits light into the network for only 20 ns (see figure 2) to prevent optical interference of the light carrier at the combiners.

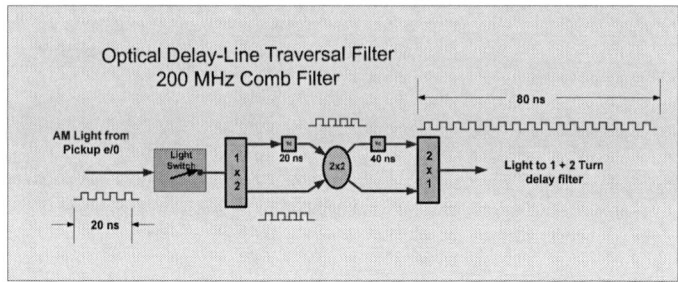

Figure 5. The fiber optic network that stretches the drive signal to the cavities to 80 ns.

Two-Turn Delay Filter for Halo Cooling

A notch filter is usually used for momentum cooling, but if two notch filters are cascaded a larger momentum spread can be cooled.[3] This is beneficial in the RHIC application where preventing de-bunching is the primary goal. The delays are realized in fiber optics before converting to electrical. Adjustments for drifts in delay are corrected with motorized optical trombones.

RESULTS WITH BEAM

The cooling loop was tested by measuring the open-loop system transfer function, shown in figure 6 with a 200 kHz span. The revolution frequency bands are spaced at 78 kHz. The top traces are the magnitudes, with and without the notch filter.

The bottom traces are the real and imaginary parts of the transfer function with the two-turn delay notch filter. The magnitude measurement (after correcting for the bunch duty factor) shows that the kicker provides sufficient voltage and the proper function of the notch filter causes the real part of the transfer function to be anti-symmetric about the revolution frequencies.

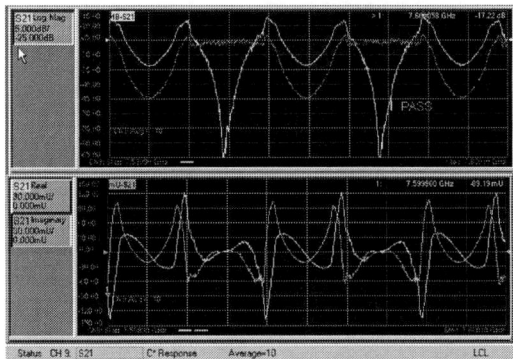

Figure 6. Cooling system open-loop transfer function at 7.6 GHz with 200 kHz span. Only one of the high-Q cavity kickers is driven. The top traces are log magnitude, with and without notch filter. Bottom traces are real and imaginary parts with filter. The real part is anti-symmetric about the revolution frequency.

Signal Suppression

When the cooling loop is closed the observed Schottky spectrum is modified by the mechanism known as signal suppression [4]. Figure 7 compares spectra with the loop open and closed, for the cases of a one-turn and two-turn delay notch filters.

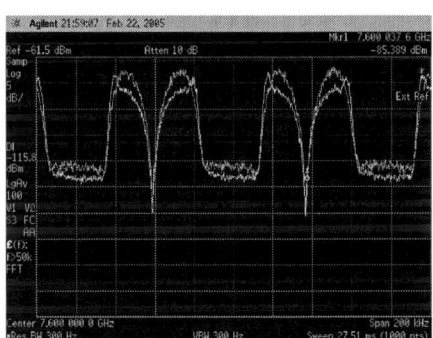 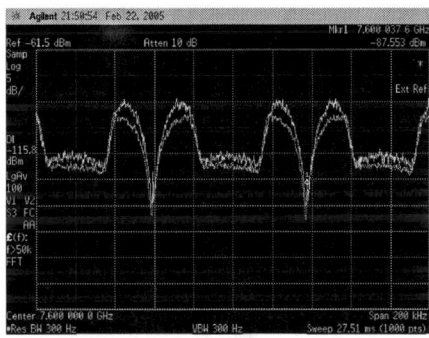

Figure 7. Schottky spectra at 7.6 GHz showing signal suppression. The observed spectra are reduced by about 5 dB when the loop is closed. Left is with one-turn delay filter. Right is with two-turn delay filter.

REFERENCES

1. J.M. Brennan, M. Blaskiewicz, J. Wei, Proceedings Cool03 Nuclear Instrumentation and Methods in Physics Research A 532 (2004). M. Blaskiewicz, these proceedings. G. Jackson, et al, PAC 1991, pp1758
2. D. Boussard, CERN Accelerator School, 87-03 Vol II, p423
3. M. Blaskiewicz, PAC 2005, Knoxville TN
4. D. Mohl, CERN Accelerator School, 87-03 Vol II, p500. J. Marriner, Proceedings, Workshop on Beam Cooling and Related Topics, Montreux, 1993.

Cooling Scenario for the HESR Complex

H. Stockhorst, D. Prasuhn, R. Maier, B. Lorentz

Forschungszentrum Jülich GmbH, D-52425 Jülich, Germany

Abstract. The High-Energy Storage Ring (HESR) of the future International Facility for Antiproton and Ion Research (FAIR) at GSI in Darmstadt is planned as an anti-proton cooler ring in the momentum range from 1.5 to 15 GeV/c. An important and challenging feature of the new facility is the combination of phase space cooled beams with internal targets. The required beam parameters and intensities are prepared in two operation modes: the high luminosity mode with beam intensities up to 10^{11} anti-protons, and the high resolution mode with 10^{10} anti-protons cooled down to a relative momentum spread of only a few 10^{-5}. Consequently, powerful phase space cooling is needed, taking advantage of high-energy electron cooling and high-bandwidth stochastic cooling. Both cooling techniques are envisaged here theoretically, including the effect of beam-target interaction and intra-beam scattering to find especially for stochastic cooling the best system parameters.

Keywords: FAIR, electron cooling, stochastic cooling
PACS: 29.20.Lq

INTRODUCTION

The High-Energy Storage Ring (HESR) [1] of the future International Facility for Antiproton and Ion Research (FAIR) at GSI in Darmstadt [2] is planned as an anti-proton cooler ring in the momentum range from 1.5 to 15 GeV/c. The basic racetrack layout of the HESR is shown in figure 1. The circumference of the ring is 574 m with two arcs of length 155 m each. The long straight sections each of length 132 m contain the electron cooler solenoid and on the opposite side the Panda experiment. The stochastic cooling tanks will be located in the straight sections. Two diagonal signal paths are foreseen for horizontal and vertical cooling. One of the systems will be used for longitudinal cooling. Two injection lines are envisaged, one coming from the RESR [2] to inject cooled anti-protons with 3 GeV kinetic energy and the other one to inject protons from SIS 18.

An important feature of the new facility is the combination of phase space cooled beams with internal targets. The desired beam quality and intensity is prepared in two operation modes: the *high luminosity mode (HL)* with a luminosity of $2 \cdot 10^{32}$ cm^{-2} s^{-1}. The HL-mode is attained with 10^{11} anti-protons and a target thickness of $4 \cdot 10^{15}$ atoms cm^{-2}. The HL-mode has to be prepared in the whole energy range of the HESR and beam cooling is needed to particularly prevent beam heating by the beam-target interaction. Much higher requirements are necessary in the *high resolution mode (HR)* with a luminosity of $2 \cdot 10^{31}$ cm^{-2} s^{-1} that can be attained with 10^{10} anti-protons and a target thickness of $4 \cdot 10^{15}$ atoms cm^{-2}. In this mode that is requested up to 8.9 GeV/c an anti-proton beam with a relative momentum spread down to $1 \cdot 10^{-5}$ has to be furnished.

To accomplish these goals this contribution discusses a possible scenario from a theoretical point of view in which electron cooling is applied in the HR-mode while a high bandwidth stochastic cooling system is utilized to provide the HL-mode over the entire momentum range of the HESR. The challenging and tough tasks in designing a 2 MeV electron cooler system at the cooler synchrotron COSY as an intermediate energy step towards the future high-energy magnetized cooler concept at the HESR is outlined in [3].

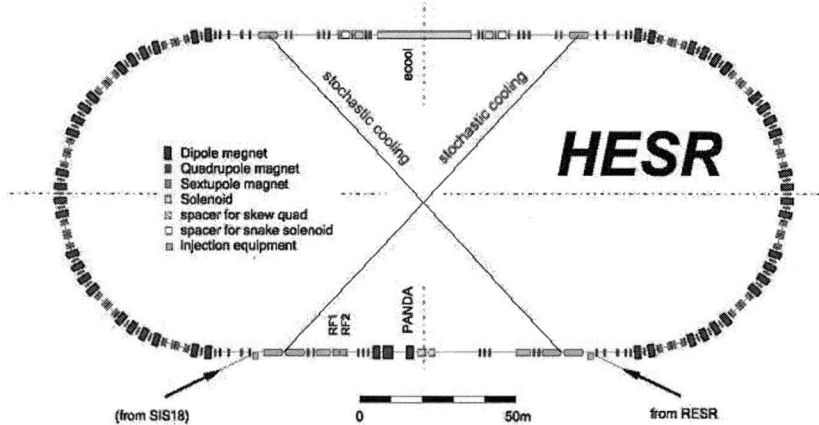

FIGURE 1. Basic layout of the HESR showing the electron cooler location and the signal paths for the stochastic cooling system. The PANDA experiment is located on the opposite side to the electron cooler.

COOLING MODELS

Stochastic and electron cooling simulations has been carried out with the computer code BETACOOL [4]. In collaboration with the Joint Institute for Nuclear Research at Dubna, JINR, stochastic cooling simulation has been now added in the program. Transverse stochastic cooling [5] is applied in accordance with the theory that was published in [6]. The longitudinal Filter cooling technique [5] has been chosen for stochastic momentum cooling in the HESR. A revised theory that was originally outlined in [7] has been prepared and integrated in the program. The transfer functions of quarter wave loop pickups and kickers are implemented in the model. These functions are discussed in detail in [8, 9] and can be expressed analytically. The stochastic cooling model calculations allow to optimize the electronic gain to achieve the desired beam parameters in the HR- or HL-mode in the presence of an internal target and intra beam scattering. The number of pickup and kicker loop pairs can be optimized with respect to the available electronic power and space for the pickup and kicker tanks installations. The model also allows to optimize the system parameters with respect to the desired equilibrium beam parameters with analytical formulas.

All cooling simulations include intra-beam scattering (IBS) according to the Martini model [4]. The HESR lattice [10] that has been used throughout has an imaginary transition energy with $\gamma_{tr} = 6.5\,i$. The target-beam interaction is treated in the formalism as outlined in detail in [11, 12].

Electron cooling simulations utilize the cooling force formula as evolved by V.V Parkhomchuk. The time development of tails which arise during electron cooling is accounted for by choosing the model beam option in BETACOOL [4]. The main electron cooler parameters as adopted in the HESR technical report [1] are listed in table 1.

TABLE 1. Electron Cooler Parameters

Beta function at cooler (h/v):	100	m
Cooling section length:	30	m
Electron beam radius:	5	mm
Electron beam current:	1	A
Electron beam density:	$2.7 \cdot 10^8$	cm^{-3}
Field homogeneity:	$1 \cdot 10^{-5}$	
Transverse electron temperature:	0.1	eV

During stochastic cooling the time development of the rms-emittances and the rms relative momentum spread of a nearly DC-beam is determined for a 2 GHz bandwidth system operating in the frequency range (2 – 4) GHz. The mixing from pickup to kicker is adjusted so that it plays no significant role. The basic parameters of the stochastic cooling system are summarized in table 2.

TABLE 2. Main Stochastic Cooling System Parameters

Beta function at pickup and kicker (h/v):	75	m
Number of PU and KI loop pairs:	64	
Electrode length, gap height and width:	2.5	cm
PU/KI length:	≈ 3	m
Impedance:	50	Ω
Cooled structures and eq. amp. temperature:	80	K

For longitudinal stochastic cooling an optical notch filter will be implemented in the signal path where the pickup and kicker loops are combined in the sum mode. Including safety margins the necessary electronic power is less than 500 W per plane. It should be pointed out that, due to band overlap, the longitudinal filter cooling method is only feasible for momenta above 3.8 GeV/c.

HIGH RESOLUTION MODE

In general, the cooled anti-proton beam is injected at 3.8 GeV/c from the RESR. The beam is then electron cooled and accelerated to the desired experiment energy. Figure 2 shows a result of an electron cooling simulation for the HR-mode at p = 3.8 GeV/c (T = 3 GeV). The inititial rms-emittance is 0.092 mm mrad in both planes and the relative rms-momentum spread is $1.5 \cdot 10^{-4}$. It is shown that the relative rms-momentum spread attains an equilibrium value of about $3.5 \cdot 10^{-5}$ after 10 s of cooling. At the same time the emittances reach an equilibrium. Note the effect of IBS. Initially, both emittances and momentum spread are decreasing during cooling. At about 5 s IBS becomes visible and leads to a minimal value for the relative momentum spread which slightly increases to the final value while the emittances attain an equilibrium. The horizontal equilibrium emittance, 0.0024 mm mrad, is larger than the vertical one, 0.00063 mm mrad, due to the fact that the mean horizontal beta function in the HESR lattice is smaller than the vertical one [1, 10]. The target is switched on at 22 s and has no visible effect on the equilibrium values. This means that IBS is the dominant mechanism at these low emittances.

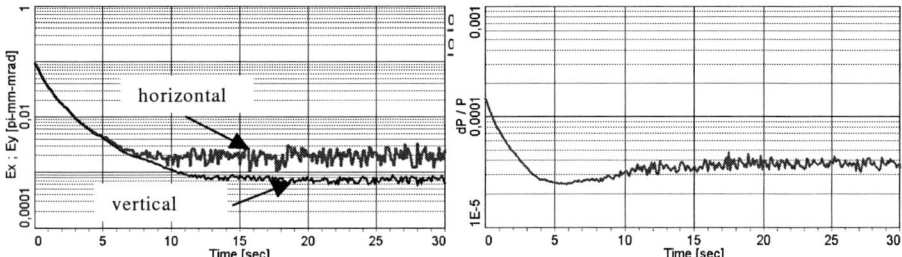

Figure 2. Time development of the rms-emittances (left figure) and relative rms-momentum spread (right figure) at T = 3 GeV. The equilibrium values are IBS dominated. A stable equilibrium relative momentum spread of about $3.5 \cdot 10^{-5}$ after 15 s of cooling is achieved.

The result of electron cooling at T = 8 GeV is drawn in figure 3. The beam is pre-cooled with electron cooling at injection energy. It is assumed that adiabatic shrinking of phase space during acceleration can be applied. The target is switched on at the experiment energy. A stable equilibrium relative rms-momentum spread of about $3 \cdot 10^{-5}$ is found in this simulation.

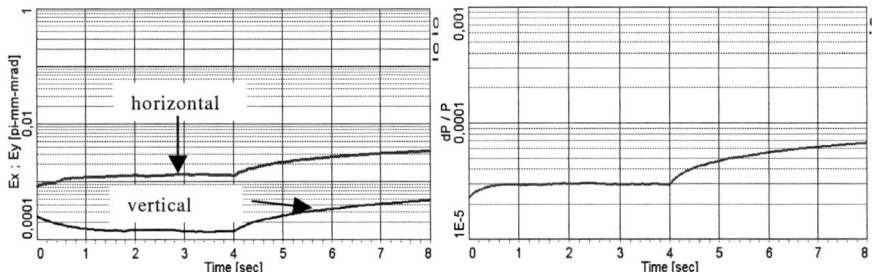

Figure 3. Time development of the rms-emittances (left figure) and relative rms-momentum spread (right figure) at T = 8 GeV. The equilibrium values are IBS dominated. A stable equilibrium relative momentum spread of about $3.0 \cdot 10^{-5}$ is achieved with electron cooling on. Beam heating is visible when electron cooling is switched off after 4 s.

In all cases above the final emittances attain tremendously small equilibrium values resulting in a beam that can be expected to be highly sensitive to perturbations. Moreover, to attain the necessary beam-target overlap with a Pellet target the beam emittance should be 1 mm mrad if the beta function at the target point is 1 m. Since electron cooling acts always on all phase planes simultaneously one possibility to circumvent such small emittances could be transverse beam heating. However simulations have shown that switching on a transverse noise source may lead to a significant beam loss due to an increasing transverse beam emittance.

HIGH LUMINOSITY MODE

Simultaneously transverse and longitudinal stochastic cooling of a DC-beam at injection momentum 3.8 GeV/c with an internal target is shown in figure 4.

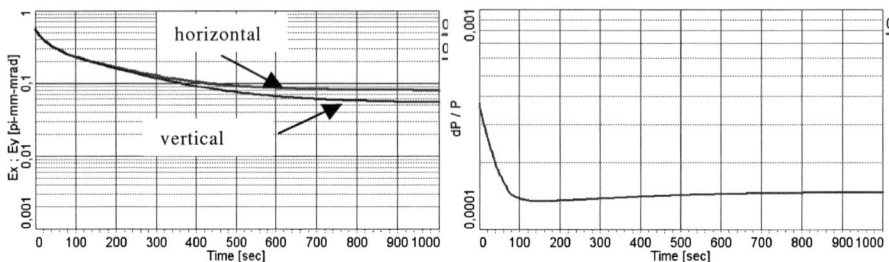

Figure 4. Time development of the rms-emittances (left figure) and relative rms-momentum spread (right figure) at T = 3 GeV. A stable equilibrium equilibrium momentum spread is attained after about 150 s. The rms-emittances becomes stable after about 800 s. The momentum acceptance of the machine is set to infinity.

The electronic gains have been adjusted independently to achieve the smallest equilibrium values. The figure demonstrates that target-beam heating can be significantly suppressed by stochastic cooling. The effect of IBS is also here visible however it is less pronounced as in case of electron cooling. Nevertheless it leads to slightly different final horizontal and vertical emittances. Transverse cooling is slower than the longitudinal one because of the reduction in momentum spread during transverse cooling. The wanted mixing from kicker to pickup is thereby reduced.

In the frame of this work detailed studies of small angle and energy straggling [11, 12] in the HESR energy range have shown that the dominant beam heating mechanism above 8.9 GeV/c due to an internal target is caused mainly by energy loss straggling. The emittance increase within a typical experiment run of one hour is less than a factor of two. Consequently it seems to be sufficient only to cool the longitudinal phase space above 8.9 GeV/c.

Table 3 lists the equilibrium relative momentum spreads that can be achieved for three different energies. Two cases have been studied. First, the momentum acceptance of the machine was assumed to be infinity, $\delta_{Cut} = \infty$, as in figure 4. In the second case the more realistic acceptance $\delta_{Cut} = 2 \cdot 10^{-3}$ was assumed. The simulation clearly demonstrates the sufficient damping of the target-beam interaction and IBS with stochastic cooling.

The table lists also the normalized gains $\sqrt{n_P n_K} G_A$ where n_P, n_K, G_A, denote the number of pickup and kicker loop pairs and the electron gain, respectively. Given the number of pickup loop pairs for a good signal-to-noise ratio at the pickup output the normalized gain allows to determine the number of kicker loop pairs if the power is limited. The table also contains the normalized gain expressed in the technical unit *dB*. $\delta_{rms,eq}$ denotes the equilibrium rms-momentum spread that is reached in approximately t_{eq} seconds.

TABLE 3. Stochastic Cooling for the HL-Mode

		3.9 GeV/c	8.9 GeV/c	14.9 GeV/c
Initial relative momentum spread:	$\delta_{rms,Ini} \cdot 10^4$:	3.8	2.5	1.8
$\delta_{Cut} = \infty$	$\delta_{rms,eq} \cdot 10^4$:	1.3	1.6	1.8
	t_{eq} [s]:	≈ 150	≈ 150	-
	$\sqrt{n_P n_K} G_A$:	$0.64 \cdot 10^7$	$2.0 \cdot 10^7$	$3.6 \cdot 10^7$
	dB:	136	146	151
$\delta_{Cut} = 2 \cdot 10^{-3}$	$\delta_{rms,eq} \cdot 10^4$:	1.0	0.94	0.9
	t_{eq} [s]:	≈ 250	≈ 400	≈ 500
	$\sqrt{n_P n_K} G_A$:	$0.36 \cdot 10^7$	$0.9 \cdot 10^7$	$1 \cdot 10^7$
	dB:	131	139	140

SUMMARY AND OUTLOOK

A possible cooling scenario has been discussed with respect to the desired high resolution and high luminosity modes at the future facility HESR. With electron cooling the HR-mode can be nearly fulfilled with respect to the desired momentum resolution. Further investigations however are needed to achieve the necessary beam-target overlap when a pellet target is used. In case of stochastic cooling it seems feasible to accomplish the HL-mode for kinetic energies above 3 GeV. A stochastic cooling system operating in the frequency range 2 to 4 GHz with a moderate power level seems to be sufficient. Since an emittance increase due to the beam-target interaction is less important above 8 GeV only longitudinal cooling is necessary in this energy range.

REFERENCES

1. HESR, *Technical Report*, to be published.
2. M. Steck, "The FAIR Project", this conference
3. J. Dietrich et al., "The Proposed 2 MeV Electron Cooler for COSY", this conference
4. A.Smirnov, A.Sidorin, G.Trubnikov, JINR, "Description of software for BETACOOL program based on BOLIDE interface"
5. D. Möhl et al., Phys. Rep. Vol. 58, No.2, Feb. 1980
6. B. Autin, "Fast Betatron Cooling in an Antiproton Accumulator", CERN/PS-AA/82-20
7. T. Katayama and N. Tokuda, Part. Acc., 1987, Vol. 21
8. Design Report Tevatron 1 Project, Sept. 1984, FNAL
9. G. Lambertson, "Dynamic Devices – Pickup and Kickers", in AIP Conf. Proc. 153, p. 1413
10. Y. Senichev et al., "Lattice Design Study for HESR", Proc. of the European Accelerator Conf. EPAC 04, Lucerne, (653)2004
11. F. Hinterberger and D. Prasuhn, Nucl. Instr. and Meth. A279(1989)413
12. F. Hinterberger, INTAS Workshop, June 3rd, 2005, GSI, to be published

Stacking of 3 GeV Antiprotons with a Moving Barrier Bucket Method at the GSI-RESR

T. Katayama, P. Beller, B. Franzke, I. Nesmiyan, F. Nolden, M. Steck (GSI), D. Möhl (CERN) and T. Kikuchi (Utsunomiya Univ.)

*Gesellschaft fur Schwerionenforshung(GSI),
Planckstrasse 1, D-64291 Darmstad, Germany*

Abstract. At the FAIR project at GSI, 3GeV pbar beams are to be accumulated in the RESR ring up to the intensity of 1.0e11. Every 5 seconds a new batch of 1.0e8 pbars is transferred from a Collector Ring (CR) where the pbar beams are pre-cooled to the momentum spread of 0.1% with stochastic cooling. The main task of the RESR is the accumulation of 1000 batches from the CR. In the classical way established at CERN's AA/AC, a new batch is injected on the injection orbit and is stacked in the radially separated stacking region. In the present paper, an alternative way, azimuthal separation with barrier bucket is proposed. The process is simulated up to 1000 injections and the emittance growth and the intra-beam scattering effects are evaluated.

Keywords: **barrier bucket, stochastic stacking**
PACS: 29.20.Dh, 29.27.Ac

INTRODUCTION

The important function of the storage ring RESR of the FAIR project at GSI [1], is the accumulation of 3 GeV antiprotons to obtain a high-density beam of 1×10^{11} antiprotons within a small momentum spread, for the experiments which will be performed at the subsequent storage rings HESR and NESR. The accumulation of antiprotons can be obviously accomplished in the way developed and established at the CERN ACOL/AA and the FNAL AS, namely with the stochastic stacking method [2]. This method requires a fairly large dispersion section in the ring for the separation of orbits of the injected beam and the stacked beam. This radial separation requires a large horizontal acceptance, and hence large aperture magnets. The barrier bucket (BB) method [3, 4, 5] is a new way of beam accumulation where part of the ring circumference is reserved for the injected beam while the other azimuthal part is used for the beam accumulation. This separation is performed with barrier buckets provided by short RF-pulses. Whereas the classical method uses the radial aperture to manipulate the injected and stacked beams, the barrier bucket method uses the azimuthal aperture.

The BB method described in this report to accumulate many batches, say 1000, is a new way of stacking (only 10 or so pulses are accumulated in the BB method of ref [3]). In contrast to the classical stacking method, it needs less aperture because a new batch is injected onto the central orbit with a full aperture kicker. The momentum spread of the injected and the stacked beams are less than +-0.3% during the whole process and then the aperture due to the dispersion can be small even if we have a lattice of large dispersion at the cooling pick-ups to avoid interference of the stack with the high gain cooling system for the injected batch. On the other hand the BB-scheme requires a pair of rectangular (or sine wave) pulses, and therefore cavities with low Q value. In addition the dynamic movement of the barriers needs the precise control of timing with only small jitters. Especially for a small ring such as RESR, the required pulses are short and hence the rise and fall time of the pulses is stringent.

OPERATIONAL SCHEME OF BB ACCUMULATION

The movement of the barrier pulses proceeds as follows (Fig. 1). After injecting the beam in the azimuthal injection area (about one half of the circumference), one of the barrier pulses, P2 is adiabatically moved to P1. The energy spread of trapped particles between these two pulses, becomes large as the distance of the two pulses shrinks. Eventually ΔE becomes large enough so that particles overflow the barriers, and move into the accumulation area. When P2 reaches P1, all the injected beam overflows into the accumulation area, and the barrier pulses are turned off. At this moment the beam is coasting all around the ring with two streams which are separated by the energy +- ΔE determined by the barrier bucket height whereas the stack remains in the middle, around $E=0$. Then the stochastic cooling stacking is applied to these coasting beams. After 5 sec, stochastic cooling is stopped and P1 and P2 are adiabatically turned on at the same azimuth. Then P2 is again moved back to the original position. The barrier pulses compress the cooled/stacked beam into the accumulation area and the injection area is now empty. Now a new pre-cooled batch is injected from the Collector Ring (CR). This process will be repeated every 5 sec, depending upon the cooling time in CR, and will be continued until the beam extraction from RESR. The accumulated antiproton number will reach to 1×10^{11} after ~2 hours stacking.

In the Fig. 1, horizontal scale is the time of revolution (+-0.4 μsec), and vertical ones are pulse voltages (blue colors) and the deviation of energy of particles from the synchronous energy. Red points represent the distribution of particles in longitudinal phase space. Sequences are from top left to bottom right. (1) injection. (2) P2 pulse (right hand side pulse) is approaching to P1(left hand side) pulse. (3) beam cooling. (4) P2 pulse is moved back to the original position.

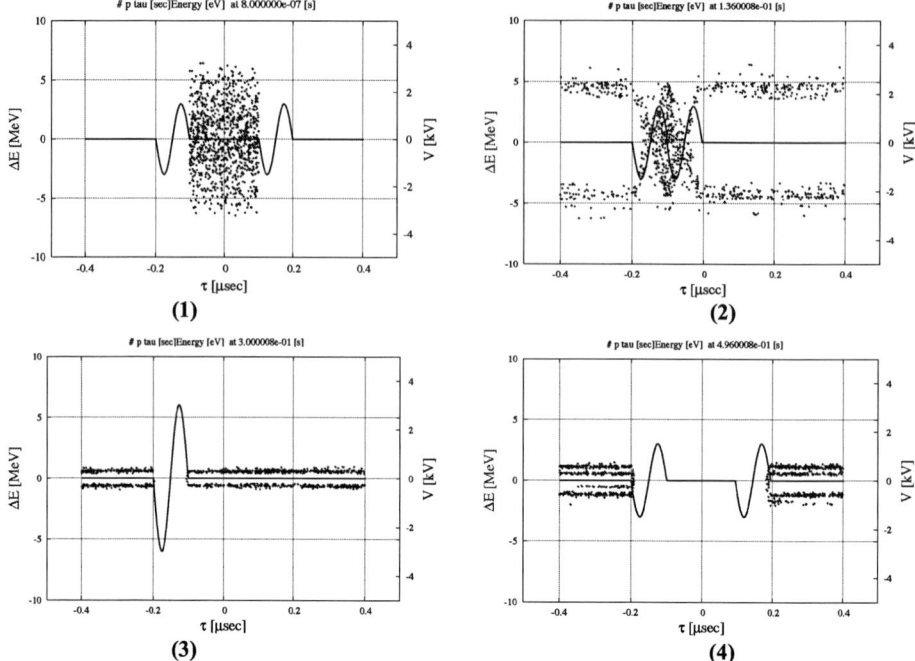

FIGURE 1. The scheme of Barrier Bucket injection and cooling during one cycle. Horizontal scale is time, and vertical one is energy deviation. The curves (blue) represent the barrier voltages. The red points represent the distribution of particles in longitudinal phase space. (1) beam injection, (2) the barrier pulses are approaching each other (3) beam cooling. (4) the barrier pulses are moved back to their original position.

The barrier voltage is taken as 3.0 kV. It is determined by the need to separate the two streams of coasting beam by +-8 MeV energy difference, and secondly to contain the injected beam in the injection area, and to confine the stacked beam in the accumulation area. From the requirement of adiabaticity, the moving time of pulse P2 towards P1 should be much longer than 5 msec. Here we assume as 200 msec.

TABLE 1. Parameters of RESR and barrier bucket pulses.

Circumference	245.5 m
Revolution time of 3 GeV pbars	820 nsec
Transition γ	3.62
Slipping factor η	0.020
BB Pulse voltage	+-3.0 kV
BB Pulse width	100 nsec (10 MHz)
Pulse movement speed(time to Move P2 to P1)	200 msec

The principle of stochastic cooling and stacking of the two coasting streams in Fig.1 (3), will be similar to the classical stacking method. The two coasting beams have initially the particle density of 10 pbar/eV, and are moved to the central core by the coherent force of stochastic momentum cooling. The Schottky noise from other particles and the amplifier noise, act as diffusion force and push back the particles from the dense part to lower density region. The Schottky noise term is proportional to the square of the coherent term and we can adjust the proper coherent strength to produce the maximal particle flux from the injection orbit to the central region. The particles accumulate in the stacking area at successive injection and cooling. Within 5000 seconds, 7.5×10^{10} antiprotons are accumulated in the momentum spread of less than +-1.0e-3.

TABLE 2. Summary of antiproton stacking parameters.

Injected pulse		
	Number of pbar/batch	1×10^8
	$\Delta p/p$ (1σ)	1×10^{-3}
	Horizontal and vertical emittance	5π mm-mrad
	Repetition cycle	5 sec
	Particle flux	2.0×10^7 /sec
Final stack		
	Number of pbars	1×10^{11}
	$\Delta p/p$ taken for stack(full width)	$<3 \times 10^{-3}$
	Horizontal and vertical emittance	$< 5\pi$ mm-mrad
	Total stacking time	5000 sec

RESULTS OF SIMULATION OF BB STACKING

Fig. 1 shows, that the injected batch is split into two coasting streams with a separation in energy of +- 4 MeV after the debunching. The separation is determined by the height of the barrier pulses and thus well controllable by the voltage on the barrier cavities. In Fig. 2, the spectrum just after debunching is illustrated for the different barrier bucket voltages, 1kV(red colors), 2kV(green) and 3kV(blue), respectively. The energy separation is proportional to the square root of the voltage as expected.

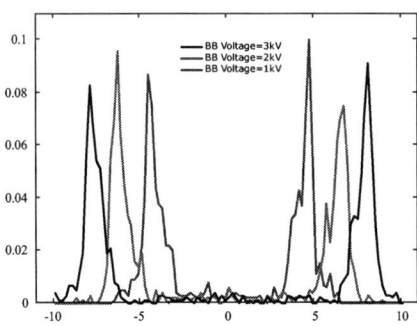

FIGURE 2. Spectrum just after the debunching process for different barrier bucket voltages. 1kV(red colors), 2kV(green) and 3kV(blue). The energy separation between two coasting beams is proportional to the square root of barrier voltage.

The energy separation is an important parameter because it determines the phase space area for cooling and stacking.

The rms longitudinal emittance is numerically calculated at each step during the stacking process and is given in Fig. 3. In this calculation, the cooling period is assumed as 0.1 sec (instead of ~5sec) and only 10 batches (instead of 1000) are stacked to save the computing time. Red color shows the variation of emittance and green the particle number. During each debunching process, the rms emittance becomes large due to the particle acceleration and deceleration by barrier voltages. In the process of cooling, the emittance becomes small. After the new batch is injected and transformed into the two streams, the overall emittance is again increased. As the stacking progresses, the corresponding peaks in Fig. 3 become smaller as the number of particles in the core increases.

Fig. 3.8 Evaluation of RMS Longitudinal Emittance

$$\varepsilon_z = \left[\left\langle (\Delta E - \Delta E_0)^2 \right\rangle \left\langle (\tau - \tau_0)^2 \right\rangle - \left\langle (\Delta E - \Delta E_0)(\tau - \tau_0) \right\rangle^2 \right]^{\frac{1}{2}}$$

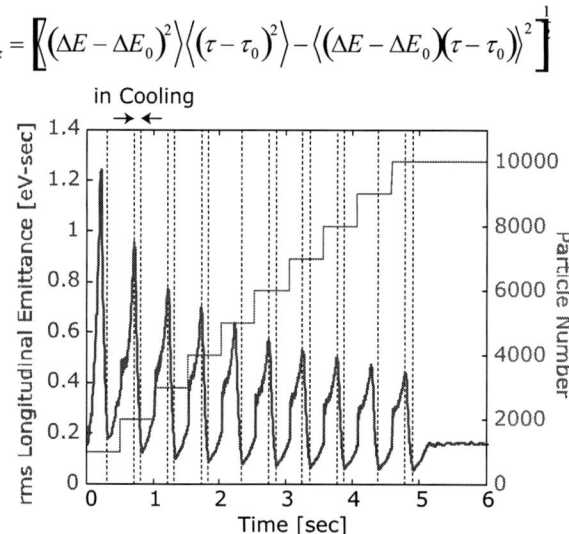

FIGURE 3 Variation of the longitudinal rms emittance (red) and particle number (green). Here only batches of 1000 particle each are injected and the cooling time is 0.1s. During the debunching process, the rms emittance becomes large due to the particle acceleration and deceleration by barrier voltages. In the process of cooling, emittance becomes small. After the new batch is added, the emittance increases during the new debunching process As stacking progresses the peaks become smaller because the number of particles in the core increases.

The stacking requirements are summarized in Table 2. Cooling system parameters obtained in the similar way as for the classical stacking in the CERN AA are given in Table 3.

TABLE 3. Numerical values of cooling stacking system.

Ring dispersion η	0.020
Barrier Voltage V_0	+-3kV
Energy separation of the two streams ΔE_b	+-8MeV
Corresponding momentum separation $\Delta p/p$	+-2.15x10^{-3}
Characteristic energy E_d	1.156 MeV
Maximal incoming flux Φ_{max}	1.0 x10^7/sec
Band width W	1.5 GHz
Cooling system voltage for injected batch ΔE_c	0.533 eV/turn
Ratio of cooling voltage for injected batch and stack top	1.01 x10^3
Number of Pbar in core region within $\Delta p/p$=+-1.0e-3	7.5 x10^{10}

In Fig. 4, the distribution of 3 GeV antiprotons calculated for these parameters is illustrated as a function of time after the start of stacking. Every 5 seconds 1.0x10^8 antiprotons are injected from the Collector Ring where antiprotons are pre-cooled to the momentum spread 0.1 %(1σ) and emittance of 5π mm-mrad. As explained in the preceding chapter, the movement of barrier buckets produces two streams of coasting beam in the RESR.

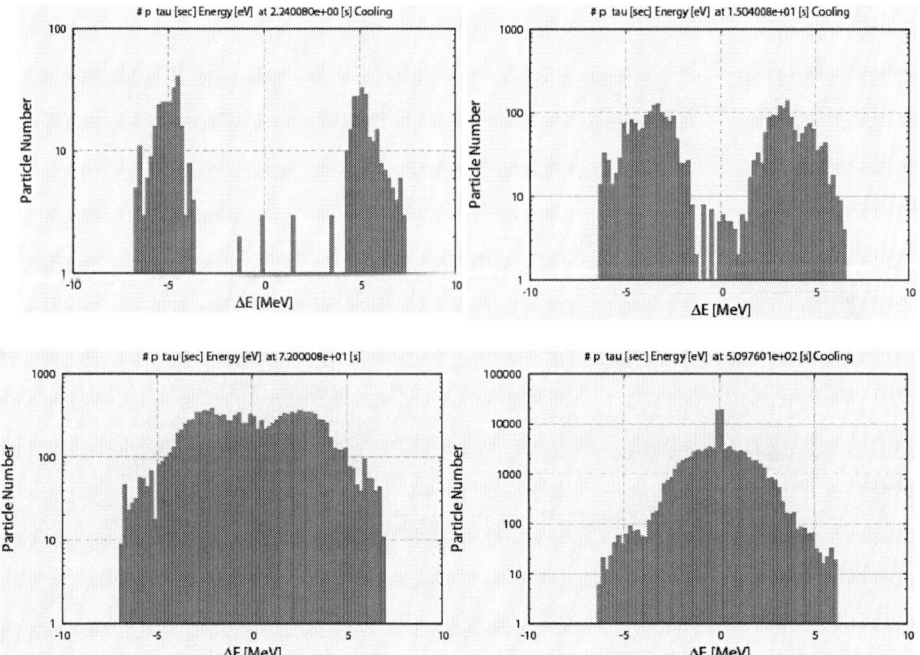

FIGURE 4 Spectrum of the cooled and stacked beam at different times after the start of stacking. The plots from top left to bottom right correspond to the time of 22 sec, 150 sec, 720 sec and 5000 sec, respectively. The number of particles (vertical scale) is in units of 1.0e6.

The figure reveals the high density core produced in the central region during the 5000 sec (1000 injections). The sharp peak in the center is due to the pickup's anti-symmetric gain function with non zero value at the origin. However in the real stacking process, such a high-density peak will be smeared out due to the strong heating due to intra-beam scattering. The number of pbars in the central core region defined by $|\Delta p/p|<1.0 \times 10^{-3}$, is 7.5×10^{10}.

The variation of rms longitudinal emittance during the accumulation is illustrated in Fig. 5. After 1000 injection cooling and stacking cycles, the emittance becomes 2.5 times larger than the initial emittance, and hence the density in phase space increases by 400.

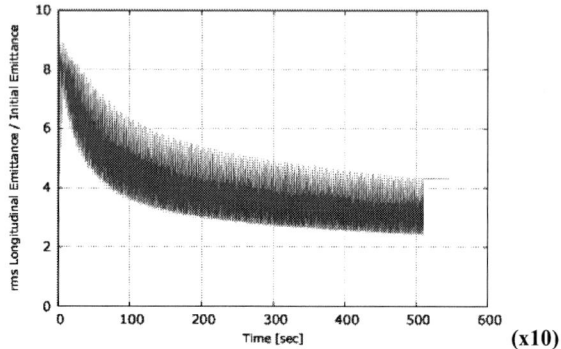

FIGURE 5: Rms longitudinal emittance during the process of cooling and stacking of 1000 batches. Horizontal scale is the time after the start of stacking (in units of 10 sec). Scale should be 10 times. Vertical scale is the ratio of rms emittance to emittance of incoming batch.

One of the critical subjects of the barrier system is a ringing of voltage. The predominant ringing frequency will be the frequency of main voltage, namely 10 MHz in the present case. Fig. 6 shows the longitudinal rms emittance variation for no ringing voltage (a), 11% ringing voltage (b) and 12 % ringing voltage(c), respectively. The cooling force is assumed to correspond to the e-folding time of 2 sec for 10^8 particles. At 60 sec after the start of stacking, (after 12 injections), the emittance is abruptly increased for the ringing voltage of 12 %. A typical example of particle distribution in the phase space at this threshold value, are illustrated in Fig. 6 (d).

We can say that the threshold ringing voltage for large blow up is 12 % of main barrier bucket voltage, and for smaller voltages particles are safely cooled and stacked. For a well designed BB pulse system, one expects less than 5 % ringing and thus no serious problem for the present stacking scheme.

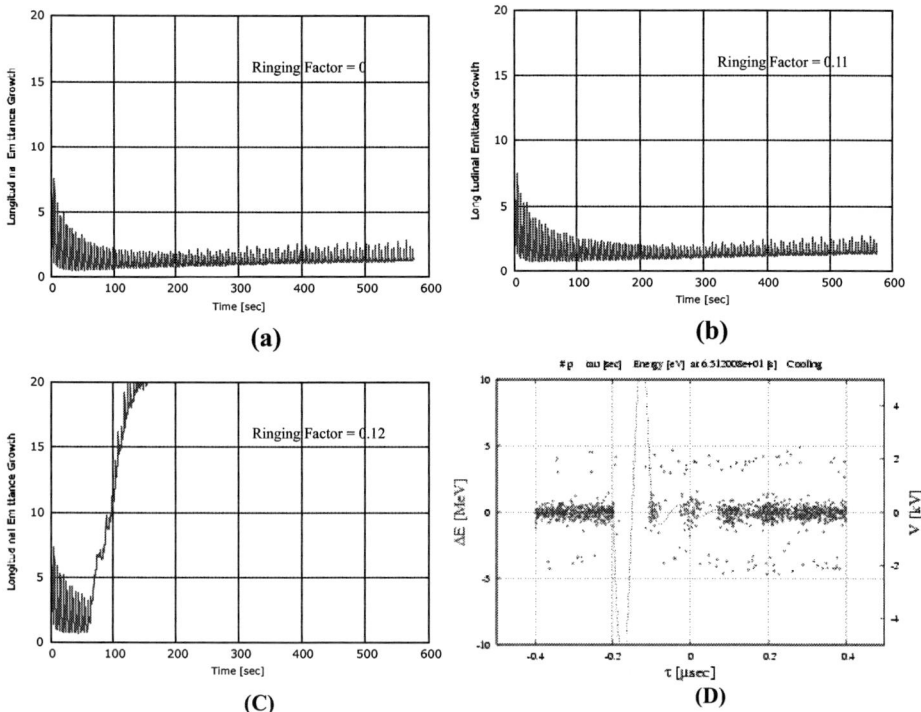

FIGURE 6 Emittance variation during 100 injection and cooling cycles. The e- folding cooling time is 2 sec for 10^8 particles. Every new batch contains 10^8 particles and the cooling time is assumed to be inversely proportional to the number of particles. The ringing voltage is 0 % (a), 11% (b) and 12 % (c). In (d) the particle distribution for the ringing voltage of 12%, is illustrated.

CORE COOLING

The purpose of core cooling is to keep the momentum spread of stacked beam in the core region as small as possible, less than +-0.1 % even after many batches are injected and stacked. Also it is aimed to suppress the emittance growth due to different sources such as Intra Beam Scattering (IBS) and amplifier noise. There are two possible methods of core cooling, the notch filter method and the Palmer method. For the present core cooling, the notch filter method will be simple and effective enough because the momentum spread of the core beam is less than +- 0.1% and the band overlapping is not the problem

In fact the frequency where overlapping starts is calculated as 30 GHz which is high enough to use high cooling band width. The allowable band width is also restricted by the phase shifts in the signal propagation lines. Here we assume the band of 2~4 GHz. Then the cooling time of 10^{11} anti-protons is around 100 sec for a system gain of 120 dB.

INTRA-BEAM SCATTERING EFFECTS

Coulomb scattering between particles in the coasting beam could be one of the main reasons of emittance growth in the stack core, where the high density beam is produced. The growth rate of transverse emittance and relative momentum spread are given by the following formula for the coasting beam [6],

$$\frac{1}{\tau_i} = 4\pi A F(\alpha_{i,j}, \beta_{i,j}, \eta, \varepsilon_{i,j}, \sigma_p) \quad (1)$$

where A is defined as

$$A = \frac{r_0^2 c N}{16\sqrt{2\pi^3} C \beta^3 \gamma^4 \varepsilon_x \varepsilon_y \sigma_p} \quad (2)$$

r_0 is the classical proton radius, c the speed of light, N the number of particles in the ring, C the ring circumference, β, γ the relativistic factor. ε_x, ε_y are transverse emittances and σ_p the relative energy spread. F is a complicated function of Twiss parameters, emittance and momentum spread, and should be determined numerically. It should be noted that the growth rates can not be simply scaled with the particle density in 6 dimensional phase spaces.

In Fig. 7 the growth rate of the longitudinal emittance is given as a function of the energy spread of the beam. We assume that the antiproton energy is 3 GeV, the initial 1σ emittance 5.0 π mm-mrad for the horizontal and vertical directions, the number of particles 1.0×10^{11} and no coupling between horizontal and vertical emittances.

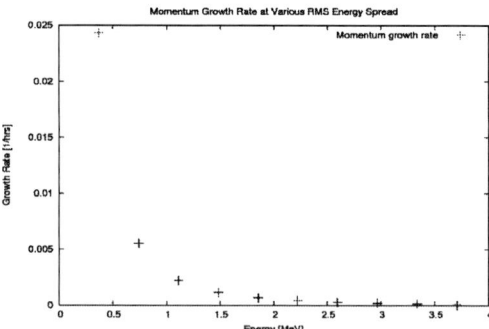

FIGURE 7 Growth rate of the longitunal emittance in RESR. The horizontal scale is the rms energy spread, the vertical scale gives the growth rate.

Note that we have an overall energy spread of around +- 8MeV during the BB stacking process. Most critical is probably the dense core. Assuming an rms energy spread of 0.5 MeV for this region, the growth time for the longitudinal emittance is 40 hrs. From these results, we can deduce that all growth times are much longer than 10 hrs. We conclude that the IBS in the RESR is of no concern because we are aiming to accumulate a fairly small number of particles 1e11 within around two hours.

CONCLUSION

We have made a feasibility study of the application of the barrier bucket method to the cooling and stacking in the RESR, based mainly on simulation work. From the results We conclude that the moving barrier bucket method can be used to stack 3GeV pbar up to 10^{11} within a momentum spread <+-0.3 %. Compared with the classical stacking method this scheme requires smaller aperture because injection and stacking are both performed on the central orbit rather then on orbits separated by momentum and dispersion. However details of technical issues have to be further investigated to be safe of "hidden flaws". In this sense, presently the classical cooling and stacking method is the preferred candidate for the RESR. The application of the BB scheme to the stacking at other storage rings, for example, heavy ion storage rings with electron cooling, could also be useful.

ACKNOWLEDGMENTS

The authors would like to express their thanks to F. Caspers, H. Eickhoff, K. Horioka, S. Kawata, Y. N. Rao and L. Thorndal for their valuable discussions.

REFERENCES

1. B. Franzke et al., "Fast cooling of antiproton and radioactive ion beams in future storage rings at GSI", Nuclear Instruments and Method in Physics Research A **532**, 2004, pp. 97-104
2. S. van der Meer, " Stochastic Stacking in the Antiproton Accumulator", CERN/PS/AA/78-22, 1978.
3. Fermilab recycler ring technical design report, FNAL report Nov. 1996, http://www-lib.fnal.gov/archive/1997/tm/TM-1991.html
4. J. Griffin et al., IEEE Trans. Nucl. Sci. **30**, 1983, pp. 3502-3504
5. S. Y. Lee and K. Y. Ng, Phys. Rev. E 55, 1998, pp. 5992-6001
6. Y. N. Rao and T. Katayama "Intra Beam Scattering" Particle Accelerators **59**, 1998, p. 251

Bunched Beam Stochastic Cooling and Coherent Lines

M. Blaskiewicz* and J.M. Brennan*

BNL 911B Upton NY 11973

Abstract. Strong coherent signals complicate bunched beam stochastic cooling, and development of the longitudinal stochastic cooling system for RHIC required dealing with coherence in heavy ion beams. Studies with proton beams revealed additional forms of coherence. This paper presents data and analysis for both sorts of beams.

Keywords: stochastic cooling, coherence
PACS: 41.75.Ak,52.35.Lv,52.35.Sb

INTRODUCTION

Intrabeam scattering causes emittance growth during heavy ion physics stores in RHIC. To combat this growth a longitudinal stochastic cooling system has been prototyped [1, 2, 3, 4, 5, 6] and an operational system is under construction [7]. During prototyping we encountered strong coherent signals in the heavy ion beams. Studies with proton beams also showed strong coherence and previous attempts to cool proton beams reported similar phenomena [8, 9, 10, 11, 12, 13].

COHERENCE IN HEAVY ION BEAMS

Heavy ion bunches are captured in 28 MHz RF buckets with harmonic number $h = 360$ and accelerated to top energy. At top energy the beam is "rebucketed" into 197 MHz buckets with $h = 2520 = 7 \times 360$. Coherent signals are strongest after rebucketing. Low level stochastic cooling signals for a single bunch are summarized in Fig 1A while the average and standard deviation of signals from five different bunches are shown in Fig 1B. The Schottky signal contributes to the standard deviation but not the average. The integrated energy ($\int V^2 dt$) for the standard deviation signal is about $1/3$ of the integrated energy in the average signal. Assuming that the standard deviation is totally due to Schottky implies that the peak voltage in the average signal is about 4 times larger than the rms voltage from the Schottky signal.

Since the average is significantly larger than the rms, the bulk of the coherent signal is due to the time average bunch shape. Frequency domain data confirm this conjecture with the coherent power occuring right at revolution lines. A significant amount of effort went into developing filtering schemes that could deal with this coherence [7].

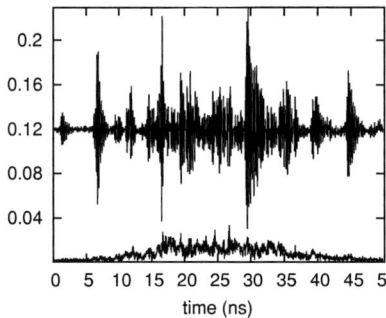
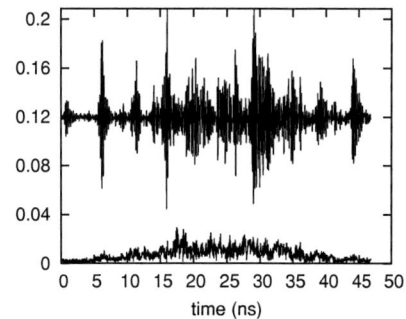

FIGURE 1. Average (top) and standard deviation (bottom) of time domain traces. The average has been offset by 0.12 for clarity. The left panel (A) is for one bunch at 7 different times while the right panel (B) is for 5 diffferent bunches.

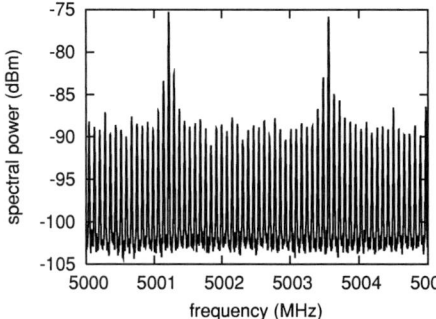
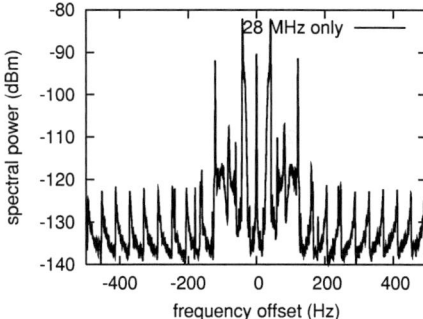

FIGURE 2. High frequency signals for 28 proton bunches in a 30 bunch fill pattern. The broad band signals (left panel, A) exhibit the 2.36 MHz bunching frequency. A narrow band signal centered on the strong right peak is shown in the right panel (B).

COHERENCE IN PROTON BEAMS

Figure 2 shows high frequency signals for 28 proton bunches with a 30 bunch fill pattern and $h = 360$. The $30/T_{rev} = 2.36$ MHz bunching frequency is clearly visible in the left panel. Such signals would be present if the spectrum of the average bunch shape dominated the Schottky signal. However, Fig. 2B shows that the power is dominated by lines centered at $\pm f_{synch}$ where $f_{synch} \approx 40$ Hz is the synchrotron frequency. The contribution from the revolution line is comparable to the contribution from lines at ± 120 Hz. Upon numerical integration of the spectrum one finds that the sum of all three accounts for about 10% of the signal. The true Schottky signal dominates for frequency offsets larger than 200 Hz and accounts for $\lesssim 1\%$ of the total power.

To explain the data in Fig 2 consider the effect of low frequency phase noise, $\phi_0(t)$, in the RF clock. Let $\phi(t)$ denote the RF phase of a proton and ω_{s0} denote the small amplitude, angular synchrotron frequency. In the smooth approximation the single particle equation of motion is

$$\frac{d^2\phi}{dt^2} + \omega_{s0}^2 \sin[\phi - \phi_0(t)] = 0, \quad (1)$$

where ϕ is measured with respect to a perfect clock. Define the position-like coordinate $z = \phi - \phi_0(t)$ and scale time so that $\tau = \omega_{s0} t$. The conjugate momentum is $p = dz/d\tau \equiv \dot z$ and the equations of motion are generated by the Hamiltonian

$$H(z, p, \tau) = p^2/2 + z\ddot\phi_0 - \cos z. \quad (2)$$

Consider action angle variables J and ψ with $z = \sqrt{2J} \sin \psi$ and $p = \sqrt{2J} \cos \psi$ so that

$$H(\psi, J, \tau) = H_0(J) + \ddot\phi_0 \sqrt{2J} \sin \psi + \text{fast terms}, \quad (3)$$

where $H_0(J)$ is the time average, unperturbed Hamiltonian. Now set $\phi_0(\tau) = \alpha \sin(\nu \tau)$, with $\nu \approx 1$. For short bunches and small ϕ_0, $\dot\psi \approx dH_0/dJ \approx 1$, so

$$H(\psi, J, \tau) = H_0(J) - \frac{\alpha \nu^2}{2} \sqrt{2J} \cos(\psi - \nu \tau) + \text{fast terms}. \quad (4)$$

Finally, set $\chi = \psi - \nu \tau$ and drop fast terms yielding

$$H(\chi, J) = H_0(J) - \nu J - \frac{\alpha \nu^2}{2} \sqrt{2J} \cos \chi. \quad (5)$$

We have performed drift-kick simulations of eq.(1) with α growing slowly from zero. Stroboscopic averaging of the phase space density at the drive frequency is compared with a contour plot of eq(5) in Fig.3. Since $\alpha = 10^{-3} \ll 1$, the difference between ϕ and z is too small to see on the plot. Similar agreement has been observed with amplitude modulation of the RF voltage [14, 15].

The stroboscopically averaged phase space density exhibits a plateau in the region bounded by the crescent shaped separatrix, which is generic in plasma physics [16]. The basic idea is that particle motion is adiabatic except for very near the separatrix. As α grows so does the separatrix. Particles cross the separatrix from both small and large actions with the net result being a flattening of the coarse grained distribution. Following this reasoning one also expects a discontinuity of the phase space density at the separatrix. These discontinuities in the phase space will produce line densities with singular derivatives (like $\sqrt{x}$ for x near zero) providing signal power up to high frequencies.

Since the synchrotron frequency is about 40 Hz it is unlikely that a single frequency source is responsible for the coherence apparent in Fig 2. Also, changing the synchrotron frequency shifts the peaks, suggesting a fairly wide band source of noise. For this case we take

$$\phi_0(t) = \int_{-\infty}^{\infty} d\Omega \alpha(\Omega) \exp[(\varepsilon - i\Omega)t], \quad (6)$$

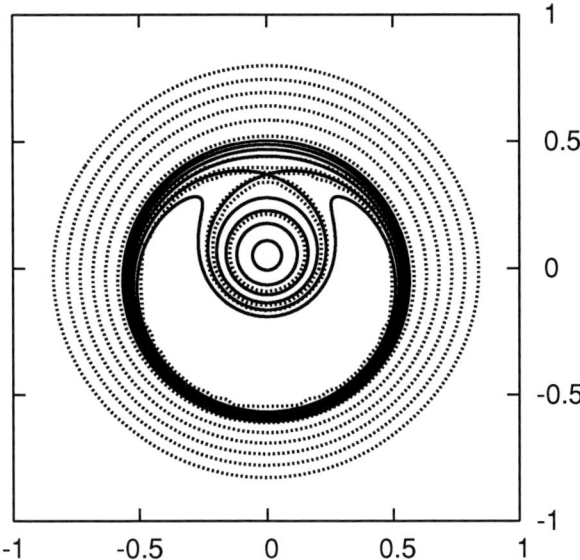

FIGURE 3. Contour plots for the stroboscopic average of the simulated phase space density (dashed lines) and for the approximate Hamiltonian (solid line). The parameters were $v = 0.99$ and $\alpha = 0.001$ with the intital, matched bunch confined to $|\phi| < 1$.

where $\varepsilon > 0$ ensures a causal response and we have gone back to regular time. Consider a single bunch and take the phase space density to be $f(\psi,J,t) = f_0(J) + f_1$. Using first order perturbation theory on the Vlasov equation yields

$$f_1(\psi,J,t) = \int_{-\infty}^{\infty} d\Omega \sum_{m=\pm 1} \frac{\exp[im\psi + (\varepsilon - i\Omega)t]}{\varepsilon - i\Omega + im\omega_s(J)} \Omega' u(\Omega) \sqrt{J/2\omega_{s0}} \frac{df_0}{dJ}, \qquad (7)$$

where $\omega_s(J)$ is the angular synchrotron frequency as a function of amplitude.

Suppose there is a pickup at azimuth $\theta = \omega_0 t - \phi(t)/h = 0$, where ω_0 is the revolution frequency. The beam current at the pickup due to the perturbation is

$$I(t) \approx Q\omega_0 \int dJ d\psi f_1(\psi,J,t) \delta_p(\omega_0 t - z(\psi,J)/h) \qquad (8)$$

where f is normalized to unity, δ_p is the periodic delta function, and Q is the bunch charge. In eq.(8) we have neglected the broad band signal due to timing modulations of $f_0(J)$ and left out the revolution lines. The power spectrum is

$$< |\tilde{I}(\omega)|^2 > = \left\langle \left| \int_0^T \frac{dt}{\sqrt{T}} \exp(i\omega t) I(t) \right|^2 \right\rangle. \qquad (9)$$

We insert eq(8) into (9), and take the ensemble average using

$$< \alpha(\Omega_1)\alpha^*(\Omega_2) > = P_\alpha(\Omega_1)\delta(\Omega_1 - \Omega_2)$$

where the $*$ denotes complex conjugate. Next we take the limit $T \to \infty$ and correct for bunch fill pattern yielding

$$< |\tilde{I}(\omega)|^2 > = \sum_{k=-\infty}^{\infty} |I_k|^2 \frac{P_\alpha(\omega - k\omega_0)}{4\omega_{s0}^2} \left| \int dJ \frac{df_0}{dJ} \sum_{m=\pm 1} \frac{m(\omega - k\omega_0)^2 \hat{\phi} J_1(k\hat{\phi}/h)}{\varepsilon - i(\omega - k\omega_0) + im\omega_s(J)} \right|^2 \quad (10)$$

where $\hat{\phi} = \sqrt{2J/\omega_{s0}}$, I_k is proportional to the Fourier component of the charge distribution for the actual fill, and J_1 is the Bessel function. Equation(10) is simply the square of the dipole beam transfer function. Sextupole and higher order modes are not present due to the approximation in (3).

Both equation (10) and the phase space densities implied by Fig. 3 would yield qualitatively similar power spectra to those shown in Fig. 2. Since the phase space density in (7) is singular it is likely that the very high frequency behavior of (10) will be an overestimate. On the other hand, the phase space density of Fig. 3 would yield a delta function spectrum for the coherent lines. Perhaps letting $\varepsilon \to \varepsilon_k[P_\alpha]$ in (10) would improve things, but space does not allow.

ACKNOWLEDGMENTS

We thank Gene Raka for helpful conversations. This work was performed under the auspices of the United States Department of Energy.

REFERENCES

1. J.M. Brennan, M. Blaskiewicz, P. Cameron, J. Wei, 308, EPAC02
2. M. Blaskiewicz, J.M. Brennan, P. Cameron, J. Wei, PAC03, p394.
3. M. Blaskiewicz, J.M. Brennan, J. Wei, EPAC04, p2861, (2004).
4. J.M. Brennan, M. Blaskiewicz, J. Wei, NIMA, **532**, p335, (2004)
5. M. Blaskiewicz, C-AD note #189 (2005).
6. M. Blaskiewicz, PAC05,toac003.
7. J.M. Brennan, *these prodeedings*.
8. D. Boussard, Proc. Joint US-CERN Acc. School, Texas, 1986, Lecture Notes on Phys. 296, p289
9. G. Jackson, PAC91, p2532, 1991.
10. G. Jackson, J. Marriner, D. McGinnis, R. Pasquinelli, D. Peterson, Workshop on Advanced Beam Instrumentation, Tsukuba, p312, (1991).
11. R.J. Pasquinelli, PAC95 Dallas, p2379 (1995).
12. F. Caspers, D. Mohl, XVII International Conference on High Energy Accelerators, Dubna, Russia, (1998). also CERN/PS 98-051 (DI).
13. G. Jackson, Workshop on Beam Cooling, Montreux, 1993, CERN 94-03, (1994).
14. M. Bai, K.A. Brown, W. Fischer, T. Roser, N. Tsoupas, J. van Zeijts, PRSTAB, **3**, 064001 (2000).
15. K.A. Brown, M. Bai,W. Fischer, T. Roser, PRSTAB, **4**, 014001, (2001).
16. R.Z. Sagdeev and A.A. Galeev, "Nonlinear Plasma Theory", Benjamin, (1969).

Applications of Schottky Spectroscopy at the Storage Ring ESR of GSI

F. Nolden*, K. Beckert*, P. Beller*, B. Franzke*, V. Gostishchev*,
C. Kozhuharov*, Y. A. Litvinov*, A. Schwinn* and M. Steck*

GSI, Darmstadt, Germany

Abstract. Schottky spectroscopy is widely used at the ESR storage ring for the diagnosis of longitudinal and transverse beam parameters. Furthermore, it plays a decisive role for the precision mass measurements of rare isotope nuclei as well as for lifetime measurements of these nuclei. These measurements are performed at low-intensity beams with longitudinal ordering. The paper discusses theoretical preliminaries of Schottky spectra, hardware considerations and aspects of digital data analysis, especially with respect to time-resolved measurements of very low-intensity rare isotope beams.

Keywords: Storage rings, Beam characteristics
PACS: 29.20.Dh, 29.27.Fh

BASIC PRELIMINARIES

Coasting Beam Pick-up Response

An introduction to the theory of Schottky spectra is presented in [1]. Schottky signals arise from both coasting and bunched beams. Here we restrict ourselves to coasting beams.

The beam current density in a coasting beam is

$$j(x,y,t) = qe \sum_{n=1}^{N} \sum_{m=-\infty}^{\infty} \delta(x-x(n,m))\delta(y-y(n,m))\delta(t-t_n-mT_n) \quad (1)$$

The beam consists of N particles. Each of them has a constant revolution period T_n. Its longitudinal position in the storage ring is parametrized by a constant t_n.

Betatron motion and the transverse position are described by

$$x(n,m) = D(s_p)\frac{\delta p}{p} + \sqrt{2J_x\beta_x(s_p)} \sin\left(\mu_{x0}(n) + \mu_x(s_p) + 2\pi mQ_x\right) \quad (2)$$

$$y(n,m) = \sqrt{2J_y\beta_y(s_p)} \sin\left(\mu_{y0}(n) + \mu_y(s_p) + 2\pi mQ_y\right) \quad (3)$$

It is assumed that the pick-up is located at some azimuthal position s_p and that the effect of the azimuthal variation of the lattice functions is negligible. J_{xy} is the constant transverse action, $\beta_{xy}(s_p)$ is the betatron function, $\mu_{xy0}(n)$ describes the betatron phase at revolution $m = 0$, $\mu_{xy}(s_p)$ is the betatron phase advance at position s_p, and Q_{xy} is the number of betatron oscillations per turn. For the sake of simplicity, we have assumed zero chromaticity. The measurement of chromaticity in transverse Schottky spectra is concisely discussed in [2].

The signal arising from this current density depends on the sensitivity $S(x,y,t)$ of the Schottky pick-up. If the pick-up is a linear device, and if the output port is a coaxial or other TEM wave port of line impedance Z_l, then the output signal due to the current density j is a voltage

$$U(t) = Z_l j(x,y,t) * \frac{S(x,y,t)}{2} \qquad (4)$$
$$= \frac{qeZ_l}{2} \sum_{n=1}^{N} \sum_{m=-\infty}^{\infty} S(x(n,m),y(n,m),t) * \delta(t-t_n-mT_n)$$

where $*$ denotes a convolution in the time domain. Obviously, $S(\ldots,t)$ describes the 'ringing' of the pick-up in the time domain when excited by a δ-formed pulse. The nature of the detected Schottky signal depends very much on the transverse dependence of S. A clear exposition of this relationship is given in [3] by Bisognano and Leeman.

Longitudinal Pick-up

In order to simplify the discussion we restrict ourselves to two idealized cases. The first case deals with a pick-up the sensitivity of which does not depend on the transverse coordinates $S(x,y,t) = S^L(t)$. In this case, the Fourier transform of (4) becomes

$$U^L(\Omega) = \sum_{n=1}^{N} \frac{Z_l qe\omega_n}{2} \sum_{m=-\infty}^{\infty} S^L(\Omega) e^{-i\Omega t_n} \delta(\Omega - m\omega_n) \qquad (5)$$

where $\omega_n = 2\pi/T_n$ is the angular revolution frequency. The spectrum consists of lines at each harmonic of the revolution frequency. In a coasting beam, the phases $\Omega_n t_n$ are random.

Transverse Pick-up

The simplest model for a horizontal pick-up is $S(x,y,t) = xS^H(t)$. At a dispersion-free location ($D(s) = 0$), this yields a spectrum

$$U^H(\Omega) = \sum_{n=1}^{N} \frac{Z_l qe\omega_n}{4} \sum_{m=-\infty}^{\infty} \sqrt{2J_x \beta_x(s)} S^H(\Omega) e^{-i\Omega t_n} \qquad (6)$$
$$\times \left[e^{-i(\mu_{x0}(n)+\mu_x(s))} \delta(\Omega - (m+Q_x)\omega_n) + e^{+i(\mu_{x0}(n)+\mu_x(s))} \delta(\Omega - (m-Q_x)\omega_n) \right]$$

The spectrum consists of lines at the betatron sidebands $(m \pm Q_x)\omega_n$ of the revolution frequency. In addition to the random phases which occur in the longitudinal spectrum, there is an additional random phase due to the initial betatron oscillation phase $\mu_{x0}(n)$.

In contrast to the longitudinal response, the transverse response is proportional to the betatron amplitude at the location of the pick-up.

In the case of non-zero dispersion, there is an additive contribution to the spectrum at the longitudinal harmonics which can be written by replacing S^L by $S^H D(s)\delta p/p$ in (5). Such signals are useful in stochastic cooling if the Palmer method is applied.

Revolution Frequency of Multi-Species Heavy Ion Beams

The longitudinal spectrum depends on the distribution of revolution frequencies $f = \omega/2\pi$ in the beam. We assume in the following heavy ion beams, which can consist of several different species characterized by their mass number A and their charge state Q. Because of the following expansion for the magnetic rigidity

$$\frac{\delta(B\rho)}{B\rho} = \frac{\delta(\beta\gamma)}{\beta\gamma} + \frac{\delta(A/Q)}{(A/Q)} \tag{7}$$

we use in the following the abbreviations

$$\delta = \frac{\delta(\beta\gamma)}{\beta\gamma} \tag{8}$$

$$r = \frac{\delta(A/Q)}{(A/Q)} \tag{9}$$

Note that in one-component beams one usually writes $\delta p/p$ instead of $\delta(\beta\gamma)/\beta\gamma$. We investigate the relative difference in revolution frequencies $f = \beta c/C$, where C is the orbit circumference. Relative changes of the latter depend on the momentum compaction factor $\alpha_p = C^{-1} \oint ds\, D(s)/\rho(s)$. In addition to changes of the magnetic rigidity of the particles, the orbit circumference is influenced by field ripple. We assume that there is an 'effective' overall ripple $\delta B/B$ which affects all the storage ring magnets. Then the relative frequency differences are

$$\frac{\delta f}{f} = \frac{\delta \beta}{\beta} - \frac{\delta C}{C} = \gamma^{-2}\delta - \alpha_p\left(\frac{\delta(B\rho)}{B\rho} - \frac{\delta B}{B}\right) = \eta\delta - \alpha_p\left(r - \frac{\delta B}{B}\right) \tag{10}$$

where $\eta = \gamma^{-2} - \alpha_p$ is the frequency dispersion.

The spectral information at a given frequency is not always unambiguous. This situation is called Schottky band overlap.

Single species longitudinal spectra are a first example. If the δ spectrum of the beam is limited by $\pm\delta_{max}$, the extreme revolution frequencies are denoted by f_+ and f_-. Schottky overlap begins if $(m+1)f_- = mf_+$. This occurs if

$$m \geq \left[(\eta\delta)^{-1} - 1\right]/2 \approx \frac{1}{2\eta\delta} \tag{11}$$

Limits for the mutual overlap between transverse and longitudinal spectra can be found by analogous reasoning. For multi-species heavy ion beams it is important to calculate the overlap limit of two species with a relative A/Q difference r. We assume that both species have the same width $\pm\delta$. Then two peaks overlap if

$$r < 2\left(|\eta\delta| + \left|\alpha_p \frac{\delta B}{B}\right|\right) \tag{12}$$

This condition does not depend on the harmonic number.

Power Spectra

Statistical Properties of Coasting Beam Power Spectra

Due to the random phases in (5) and (6) the expectation value of the voltages U_L and U_H vanishes. However, the power spectrum is non-zero. The power spectrum of a voltage U is defined as the expectation value $\langle U(\Omega)U(\Omega')\rangle$. In a coasting beam, U is a so-called stationary process (this is discussed in [1]. Stationary processes are discussed in textbooks on signal theory). In stationary processes, one has $\langle U(\Omega)U(\Omega')\rangle = C_U(\Omega)\delta(\Omega+\Omega')$ (Wiener-Khintchine-theorem) and it is sufficient to deal with C_U. On the other hand, C_U is the Fourier transform of the autocorrelation function $R(\tau) = \langle U(t+\tau/2)U(t-\tau/2)\rangle$, which is independent of t if U is stationary.

Longitudinal and Transverse Power Spectra

The coasting beam is characterized by distributions $\Psi_{A,Q}^{3D}(\delta, J_x, J_y)$, which are normalized to the numbers $N(A,Q)$ of each species (characterized by A and Q) stored in the ring. In the context of our zero-chromaticity model, the Schottky power spectra depend on the integral density

$$\Psi_{A,Q}(\delta) = \int dJ_x dJ_y \Psi_{A,Q}^{3D}(\delta, J_x, J_y) \tag{13}$$

and on the mean emittance per rigidity

$$\langle J_{x,y}\rangle(\delta) = \frac{\int dJ_x dJ_y J_{x,y} \Psi^{3D}}{\int dJ_x dJ_y \Psi^{3D}} \tag{14}$$

In order to simplify the writing, we restrict the discussion to the case of vanishing Schottky overlap. The ideal longitudinal spectrum can be derived from (5) under the assumption that the two-particle correlation function is a delta function [5] in all phase space variables.

$$C_L(m\omega) = \frac{(Z_l q e)^2 \omega}{8\pi |m\eta|} |S^L(\omega_n)|^2 \Psi_{A,Q}(\delta) \tag{15}$$

where ω is related to δ via (10). If there is overlap, all the possible contributions of the type (15) have to be added.

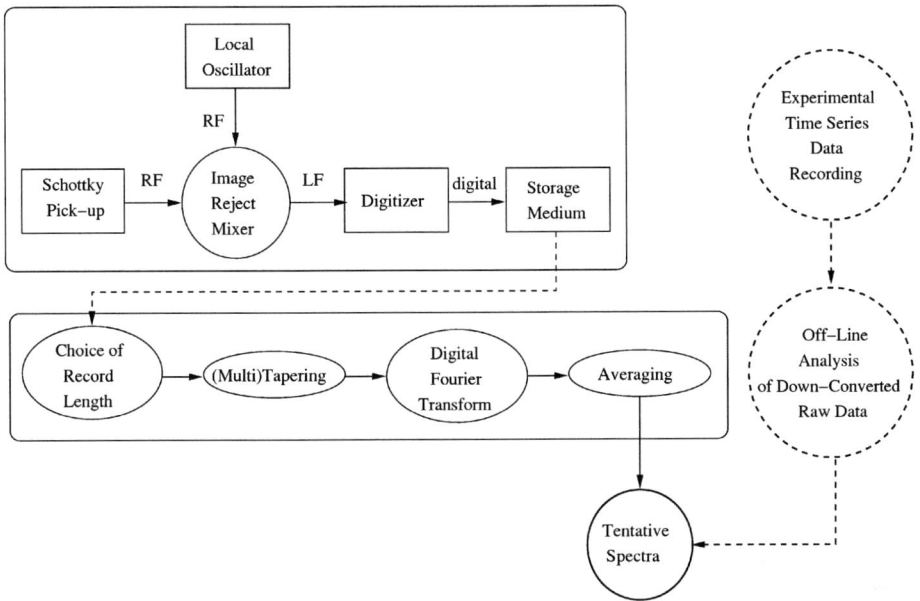

FIGURE 1. Signal processing with off-line analysis.

The ideal transverse spectrum from (6) has contributions at the sidebands of the revolution harmonics:

$$C_H((m \pm Q_x)\omega) = \frac{(Z_l q e)^2 \omega \beta_x(s) \langle J_{x,y} \rangle (\delta)}{32\pi |m\eta|} \left| S^H((m \pm Q_x)\omega_n) \right|^2 \Psi_{A,Q}(\delta) \quad (16)$$

RF Signal Processing

We discuss in the following some aspects of digital spectrum analysis. Usually Schottky spectra are recorded at one definite harmonic. A stable local oscillator with low phase noise and low spurious sidebands is needed if a high frequency resolution is required. An image reject mixer is used for down-converting the signal to the low-frequency regime (typically dc to maximum 500 kHz). A good mirror frequency rejection is needed in order to get unambiguous signals and to reduce the noise level. The low frequency signal should be sampled by use of analog-to-digital converters with an accuracy of 12 bits or higher. In order to attain flexibility in the evaluation of the data, it is beneficial to store the converted data on an adequate storage medium before transforming into the frequency domain. This excludes the application of many commercial systems.

Digital Spectral Estimation

If one applies the usual DFT technique for estimating the Schottky spectra, important parameters have to be set:

1. the time T_s between two samples
2. the record length $T_r = N_r T_s$, where N_r is the number of samples per record.

In order to avoid sidelobes, it is important to choose some *window* or *taper* function w_n. Instead of transforming the record a_n of digitiyed data directly, one transforms the sequence $a_n w_n$ ($0 \leq n < N_r$). Commercial analyzers often offer many different tapers, e.g. the Hanning (named after the engineer *von Hann*), Hamming or Blackwell tapers.

As the tapers decrease the efficient record length, one often applies overlapping time records with typically 50 % overlap. If the signal is noisy, it is beneficial to average over N_{avg} such spectra (overlapping sample averaging).

Multitapering [4] is a more recent algorithmic technique, which is used successfully in scientific fields where spectral analysis is confronted with the problem of large underlying noise (geoseismology, helioseismology, gravitational waves, neurophysiology). The original technique is based on DPSS (*D*iscrete *P*rolate *S*pheroidal *S*equences) tapers. These orthogonal function families have analytic properties which make them unique for minimizing the power in the sidelobes. Therefore it is advisable to apply multitapers to Schottky spectra, as well.

The DFT yields a frequency sequence f_n ($0 \leq n < N_f$), where the number of points N_f in the frequency domain is not only limited by the Nyquist limit $N_f \leq N_r/2$, but is still smaller due to the windowing procedure, e.g. $N_f = N_r/2.56$.

The frequency resolution depends only on the record size, with windowing it is roughly $\delta f = 2.56/T_r$. This relation is most important for the choice of the right Schottky harmonic. As the frequency difference between Schottky lines incrases linearly with the harmonic number, the record length needed to separate different Schottky lines is inversely proportional to the harmonic number. However, the power density has exactly the same dependence. Therefore one needs to ascertain a minimum signal to noise ratio at very high harmonics. This can be done by using resonant structures. Their practical high freqency limit appears to be related to the cut-off frequency of the vacuum chamber.

The maximum frequency $f_{max} = f_s/2.56$ depends only on the sampling rate.

At GSI, both commercial analyzers and a dedicated system for acquiring very long time sequences have been used, mainly for the purpose of Schottky Mass Spectrometry. The off-line spectral analysis enhances strongly the versatility of the choice of the number of averages and of the record length.

SCHOTTKY MASS SPECTROMETRY OF COLD HEAVY ION BEAMS

Mass Measurements

An important application of longitudinal Schottky spectra is the determination of nuclear masses [6] [7] and the study of decay properties [8] [9].

If there is a mixture of fragments with a relative spread in δ and a relative mass over charge difference r (see (8) and (9)), then their relative frequency deviation is given by (10).

There are two principle methods to use this relationship in order to determine r experimentally

1. make δ as small as possible by cooling.
2. make η as small as possible by tuning the storage ring exactly on transition over the full range of accepted orbits filled by the injection process [10] [11].

In any case, the power supply ripple has to be very small.

Both methods are being applied at the storage ring ESR of GSI. The second method, however, uses presently a time of flight technique rather than a Schottky spectrometry method and is therefore not discussed here.

With the first method [12] the utmost precision is reached by detecting electron-cooled low intensity beams, the properties of which indicate longitudinal ordering [13] – [16].

With these beams, the fwhm width $\delta f/f$ of single lines is typically $4 \cdot 10^{-7}$. The position of these lines can be determined with even better accuracy. The 1σ error of this procedure is dominated by systematic errors, amounting to typical values of 28 keV for nuclei with masses between 79 and 206. The relative accuracy in the latter case is $1.5 \cdot 10^{-7}$.

It is well worth noting that these results depend strongly on the stability of the magnet power supplies in the ESR. From (10) a current ripple amplitude below $1 \cdot 10^{-6}$ for $\alpha_p \approx 0.18$ can be inferred.

Time Resolved Schottky Spectra

Furthermore, the measurement of spectral lines as a function of time allows to derive nuclear decay properties. The temporal resolution of these spectra is closely linked to the required spectral resolution and to the signal-to-noise ratio of the spectral lines. Let us assume that one uses N_{avg} overlapping samples with a taper as discussed above. Then a spectral resolution δf requires an overall measurement time of $T_{\text{avg}} \approx 1.28 N_{\text{avg}}/\delta f$. In order to avoid spreading the Schottky energy over too large a frequency band, it is preferable to resolve the relative width $4 \cdot 10^{-7}$ as it is available. Measurements at a frequency of 60 MHz (as in the ESR) therefore yield $T_{\text{avg}} \approx N_{\text{avg}} \times 46$ ms.

Figure 2 shows an example of the time resolved measurement of the electron capture decay of $^{140}\text{Pr}^{58+}$ in the ESR. The daughter nucleus $^{140}\text{Ce}^{58+}$ is stored in the storage ring, as well. Up to the difference in binding energy (3.388 MeV), it has the same

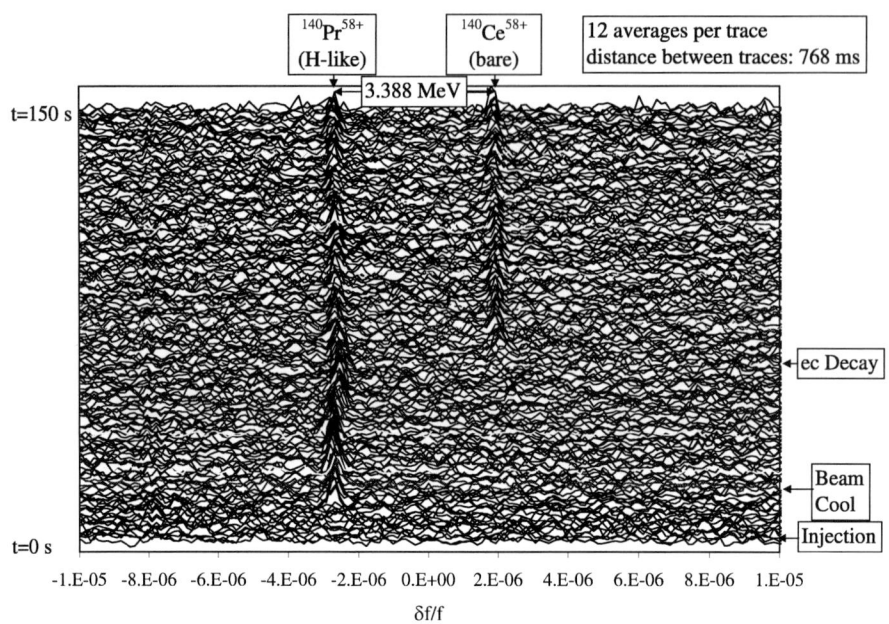

FIGURE 2. Fate of two praseodyme ions (waterfall plot with 12 overlapping averages per record).

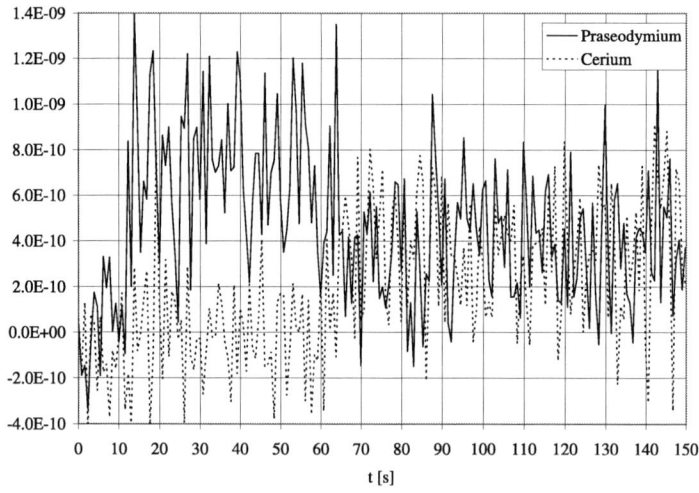

FIGURE 3. Schottky area corresponding to mother and daughter nuclei after background subtraction.

mass over charge ratio m/q as the mother nucleus. The Schottky lines of both species are separated by $\delta f/f \approx 5 \cdot 10^{-6}$, about an order of magnitude above the individual linewidth.

The record length of these measurements was 128 ms; applying 12 averages with 50 % overlap one gets consecutive spectra at a distance of 768 ms. After injection at $t = 0$ stochastic cooling was used for 8 s followed by electron cooling. After about 13 seconds, a line appears which is due to *precisely two praseodymium ions*. The following convincing arguments support this interpretation:

1. It agrees with the value one gets from scaling the Schottky area of a beam which is sufficiently intense to be measured simultaneously by means of a beam current transformer.
2. There is an obvious decay of one of the mother nuclei. The decay is accompanied by the sudden and simultaneous emergence of the daughter nucleus. The areas of both peaks are the same and they are half as large as the original line with two mother nuclei.

Figure 3 illustrates these considerations.

Although beams with one-dimensional ordering obviously offer unique experimental opportunities, their Schottky spectra deserve theoretical investigations not yet performed so far. An interesting problem is the statistical description of the azimuthal particle distribution. One should not misunderstand one-dimensional ordering as a regular string of equally spaced particles. Ordering is due to the Coulomb force at short distances, but most of the time the particles move almost freely on their closed orbit. Therefore it is appears to be worthwhile investigating whether there are differences between ordinary Schottky spectra and those of one-dimensional ordered beams.

REFERENCES

1. S. Chattopadhyay, Some fundamental aspects of fluctuations and coherence in charged-particle beams in storage rings, *CERN Yellow Report* **84-11** (1984).
2. F. Nolden, Instrumentation and Diagnostics Using Schottky Signals, *Proc. DIPAC 2001*, 6-10 (2001)
3. J. Bisognano, C. Leemann, "Stochastic Cooling", in *Physics of High Energy Accelerators*, Fermilab School 1981, *AIP Conf. Proc.* **87** (1982).
4. D.B. Percival and A.T. Walden, *Spectral Analysis For Physical Applications: Multitaper And Conventional Univariate Techniques*, Cambridge University Press, Cambridge, 2002.
5. F. Nolden, Zur stochastischen Vorkühlung am ESR, *PhD Thesis Technische Universität München* (1996)
6. T. Radon et al., Schottky Mass Measurements of Stored and Cooled Neutron-Deficient Projectile Fragments in the Element Range of $57 \leq Z \leq 84$, *Nucl. Phys. A*, **677** (2000), 75-99.
7. Yu.A. Litvinov et al., Mass Measurements of Cooled Neutron-Deficient Bismuth Projectile Fragments with Time-Resolved Schottky Mass Spectrometry at thr FRS-ESR Facility *Nucl. Phys. A*, **756** (2005), 3-38.
8. M. Jung et al., First Observation of Bound State β^- Decay, *Phys. Rev. Lett.* **69** (1992) 2164-2167
9. F. Nolden et al., Half-Life Measurement of the Bound State Beta Decay of $^{187}\text{Re}^{75+}$, *Nucl. Phys. A*, **621** (1997), 297c-304c.
10. M. Hausmann et al., First isochronous mass spectrometry at the experimental storage ring ESR, *Nucl. Inst. Meth. A*, **446** (2000), 569-580.
11. J. Stadlmann et al., Direct mass measurement of bare short-lived ^{44}V, ^{48}Mn, ^{41}Ti, and ^{45}Cr ions with isochronous mass spectrometry, *Phys. Lett. B*, **568** (2004), 27-33.

12. B. Schlitt et al., Schottky mass spectrometry at the ESR, a novel tool for precise direct mass measurements of exotic nuclei *Nucl. Phys. A*, **626** (1997), 315c-325c.
13. M. Steck et al., Anomalous Temperature Reduction of Electron-Cooled Heavy-Ion Beams in the Storage Ring ESR, *Phys. Rev. Lett.*, **77** (1996), 3803-3806.
14. M. Steck, Diagnostic methods to detect the properties of cooled heavy-ion beams in storage rings, *Nucl. Phys. A*, **626** (1997), 473c-483c.
15. M. Steck et al., New evidence for one-dimensional ordering in fast heavy ion beams, *J. Phys. B: At. Mol. Opt. Phys*, **36** (2003), 991-1002.
16. M. Steck et al., Electron Cooling Experiments at the ESR, *Nucl. Inst. Meth. A*, **532** (2004), 357-365.

Pick-Up and Kicker Electrodes for the CR

C. Peschke*, F. Nolden* and L. Thorndahl[†]

*Gesellschaft für Schwerionenforschung, Planckstraße 1, D-64291 Darmstadt, Germany
[†]CERN European Organization for Nuclear Research, CH-1211 Genïɟe 23, Switzerland

Abstract. The collector ring (CR) of the proposed GSI project FAIR includes a fast stochastic cooling system for exotic nuclei and antiprotons. To reach a good signal to noise ratio of the pick-up even with a low number of particles, a novel pick-up and kicker electrode system based on slotlines is presented. The sensitivity and noise properties of electrode models are calculated. These are compared with other types of electrodes. Different options for signal processing and layout of a pick-up or kicker with many electrodes for different beam velocities are discussed.

Keywords: heavy ions, storage rings, stochastic cooling
PACS: 29.20.Dh, 41.75.-i

SLOTLINE ELECTRODE DEVELOPMENT

The main task of the collector ring (CR) is stochastic cooling of rare isotope beams (RIBs, $\beta = 0.83$) and antiprotons ($\bar{p}$, $\beta = 0.97$). The CR should achieve a phase space volume reduction of $1.6 \cdot 10^4$ in 5 s for $\bar{p}$ and $1.3 \cdot 10^6$ in 1 s for RIBs. Therefore the pick-ups must have a large bandwidth and a high S/N ratio in spite of their large aperture of $120 \cdot 120$ mm^2. A new planar electrode is developed [1] to meet these requirements.

The electrode consists of a slotline perpendicular to the beam (Fig. 4, top, left), and a microstrip circuit on the rear side of the planar substrate (top, right). The mirror currents induce traveling waves in both directions of the slotline. At $\lambda/4$ from the end of the slotline, the signal is coupled out to the microstrip line. The $\lambda/4$-section at the beginning of the microstrip is a virtual short to one of the two conductors of the slotline. An alternative is a direct feedthrough. Both constructions have different frequency responses. With feedthrough one gets one transmission maximum. The $\lambda/4$ microstrip can be dimensioned to have two slightly lower maxima and a flatter frequency response. A Wilkinson combiner transforms both microstrip lines to a single output.

An important performance issue is the selection of the substrate and the type of the slotline. The substrate must be UHV compatible and must have good dielectric properties. Hence, only few materials are suitable. Two possible choices are alumina (Al_2O_3) or just vacuum. The substrate has to fulfill different conflicting demands. The slotline impedance should be high and the field should not be too much concentrated in the gap. This requires a wide gap and a thin substrate. On the other hand, the microstrip line should have a high impedance requiring a thick substrate. Slotline and beam width must be comparable, leading to a low effective dielectric constant (ε_{eff}).

An ideal slotline consists of a narrow gap in the conductive coating of an infinitely wide dielectric. In practice, the conductors are finite and connected by a metal shielding. With this, the line acts more like a loaded waveguide with a non-zero cut-off frequency

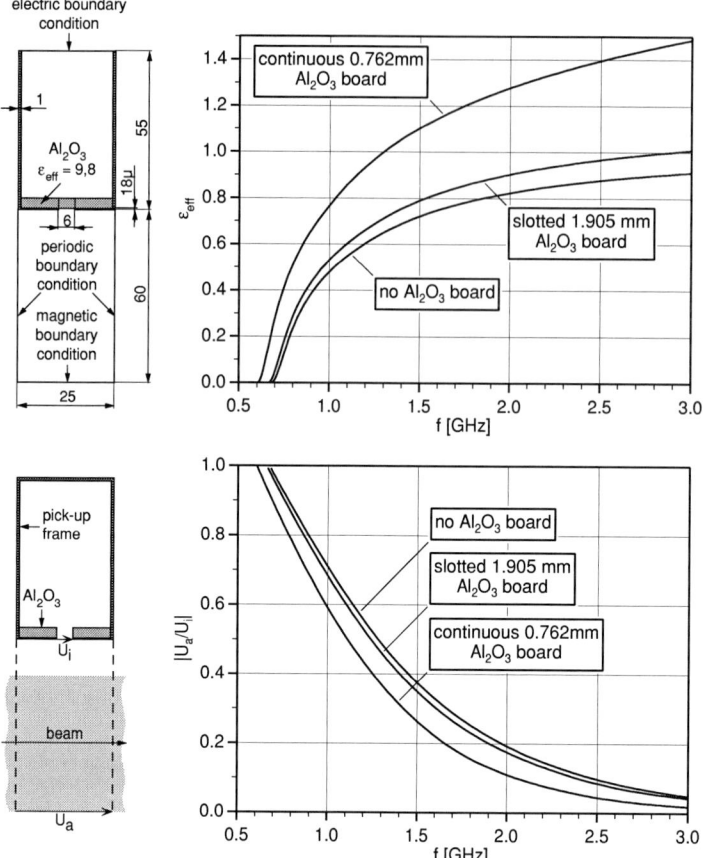

FIGURE 1. Effective dielectric constant of different pick-up lines versus frequency (top) and voltage in the middle of the beam relative to the voltage at the electrode surface versus frequency (bottom).

and a higher dispersion. Fig. 1 shows properties of different line types. The numeric calculations were done using an eigenmode solver [2]. The height of the shielding is 55 mm, resulting a cut-off frequency below the operating band (1-2 GHz). The distance from the slotline to the center of the beam pipe is 60 mm. Symmetrically driven electrodes on each side are simulated by a magnetic boundary condition.

The vacuum line has the lowest ε_{eff} and the lowest field concentration in the gap, but an ambitious mechanical construction. The second line uses a thin contonous Al_2O_3-substrate. This line is easy to manufacture, but the ε_{eff} as well as the field concentration are much higher. The third line as compromise of the above uses a slotted Al_2O_3-substrate. The ε_{eff} and the field concentration resembles the vacuum line.

We have analysed the behavior of the slotline electrodes and compared its performance with alternative designs. A major result is that the novel electrodes (vacuum line,

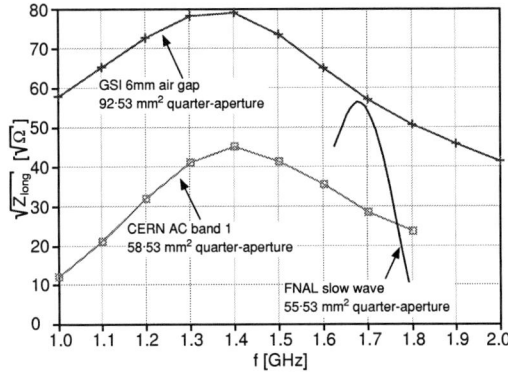

FIGURE 2. Comparison between different longitudinal kicker electrodes.

direct feedthrough) are superior to $\lambda/4$-superelectrodes and slow-wave couplers. Fig. 2 shows the square root of the longitudinal impedance (for a beam in chamber center, and longitudinal kicker mode) of a 2 m long array of 80 slotline electrodes, a 2 m long array of 23 CERN AC band 1 superelectrodes, and a FNAL slow-wave structure scaled to 80 cells and a length of 32.5 cm without coupler. The $\lambda/4$-superelectrodes have a similar bandwidth, but a lower impedance. The advantage of the slow-wave structure is simplicity without combiners and β-switches. Disadvantages are the need to use five sub-bands with signal processing per octave, and the dependance of the center frequencies from β and plunging position. Plunging is needed due to the large phase space reduction.

SIGNAL COMBINATION AND NOISE TEMPERATURE

The two velocities make it necessary to use switchable delays for the phase consistent combination. In order to reduce the number of switches and feedthroughs, we plan to combine a small group of electrodes with fixed delays, designed for antiprotons with $\beta = 0.97$ and combine the groups outside the vacuum with switchable delays. The RIBs deliver much stronger signals, making a slight performance degradation due to the fixed delay acceptable. As dicussed in [1], the amplitude degradation is only 4.6 % at 1.5 GHz with an electrode distance of 25 mm and eight electrodes per group.

Fig. 3 shows the circuit of a pick-up module with eight electrodes. Simple Wilkinson-combiners with small printed resistors can be used for the signal combination. In the kicker modules, there will be a large power dissipation in the resistors, due to the different delays. Two port power resistors would have much higher parasitic effects. So, the Wilkinson-splitters will be replaced by 180°-hybrids and one port power resistors.

Fig. 4 shows a preliminary pick-up module. The pick-up board with eight slotlines and first combiners is at the bottom. The board above establishes the connection between the pick-up board and the combiner board on top. It also contains a small test signal antenna. The connection board can optionally contain the first low noise amplifier. The module is vertically plungeable to follow the cooled beam. Two BeCu springs establish the thermal

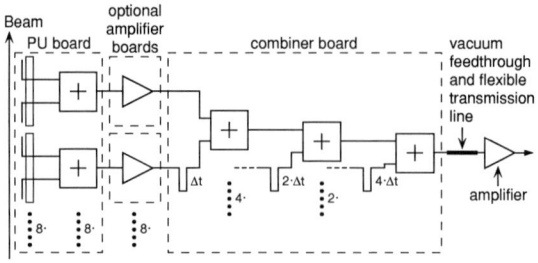

FIGURE 3. RF-circuit of a pick-up module.

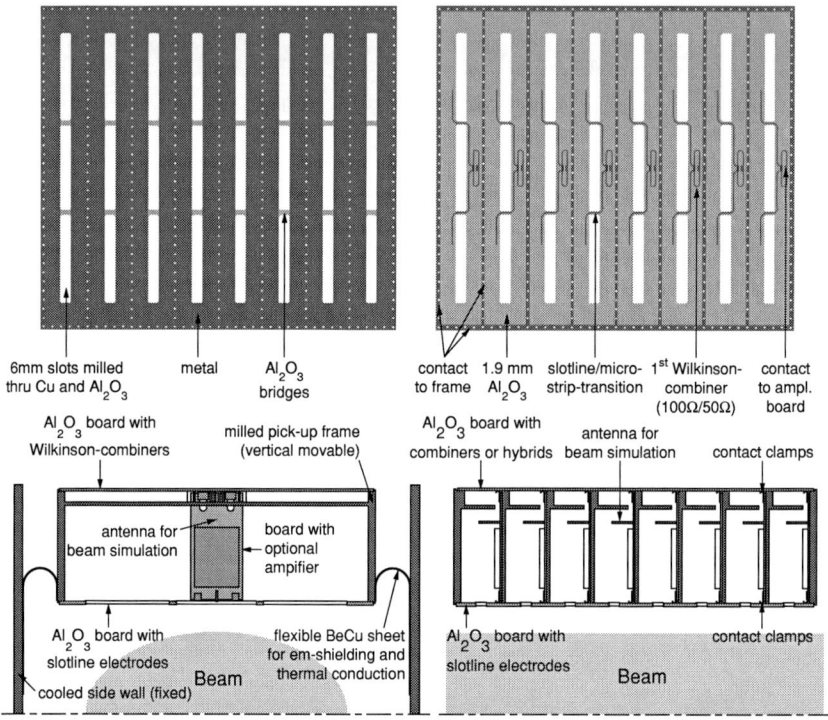

FIGURE 4. A possible layout of a pick-up module.

connection to the fixed cooled side walls and shield the electrodes. An alternative to the mechanically complicatad wide springs could be a simple gap with damping material and narrower springs for thermal conduction. A tank will probably consist of eight modules. Both ends will be completed with damping material and a thermal shielding.

One possibility to reduce the noise is to cool the modules with liquid nitrogen (80 K) or with helium to approximately 30 K. Another possibility is to terminate the slotlines with electrically cold loads. An electrode with the first combiner has a high reflection

TABLE 1. Parameters for effective noise temperature calculations.

transmission of the slotline coupler	-0.92 dB	(≡ 0.9)
transmission of one Wilkinson combiner	-0.2 dB	
transmission of the vacuum feedthrough with flexible line	-0.5 dB	
transmission of the amplifier	10 dB	
noise figure of the amplifier	0.25 dB	($\hat{=}$ 17.2 K)

factor. An amplifier at this point is terminatad with its own electrically cold input. Due to the different delays in Fig. 3, amplifiers behind the module are terminated by combiner resistors instead of their inputs. The disadvantage of these amplifiers is having to be mounted inside the unaccessable vacuum tank. Hence the amplifiers must be indiviually switchable, a failed device decreasing the performance only slightly. To avoid outgassing, the amplifiers have to be hermetically sealed. To compare the different cases, a simple equivalent circuit of the 1-2 GHz pick-up with 25 mm electrode spacing has been calculated, assuming that all components in Fig. 3 have the same temperature, are perfectly matched, and have the parameters listed in Table 1. The result of the calculation is shown in Table 2. The noise temperature with the amplifier outside at 80 K is probably unacceptable. The case with amplifier inside at the same temperature looks much better. A temperature of 30 K with amplifiers inside is slightly worse. With amplifiers inside, the lower temperature does not help as much as with amplifiers outside.

TABLE 2. Comparison of noise temperatures.

	noise temperature	rel. S/N-ratio
amplifiers inside (30 K)	28.1 K	5.4 dB
amplifiers inside (80 K)	36.4 K	4.3 dB
amplifiers outside (30 K)	47.0 K	3.1 dB
amplifiers outside (80 K)	96.6 K	:= 0 dB

SUMMARY AND OUTLOOK

The slotline structures have a satisfying impedance. Both a slotted Al_2O_3 line and a vacuum line are possible. A tank will contain eight modules having eight slots each. The microstrip to slotline transition will be optimised using equivalent circuits. A prototype of a module pair will be build for bench measurements and in the next step for testing in the existing experimental storage ring (ESR) of the GSI. From the point of view of noise temperature, systems with internal amplifiers at 80 K look attractive, however they are challenging in a UHV environment. Further investigations are nessecary.

REFERENCES

1. C. Peschke, F. Nolden, M. Balk: "Planar Pick-Up Electrodes for Stochastic Cooling"; Nuclear Instruments and Methods A 532 (2004), pages 459–464
2. Program "Microwave Studio", CST GmbH, Bad Nauheimer Straße 19, D-64289 Darmstadt, Germany, http://www.cst.de

Fermilab Recycler Stochastic Cooling for Luminosity Production

D. Broemmelsiek* and C. Gattuso*

Fermi National Accelerator Laboratory, Batavia, IL, U.S.A.

Abstract. The Fermilab Recycler began regularly delivering antiprotons for Tevatron luminosity operations in 2005. Methods for tuning the Recycler stochastic cooling system are presented. The unique conditions and resulting procedures for minimizing the longitudinal phase space density of the Recycler antiproton beam are outlined.

Keywords: stochastic cooling, Recycler
PACS: 41.75.Lx

INTRODUCTION

The Fermilab Recycler is a fixed 8 GeV kinetic energy storage ring in the Fermilab Main Injector tunnel near the ceiling. The Recycler's role in the Tevatron Collider program is to store antiprotons from the Fermilab Antiproton Accumulator. Stochastic cooling [1] is used in the Recycler to preserve and cool the antiproton emittances for transfer to the Tevatron. Table 1 gives some relevant parameters of the Recycler storage ring.

Stochastic cooling of bunched beams at the Recycler presents several difficulties. Stochastic cooling is not only limited by intensity and bandwidth, but also the introduced diffusion when the feedback loop is closed. One of the dominant beam-based sources of diffusion in the Recycler is intrabeam scattering. The Recycler bunch length is continuously changed to adjust the momentum width of the beam to minimize IBS diffusion and achieve the required longitudinal emittance necessary for the Tevatron. Therefore, the stochastic cooling system has to remain optimized for a wide variety of beam intensities and currents.

TABLE 1
Recycler Parameters

Circumference	C	3320	m
Momentum	p	8.9	GeV/c
Slippage factor	η	-0.0086	
Emittance(n, 95%)	ε_n	$\sim 3 - 5$	μm
Average β-function	β_{avg}	30	m

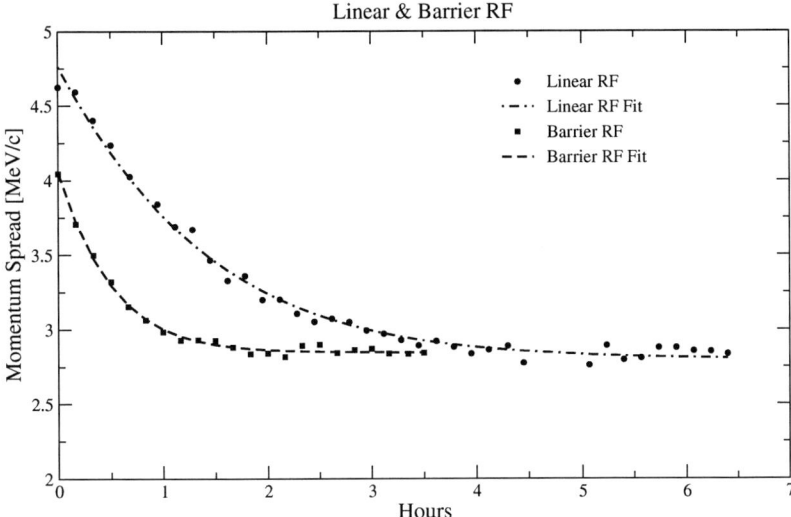

FIGURE 1. Evolution of the beam rms momentum spread for parabolic and barrier RF buckets. Points are data, lines are fits.

STOCHASTIC COOLING LIMITATIONS

The design of the Recycler lattice prohibits the use of Palmer cooling [2]. Filter cooling is implemented in the frequency bands 0.5-1 GHz and 1-2 GHz. Experiments [3] have shown, Figure 1, that for fixed RF and stochastic cooling gain, the asymptotic momentum spread is approximately the same. The asymptotic momentum spread indicates a large diffusion introduced by the stochastic cooling system itself and, more importantly for operations, a limit to the attainable longitudinal emittance for a fixed bunch.

INTRABEAM SCATTERING

Intrabeam scattering is expected to be the dominant beam-based diffusion mechanism for the intense antiproton beams in the Recycler. A detailed IBS model [4] using the measured Recycler lattice functions predicts a momentum spread that minimized the normalized 6-dimensional IBS diffusion rate. Because there is a small non-zero dispersion, there is no momentum spread at which the normalized 6-d diffusion rate is zero. Figure 2 shows normalized diffusion rates for 40×10^{10} particles with a fixed transverse emittance of 3.2π mm mrad.

To verify this IBS model, experiments [5] at the Recycler were preformed to demonstrate sympathetic IBS cooling. The measurements were made with a bunched antiproton beam of 25×10^{10} particles which were initially cooled transversely to a very small emittance ($< 2\pi$ mm mrad). The momentum spread was increased above 4.5 MeV/c by

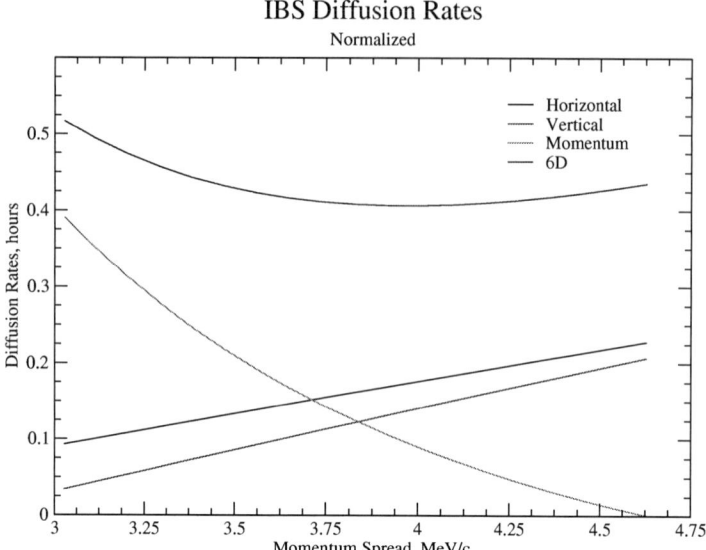

FIGURE 2. Recycler IBS model calculations for normalized diffusion rates.

compressing the beam with the RF. The experiment successfully demonstrated sympathetic momentum cooling.

STOCHASTIC COOLING

Given these limitations, a strategy for attaining the longitudinal emittances required by the Tevatron Collider program is needed. The rms momentum spread is kept constant by decreasing the bunch length. Since the beam is confined inside a barrier-bucket, bunch length is decreased by decreasing the separation of the barrier voltages.

Since the RF beam structure is constantly being adjusted to compensate for the IBS diffusion rate, the stochastic cooling systems also need to be constantly adjusted. As the duty cycle and peak voltages are changing, consistent and appropriate methods for quickly tuning the stochastic cooling systems are needed to keep optimal performance. Because of the arbitrary nature of the beam conditions in the Recycler, a simple table of settings becomes cumbersome.

Avoiding amplifier compression is critical for optimal performance in the Recycler stochastic cooling systems. The signal to noise ratio measurements shown in Figure 3 are done using the open loop Schottky spectral measurement. A side-band analysis is performed to calculate the signal to noise ratio.

Single band signal suppression measurements are used to set the betatron cooling system gain. The peak signal suppression per band suffers from noise induced by the Main Injector ramps. The average of several zero span signal suppression measurements, normalized to the resolution bandwidth used, is fast and gives a low noise result proportional

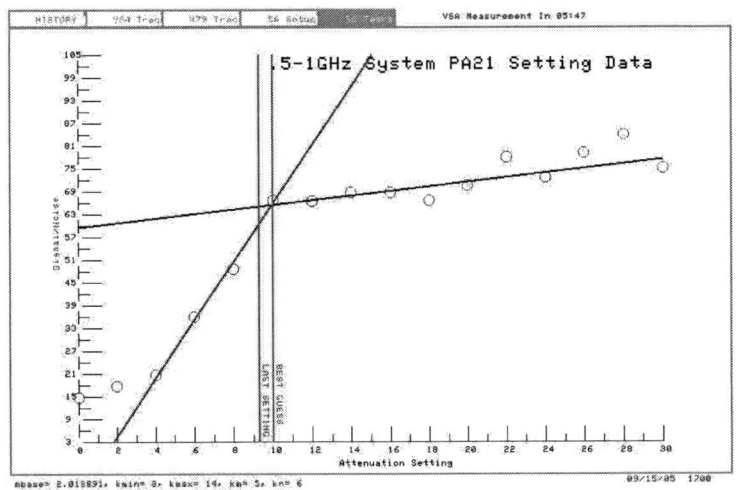

FIGURE 3. Signal-to-Noise measurements used to power balance stochastic momentum cooling bands.

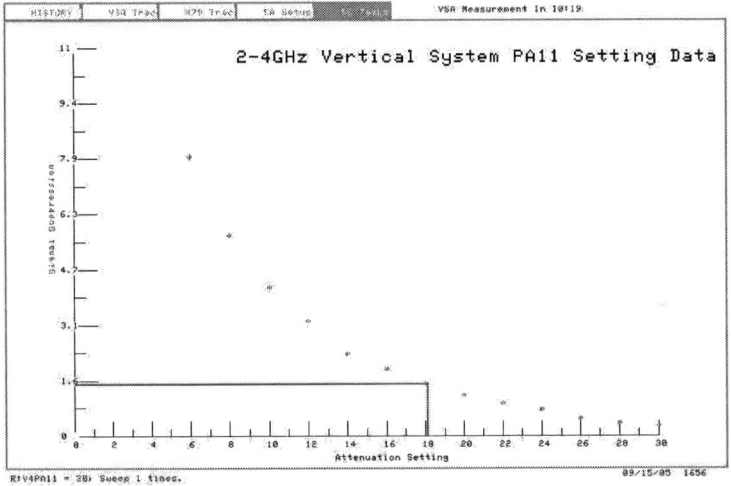

FIGURE 4. Signal suppression measurement versus attenuation setting.

to the peak signal suppression. Figure 4 shows one such measurement.

These measurements allow operations to optimize the stochastic cooling systems for current conditions. Appropriate feedback is provided within minutes, giving operators the opportunity to improve beam conditions rapidly and thereby retaining antiproton intensity.

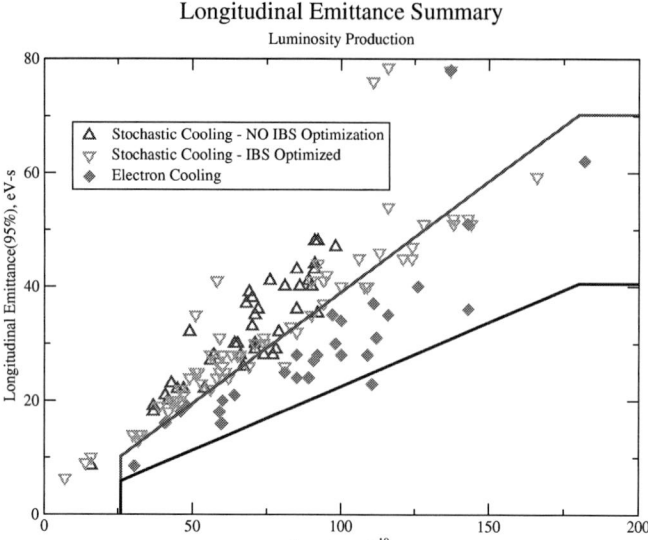

FIGURE 5. Attained longitudinal emittance of antiproton beams before extraction from the Recycler for collider luminosity production.

SUMMARY

Stochastic cooling in the Recycler has been shown to be limited by diffusion processes. A precise IBS model has been developed for the Recycler. This model has been verified with beam measurements made during storage and cooling of the antiprotons before extraction to the Tevatron. Longitudinal IBS heating and cooling effects have been demonstrated by designing the beam parameters based on the IBS model. Minimization of IBS diffusion has been incorporated in the storage and cooling procedures for antiproton beams for Tevatron collider luminosity generation. Operational data analysis and tuning procedures have been developed for the stochastic cooling systems. The Recycler has successfully increased the number of antiprotons available for Tevatron collider operations for the past year.

REFERENCES

1. J.P. Marriner, Fermilab MI Note 168 & 169, 1996.
2. D. Broemmelsiek, R.J. Pasquinelli, "Fermilab Recycler Stochastic Cooling Commissioning and Performance," PAC'03, Portland, Oregon, May 2003, p. 3431.
3. D. Broemmelsiek, D. Neuffer, "Bunched Beam Cooling in the Fermilab Recycler," PAC'05, Knoxville, Tennessee, May 2005, FPAE009.
4. S. Nagaitsev, "Intrabeam scattering formulas for fast numerical evaluation," Phys. Rev. ST Accel. Beams 8, 064403 (2005).
5. Martin Hu, Sergei Nagaitsev, "Observation of Longitudinal Diffusion and Cooling Due to Intrabeam Scattering at the Fermilab Recycler Ring," PAC'05, Knoxville, Tennessee, May 2005, FPAE017.

Stochastic Cooling with Schottky Band Overlap[*]

Valeri Lebedev

Fermilab, P.O. Box 500, Batavia, IL 60510

Abstract. Optimal use of stochastic cooling is essential to maximize the antiproton stacking rate for Tevatron Run II. Good understanding and characterization of the cooling is important for the optimization. The paper is devoted to derivation of the Fokker-Plank equations justified in the case of near or full Schottky base overlap for both longitudinal and transverse coolings.

Keywords: Stochastic cooling, band overlap, antiproton stacking.
PACS: 29.20.-c.

INTRODUCTION

The Schottky band overlap compromises the performance of stochastic cooling. Therefore all practical cooling systems are designed and built to avoid the band overlap. Nevertheless, operating cooling systems are frequently used in a regime when bands are close to overlap or slightly overlapped. In this case the band overlap need to be taken into account if detailed description of the cooling is required. The stochastic cooling theory with no band overlap is well developed [see Ref. 1 and 2 and included bibliography]. In this paper we extend this theory to the case of arbitrary band overlap. First, we derive expressions for the beam permeabilities of the longitudinal and transverse coolings and, then, proceed to derivation of the Fokker-Plank equations describing transverse and longitudinal coolings.

BEAM PERMEABILITY FOR LONGITUDINAL COOLING

Usually, a calculation of the beam permeability is based on azimuthal harmonics. It does not work well if bands are close being overlapped because the amplitudes of the harmonics are changed within one revolution. In this paper we limit ourselves to the case of the beam with sufficiently small intensity so that the beam interaction with vacuum chamber could be neglected. That allows us to reduce the problem from one of finding the entire ring distribution function to one of finding the local distribution functions in the pickup and kicker. Let $f_1(x,t)$ be the distribution function immediately after the kicker, $f_2(x,t)$ be the distribution function in the pickup, and $f_3(x,t)$ be the distribution function just before the kicker. Taking into account that the particle momentum is changed only in the kicker one can write the equations binding up these functions:

---

[*] Work supported by the Universities Research Assos., Inc., under contract DE-AC02-76CH03000 with the U.S. Dept. of Energy.

$$f_2(x,t) = f_1(x, t - T_1(1+\eta_1 x_0)) \ ,$$
$$f_3(x,t) = f_2(x, t - T_2(1+\eta_2 x)) \ , \qquad (1)$$
$$f_1(x,t) = f_3(x - \delta p(t)/p_0, t) \ .$$

Here $x = (p - p_0)/p_0$ is the relative momentum deviation, T_1, T_2 and $T_0 = T_1 + T_2$ are the kicker-to-pickup, pickup-to-kicker and revolution times for the reference particle, $\eta = \alpha - 1/\gamma^2$ is the slip factor, η_1 and η_2 are the partial kicker-to-pickup and pickup-to-kicker slip factors so that $\eta T_0 = \eta_1 T_1 + \eta_2 T_2$, and $\delta p(t)$ is the particle momentum change by the kicker. Expressing the distribution function through its equilibrium value and the perturbation, $f_k(x,t) = f_0(x) + \tilde{f}_k(x,t)$, $k = 1,...3$, and leaving only the first order addend in the Taylor expansion of the third equation in Eq. (1) one obtains:

$$\tilde{f}_2(x,t) = \tilde{f}_1(x, t - T_1(1+\eta_1 x_0)) \ ,$$
$$\tilde{f}_3(x,t) = \tilde{f}_2(x, t - T_2(1+\eta_2 x)) \ , \qquad (2)$$
$$\tilde{f}_1(x,t) = \tilde{f}_3(x,t) - \frac{\delta p(t)}{p_0}\frac{df_0(x)}{dx} \ .$$

Looking for a solution in the form $\tilde{f}_k(x,t) = \tilde{f}_{k\omega}(x)e^{i\omega t}$, $\delta p(t) = \delta p_\omega e^{i\omega t}$ and excluding $\tilde{f}_{1\omega}(x)$ and $\tilde{f}_{3\omega}(x)$ from the resulting equations we obtain:

$$\tilde{f}_{2\omega}(x)\exp(i\omega T_1(1+\eta_1 x)) = \tilde{f}_{2\omega}(x)\exp(-i\omega T_2(1+\eta_2 x)) - \frac{df_0(x)}{dx}\frac{\delta p_\omega}{p_0} \ . \qquad (3)$$

Let the momentum kick be determined by the sum of amplified pickup signal and an external harmonic perturbation so that:

$$\delta p_\omega / p_0 = \int dx \tilde{f}_{2\omega}(x) G(x,\omega) e^{-i\omega T_2}\left[1 - A(\omega)e^{-i\omega T_0}\right] + \Delta p_{ext\omega}/p_0 \ . \qquad (4)$$

Here the term $e^{-i\omega T_2}$ takes into account the delay in signal propagation from the pickup to the kicker, $\tilde{f}_2(p, t - T_2) \to \tilde{f}_{2\omega}(p)e^{-i\omega T_2}$. The total system gain, $G(x,\omega)\left[1 - A(\omega)e^{-i\omega T_0}\right]$, is chosen so that it would describe both Palmer and momentum cooling. For Palmer cooling $A(\omega) = 0$ and the pickup signal depends on the particle momentum due to non-zero dispersion in the pickup. For filter cooling the pickup signal does not depend on particle momentum, $G(x,\omega) \to G(\omega)$, and the cooling signal is formed by the notch filter, $A(\omega) \approx 1$. Its delay is equal to the revolution time for the reference particle, T_0. Taking into account the distribution function normalization, $\int f_0(x)dx = 1$, and introducing the impedances of pickup, Z_p, and kicker, Z_k, so that the pickup voltage is

$$U_{pickup_\omega} = I_0 \int Z_p(x,\omega) f_{2\omega}(x,\omega) dx \ , \qquad (5)$$

and the energy gain in the kicker is

$$\delta E_{kicker_\omega} = e\frac{Z_k(\omega)}{Z_{ampl}} U_{kicker_\omega} \ , \qquad (6)$$

we obtain that the system gain is:

$$G(x,\omega) = \frac{eI_0 Z_p(x,\omega) Z_k(\omega)}{\gamma \beta^2 mc^2 Z_{ampl}} K(\omega) \ . \qquad (7)$$

Here I_0 is the beam current, $Z_{ampl} = 50\ \Omega$ is the impedance of power amplifier, $K(\omega)$ is the total electronic amplification of the cooling system, c is the speed of the light, e and m are the particle charge and mass, and β and γ are the relativistic factors.

Substitution Eq. (4) into Eq. (3) yields:

$$\tilde{f}_{2\omega}(x)\left[e^{i\omega T_1(1+\eta_1 x)} - e^{-i\omega T_2(1+\eta_2 x)}\right] + \frac{df_0(x)}{dx}\left[\frac{\Delta p_{ext\omega}}{p_0} + e^{-i\omega T_2}\left[1 - A(\omega)e^{-i\omega T_0}\right]\int dx'\tilde{f}_{2\omega}(x')G(x',\omega)\right] = 0 \quad . \tag{8}$$

Dividing both addends by $e^{i\omega T_1(1+\eta_1 x)} - e^{-i\omega T_2(1+\eta_2 x)}$, multiplying them by $G(x,\omega)$ and integrating we obtain:

$$S_\omega + \frac{\Delta p_{ext\omega}}{p_0}\int\frac{df_0(x)}{dx}\frac{G(x,\omega)dx}{e^{i\omega T_1(1+\eta_1 x)} - e^{-i\omega T_2(1+\eta_2 x)}} + e^{-i\omega T_2}\left[1 - A(\omega)e^{-i\omega T_0}\right]S_\omega\int\frac{df_0(x)}{dx}\frac{G(x,\omega)dx}{e^{i\omega T_1(1+\eta_1 x)} - e^{-i\omega T_2(1+\eta_2 x)}} = 0 \quad , \tag{9}$$

where $S_\omega = \int dx'\tilde{f}_{2\omega}(x')G(x',\omega)$. Solving Eq. (9) relative to S_ω we finally obtain the system response at the pickup location due to the external harmonic perturbation:

$$S_\omega = -\frac{1}{\varepsilon(\omega)}\frac{\Delta p_{ext\omega}}{p_0}\int\limits_{\delta\to 0_+}\frac{df_0(x)}{dx}\frac{G(x',\omega)e^{i\omega T_2(1+\eta_2 x)}}{e^{i\omega T_0(1+\eta x)} - (1-\delta)}dx \quad , \tag{10}$$

where $\varepsilon(\omega)$ is the beam permeability

$$\varepsilon(\omega) = 1 + \left(1 - A(\omega)e^{-i\omega T_0}\right)\int\limits_{\delta\to 0_+}\frac{df_0(x)}{dx}\frac{G(x,\omega)e^{i\omega T_2\eta_2 x}}{e^{i\omega T_0(1+\eta x)} - (1-\delta)}dx \quad . \tag{11}$$

In the above equations the rule to traverse the poles, $\delta \to 0_+$, follows from the fact that for the complex Laplace transform ω is shifted to the lower complex plane.

Far away from Schottky band overlap the exponent in the denominator of Eq. (11) can be expended near revolution harmonic, $\omega = n\omega_0 + \delta\omega$, $\omega_0 = 2\pi/T_0$ and we arrive to the standard formula for the permeability[1].

BEAM PERMEABILITY FOR TRANSVERSE COOLING

Similar to the method used above for the longitudinal cooling the beam evolution is considered at three points: (1) after kicker, (2) in the pickup, and (3) before the kicker. The beam dipole moment at each point is

$$d_k(t) = \frac{I_0}{c\beta}\int y_k(x,t)f_0(x)dx \quad , \quad k = 1,2,3. \tag{12}$$

Here $f_0(x)$ is the distribution function over momentum, and $y_k(x)$ is the average transverse beam displacement for particles with relative momentum deviations equal to x. Normalizing the beam displacements, $y_k(x)$, and angles, $\theta_k(x)$, by the beta-functions so that $\tilde{y}_k = y_k/\sqrt{\beta_k}$ and $\tilde{\theta}_k = \theta_k\sqrt{\beta_k} + \alpha_k x_k/\sqrt{\beta_k}$ one can write the system of equations binding up the beam displacements after and before the kicker:

$$\tilde{y}_3(x,t) = c(x)\tilde{y}_1(x,t-T_0(1+\eta x)) + s(x)\tilde{\theta}_1(x,t-T_0(1+\eta x)) \ ,$$
$$\tilde{\theta}_3(x,t) = -s(x)\tilde{y}_1(x,t-T_0(1+\eta x)) + c(x)\tilde{\theta}_1(x,t-T_0(1+\eta x)) \ .$$
(13)

Here $c(x) = \cos(2\pi(\nu + \xi x))$, $s(x) = \sin(2\pi(\nu + \xi x))$, ν is the betatron tune, and ξ is the tune chromaticity. Passing the kicker changes the beam angle but does not change beam coordinate so that

$$\tilde{y}_1(x,t) = \tilde{y}_3(x,t) \ ,$$
$$\tilde{\theta}_1(x,t) = \tilde{\theta}_3(x,t) + \delta\tilde{\theta}(t) \ .$$
(14)

We look for a solution in the form $\tilde{y}_k(x,t) = \tilde{y}_{k\omega}(x)e^{i\omega t}$ and $\delta\theta(t) = \delta\theta_\omega e^{i\omega t}$. Substituting it into Eqs. (13) and (14) and solving obtained equations relative to $\tilde{y}_{1\omega}(x)$ and $\tilde{\theta}_{1\omega}(x)$ we obtain

$$\tilde{\theta}_{1\omega}(x) = -\frac{(c(x)-\exp(i\omega T_0(1+\eta x)))\exp(i\omega T_0(1+\eta x))}{\exp(2i\omega T_0(1+\eta x)) - 2c(x)\exp(i\omega T_0(1+\eta x)) + 1}\delta\tilde{\theta}_\omega \ ,$$
$$\tilde{y}_{1\omega}(x) = \frac{s(x)\exp(i\omega T_0(1+\eta x))}{\exp(2i\omega T_0(1+\eta x)) - 2c(x)\exp(i\omega T_0(1+\eta x)) + 1}\delta\tilde{\theta}_\omega \ .$$
(15)

Taking into account the relationship between coordinates and angles of points 1 and 2,

$$\tilde{y}_2(x,t) = c_1(x)\tilde{y}_1(x,t-T_1(1+\eta_1 x)) + s_1(x)\tilde{\theta}_1(x,t-T_1(1+\eta_1 x)) \ ,$$
$$\tilde{\theta}_2(x,t) = -s_1(x)\tilde{y}_1(x,t-T_1(1+\eta_1 x)) + c_1(x)\tilde{\theta}_1(x,t-T_1(1+\eta_1 x)) \ ,$$
(16)

and transforming the time dependent values in Eq. (16) to their Fourier harmonics we obtain for the beam displacement in the pickup

$$\tilde{y}_{2\omega}(x) = \frac{(s_2(x)+s_1(x)e^{i\omega T_0(1+\eta x)})e^{i\omega T_2(1+\eta_2 x)}}{e^{2i\omega T_0(1+\eta x)} - 2c(x)e^{i\omega T_0(1+\eta x)} + 1}\delta\tilde{\theta}_\omega \ .$$
(17)

Here $c_{1,2}(x) = \cos(2\pi(\nu_{1,2}+\xi_{1,2}x))$, $s_{1,2}(x) = \sin(2\pi(\nu_{1,2}+\xi_{1,2}x))$, $2\pi\nu_1$ and $2\pi\nu_2$ are the betatron phase advances between pickup and kicker so that $\nu_1+\nu_2 = \nu$, and ξ_1 and ξ_2 are the partial tune chromaticities so that $\xi_1+\xi_2 = \xi$.

Similar to Eq. (4) the beam kick is determined by the sum of amplified pickup signal and an external harmonic perturbation so that:

$$\delta\tilde{\theta}_\omega = \int dx f_0(x)\tilde{y}_{2\omega}(x)G_\perp(\omega)e^{-i\omega T_2} + \Delta\tilde{\theta}_{ext\omega} \ .$$
(18)

We introduce the impedances of pickup, $Z_{p\perp}$, and kicker, $Z_{k\perp}$, so that the pickup voltage is

$$U_{pickup\,\omega} = I_0 Z_{p\perp}(\omega)\overline{y_\omega} = I_0 Z_{p\perp}(\omega)\int y_\omega(x)f_0(x)dx \ ,$$
(19)

and the transverse angle obtained by a particle in the kicker is

$$\delta\theta_{kicker_\omega} = \frac{e}{mc^2\gamma\beta^2}\frac{Z_{k\perp}(\omega)}{Z_{ampl}}U_{kicker_\omega} \ .$$
(20)

That yields that the system gain is:

$$G_\perp(\omega) = \frac{eI_0 Z_{p\perp}(\omega)Z_{k\perp}(\omega)}{\gamma\beta^2 mc^2 Z_{ampl}}\sqrt{\beta_p\beta_k}K(\omega) \ ,$$
(21)

where β_p and β_k are the beta-functions in the pickup and kicker.

Substituting Eq. (18) into Eq. (17) we obtain:
$$\tilde{y}_{2\omega}(x) = \frac{(s_2(x) + s_1(x)e^{2i\omega T_0(1+\eta x)})e^{i\omega T_2(1+\eta_2 x)}}{e^{2i\omega T_0(1+\eta x)} - 2c(x)e^{i\omega T_0(1+\eta x)} + 1}\left(G_\perp(\omega)e^{-i\omega T_2}\int dx\, f_0(x)\tilde{y}_{2\omega}(x) + \Delta\tilde{\theta}_{ext\omega}\right). \quad (22)$$

The solution is similar to the solution carried out in the previous section. The result is:
$$\overline{\tilde{y}}_{2\omega} \equiv \int dx\, f_0(x)\tilde{y}_{2\omega}(x) = \frac{\Delta\tilde{\theta}_{ext\omega}}{\varepsilon_\perp(\omega)}\int dx\, f_0(x)\frac{(s_2(x) + s_1(x)e^{i\omega T_0(1+\eta x)})e^{i\omega T_2(1+\eta_2 x)}}{e^{2i\omega T_0(1+\eta x)} - 2c(x)e^{i\omega T_0(1+\eta x)} + 1}. \quad (23)$$

where the beam permeability is:
$$\varepsilon_\perp(\omega) = 1 - \frac{G_\perp(\omega)}{2}\int_{\delta\to 0_+}\frac{[e^{-i\omega T_0(1+\eta x)}\sin(2\pi(v_2 + \xi_2 x)) + \sin(2\pi(v_1 + \xi_1 x))]e^{i\omega T_2 \eta_2 x}}{\cos(\omega T_0(1+\eta x)) - \cos(2\pi(v + \xi x)) + i\delta\sin(\omega T_0(1+\eta x))}f_0(x)dx. \quad (24)$$

FOKKER-PLANCK EQUATIONS

Evolution of the beam longitudinal distribution function is described by:
$$\frac{\partial f}{\partial t} + \frac{\partial}{\partial x}(F(x)f) = \frac{1}{2}\frac{\partial}{\partial x}\left(D(x)\frac{\partial f}{\partial x}\right). \quad (25)$$

The drag force is created by the particle self-interaction and therefore is not directly affected by the band overlap but is affected by screening of the particle signal. The result is well-known[1]:
$$F(x) \equiv \frac{dx}{dt} = \frac{1}{T}\sum_{n=-\infty}^{\infty}\frac{G_1(x,\omega_n)}{\varepsilon(\omega_n)}(1 - A(\omega_n)e^{-i\omega_n T_0})e^{i\omega_n T_2 \eta_2 x}, \quad \omega_n = \omega_0(1+\eta x)n. \quad (26)$$

Here $G_1(x,\omega) = G(x,\omega)/N$ is the single particle gain, N is the particle number in the beam and $\varepsilon(\omega_n)$ in the denominator takes into account particle screening[3].

The diffusion is created by noise in the kicker voltage:
$$D(x) = \frac{2\pi e^2}{T_0^2(\gamma\beta^2 mc^2)^2}\sum_{n=-\infty}^{\infty}P_U(\omega_n), \quad (27)$$

where $P_U(\omega)$ is the spectral density of kicker voltage consisting of two contributions. The first one is related to the noise of the electronics, P_{Unoise}, and the second one is related to the particle shot noise. Note that we normalize all spectral densities so that $\int_{-\infty}^{\infty}P_x(\omega)d\omega = \overline{x^2}$. The beam current shot noise for non-interacting particles is equal to:
$$P_I(\omega) = \frac{e^2 N}{2\pi T_0}\sum_{k=-\infty}^{\infty}\frac{1}{|k\eta|}f\left(\frac{\omega - k\omega_0}{\eta k\omega_0}\right). \quad (28)$$

Combining Eqs. (27) and (28) and simplifying one obtains:
$$D(x) = \sum_{n=-\infty}^{\infty}\left[\frac{2\pi e^2 P_{Unoise}(\omega_n)}{T_0^2(\gamma\beta^2 mc^2)^2} + \frac{N}{T_0}\left|\frac{G_1(x,\omega_n)(1 - A(\omega_n)e^{-i\omega_n T_0})}{\varepsilon(\omega_n)}\right|^2\sum_{k=-\infty}^{\infty}\frac{1}{|k\eta|}f\left(\frac{(1+\eta x)n - k}{\eta k}\right)\right]. \quad (29)$$

Natural variables for transverse cooling description are the action-phase variables (I, ψ). We determine the action so that $I = (\beta_y\theta^2 + 2\alpha_y y\theta + (1+\alpha_y^2)y^2/\beta_y)/2$, where β_y and α_y are the beta- and alpha-functions of the ring. We assume that there is no x-y coupling in the lattice, and the cooling is linear in betatron amplitude. That yields that the beam distribution function can be described by the following equation:

$$\frac{\partial f_\perp}{\partial t} + \lambda_\perp(x)\frac{\partial}{\partial I}(If_\perp) = D_\perp(x)\frac{\partial}{\partial I}\left(I\frac{\partial f_\perp}{\partial I}\right) . \tag{30}$$

Here $f_\perp \equiv f_\perp(x,I,t)$ is the distribution function normalized so that $\int f_\perp(x,I,t)dI = f_0(x)$ and the same as above $\int f_0(x)dx = 1$, $\lambda_\perp(x)$ is the cooling decrement, and $D_\perp(x)$ is the diffusion coefficient. $\lambda(x)$ and $D_\perp(x)$ do not depend on I because of system linearity in the transverse plane.

Similar to the longitudinal cooling the transverse cooling is created by the particle self-interaction and therefore is not directly affected by the band overlap but still affected by screening. The result is[1]:

$$\lambda_\perp(x) = \frac{1}{T_0}\sum_{n=-\infty}^{\infty} \text{Re}\left(i\frac{G_{\perp 1}(\omega_n)}{\varepsilon_\perp(\omega_n)}e^{i\omega_n T_2 \eta_2 x - 2\pi i v_2}\right), \quad \omega_n = \omega_0(n(1+\eta x) - (v+\xi x)), \tag{31}$$

where $G_{\perp 1}(\omega) = G_\perp(\omega)/N$ is the single particle gain.

The diffusion coefficient is:

$$D_\perp(x) = \frac{\pi\beta_k}{2T_0^2}\sum_{n=-\infty}^{\infty} P_\theta(\omega_n) , \tag{32}$$

where $P_\theta(\omega)$ is the spectral density of the angle kicks produced by the kicker. $P_\theta(\omega)$ consists of two contributions: the spectral density of amplifier noise, $P_{\perp Ua}(\omega)$, and the amplified shot noise of the beam. The shot noise of the beam at the pickup is

$$P_{\perp Up}(\omega) = \frac{e^2|Z_{p\perp}(\omega)|^2 y^2}{T_0^2} N\sum_{m=-\infty}^{\infty}\frac{1}{\|\omega_0(\xi+\eta m)\|}f\left(\frac{\omega-\omega_0(v+m)}{\omega_0(\xi+\eta m)}\right), \tag{33}$$

Substituting Eq. (33) into Eq. (32) and using definition of the single particle gain we obtain:

$$D_\perp(x) = \sum_{n=-\infty}^{\infty}\left(\frac{\pi\beta_k}{2T_0^2}\left(\frac{e|Z_{k\perp}(\omega_n)|}{mc^2\beta^2\gamma Z_{ampl}}\right)^2 P_{\perp U}(\omega_n) + \left|\frac{G_{\perp 1}(\omega_n)}{\varepsilon_\perp(\omega_n)}\right|^2 \frac{\overline{I(x)}N}{2T_0}\sum_{m=-\infty}^{\infty}\frac{f\left(\frac{n-m+(\xi+\eta m)x}{\xi+\eta m}\right)}{|\xi+\eta m|}\right), \tag{34}$$

where $\overline{I(x)} = \int f_\perp(x,I,t)I dI$ is the average action for given momentum deviation x.

ACKNOWLEDGMENTS

The author would like to thank A. Burov and J. Bisognano for many fruitful discussions.

REFERENCES

1. J. Bisognano and C. Leemann, "Stochastic Cooling" in *1981 Summer School on high Energy Particle Accelerators,* edited by R. A. Carrigan et al., AIP Conference Proceedings 87, American Institute of Physics, Melville, NY, 1982, pp. 584-655.
2. D. Möhl, "Stochastic Cooling" in *CERN Accelerator School, Fifth Advanced Accelerator Physics Course,* edited by S. Turner, CERN, Geneva, Switzerland, 1995, pp. 587-671.
3. V. V. Parkhomchuk and D. V. Pestrikov, *Sov. Phys. Tech. Phys,* **25**(7), 818 (1980).

Debuncher Cooling Performance

P.F. Derwent, David McGinnis, Ralph Pasquinelli, David Vander Meulen, Steven Werkema

Fermi National Accelerator Laboratory, P. O. Box 500, Batavia IL 60510-0500

Abstract. We present measurements of the Fermilab Debuncher momentum and transverse cooling systems. These systems use liquid helium cooled waveguide pickups and slotted waveguide kickers covering the frequency range 4-8 GHz.

Keywords: Stochastic Cooling, Antiproton Beams
PACS: 41.75.Lx

THE FERMILAB DEBUNCHER

The Fermilab Debuncher is an 8 GeV ring designed for the collection, RF debunching, and storage of anitprotons. The Tevatron Collider program requires 1e13 antiprotons for the study of proton-antiproton collisions at $\sqrt{s} = 1.96$ TeV. Antiprotons are produced by impinging a 120 GeV proton beam on an nickel alloy target and collected through a lithium focussing lens and the Debuncher ring then stochastic stacked in the Fermilab Accumulator [1, 2]. The momentum acceptance of the Debuncher is about 4% (350 MeV). The large momentum spread and the short bunches of the antiproton beam are exchanged with an RF bunch rotation . After the bunch rotation, the coasting beam has a momentum spread of about 0.3-0.4%. The bunch rotation and adiabatic debunching takes less than 100 msec, leaving additional time for stochastic cooling of the beam in all 3 planes. Because of the low beam current (few $\times 10^8$ antiprotons per pulse), the cooling rate of these systems is limited by the power available to the kickers. To maximize signal to noise, all pickups and front end amplifiers are cryogenically cooled with liquid Helium to approximately 8 K.

PERFORMANCE REQUIREMENTS

The Debuncher accepts a few $\times 10^8$ antiprotons every 2 seconds. The input beam fills the transverse aperture of the beam, consistent with a transverse emittance of 320π mm mr (95% unnormalized). At the end of the 2 second cycle, the beam is required to have transverse emittance less than 45π mm mr (95% unnormalized) in both planes (factor of 7). After bunch rotation, the 95% momentum width is approximately 60 MeV/c. At the end of the 2 second cycle, the 95% momentum width of the beam is required to be less than 6 MeV/c (factor of 10). These requirements correspond to a 6-dimensional phase space density ($\rho_{6d} = \frac{N_{particles}}{\varepsilon_l \varepsilon_h \varepsilon_v}$) increase of a factor of 500.

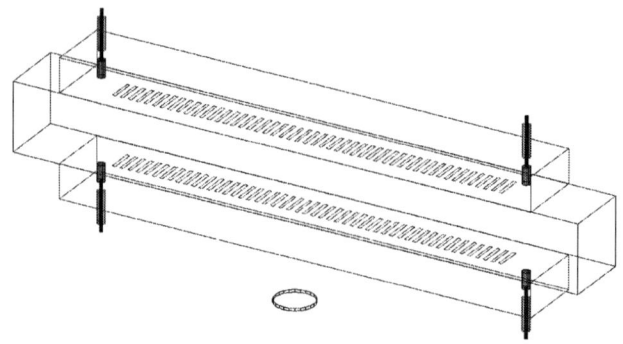

FIGURE 1. A drawing of slotted waveguide pickups. For comparison, a quarter is drawn below the array.

SYSTEM ARRANGEMENT

The Debuncher cooling systems make use of slot coupled "slow wave" waveguide structures (see figure 1 as pickups and kickers [3]. Due to the large transverse aperture of the Debuncher and value of the β functions in the region of the pickups and kickers, the beam pipe allows for waveguide modes in the frequency range of the pickups and kickers. These modes limit the fractional bandwidth of a particular pickup/kicker. As a result, the system was built with 8 narrow pickup bands and 4 slightly wider kicker bands, as shown in figure 2. Narrow band transversal filters are used to prevent band overlap. As the kicker bands are wider than the pickup bands, we sum 2 pickup bands into one kicker band. The pickups and front end amplifiers are cryogenically cooled by liquid Helium. The effective front end temperature of the system, including noise, cabling, and amplifier noise figure, is 25 K [4]. We use a broadband notch filter [6] with an optical delay line to put a null in the system response at the desired frequency. Over the 4-8 GHz band, the filter has an RMS spread of X Hz and average depth greater than 30 dB [5].

MOMENTUM COOLING PERFORMANCE

To measure the performance of the momentum cooling, we down convert a 5.2 GHz longitudinal schottky signal (near the peak of the response in band 2) and use a Vector signal analyzer to record 10 seconds worth of traces. We are then able to calculate the mean, RMS, and 95% momentum width of the antiproton beam. The beam cools quickly, with $1/e$ cooling times of 0.3 seconds, and reaches an asymptotic final width within 2-3 seconds. The asymptotic width has been found to be a function of both beam current and system gain. As the beam cools longitudinally, the notches fill up with the beam signal and beam heating becomes important. In addition, at high power levels, intermodulation

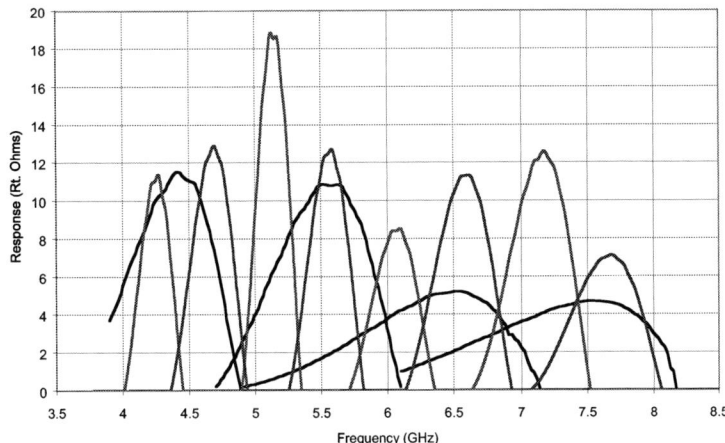

FIGURE 2. The impedance response (in $\sqrt{\Omega}$) vs frequency for the 8 pickup bands (red and blue) and the 4 kicker bands (black). The pickup half bands are summed before the kicker. Transversal filters are used to prevent band overlap.

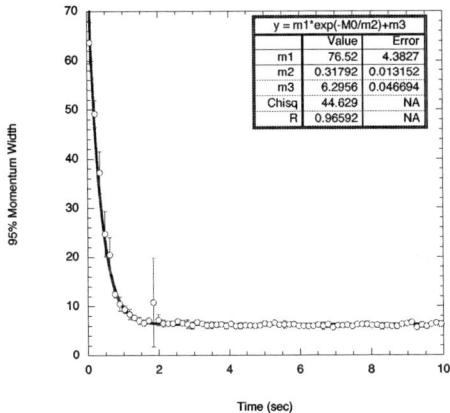

FIGURE 3. The 95% momentum width in the Debuncher as a function of cooling time. The data is the average of 5 pulses, with a fit to an exponential plus a constant overlaid.

distortion in the final stage traveling wave tube amplifiers causes the notches to fill up with noise power. Both of these problems would point toward lowering the system gain. However, the initial cooling rate depends strongly on the gain, which points toward maximum system gain at the start of the cycle. We have built automatic gain ramps into the system, allowing a 6 dB change in the gain during the 2 second cooling cycle, with maximum gain at the beginning of the cycle. With the cooling ramps, we have a cooling time of 0.32±0.01 seconds, with an asymptotic final width of 6.30±0.05 MeV/c. The gain ramp lowers the asyptotic final width by 20%. In figure 3, the 95% momentum width is plotted versus time, along with a fit to a falling exponential plus a constant.

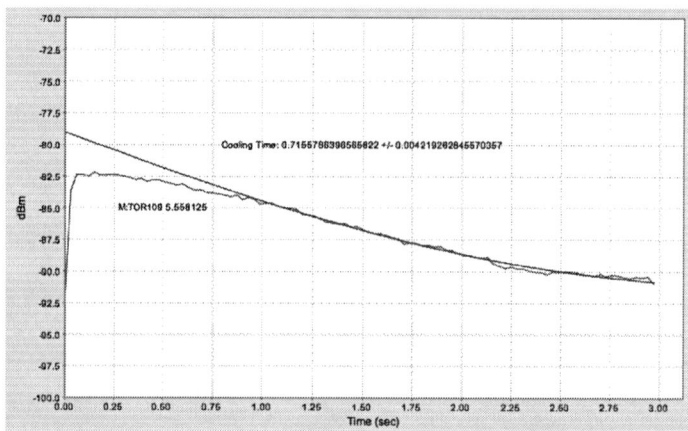

FIGURE 4. The horizontal schottky power in the Debuncher as a function of cooling time. The data is for a single pulse (red trace), with a fit to an exponential plus a constant overlaid (blue trace). The fit only starts after 1 second, as during the first second the beam is cooling into the resolution bandwidth of the analyzer.

TRANSVERSE COOLING PERFORMANCE

To measure the performance of the transverse cooling, we look at the power in a transverse schottky sideband, which is proportional to the average dipole moment of the beam, which in turn is proportional to the beam emittance. As the width of the sideband depends upon the momentum width of the beam and we are using a spectrum analyzer with a given resolution bandwidth setting, we only fit over the time period where the entire sideband is well within the resolution bandwidth. For the current performance, that is after 1 second of beam cooling. We have also implemented gain ramping for the transverse systems. As these systems are power limited during the entire cycle, we ramp the gain up (again, roughly 6 dB during the 2 second cycle) to keep the power at the maximum value. For beam currents of 1×10^8 antiprotons, we measure cooling times of 0.69 ± 0.03 seconds for the horizontal system and 0.74 ± 0.03 seconds for the vertical system. Figure 4 shows the power vs time for a single pulse. These values are in very good agreement with predictions made during the design phase [7].

CONCLUSIONS

We have presented measurements of the momentum and transverse cooling systems of the Fermilab Debuncher. For 1×10^8 antiprotons, the 95% momentum width of the beam is compressed by a factor of 12.8 in 2 seconds, limited by the asymptotic width, not the cooling time. For the transverse planes, the systems cool by a factor of 17 in 2 seconds. As the intensity goes up, we expect a factor of 12 with 2×10^8 antiprotons as we are limited by available power.

REFERENCES

1. D. McGinnis, "Antiproton production rate increase", these proceedings.
2. P.F. Derwent, *et al.*, "Performance and upgrades of the Fermilab Accumulator Stacktail Stocastic Cooling", these proceedings.
3. D. McGinnis, "The 4-8 GHz Stochastic Cooling Upgrade for the Fermilab Debuncher", Contributed to IEEE Particle Accelerator Conference (PAC 99), New York, "New York 1999, Particle Accelerator Conference, vol. 3", 1713-1715.
4. R.J. Pasquinelli, "Noise Performance of the Debuncher Stochastic Cooling Systems", Fermilab Pbar Note 661 (unpublished), http://www-bdnew.fnal.gov/pbar/documents/pbarnotes/pdf_files/PbarNote661.pdf.
5. R.J. Pasquinelli, Need reference from Ralph....
6. D. Möhl, *et al.*, "Physics and Technique of Stochastic Cooling", *Physics Reports*, **58**, No. 2 (1980).
7. J. Marriner, "Debuncher Stochastic Cooling Upgrade for Run II and Beyond", Fermilab Pbar Note 573 (unpublished), http://www-bdnew.fnal.gov/pbar/documents/pbarnotes/pdf_files/PB573.pdf.

Performance and Upgrades of the Fermilab Accumulator Stacktail Stochastic Cooling

P.F. Derwent, Ed Cullerton, David McGinnis, Ralph Pasquinelli, Ding Sun, David Tinsley

Fermi National Accelerator Laboratory, P. O. Box 500, Batavia IL 60510-0500

Abstract. We report on the performance and planned upgrades to the Fermilab Accumulator Stacktail Stochastic Cooling System. The current system has achieved a maximum flux of 16.5e10/hour, limited by the input flux of antiprotons. The upgrades are designed to handle flux in excess of 40e10/hour.

Keywords: Stochastic Cooling, Antiproton Beams
PACS: 41.75.Lx

THE FERMILAB ACCUMULATOR

The Fermilab Accumulator is an 8 GeV ring designed for the collection, cooling, and storage of anitprotons. The Tevatron Collider program requires 1e13 antiprotons for the study of proton-antiproton collisions at $\sqrt{s} = 1.96$ TeV. Antiprotons are produced by impinging a 120 GeV proton beam on an nickel alloy target and collected through a lithium focussing lens and the Fermilab Debuncher ring [1]. As the number of antiprotons collected per pulse (which occur every 2 seconds) is on the order of a few $\times 10^8$, it is necessary to have a storage ring to collect and compress the phase space. For the last 20 years of the Tevatron Collider complex, the Fermilab Accumulator has served as that ring. With the advent of electron cooling in the Fermilab Recycler [2], the Accumulator will not be the final storage ring but will still play a significant role in the compression of the antiproton phase space.

THEORY OF STOCHASTIC STACKING

Stochastic stacking with a constant flux is achieved by designing a system with gain as a function of energy that falls exponentially, with characteristic energy E_d. The resulting density distribution then rises exponentially with the same characteristic energy [3]. The resulting maximum flux Φ can be expressed as:

$$\Phi = \frac{W^2|\eta|E_d}{f_0 p \ln(\frac{F_{max}}{F_{min}})} \quad (1)$$

where W is the electronic bandwidth of the system, η is the phase slip factor, f_0 is the beam revolution frequency, p is the beam momentum, and F_{max} and F_{min} are the maximum and minimum frequencies in the system electronic bandwidth. Planar pickups

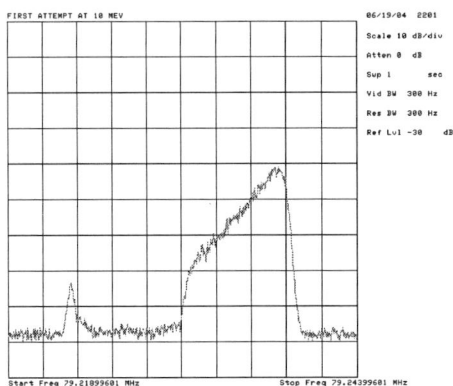

FIGURE 1. Current stacktail profile, which fits to a E_d of 10 MeV.

have a response that follows $exp(-\frac{\pi x}{d})$ where x is the transverse distance from the center of the pickup and d is the vertical aperture. If the pickups are located in a region of high momentum dispersion, a system can be designed where the gain response falls off exponentially with energy. The region of exponential density increase is called the stacktail and the region where beam accumulates is called the core. As the revolution frequency changes with energy so does the flight time between pickup and kicker. The delay time through the electronics is a constant, hence, it is necessary to use multiple sets of pickups with different gains and delays to build the gain slope across the aperture. Correlator notch filters are used to null the signal at the core.

CURRENT PERFORMANCE

The present Accumulator provides a good example of the basic principles. Figure 1 shows the antiproton density distribution as a function of the beam revolution frequency, using a longitudinal Schottky pickup. We have 2 sets of pickups, located in a region with 10 m of dispersion, separated by 15 mm radially. The pickups and front end amplifiers are cooled by liquid Nitrogen, with an effective noise temperature of 125K. There are 256 pickup loops at 15 MeV (with respect to the central energy of the Accumulator) and 48 pickup loops at -8 MeV. There are 128 kicker loops in 8 tanks, with 4 TWTs per tank. There is approximately 150 dB of gain from pickup to kicker. By adjusting the relative gain and phase of the two sets of pickups, we have achieved an exponential gain slope of 10 MeV. The system bandwidth, accounting for phase variation as a function of frequency, is measured to be 1.2 GHz. With the machine parameters of the Accumulator, these values support a maximum flux of 29.5e10/hour. Integrating over an hour, our best performance has been the accumulation of 16.5e10. The system is still limited by the input flux.

UPGRADE DESIGN APPROACH

Of the parameters appearing in Eq. (1), E_d, W, and η are the only ones that can reasonably be considered as changeable. The simplest approach to maximize the flux is to increase E_d. This approach sacrifices the amount of density compaction achievable, since the density grows as $exp(\frac{E}{E_d})$, but has fewer implications for other systems in the Accumulator. Increasing the bandwidth clearly increases the maximum flux. Both approaches will be taken in this upgrade. Changes to η are not being considered at this time.

A two-stage upgrade is planned to handle increased input flux [1]. In the initial stage, the characteristic energy E_d is increased from 10 MeV to 18 MeV. This change can be implemented with a minimal change in hardware through changes to pickup position (moving tanks radially in the Accumulator tunnel) and electronic gain and phase settings. The second stage requires additional pickups, electronics, and kickers; all covering the frequency range 4-6 GHz.

DESIGN CONSTRAINTS

There are drawbacks to increasing E_d. The Accumulator has a finite momentum aperture. It is therefore necessary to stop the flux at some point and accumulate it in a 'core'. The gain function will then deviate from a pure exponential and other considerations come to the fore. It is necessary to match the stacktail system gain to the core system gain to have a smooth transition in the gain profile. As the density increases for a given value of the gain, diffusive beam heating from other particles (through the cooling systems) eventually dominates the cooling term and the system no longer is able to effectively increase the density. It is generally true that as the density of the core increases it becomes necessary to decrease the system gain to maintain some margin between the cooling and diffusive terms in the Fokker-Planck equation.

Another limitation is the assumption of constant input flux. The input flux is a transient, with large pulses coming every 2 seconds. It is necessary for the input pulse to move completely into the stacktail region before the next pulse arrives or it will be phase displaced by the RF bucket moving the new pulse onto the deposition orbit. The fraction of the input pulse that moves across the aperture is a function of the gain of the system and the momentum distribution of the input pulse. The larger the gain, the more efficiently the input pulse moves off of the deposition orbit. The large gain necessary for effective stacking of the input pulse is also detrimental (for reasons given above) to accumulating large amounts of beam in the core.

DESIGN CALCULATIONS

Using a numerical integration of the Fokker-Planck equation, including models of pickup and kicker response, amplifier and notch filter performance, and a full implementation of the beam feedback terms [4], both stages of the upgrade have been simulated. The simulation designs for stage 1 (2-4 GHz bandwidth, 18 MeV characteristic energy) and stage 2 (2-6 GHz bandwidth, 10 MeV characteristic energy) are complete. For stage 1, the following changes are necessary:

- Move 256 pickups 1mm radially outward from current location (an energy change of 1 MeV)

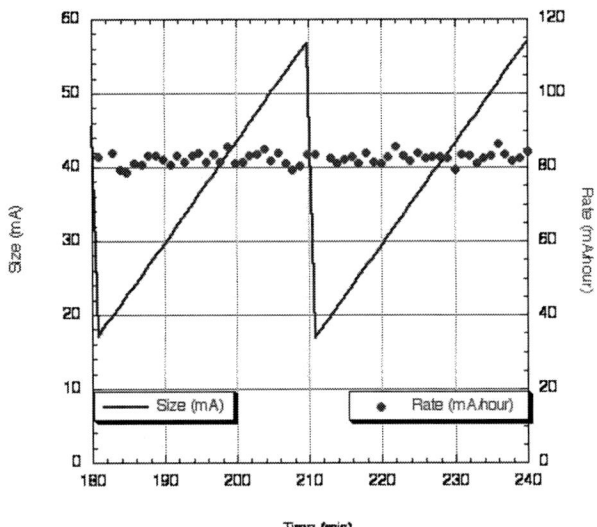

FIGURE 2. Stage 2 simulation results. The green dots are the accumulation rate per minute, the blue line is the beam current in the Accumulator. The input flux is approximately 80e10/hour, transfers to the Recycler occur every 30 minutes and take 1 minute to complete.

- Move 48 pickups 7 mm radially outward from current location (an energy change of 8 MeV)
- Adjust system gains and phases

With these changes, the stacktail can sustain a stack rate of >55e10/hour for 30 minutes.

For stage 2, because of the increased bandwidth of the system, it is necessary to decrease the total energy change in the system to avoid Schottky band overlap. Hence, the core energy is closer to the stacktail. Half the 2-4 GHz pickups and kickers are removed and replaced with 4-6 GHz pickups and kickers. With the increased bandwidth, the characteristic energy is lowered to approximately 15 MeV. The resulting system can sustain 80e10/hour for 30 minutes. The accumulation rate and beam current for this simulation are shown in figure 2.

UPGRADE STATUS

The current configuration of the stacktail is able to handle the current maximum input flux to the Accumulator. With the coming improvements [1], we anticipate that we will make use of the staged approach. The 4-6 GHz pickups and kickers have been designed and measured with a stretched wire, with the results agreeing well with microwave structure simulations. We are currently building a prototype tank to house several design iterations of the pickup loops. This tank will be installed in the Fermilab Debuncher this fall for beam tests. We anticipate assembly of the 4-6 GHz pickup and kicker tanks to take place during the 2006 calendar year, with installation during an appropriate accelerator shutdown period.

CONCLUSIONS

The Fermilab Accumulator stacktail design and performance has been described. The current arrangement can support a theoretical maximum flux of 29.5e10/hour, though the maximum value achieved has been limited by the input flux and is 16.5e10 in one hour. A two stage design approach has been presented, to handle increases of the input flux to 80e10/hour. Beam measurements are planned and we anticipate installation as the input flux outperforms the current stacktail cooling system..

REFERENCES

1. D. McGinnis, "Antiproton production rate increase", these proceedings.
2. S. Nagaitsev, "Antiproton cooling in the Fermilab Recycler", these proceedings.
3. D. Möhl, *et al.*, "Physics and Technique of Stochastic Cooling", *Physics Reports*, **58**, No. 2 (1980).
4. J. Marriner & V. Visnjic, "Fermilab Stochastic Cooling Code User's Guide", Fermilab Pbar Note 498 (unpublished), `http://www-bdnew.fnal.gov/pbar/documents/pbarnotes/pdf_files/pb498.pdf`. The Fortran code has been rewritten in C++ for these simulations.

/*ELECTRON COOLING*/

ELECTRON COOLING

Development of a New Generation of Coolers with a Hollow Electron Beam and Electrostatic Bending

V.V. Parkhomchuk

BINP, Novosibirsk-90,630090,Russia

Abstract. The basic features and design of a new generation coolers made for CSRm,CSRe (Lanzhow, IMP) and for LEIR (Geneva, CERN) will discussed. The hollow profile electron beam help suppress recombination at the accumulation zone. The low electron beam density at the core of the intensive ion beam decrease the amplitude coherent electron-ion beam oscillations (so called electron heating effect). The electrostatic bending made the recuperation loss electron beam current less then 1 mkA for 1-2 Amp the main electron beam current. Decreasing out gassing by the electrons desorption the vacuum chamber cooler open perspective for obtain the high vacuum at cooler on level 1E-12 Torr (for LEIR cooler).

Keywords: Beam; Cooling, Electron gun
PACS: R29.20.Dh; 29.27.Bd

ELECTRON COOLING STAGES

The electron cooling process was invented G.I.Budker at 1965 as method for preparing beams for hadron collider. The electron beam moving with the same average velocity as proton beam absorbed the kinetic energy heavy particles (protons or ions). The first experiments carried out in Novisibirsk in 1974 demonstrated the high efficiency of this method and triggered all the development of heavy particle cooling methods. First electron cooler have cooling length 1 m made for proof the principle of cooling. In course of the experiments carried out at the NAP-M facility it was discovered that the time required for cooling, expected to be several seconds, in fact turned out to be 0.1 s. Such an abrupt increase in cooling efficiency was the result of magnetization of the electron beam and low effective electron beam temperature..

For a detailed study of the kinetics of cooling under the conditions of strong magnetization, an installation called MOSOL, with a field of 4 kG and a very good field straightens $<10^{-5}$ was built. The new economic situation at Russia pushed BINP for production cooler according contract with other laboratories. At 1998 the low energy electron cooler designed and built (magnet system) at BINP for heavy synchrotron SIS at GSI (Germany). The investigation of electron cooling at different laboratories open same problems for cooling the intensive ion beams. For the optimization cooling should be used technique for more ease control the equilibrium ion beam emittance and momentum spread. From 2000 to 2005 was made new coolers for CSRm 35 kV (EC35) and CSRe project 300 kV (EC300) (IMP,China) and for

LEIR ring CERN 40 kV (EC40). This new coolers was appointed electron gun with variable profile electron beam and the electrostatic bending plate for the electron and ion beams convergence. At LEIR cooler was used the very powerfully vacuum system pumping base on NEG absorbers.

TABLE 1. Basic parameters electron coolers.

	EC35	EC300	EC40
Max. voltage (kV)	35	300	40
Max. electron current (A)	3	3	3
Cooling length (m)	4	4	2.5
Magnet field at cooling (kG)	0.5-1.5	0.5-1.5	0.5-1.5
Vacuum at cooler (Torr)	2E-11	2E-11	5E-12
Storage ring	CSRm	CSRe	LEIR

PROBLEMS OF COOLING INTENSIVE ELECTRON BEAM

The electron cooling base on the transfer kinetic energy of random (thermal) motion ions to the electron beam moving with the same average velocity as the ion beam. The cooling process continues to equalization the ion temperature to the electron beam temperature. The concept of the electron beam temperature at condition strong magnet field at cooler section is complicate and it value is function of process. For electron ion recombination temperature is just the kinetic energy of the electrons at the beam rest system. But for the cooling at strong magnet field the transverse motion of the electrons magnetized and the effective temperature electrons is just longitudinal kinetic energy and transverse motion connected with transverse field at cooling section. It can be transverse component the magnet field at cooling section and the space charge field of electron beam. After acceleration from the electron gun kinematics spread of longitudinal velocity very small and effective temperature determent the repulsing of electrons [1]:

$$T_{eff} = 2e^2 n_e^{1/3} . \qquad (1)$$

Usually for the low energy cooler the electron density is $n_e=10^8$ $T_{eff}=1.2\times10^{-4}$ eV$\approx$1 K° and the low intensive ion beam cooled practically to few Kelvin degree temperature. But cooling the high intensive ion beams demonstrate effects of so called "electron heating" when instead cooling the ion beam heated and show fast losses intensity [3,4]. This phenomena close to beam-beam effects at a colliders when the emittances of the intensive beams increased without clear visible coherent signals. The simplest model of this phenomena base on generation fluctuation (plasma oscillations) at beams by ahead moving ions [3,4]. Moving ions generate at the electron beam the friction electric field (cooling force) (for non magnetized cooling) equal:

$$\vec{F}(\vec{V}) = \frac{4\pi e^4 Z^2 n_e}{m} \frac{\vec{V}}{(\sqrt{(V^2+V_{eff}^2)})^3} \ln(\frac{\rho_{max}}{\rho_{min}}) , \qquad (2)$$

where impact parameters can be written at the simplest form:

$$\rho_{max} = V\tau,$$

$$\rho_{min} = \frac{Ze^2}{mV^2} = \frac{Zr_e}{(V/c)^2}, \quad r_e = \frac{e^2}{mc^2}, \quad (3)$$

$$\rho_L = \frac{V_{et}mc}{eB}$$

where Z is charge of the ions, τ is time of flight cooling section (all values at beam reference system), V_{eff} is the effective velocity. For non magnetize cooling V_{eff} there is spread of velocity $V_{e\perp}$ at the beam system of reference. But for magnetized cooling at strong magnet field B, V_{eff} is spread of velocity of center Larmor cycles but not the electron velocity. At this magnetized case V_{eff} determined the electrical field from space charge of the electron beam and the transverse component magnet field at cooling section:

$$V_{eff} = c\frac{2\pi e n_e x}{B} + c\beta\gamma\theta_B, \quad (4)$$

where θ_B is the deflection of direction the magnet line from center axis of ion beam orbit. The losses of momentum ion after flight cooling section is:

$$\Delta p^2 = -2F\tau \times p + (F\tau)^2 n_i \frac{4\pi}{3}\rho_{max}^3. \quad (5)$$

The first term is cooling and the second one is the heating by presents at the zone radius ρ_{max} other ions with density n_i. Other ions with number $N = n_i \rho_{max}^3 4\pi/3$ produce random kick increasing momentum spread. Using equations (2) the losses momentum can be written as:

$$\left(\frac{\Delta p}{p}\right)^2 = -2\frac{F\tau}{p}(1 - \omega_e^2 \omega_i^2 \tau^4 g) \quad (6)$$

where ω_e, ω_i plasma frequencies at the electron and the ion beams:

$$\omega_e = \sqrt{\frac{4\pi e^2 n_e}{m_e}} = c\sqrt{4\pi r_e n_e},$$

$$\omega_i = \sqrt{\frac{4\pi e^2 Z^2 n_i}{M_i}} = c\sqrt{4\pi r_i n_i} \quad (7)$$

g is numerical factor near 1-3 is result of averaging friction force over interaction zone of the ion. This number g can be calculated by numerical simulation of the interaction electrons and ions. The fig.1 demonstrate example of the calculation the electric field around Bi^{+67} ion moved with the velocity 4.6×10^6 cm/s at the electron beam with density 1×10^6 1/cm^3 distance 3 mm. This calculation was made with neglecting interaction between electrons by short time interaction (time of flight less period of electron plasma oscillations) and the low electron beam density.

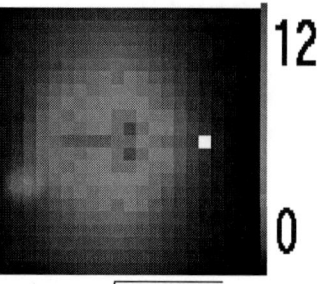

Figure 1. Map of the electric field at units $\sqrt{(E_x^2 + E_y^2)}/(F/q)$ after passing ion Bi^{+67} from –0.15 cm to +0.15 cm at electron gas with density 10^6 1/cm^3., white point is position of ion at moment calculation electric fields. Palette line (right) shows corresponding color with value number from 0 - black to 12 -red.

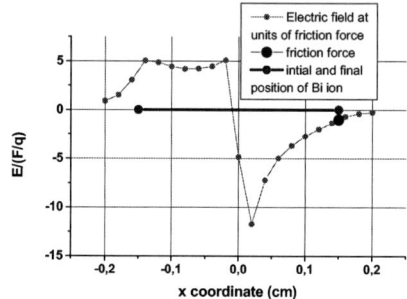

Figure 2. The electric field along of the ion trajectory at units F/qZ.

From fig. 1 and 2 clear see that maximum electric field is not at position of ion but more close to center of exited zone. The maximum value field is at 10-12 times high that friction force F/qZ that was surprise for author. The coincidence of the electric field with calculation the friction force according equation (2) (E/(F/q/Z)=-1 at x=0.15 cm. (at ion position)) shows self consistence of this simple model. Averaging kick for all other ions gives numerical value for this case equal to:

$$g = \int E^2 dV /(4\pi\rho_{max}^3/3)/(F/qZ) \approx 1.4 . \quad (8)$$

Cooling measured the value of $\delta=F\tau/p$ dimensionless decrement and the normal cooling rate (at lab. system) is $\lambda=\delta f_0$ (f_0 is the ion revolution frequency at ring). At case high ion velocity the cooling rate for the magnetization cooling is:

$$\lambda = f_0 \frac{4r_e r_i n_e c \tau}{(V/c)^3} \ln(\frac{\rho_{max}+\rho_L+\rho_{min}}{\rho_{min}+\rho_L}) = \\ \frac{4r_e r_i \eta_e}{\gamma^5 \beta^4 \theta^3} \frac{J_e}{q\pi a_e^2} Ln_c = \frac{4r_e r_i \eta_e \beta_{cool}^3}{\gamma^5 \beta^4 a_i^3} \frac{J_e}{q\pi a_e^2} Ln_c \quad (9)$$

where η_e=lcool/P is the fraction electron beam at ion orbit, J_e is electron beam current, q is electron charge, θ is angle spread ion beam at cooling section, a_i is amplitude of ion oscillations. From equation (9) clear that at the process of cooling decreasing of amplitude ion oscillations leads to fast increasing cooling rate (as $(1/a_i)^3$). Practically there is reserve for optimization cooling. If the electron beam density on axis decrease with amplitude so that all ions cools with the same cooling rate the ion beam profile shape is not change at time and cooling looks optimal. But at this case the electron beam density can decreased at center. For flat electron beam after first injection of ions at the ring the cooling central port of ion beam lead the formation at center very intensive ion beam core containing only small fraction of beam. The plasma oscillations at this beam core produce the electric fluctuations that heat the slow cooled ions with large amplitude. Most systematically the phenomena electron heating was study at CELSIUS ring cooling experiments [3,4]. Main result are that after injection intensive proton beam so high that after cooling only small part of initial intensity of the ion beam cooled. Injection the low intensive ion beams shows at many times low losses rate. Many experiments show that the minimal emittance of ion beam limited the Lasslet betatron tune shift by ion beam space charge with value near $\Delta Q \approx 0.1$ or at units plasma frequency:

$$\Delta Q = \frac{Nr_i}{4\pi\varepsilon_i \gamma^3 \beta^2} = (\omega_i \tau)^2 \frac{\beta_{cool}}{8\pi l_{cool} \eta_e} \quad (10)$$

From equation (10) is clear that maximal density ($\omega_i^2 \tau^2$) can reach for very low initial intensity of beam: for low intensity less equilibrium emittance. The threshold electron current when $\omega_e^2, \omega_i^2 \tau^4 = 1$ (for flat profile electron beam with radius a_e) is equal:

$$J_{th}/q = \frac{c\beta_{cool} a_e^2}{\Delta Q 32\pi r_e \eta_e l_{cool}^3} \beta^3 \gamma^3 \quad (11)$$

For parameters of CELSIUS experiments [3] this threshold current is 0.12 A and 0.2 A used at experiments looks up threshold current. Really there is limit for electron beam density $j = Je/(\pi a_e^2)$ and we can decrease density at center for made better condition for the stacked ion beam. The rate of electron cooling with magnetized electrons cooling becomes very high for small amplitude of ion oscillations a :

$$\lambda(a) = \lambda_0 \frac{a_{eff}^3}{(a_{eff}^2 + a^2)^{3/2}}, \quad (12)$$

where λ_0 is cooling rate for small amplitude a<<a_{eff}, a_{eff} is effective amplitude. So the cooling rate is high for ions at the center but rather low for distant ones with a>> a_{eff}. For example, the ions at stack with a =0.3cm are cooled in 1 ms but ions with amplitude a =3 cm are cooled in 1 s. While waiting new injection stacked ions are intensively recombined with electrons. If the transverse density distribution of electron beam is formed according to the following law all ions will be cooled in the same period of time:

$$n_e(r) = n_{max} \frac{(a_{eff}^2 + r^2)^{3/2}}{(a_{eff}^2 + a_{max}^2)^{3/2}}, \quad (13)$$

where a_{max} is maximal radius of electron beam, n_{max} is the density of electrons distribution near the beam edge.

Advantages of this profile are:

The cooling rate will be constant and the ion beam will shrink without changing profile shape. It gives an opportunity to avoid this kind of the electron heating (development of plasma oscillations inside dense ion beam results in those ones in electron beam).

The life time of stacked ions due to the recombination with electrons increases by decreasing density of electron beam at the center. The cooling rate of new injected ions is proportional to $<n(r)>$ the average electron beam density but losses of stacked ions are proportional to the density at the center $n(0)$. This factor can be rather high $n(0)/n_{max} = (a_{eff}/a_{max})^3$ is for hollow electron beam.

The hollow electron beam cool only tails of ion beam in experiments with internal target. Ions oscillate with large amplitude and move in intensive electron beam after scattering on atoms of target. The intensive cooling for high amplitude without recombination of stacked ions with small amplitude allows achieving the high luminosity for this type experiments. Close to central axis can be empty from electrons zone for control the equilibrium ion beam emitance. This zone will used for stopping over cooling the transverse beam size and decrease Intra Beam Scattering that heat the momentum spread beam. Let to illustrate the cooling intensive ion beam for LEIR cooler. The electron beam with radius ae=3 cm cool Pb^{+54} ion beam with emittance 1 π mm*mrad on energy 4 MeV/n. Using equations (6) and (9) calculation of cooling rate as:

$$\lambda = \frac{4 r_e r_i \eta_e \beta_{cool}^3 (J_e/q) Ln_c}{\pi \gamma^5 \beta^4 a_i^3 a_e^2} *$$
$$(1 - k \frac{8 r_e r_i (J_i/Zi/q)(Je/q) l_c^4}{a_i^2 a_e^2 c^2 \beta^6 \gamma^6}) \quad (14)$$

where Ji is ion current, $k = n_e(0)/<n_e>$ is ration the electron beam density at center to average over ion distribution. For k=1 flat electron beam maximum cooling decrement for Ni=9E8 will be 5 1/s for Je=0.025A that far from requirements. For hollow electron beam (k=0.1) it is possible to reach requirements ($\lambda_{max} \approx 30$) on electron current 0.2 A.

ELECTRON GUN WITH VARIABLE PROFILE ELECTRON BEAM

The electron gun with a control electrode was designed to produce hollow electron beams [5]. The electron gun under consideration is shown on Fig. 3. By digits on the figure are marked: 1 – cathode, 2 – forming electrode, 3 – control electrode, 4 – anode. The gun is immersed into the longitudinal magnetic field of 700-1000 Gs. Convex oxide cathode Ø29 mm is used. The control electrode is situated near the cathode edge, so its potential strictly influences on the emission from this area. By varying the potential of this electrode it is possible to obtain on the gun output the beam with parabolic, flat or hollow profile. The potential of the forming electrode is equal to the cathode potential; the purpose of this electrode is to dump exceeding emission from the cathode edge.

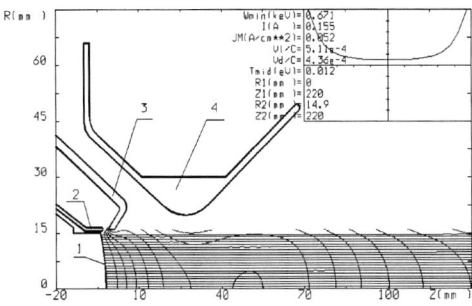

Figure 3. Electron gun calculation.

The negative voltage on control electrode suppress emission near edge and decrease electron beam size- the positive voltage increase emission near edge but space charge suppress emission at centre. Fig.4 show changing of the shape electron beam profile from small size with parabolic form (-0.1/0.9) to flat (+0.1/0.9) and hollow beam profile (+0.6/0.9).

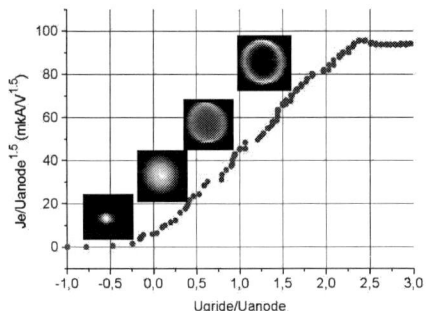

Figure 4: Perveance of the electron gun versus ratio voltage on control (grid) electrode to anode voltage. Profile figures show the corresponding of shape the electron beam for different perveance.

The very high value of perveance connected with small thick the electron beam cylinder moved inside anode close to inner diameter of the anode aperture. At this case normal perviance about 3-5 mkA/V$^{3/2}$ increase at ratio $2\pi a_{anode}/\Delta$ (a_{anode} is radius anode aperture, Δ is space from anode to the external electron beam surface (see. fig 3))

ELECTROSTATIC BENDING

The electron and the ion beams convergence before entrance at cooling section base on bending at electric field. Advantage of using electric field instead of usually used transverse magnet field is compensation of drift shift at bending for the both direction of moving electrons from cathode to collector and reflected electrons moves from collector to gun. First experimental testing of this idea was made Tim Ellison at Indiana University cooler. Fig.5 show practical design of this idea at BINP coolers.

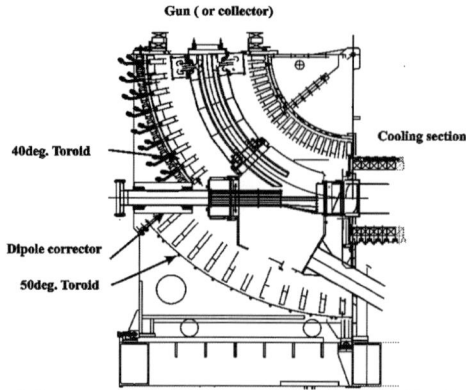

Figure 5. Design of electrostatic bending plates for electron and ion beams convergence.

Absent drift motion for electrons oscillated between electron gun and collector results to very high efficiency capture its at collector after few attempts. The force on bending orbit (inside toroid) is:

$$F = \frac{\gamma\beta^2 m_e c^2}{R} = \frac{q[B \times V]}{c} + qE, \quad (14)$$

where B,E magnet and electric field for moving electron at trajectory with radius R. It is possible to have pure magnet bending $B=B_{max}$, $E=0$ and pure electrostatic bending $B=0$, $E=E_{max}$. At the case magnet bending electrons reflected from collector have twice large drift from geometric trajectory by action centrifugal and magnet force at the same direction instead compensation. The figure 6 show losses current versus voltage on electrostatic bending plates ($E=0 \rightarrow 2kV$). The magnet field at this experiment simultaneously change ($B=Bmax \rightarrow 0$) so that orbit of the primary electron beam is not change (F=const from (14)). From this figure clear see that the recuperation efficiency change from 4×10^{-4} for pure magnet bending to $6-7 \times 10^{-7}$ for pure electrostatic bending. The minimization of the electron beam losses current differ for electrostatic and magnet bending (fig.7). At the case of magnet bending using the suppresser electrode for more effective capture electron help decrease relative losses from 2×10^{-3} to 8×10^{-5} when voltage on the suppresser change from the collector voltage 1.25 kV to –0.7 kV. At the voltage less them –0.7 kV the tails of the primary beam reflected from collector and losses very sharp increased. But at case the electrostatic bending more impotent to have free enter at the collector then suppress reflection from the surfer collector electrons. The multiply coming the electrons at the collector made requirements on the collector capture efficiency not so significant. These phenomena open the new possibility for the operation cooler with the low voltage on the collector without strong suppression at the entrance. Decreasing bombarding the vacuum chamber the electrons with high energy on the few order magnitude decrease out gassing and suppress the radiation emitting for the high energy cooler. For example if we need vacuum 10^{-12} Torr and have losses current 400 mkA the pumping power of the cooler vacuum system should be 50000 l/s ($\eta_{des}=10^{-3}$) but for electrostatic bending and loss current 1 mkA we need only 130 l/s pumping power that at many times easy. Figure 8 show experiments with measuring vacuum pressure at LEIR cooler (at time of commission at BINP with low power pumping system) with switch of ion pumps. At this case gas components with weak pumping by NEG (mainly CO) produced fast increasing pressure at cooler with rising time near 0.6E-9 Torr/(100 s).

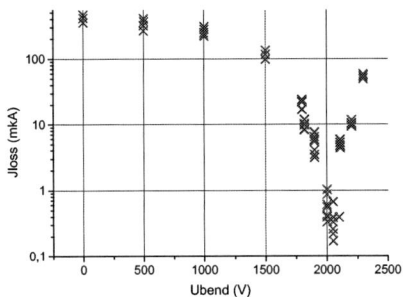

Figure 6. Losses current at EC-35 versus voltage on electrostatic bending plates. (The electron beam 15 keV*1A.).

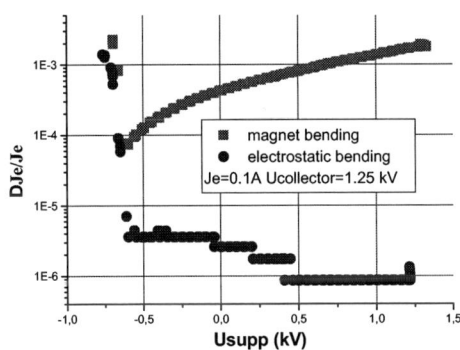

Figure 7. Relative losses electron current versus suppresser voltage for magnet and electrostatic bending.

The switch on the electron beam give high pumping and pressure go down to 4×10^{-10} very fast. Estimation of pumping give value near 140 l/s/A that correspond destroying molecule the electron beam $dV/dt = \sigma \times l_{beam} \times (J_e/q)$ with cross section about 0.5×10^{-16} cm^2. The combination the low voltage on collector and the low losses the electron current open perspective to have good vacuum at the electron cooler. After relatively short time operation with the electron beam the out gassing inside cooler becomes so low that electron beam switch on improve vacuum at cooler.

The radiation level is proportional of loss current and equal to 25 µRem/mkA (for 260 keV). Fast increasing the radiation with increasing cooler voltage made this problem important for the next generation high voltage coolers.

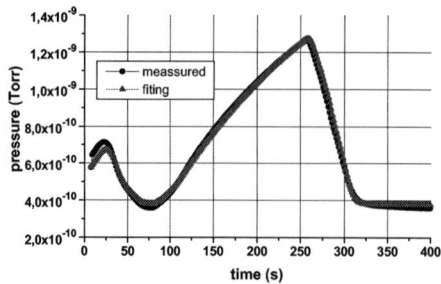

Figure 8. Pressure versus time with switch off ion pump (electron current 0.5 A was switch off in interval 75-260 s).

CONCLUSION

The commissioning of coolers with electron beam demonstrated high performance of ne electron coolers. The electron current up to 3 A was obtain that few times high that can used for the cooling ion beams. The nearest future these coolers will tested at experiments with cooling ion beam and it can open many interesting physics pheromones. Most interesting will be optimization of cooling the high intensive ion beam. It will be very interesting for next cooler for the high luminosity hadron colliders as RHIC or HESR.

ACKNOWLEDGMENTS

The development new generation of the electron coolers are the result of effort high intellectual team of sciences, engineers, workers at BINP with using world experience technique of the electron cooling. Especially I want mention important contribution to cooler techniques the general designer of these coolers Boris Smirnov. He was pioneer the electron cooler design from the first cooler practically up to the last LEIR cooler.

REEFERENCES

1. V.V. Parkhomchuk., A.N. Skrinsky Report on progress in physics 1991, v. 54,n.7, p.919-947.
2. V.V. Parkhomchuk, A.N. Skrinsky, Physics-Uspekhi 43(5) 433-452 (2000)
3. D. Reistad et al., in Proc. Workshop on Beam cooling and Related Topics (Montreux, Switzerland, 4-8 Oct. 1993) (CERN (Series), 94-03, Ed. J.Bosser) (Geneva: European Aorganization for Nuclear Research, 1994) p.183
4. Hermanssson L., Reistad D., NIM in Physics Research A 441 (2000) 140-144
5. http://accelconf.web.cern.ch/AccelConf/e02/PAPERS/WEPRI049.pdf
6. http://accelconf.web.cern.ch/AccelConf/r04/papers/TUAI02.PDF

Cooling Force Measurements at CELSIUS

B. Gålnander[1], A.V. Fedotov[2], V.N. Litvinenko[2], T. Lofnes[1],
A.O. Sidorin[3], A.V. Smirnov[3], V. Ziemann[1]

[1]*The Svedberg Laboratory, S-75121, Uppsala, Sweden*
[2]*Brookhaven National Lab, Upton, NY 11973*
[3]*JINR, Dubna, Russia*

Abstract. The design of future high energy coolers relies heavily on extending the results of cooling force measurements into new regimes by using simulation codes. In order to carefully benchmark these codes we have accurately measured the longitudinal friction force in CELSIUS by recording the phase shift between the beam and the RF voltage while varying the RF frequency. Moreover, parameter dependencies on the electron current, solenoid magnetic field and magnetic field alignment were carried out.

Keywords: electron cooling, friction force.
PACS: 29.27.Bd; 41.75.-i;

INTRODUCTION

Electron cooling friction force has several different descriptions [1, 2, 3], which predict different cooling force and are valid at different degrees of magnetization. In order to make predictions about the cooling times at future coolers it is important to know the cooling force accurately. This is especially important for high energy cooling projects such as the RHIC-project [4] where the cooling times could be of the order of 1000 s and a prediction on the order of magnitude is not sufficient. In this paper we present measurements of the longitudinal cooling force in CELSIUS using the phase shift between the beam and the rf voltage varying the rf frequency.

EXPERIMENTS

The cooling force has a linear dependence on the relative velocity of ions and electrons at low relative velocities. One way to measure the cooling force in this linear region is the so called phase-shift method. In this method the phase difference, $\Delta\phi$, between the bunched beam and the rf voltage is measured. The phase shift results from the competition between the cooler force and the force from the rf voltage. The friction force is then given by

$$F_\| = \frac{Ze\hat{U}_{RF}\sin(\Delta\phi)}{L_{cool}} \quad (1)$$

where Z is the charge of the ion, e is the elementary charge, $\Delta\phi$ is the phase shift and L_{cool} is the interaction length of the cooler.

The relative velocity between the ions and the electrons in the beam frame is given by

$$v_{\parallel}^{*} = \beta c \frac{\Delta p}{p} = \frac{\beta c}{\eta_p} \frac{\Delta f}{f} = \frac{C}{\eta_p} \cdot \Delta f, \qquad (2)$$

where C is the circumference, η_p the slip factor, f the rf frequency and Δf the frequency shift. The phase shift method has been employed earlier in different varieties at CELSIUS [5] and at other laboratories such as IUCF [6], TSR [7] and MSL [8].

Phase Discriminator

A phase discriminator was used to measure the phase difference between the beam particles and the rf voltage. In the following this technique is described in more detail.

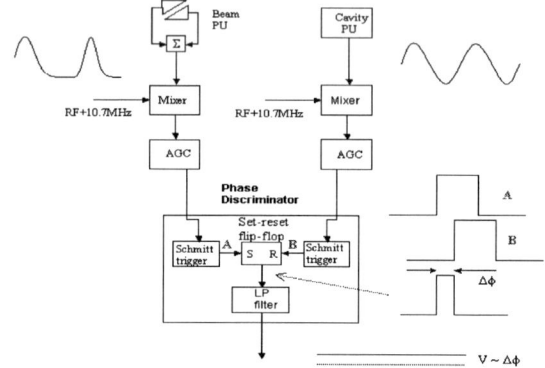

FIGURE 1. Schematic drawing of the principle of the phase discriminator.

In order to measure the phase difference between the rf cavity and the beam a phase discriminator was used as shown in Fig. 1. The signals from the beam pick-up and the cavity pick-up are sent to a phase discriminator after being up-converted to the carrier frequency (10.7 MHz) by a mixer and amplified to 10 dBm by an automatic gain control module. The phase discriminator converts the phase difference between signal (A) and (B) in Fig. 1 to a pulse length in a flip-flop. This pulse length is proportional to the phase difference, $\Delta\phi$, and is averaged in a low-pass filter to obtain an output voltage proportional to the phase difference. Since the low pass filter in the phase discriminator has a cut off frequency of 15 Hz, the output represents an average phase difference over about 20 ms. A digital Le Croy oscilloscope was used for further averaging of the signal over 2 s to get a good signal-to-noise ratio.

Measurement Accuracy

The accuracy of the measurements from different sources are summarized in Tables 1 and 2. The uncertainty in determining $v_{\parallel}$ (Table 1) is relatively small and is dominated by the uncertainty in η_p, which is estimated from optics calculations. The uncertainty in $F_{\parallel}$ (Table 2), on the other hand, is larger and dominated by the uncertainty in the effective cooler length. The field of the toroids influence the effective solenoid length, giving a field region which is different from the nominal length. Another significant source to the uncertainty is the true value of the voltage in the rf cavity. The rf voltage was measured with two different techniques. One was voltage measurement with a probe in the cavity and the other measurement of the synchrotron frequency. It was concluded that the synchrotron method was the most accurate. The synchrotron frequency is given by $f_s = \sqrt{\eta_p e U_{RF} / 2\pi T \beta^2 \gamma} \cdot f_{RF}$, and the rf voltage can thus be determined by measuring the synchrotron frequency. The uncertainty of the measurement of the synchrotron frequency was ± 3 % and η_p is known by ± 0.5% giving an accuracy in U_{RF} of ± 7 %.

TABLE 1. Parameter values and estimation of accuracy for $v_{\parallel}$, Eq. (2).

Parameter	Value	Estimated accuracy	Comment
C	81.76	± 0.1 %	Exact orbit unknown in the arcs
η_p	0.783	± 0.5 %	From optics
Δ_f	Varied around 1129.0 kHz	± 0.01 %	Determined by accuracy in the frequency generator.
$v_{\parallel}$		± 0.5 %	Total estimated accuracy.

TABLE 2. Parameter values and estimation of accuracy for $F_{\parallel}$, Eq. (1).

Parameter	Value	Estimated accuracy	Comment
U_{RF}	10.2	± 7 %	From synchrotron frequency measurements;
$\Delta\phi$	25.0 mV / 1° @ 50 Ω	± 1 %	Read as a voltage from phase discriminator. From input of a known phase difference to the discriminator
L_C	2.50 m	± 10 %	Effective cooler length. Influenced by the toroidal field.
$F_{\parallel}$		± 12 %	Total estimated accuracy.

High Voltage Ripple

The ripple of the high voltage power supply is a potential contribution to the longitudinal electron velocity spread. The ripple of the CELSIUS cooler has dominating contributions at 50 and 300 Hz while the typical cooling time is of the order of 1 s. The ripple is measured to be < 3 V rms at 26 kV voltage, thus a relative ripple $\Delta U/U < 1.1 \cdot 10^{-4}$. This ripple corresponds to a longitudinal electron velocity spread rms in the beam frame of $v^*_{\Delta U} = \beta c \dfrac{\gamma}{\gamma+1} \cdot \dfrac{\Delta U}{U} = 5.5 \cdot 10^3$ m/s.

Measurement Conditions

The experiments were performed using the phase-shift method described above. The experimental conditions can be summarized as follows.

The measurements were done with protons at the injection energy, 48 MeV, which corresponds to a cooler voltage of 26 kV. Since the phase shift method was used, the measurements were performed with a bunched beam, however using a rather low rf voltage around 10 $\hat{V}$. The phase shift was measured with a phase discriminator with integration time of 2 s to get a good signal-to-noise ratio. Changing the rf frequency instead of cooler voltage allowed us to make measurements in fine steps in relative velocity (1 Hz of 1129 kHz). The typical measurement step was 10 Hz.

We recorded transverse profiles with the magnesium-jet monitor as well as longitudinal bunch profiles. The longitudinal profiles have a distinct parabolic shape, which indicates that the beams are space charge dominated and have considerably smaller momentum spread than could be directly inferred from the bunch length [9].

RESULTS

In the following examples of results of longitudinal cooling force measurements are presented. The measurements were of different kinds: 1) Measurements for standard operational parameters of the cooler. 2) Measurements for different alignment angles between the electron and proton beams. 3) Measurements to study the influence of the non-straightness of the longitudinal magnetic field lines. 4) Measurements at different settings of the electron cooler to explore the friction force at various regimes of magnetization. In this report only the measurement data is presented and the detailed comparison with theoretical models will be presented elsewhere [10] [11].

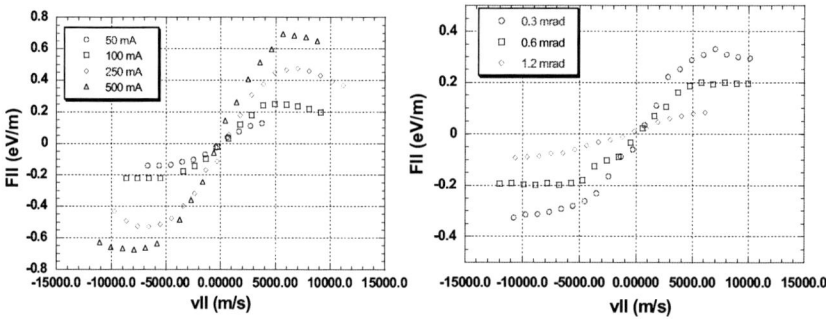

FIGURE 2. (Left) Longitudinal cooling force at standard cooler settings, B = 0.1 T and 48 MeV protons, for electron currents 50, 100, 250 and 500 mA. (Right) Longitudinal cooling force at different misalignment angles horizontally between the proton and the electron beams of 0.3, 0.6 and 1.2 mrad.

In Fig. 2 are shown results from cooling force measurements at standard cooler settings. The proton current was rather low, 40 µA, in order to reduce IBS.

In another set of experiments we measured the dependence of the cooling force on the alignment angle between the proton and the electron beam in both vertical and horizontal directions. The purpose was to increase the relative velocity in a controlled manner. In Fig. 2, right panel, are shown measurements for 0.3, 0.6 and 1.2 mrad misalignment angle horizontally. For calibration, both beam position monitors and an H^0 monitor was used. The H^0 monitor [12] is a silicon-strip detector situated 9 m from the cooler; a tilt angle of 1 mrad thus corresponds to a movement at the H^0 detector of 9 mm. The resolution of the H^0 detector is 1 mm, giving a resolution of about 0.1 mrad for the tilt angle.

In a third experiment the effects of the errors of the longitudinal solenoid field was investigated. The solenoid of the CELSIUS cooler is equipped with correction coils for correction of the solenoid field errors. Measurements of the magnetic field error in the CELSIUS cooler have been reported earlier [13] to be $\theta_e = 1$ mrad rms before corrections and $\theta_e = 0.2$ mrad rms after corrections. The measurements were carried out without corrections applied (DTCOR off) and with corrections applied (DTCOR on), see Fig. 3. It is clear that the cooling force is significantly reduced with a larger magnetic field error. The data can be used in comparisons with simulations of magnetic field errors [14].

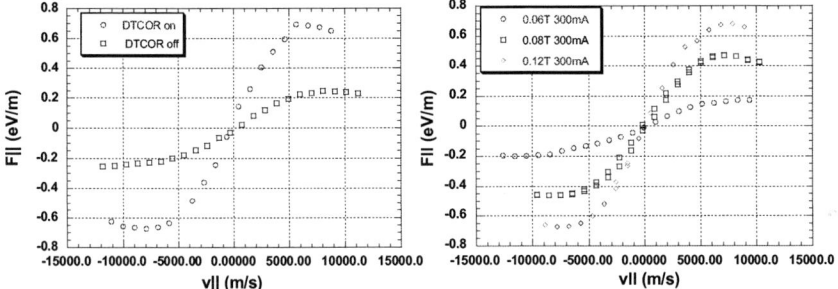

FIGURE 3. (Left) Longitudinal cooling force at different errors of the magnetic field in the cooling section. DTCOR on corresponds to an error of 0.2 mrad rms and DTCOR off corresponds to an error of 1 mrad rms. (Right) Longitudinal cooling force for different magnetic field 0.06, 0.08 and 0.12 T. The electron current was 300 mA.

We also compared the friction force for different magnetic fields as shown in Fig. 3 right panel. The idea here was to compare the cooling force at different degree of magnetization. Similar measurements were carried out for a number of different combinations of magnetic fields and electron currents.

ACKNOWLEDGMENTS

We would like to thank Dag Reistad and Ilan Ben-Zvi for numerous useful discussions and constant support during these studies. We are grateful to Oliver Boine-Frankenheim for taking an active role in planning of these experiments. A. Sidorin, A. Smirnov, and V. Ziemann acknowledge the support from INTAS grant 03-54-5584 "Advanced Beam Dynamics for Storage Rings". This work is supported by the US Department of Energy.

REFERENCES

1. Y. S. Derbenev and A. N. Skrinsky, Part. Acc. **8**, 235 (1978)
2. I. N. Meshkov Phys. Part. Nucl. **25**, 631 (1994).
3. V.V. Parkhomchuk, Nucl. Instr. Meth. A **441**, 70 (2000).
4. RHIC E-cooler, http://www.agsrhichome.bnl.gov/eCool
5. Y.-N. Rao et al. Proc. of Workshop on Beam Cooling and Related Topics, Bad Honnef, 2001.
6. D.D. Caussyn et al. Phys. Rev. E **51**, 4947 (1995).
7. M. Beutelspacher et al. Nucl. Instr. Meth. A **441**, 110 (2000).

8. H. Danared, Nucl. Instr. Meth. A **391**, 24 (1997)
9. S. Nagaitsev et al., 1993 Proc. of Workshop on Beam Cooling and Related Topics, Montreux, Switzerland, 1993, CERN 94-03, 405.
10. A.V. Fedotov et al. These Proceedings.
11. A.V. Fedotov et al. "Experiments towards high-energy cooling", submitted for publication (2005).
12. T. Bergmark et al. Nucl. Instr. Meth. A **441**, 70 (2000).
13. M. Sedlaček et al. "Design and construction of the Celsius electron cooler", 1993 Proc. of Workshop on Beam Cooling and Related Topics, Montreux, Switzerland, 1993, CERN 94-03, 235.
14. D. L. Bruhwiler et al. Proceedings of PAC 2005, 4206.

Experimental Benchmarking of the Magnetized Friction Force

A.V. Fedotov[1], B. Galnander[2], V.N. Litvinenko[1], T. Lofnes[2], A.O. Sidorin[3], A.V. Smirnov[3], V. Ziemann[2]

[1]*Brookhaven National Lab, Upton, NY 11973*
[2]*The Svedberg Laboratory, S-75121, Uppsala, Sweden*
[3]*JINR, Dubna, Russia*

Abstract. High-energy electron cooling, presently considered as essential tool for several applications in high-energy and nuclear physics, requires accurate description of the friction force. A series of measurements were performed at CELSIUS with the goal to provide accurate data needed for the benchmarking of theories and simulations. Some results of accurate comparison of experimental data with the friction force formulas are presented.

Keywords: electron cooling, beam dynamics, friction force
PACS: 29.28.Bd.,41.75.Lx

INTRODUCTION

High-energy magnetized electron cooling puts special demands on the accuracy of estimates of the cooling times [1]. For example, for parameters of the proposed RHIC cooler [2], the cooling time of Au ions at the energy of 100 GeV/u (γ=108) is of the order of 1000 s making a typical order of magnitude estimates not practical. The major goal of the present experiments was thus the accurate measurement of the magnetized friction force in order to provide the data needed for detailed benchmarking of theories and simulations. The other goal was to begin a detailed study of some features which are critical for proposed high-energy coolers [2], [3]. As part of a collaboration between BNL and European laboratories working on high-energy cooling for the RHIC-II [2] and FAIR projects [3], a variety of experiments were proposed in order to resolve the issues regarding the magnetized cooling force. The experiments, which were performed at CELSIUS [4] in December 2004 and March 2005, can be summarized as follows: 1) Accurate measurement of the longitudinal friction force for standard operational parameters of the cooler. 2) Cooling force measurements for different alignment angles between electron and ion beams. 3) Measurement of the time evolution of ion beam profiles as a result of simultaneous electron cooling and heating due to Intrabeam Scattering (IBS) before an equilibrium is reached. 4) Set-up the electron cooler to explore the friction force for various regimes of magnetization. 5) Study the effect of imperfections of solenoidal magnetic field lines. A detailed description of all the experiments performed can be found in Refs. [5, 6].

MEASUREMENT APPROACH

For low relative velocities between ions and electrons the longitudinal magnetized friction force increases linearly with velocity, reaching its maximum near the longitudinal velocity spread of the electron beam. Here, we used a measurement technique that allows very precise measurements in the region of low relative velocities, including accurately finding the maximum of the force. In this range, the simplest way of measurement is the phase-shift method. It uses a bunched ion beam and is based on measuring the phase difference between the RF system and the ion beam, resulting from the competition of the weak RF voltage and the longitudinal friction force.

Typically, the relative velocity difference is introduced by changing the energy of the electron beam. However, changing the electron acceleration voltage is usually done with a rather large voltage step. On the other hand, since this method employs bunched beams, changing the energy of the ion beam by changing the frequency of the RF cavity is more accurate. In our experiments, changing the RF frequency by a few Hz resulted in a very fine step in relative velocity. A similar technique was used successfully before, for example at IUCF [7]. In experiments reported in this paper, the longitudinal friction force curves were measured for protons at the injection energy of 48 MeV. More details on the experimental setup and accuracy of the measurements can be found in [6].

FITTING EXPERIMENTAL CURVES

The measured force for the magnetic field in the cooling section of B=0.1T and the electron currents of 500, 250, 100 and 50 mA (with good alignment of proton and electron beams) is shown in Fig. 1:

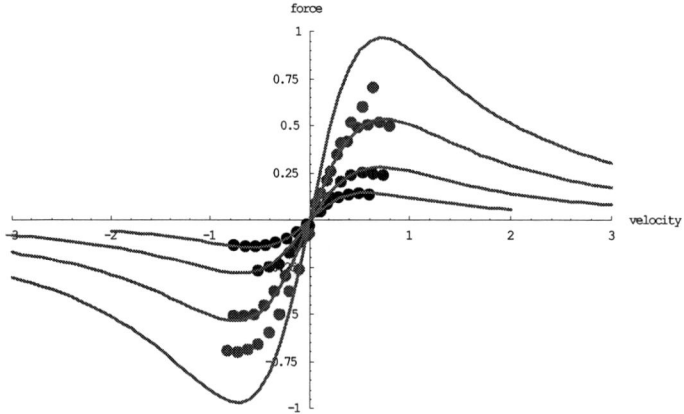

FIGURE 1. Longitudinal friction force in [eV/m] vs. velocity [10^4 m/s] for electron currents of 50 (blue, lowest set of data), 100 (red), 250 (pink), 500 mA (black, highest set of data) and magnetic field in cooling solenoid B=0.1T.

In Fig.1, we also plot the curves with fitted $\Delta_{e,eff}$ based on the empiric formula introduced by Parkhomchuk [8]

$$\vec{F} = -\vec{v}\frac{4Z^2 e^4 n_e L_M}{m}\frac{1}{(v^2 + \Delta_{e,eff}^2)^{3/2}}, \text{ with } L_M = \ln\left(\frac{\rho_{max} + \rho_{min} + \langle\rho_L\rangle}{\rho_{min} + \langle\rho_L\rangle}\right) \quad (1)$$

where v is the velocity of ion, Z is the ion charge, n_e is the electron density, e and m is charge and mass of electron, $\Delta_{e,eff}$ is an effective longitudinal velocity spread of electrons, and L_M is the Coulomb logarithm of the magnetized collisions.

For small currents of electron beam the major contribution to the effective velocity comes from the solenoid imperfections. One can see that for the electron currents of 50 (blue; lower curve), 100 (red), and 250 (pink) mA the curves based on Eq. (1) go nicely through the experimental data. The resulting effective velocity $\Delta_{e,eff}$ in these cases is about $0.75*10^4$ m/s.

For a high electron current (500 mA), the experimental data is significantly lower than the corresponding curve if the effective velocity (based only on the magnetic imperfections) comparable to the low currents is used. However, for high electron currents one needs to take into account the space-charge effect of the electron beam which results in the drift of the electrons in the crossed fields (the electric and magnetic fields of the electron beam and longitudinal magnetic field of the cooler). At a given radius r within the electron beam of uniform density distribution, the drift of the electron velocity is given by the expression:

$$v_d = \frac{2I}{B\beta\gamma^2}\frac{r}{a^2}, \quad (2)$$

where I is the current of electron beam, B is the strength of the magnetic field in the cooling section, β and γ are the relativistic parameters, and a is the radius of the electron beam.

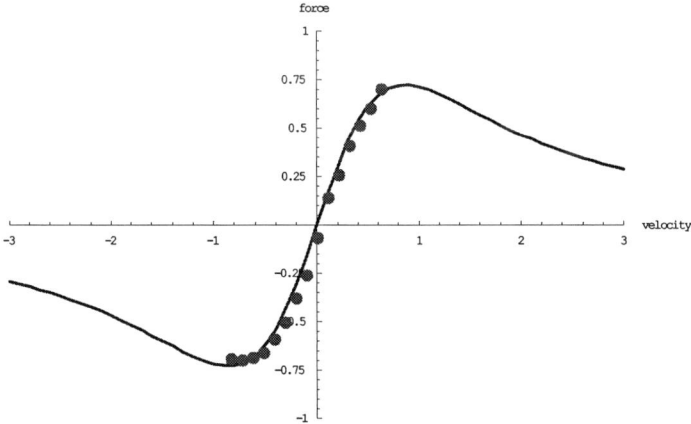

FIGURE 2. Measured friction force (points) in [eV/m] vs. velocity [10^4 m/s] for an electron current of 500 mA (B=0.1T) and the curve based on Eq. (1) with additional contribution to the effective velocity from the space charge of the electron beam.

For the set of data with 500 mA (B=0.1T) and measured parameters of the proton distribution the resulting drift velocity is around $0.7*10^4$ m/s (calculated at an rms radius of the proton beam). We can take this into account by adding an extra term to the effective velocity. This results in the solid curve shown in Fig.2 that is based on Eq. (1) and goes through the experimental data even for high electron beam currents.

The friction force with the numerical factor given in Eq. (1) agrees well when directly compared to the experimental data. However, pre-cooled proton beam, with which the friction force measurements are made, has some finite values of an rms emittance and momentum spread. To provide an accurate comparison with the experimental data one also needs to measure an rms distribution of the proton beam during the measurement of the cooling force. The single-particle friction force formula should be then averaged over the proton distribution:

$$\langle F \rangle = C \frac{4\pi Z^2 e^4 n_e}{m\sqrt{2\pi} \Delta_\perp^2 \Delta_\parallel} \int_0^\infty \int_{-\infty}^\infty \frac{v_\parallel \cdot L_M(v_\perp, v_\parallel, v_{eff})}{(v_\perp^2 + v_\parallel^2 + v_{eff}^2)^{3/2}} \exp\left(-\frac{v_\perp^2}{2\Delta_\perp^2} - \frac{(v_\parallel - v_0)^2}{2\Delta_\parallel^2}\right) v_\perp dv_\parallel dv_\perp \qquad (3)$$

where $\Delta_\perp$ and $\Delta_\parallel$ are measured rms velocities of the proton distribution, and the integrals are performed over the transverse and longitudinal velocities of the protons. In our experiments, an rms distribution of the proton beam was carefully measured for each set of the experimental curves.

In principle, there could be different approaches to a fitting procedure based on such averaging. The first one assumes that the numerical coefficient C in the expression for the single-particle force is known (for example, $C=1/\pi$ as in Eq. (1) or $1/(2\pi)^{1/2}$ as in Ref. [9] for low relative velocities), while v_{eff} is a fitting parameter. In our experiments, the measured rms velocity spread of the proton beam was rather large so that fitted v_{eff}, as a result of such averaging, became very small (v_{eff}=0.1-0.2*10^4 m/s is needed for the case shown in Fig. 3). For such low effective velocity, the single-particle friction force would have a maximum around this small value having a non-linear decrease of the force for larger velocities. As a result, significant large-amplitude oscillations for the relative velocities corresponding to the non-linear part of the force are expected [7]. However, we did not see such oscillations for the velocities in this range. In fact, we measured the maximum of the friction force by carefully recording the onset of such large-amplitude oscillations but at significantly larger velocities.

The second approach is to take v_{eff} as a known parameter based on recorded oscillations of the longitudinal profile and measured maximum of the friction force. The fitting parameter is then an unknown numerical coefficient C. In such an approach, for our parameters and the region of low relative velocities discussed here, we find some enhancement for the numerical coefficient C compared to the factor $1/\pi$ or $1/(2\pi)^{1/2}$. However, the uncertainty of the numerical factor C found by such a procedure is rather large due to significant error bars in the measurements of the proton distribution as well as the values of the effective velocity which is found from the measured distribution but needs to be used in the single-particle expression. Rather than quoting some empiric values for the coefficient C at this point, we plan to determine numerical factors for the friction force expressions using numerical studies with the VORPAL code, by studying the velocities of ion which make various angles with respect to the magnetic field lines [10]. By accurately comparing with the formulae and experimental data, we should be able to have an estimate of the numeric coefficients for the magnetized friction force with a reasonably good accuracy.

A detailed study of the friction force both for the low and high relative velocities is presently in progress [11]. A systematic benchmarking of each of the performed experiments will be reported elsewhere.

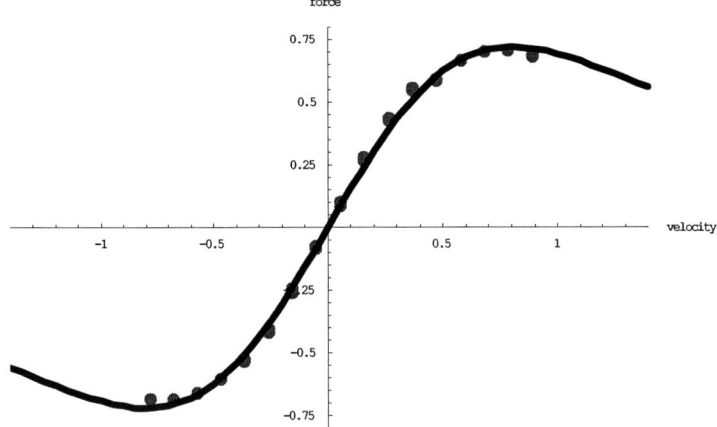

FIGURE 3. Longitudinal friction force in [eV/m] vs. velocity [10^4 m/s] for electron current of 300 mA (B=0.12T) and the force averaged according to the Eq. (3).

ACKNOWLEDGMENTS

We would like to thank Dag Reistad and the The Svedberg Laboratory for providing beam time and support during these experiments. We thank Ilan Ben-Zvi for numerous useful discussions and constant support during these studies. We are grateful to Oliver Boine-Frankenheim for taking an active role in planning of these experiments. A. Sidorin, A. Smirnov, and V. Ziemann acknowledge the support from INTAS grant 03-54-5584 "Advanced Beam Dynamics for Storage Rings". This work is supported by the US Department of Energy.

REFERENCES

1. A.V. Fedotov et al., Proceedings of PAC'05 (Knoxville, TN, 2005), TPAT090.
2. RHIC E-cooler, http://www.agsrhichome.bnl.gov/eCool
3. FAIR facility, http://www.gsi.de/GSI-Future/cdr
4. M. Sedlavcek et al., Proc. of Workshop on Cooling (Montreux, Switzerland, October 1993), CERN 94-03, 235.
5. A.V. Fedotov et al., "Experiments towards high-energy cooling", submitted for publication (2005).
6. B. Galnander et al., these proceedings.
7. D.D. Caussyn et al. Physical Review E 51, 4947 (1995).
8. V.V. Parkhomchuk, Nuc. Instr. Meth. A 441, 9 (2000).
9. Ya.S. Derbenev and A.N. Skrinsky, Particle Accelerators 8, p. 235 (1978).
10. D.L. Bruhwiler et al., "VORPAL simulation for CELSIUS parameters", unpublished (2005).
11. A.V. Fedotov et al., "Detailed studies of electron cooling friction force", these proceedings.

Electron Cooling of Intense Ion Beam

J. Dietrich[1], V.Kamerdjiev[1], Yu.Korotaev[2], R.Maier[1], I.Meshkov[2], D.Prasuhn[1], A.Sidorin[2], A.Smirnov[2], J.Stein[1], H.Stockhorst[1]

[1]*FZJ, Juelich, Germany*
[2]*JINR, Dubna, Russia*

Abstract. Results of experimental studies of the electron cooling of a proton beam at COSY (Juelich, Germany) are presented. Intensity of the proton beam is limited by two general effects: particle loss directly after the injection and development of instability in a deep cooled ion beam. Results of the instability investigations performed at COSY during last years are presented in this report in comparison with previous results from HIMAC (Chiba, Japan) CELSIUS (Uppsala, Sweden) and LEAR (CERN). Methods of the instability suppression, which allow increasing the cooled beam intensity, are described. This work is supported by RFBR grant # 05-02-16320 and INTAS grant #03-54-5584.

Keywords: Electron Cooling, Beam Stability.
PACS: 29.20.Dh; 29.27.Bd

INTRODUCTION

Electron cooling method is widely used to increase an ion beam density in the six dimensional phase space. To increase intensity of the stored beam the stacking-cooling procedure is applied: injected beam is cooled to small dimensions and thereafter a new injection can be performed into free part of the ring acceptance. Such a procedure is repeated to saturation of the stored beam current. At high phase space density one of the general limitations of the ion beam intensity is related to development of coherent instabilities of the stored beam. Different types of instabilities developing at a high beam phase density were observed at a few coolers [1]. Specific instability, which appears directly at electron cooling application, was firstly observed in CELSIUS (Uppsala, Sweden) storage ring [2]. This instability leading to decrease of the ion beam life-time in the presence of an electron beam was named "electron heating".

At present, the main application of electron cooling at COSY is the preparation of low-emittance ion beams (protons or deuterons) to be used after acceleration and extraction. The large emittance of the ion beam after stripping injection of H⁻ or D⁻ ions is strongly reduced so that the experimental requirements are much better fulfilled. However, the intensity of such low-emittance, small momentum-spread beams is governed by particle losses due to instabilities.

Both types of instabilities – "electron heating" and coherent oscillations - were observed at COSY since beginning of the electron cooling system operation. Results of the instability investigations performed at COSY during last years are presented in this report in comparison with previous results from CELSIUS, LEAR and HIMAC. At COSY the experiments were done with coasting proton beam at injection energy (45 MeV). Especial attention is paid to study of an influence of the electron beam neutralization on stability of the circulating ion beam.

"ELECTRON HEATING"

The initial particle losses are resulted from strong diffusion process which power is proportional to the electron beam current. The diffusion takes place even when electron beam energy is detuned and the electron cooling does not work (Fig. 1). The proton beam lifetime increases with decrease of the proton beam intensity (Fig. 2) and after about 10 sec reaches saturation at relatively long value.

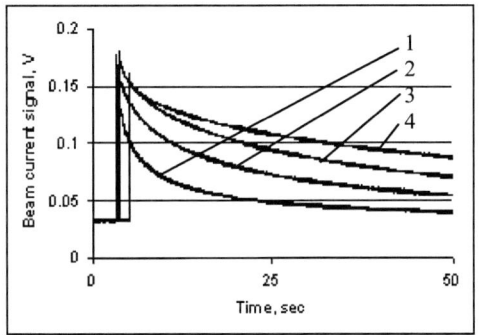

FIGURE 1. Proton beam current as a function of time at electron beam current of 243 (1), 95 (2), 45 (3) and 0 (4) mA and the electron energy detuned by 2 keV from optimum for cooling.

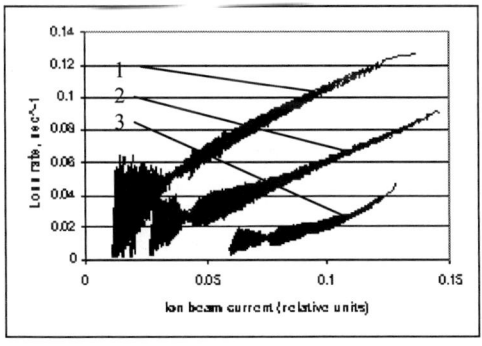

FIGURE 2. Rate of the proton losses versus the beam intensity. Electron current is 243 (1), 45 (2) and 0 (3) mA.

Decrease of the beam intensity is related mainly to a loss of the particles with amplitude of betatron oscillations larger than the electron beam radius. Nonlinearity of the electron beam self-fields can be a reason of the loss [2]. Other explanation proposed in [3] is based on assumption, that the interaction of the proton and electron beam leads to the proton beam coherent oscillations and electric field of these oscillations generate the proton beam heating. However, no coherent oscillations can be observed in COSY beam just after injection (see below).

At COSY the proton beam life-time in presence of the electron beam decreases by about 10 times, at CELSIUS this effect is even more pronounced – the proton beam at injection (which practically coincides with injection energy at COSY) decreases by up to 100 times in presence of intensive electron beam.

COHERENT INSTABILITY

Single Injection

After fast losses caused by the "electron heating" and some period of the beam cooling new particle loss can appear due to coherent instability. After injection at COSY the initial losses take place during first 5 sec of the cooling process (Fig. 3, lower curve). The cooling process is accompanied by H^0 production in the cooling section (upper curve in the Fig. 3) and H^0 count rate increases during initial particle loss. It reflects the fact that the lost particles have an amplitude of betatron oscillations larger than electron beam radius. The H^0 count rate saturates at approximately the same moment as the proton beam intensity stabilizes.

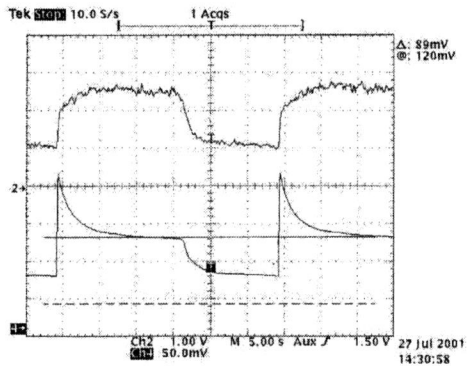

FIGURE 3. Proton beam and H^0 flux intensity vs time at COSY during repeated injection at low frequency.

Coherent oscillations of the circulating beam start after about 7 seconds of the cooling but initially do not lead to the particle loss. In the case presented in the Fig. 3 initially the dipole oscillations appeared in the longitudinal and horizontal plane. These oscillations were registered in the spectrum of the beam Schottky noise, or directly from pick-up electrodes using oscilloscope. After some period of time the horizontal oscillations were transformed to the vertical ones. The coherent oscillations of the beam in vertical plane lead to fast particle losses due to smaller value of the vertical acceptance.

Cooling - Stacking

The instability nature of the intensive proton beam in the cooling-stacking mode [4] is the same as in the case of single injection - coherent oscillations of the stored ion beam: initially in the longitudinal and horizontal degree of freedom and, after some time, the oscillations "jump" into the vertical plane. Coherent instability during stacking leads to random variations of the beam intensity near saturation and limits the stored beam intensity. More probably explanation of the instability origin is the plasma oscillations of the ion beam in accordance with the theory developed in [3]: the coherent oscillations appear when the ion beam density increase to the level corresponding to more than one plasma oscillation of the ions during passage of the electron beam.

Instability Suppression

The instabilities of cooled ion beam can be avoided by preservation of "overcooling" of the beam core. For instance at CELSIUS and later at COSY an additional external heating of the beam in longitudinal and transverse degrees of freedom and/or misalignment of the electron beam were tested for instability suppression. However, both of these methods stabilize the stored beam but do not give a substantial increase of its intensity. The external heating leads to higher value of the particle loss after injection due to decrease of dynamic aperture of the ring. A misalignment of the electron beam leads to non-Gaussian distribution in the transverse plane and, in principle, can provoke the chromatic instability. The experience of the CELSIUS cooling system operation demonstrated that more effective way to suppress the stored beam instability is artificial increase of the electron beam energy spread by its modulation. It was demonstrated that most effective is square-wave modulation at amplitude of 50 V when electron energy is of 115 keV [5].

Another method developed recently but did not tested yet in a cooler is formation in the electron gun the hollow electron beam [6].

The instability threshold depends not only on the beam current and on the magnitude of the transverse machine impedance but also on the particle momentum spread. Therefore to avoid beam losses the chromaticity should be negative below transition. Stabilization of the proton beam by adjustment of the chromaticity accomplished with sextupoles was demonstrated at COSY [7] (Fig. 4).

The injected proton beam (current of 2 mA) with 293.8 MeV/c (45 MeV) is electron cooled for 10 s and then is accelerated within 1.8 s to the flat top momentum of 2085 MeV/c. In the first cycle a strong vertical oscillation occurs and leads to beam loss soon after injection. In this case the tunes and chromaticities measured after 8 s were equal to $Q_x = 3.587$, $Q_y = 3.696$ and $\xi_x = -2.8$, $\xi_y = 0.3$, respectively. In the second cycle the sextupole family located in the arc section of the ring was powered only after injection until acceleration started with -1.7 % to shift the vertical chromaticity ξ_y from $+0.3$ to -0.6 within 100 ms. By this measure the coherent vertical betatron oscillations at injection were significantly suppressed and intensity of the accelerated beam was increased by two times (see "flat top" – FT level).

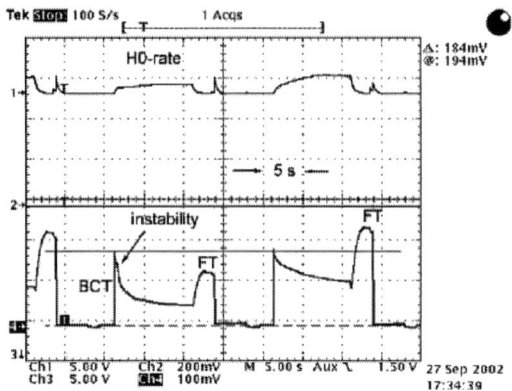

FIGURE 4. In the first cycle, without sextupole, a strong beam loss after injection due to coherent vertical betatron oscillations (instability) is visible in the beam current signal (BCT). In the next cycle this beam loss is compensated by the sextupole family. The flat top (FT) intensity is doubled. The horizontal line in the lower picture indicates the same number of injected protons in both cycles (100 mV BCT signal corresponds to 1 mA proton current).

Feed back system application allows to increase stored beam intensity, as it was demonstrated, for instance, at LEAR [1]. There strong transverse instabilities occurred once the intensity exceeds a few 10^8 protons. A large number of modes was observed at all energies accessible with electron cooling (5.3 – 50 MeV). The feedback system of the bandwidth of 0.1 – 70 MHz was implemented to stabilize the first 100 or so dipole modes. It was then possible to store up to about 3×10^9 protons with the small emittances given by the equilibrium between intrabeam scattering and cooling in the energy range accessible. Higher intensities, up to 8×10^{10} protons, could be cooled to the intrabeam scattering limit, when the stochastic cooling system of a reduced gain with a bandwidth up to 500 MHz was used as additional dipole damper.

At COSY the vertical feed back system was designed and implemented after investigations of the particle loss nature. Its bandwidth up to 70 MHz was chosen to suppress all the exiting modes. The system application made it possible to stabilize the cooled proton beam at a level of 2×10^{10} particles (1.8 mA) after a single injection. With the stacking technique a maximum of 1.2×10^{11} cooled protons (9.2 mA) at injection energy were stored without instability, which is about two times higher than without a feed back system application [8].

EQUILIBRIUM BEAM EMITTANCE

The size of the cooled proton beam can be efficiently measured with the neutral particle flux profiles originating from the recombination of protons with electrons in the cooling section (Fig. 5). Using the beta functions obtained by a lattice model, the proton beam dimensions at the position of the electron cooler can be derived (Fig. 6).

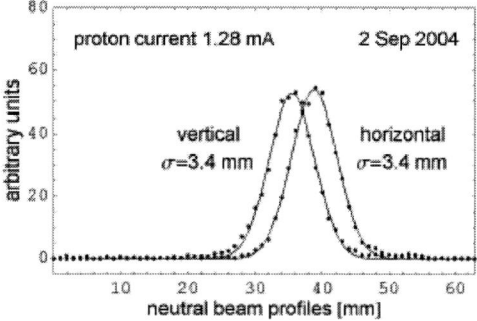

FIGURE 5. Neutral beam profiles measured with two MWPC detectors in the horizontal and the vertical planes located at the distance $D = 24.3$ m downstream the electron cooler. Wire distance 1 mm. The measured profiles are perfectly fitted by Gaussian curves. Proton beam current of 1.28 mA.

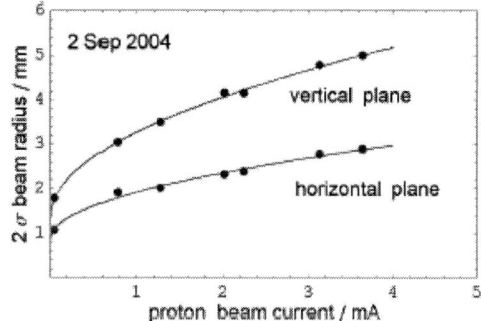

FIGURE 6. Proton beam sizes vs proton beam current at the position of the electron cooler calculated with $\beta_h = 7.5$ m, $\beta_v = 14.5$ m. Electron current 170 mA. $Q_h = 3.620$, $Q_v = 3.645$.

The beam sizes are well fit to the functional dependence

$$r_{h,v} = a_{h,v} + b_{h,v}\sqrt{I_p}, \qquad (1)$$

where a and b are constants, I_p is the proton beam current. The corresponding emittances of the measured sizes range from 0.15 to 1.1 π·mm·mrad horizontal and from 0.22 to 1.7 π·mm·mrad vertical. The square root dependence of the beam profile width dominates at higher proton currents, i.e. the emittances grow almost linearly with increasing proton intensity. That corresponds to equilibrium between cooling and heating at a constant tune shift value, i.e. general heating effect at cooling of intense ion beam is defined by tune resonances. Such resonances are often associated with beam losses, but if the resonance is weak enough it may lead to amplitude growth only that is equivalent to heating.

At low ion beam intensity the equilibrium is determined by intrabeam scattering in the ion beam and in this case the beam emittance is scaled with the particle number N as [9]

$$\varepsilon \sim N^{2/3}. \tag{2}$$

ION CLOUD IN AN ELECTRON COOLING SYSTEM

The ions of residual gas can be trapped in the electron beam in transverse direction by the electron beam electric field. When the vacuum chamber radius is varied along the cooling system from cathode to collector the condition of the ion trapping in longitudinal direction can be also met. The trapped ions partially compensate the electron beam electric field and this effect leads to so called "natural neutralization" of the electron beam. The level of the natural neutralization in COSY was measured by two independent methods and is about 37 %, that is in a good agreement with estimations based on geometry dimensions of the vacuun chamber [10].

Thus, in the cooling section the circulating ion beam interacts with the primary electron beam, secondary electrons and different species of neutralizing ions, that can lead to various types of multi-stream instabilities [11]. The stability of antiproton beam at HESR ring of FAIR project (GSI) in presence of neutralizing ions was descussed in [12] where was shown that even a few percent of the neutralization level can lead to sufficient decrease of the instability threshold. Influence of the neutralizing ions on the circulating beam stability was experimentally investigated at HIMAC [13] and later at COSY.

The ions trapped in the electron beam oscillate in the longitudinal magnetic field of the cooler solenoid and electric field of the electron beam with frequency determined by the following formula

$$\omega = \sqrt{\omega_i^2 (1-\eta_{neutr}) + \omega_B^2/4} \pm \omega_B/2, \tag{3}$$

where $\omega_B = \dfrac{ZeB}{Am_p}$ and $\omega_i^2 = \dfrac{Ze^2 n_e}{2Am_p}$ are the cyclotron frequency of the ion of the mass of Am_p and charge of Ze in the magnetic field B, and the ion plasma frequency in the electron beam of the density of n_e correspondingly. To control the neutralization level and to clear the electron beam from one of the ion spesies one can use resonant excitation of the ion oscillations with transverse sinusoidal electric field. This field is applied to pair of electrodes, so called "shaker electrodes", which are, for instance, position pick-ups in the cooling section. At shaker frequency equal to the ion oscillation frequency ω the ions leave very fast the electron beam and neutralization level is changed. That leads to a change of potential at the electron beam axis and, as result, to a change of the revolution frequency of the ions circulating in the ring. This change of the revolution frequency can be compensated by change of the electron gun cathode potential (Fig. 7).

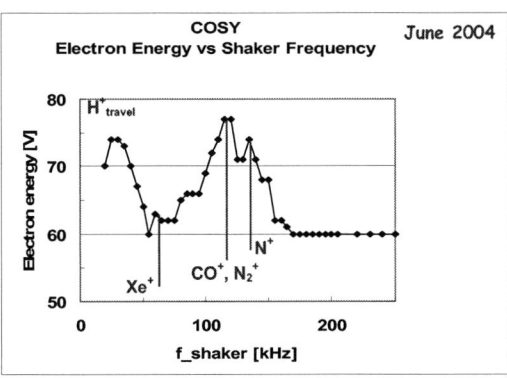

FIGURE 7. Spectrum of resonance shaker frequencies measured at COSY cooler. Transverse shaking. Electron beam current is 170 mA, electron energy is 24.5 keV, magnetic field in the cooling section is 800 G.

A few peaks in the shaker frequency range from 70 to 160 kHz correspond to Xe^+, CO^+, N_2^+, N^+ and highly charged ions. In accordance with formula (3) the peak in the frequency range from 20 to 50 kHz corresponds to ions at $A/Z > 100$. Such heavy ions are absent in the COSY vacuum chamber. To clarify nature of the peak in the small frequency range one can provide ion heating in the same frequency range using longitudinal electric field. In this case the shaker signal was applied to both plates of PU station in the same phase. At longitudinal shaking (Fig. 8) the peaks corresponding to transverse ion oscillations dissapear. Presence of the peak in small frequency range indicates that the peak corresponds to the ions traveling along the cooler. More probably that peak corresponds to H^+ ions.

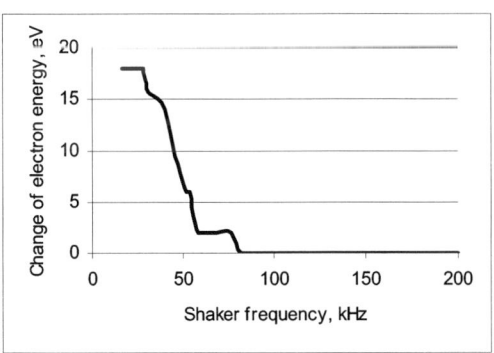

FIGURE 8. Spectrum of resonance shaker frequencies measured at COSY cooler. Longitudinal shaking.

The measurements of the frequency of the ion transverse oscillations were performed at different shaker voltage, which was varied in the range from 10 to 60 V. Width of the resonant peaks increases with increase of the voltage (Fig. 9).

The peak shape is typical for nonlinear resonance and at a high value of the shaker voltage a hysteresis behavior appears. At increase of the shaker frequency the neutralization level decreases monotonically, but at some frequency it jumps to initial value. At decrease of the shaker frequency the opposite jump of neutralization level takes place at smaller value of the frequency.

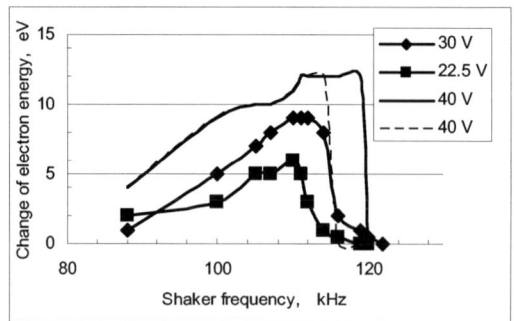

FIGURE 9. Shape of the resonant peak at 90 – 120 kHz as function of the shaker voltage amplitude. At 40 V there is a hysteresis effect – solid line corresponds to an increase of the shaker frequency, dashed line – to a decrease.

In the experiments at HIMAC the clearing of the electron beam from the residual gas ions led to increase of the stored beam intensity after stacking by about 2 times. Simultaneously the Schottky noise power of the circulating beam was of sufficiently less level [13].

Influence of the neutralization on the proton beam stability at COSY is well illustrated in the Fig. 10. Without shaker or at nonresonant shaker frequency a coherent instability leading to fast particle losses appears after approximately 25 s (Fig. 10 a, b). At clearing the electron beam from one specie of the ions (the frequency of 114 kHz, molecular nitrogen ions) the instability develops after about 70 s and the ion loss rate at instability decreases by about 3 times (Fig. 10 c).

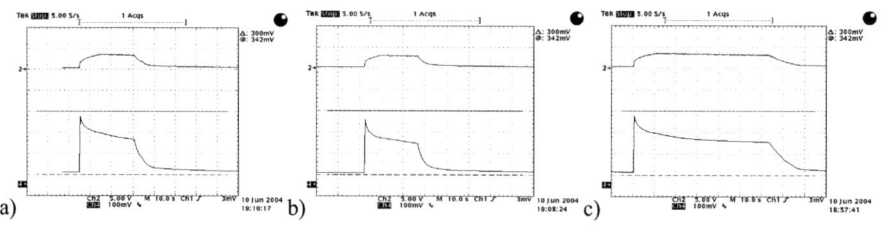

FIGURE 10. Ion current versus time: a) shaker is off, b) non resonant excitation, c) excitation on resonant frequency of 114 kHz.

This indicates that the threshold ion beam phase space density increases and instability increment decreases. However, in the cooling-stacking process the maximum beam intensity is slightly affected by the shaker in the case of monochromatic excitation. It means that at high proton beam intensity the beam stability can be determined by interaction with the ion cloud consisting of other species.

CONCLUSIONS

The experimental studies of stacking process at COSY cooler-synchrotron have shown the limitation of the ion beam intensity due to three phenomena related each to other:
1) fast losses directly after injection,
2) losses in the cooled proton beam caused by a coherent instability with transformation of horizontal oscillations into vertical ones,
3) trapping of residual gas ions in the cooling electron beam.

The fast initial losses are supposed to be a result of an influence of the electron beam field nonlinearity. Another explanation is related to plasma oscillations in the ion and electron beams, which lead to noise of big amplitude reducing the ion lifetime.

The second stage of the loss takes place when coherent oscillations appear in the cooled ion beam. The transformation of horizontal coherent oscillations into the vertical ones leads to particle losses due to smaller value of the vertical acceptance. One of the effective ways for coherent instability suppression is a feedback system application.

Comprehensive explanation of these effects was not done yet and they have to be studied in more details. Last results of the experiments at COSY and HIMAC demonstrated importance of the electron beam neutralized state control especially at cooling of intense beams. At low ion beam intensity neutralization of the electron beam can decrease the cooling time, at high intensity the ion cloud in the cooling section can provoke instability of the cooled beam.

REFERENCES

1. J.Bosser et al., Stability of cooled beams, NIM A 441 (2000), pp. 1-8.
2. D.Reistad et al., Measurements of electron cooling and «electron heating» at CELSIUS, Workshop on Beam Cooling, Montreux, 1993, CERN 94-3, pp. 183 – 187.
3. V.Parkhomchuk, D.Pestrikov, Coherent instabilities at electron cooling, Workshop on Beam Cooling, Montreux, 1993, CERN 94-3, pp. 327 – 329.
4. V.Kamerdzhiev et al., Instability phenomena of electron-cooled ion beams at COSY, NIM A 532 (2004), pp. 285-290.
5. L.Hermansson, D.Reistad, Electron cooling at CELSIUS, NIM A 441 (2000), pp. 140-144.
6. A. V. Bubley, V. M. Panasyuk, V. V. Parkhomchuk and V. B. Reva, Measuring a hollow electron beam profile, NIM A 532 (2004), pp. 413 - 415.
7. H. Stockhorst, H.J. Stein, D. Prasuhn and R. Maier Stabilization of an Electron Cooled Proton Beam at Injection with Sextupoles, IKP annual report, Juelich, 2002.
8. V. Kamerdzhiev, J. Dietrich, I.N. Meshkov, I. Mohos, A.O. Sidorin, H.J. Stein, Application of Transverse Feedback for Electron Cooled Ion Beams at COSY, IKP annual report, Juelich, 2003.
9. M. Steck, K. Beckert, F. Bosch, et. al., Electron cooling of highly charged heavy ions at the ESR, in Proceedings of Workshope on Beam Cooling and Related Topics, Montreux, 1993, CERN 94-3, pp. 395 – 399.
10. A. Sidorin, I.N. Meshkov, H.J. Stein, H. Stockhorst Natural Neutralization in the Electron Beam of the COSY Electron Cooler, IKP annual report, Juelich, 2001.
11. A.V.Burov, Secondary particle instability in storage rings with electron cooling, in Proceedings of Workshope on Beam Cooling and Related Topics, Montreux, 1993, CERN 94-3, pp. 230 – 234.
12. P. Zenkevich, A. Dolinskii and I. Hofmann, Dipole instability of a circulating beam due to the ion cloud in an electron cooling system, NIM A 532 (2004), pp. 454 – 458.
13. E.Syresin, K.Noda, T.Uesugi, I.Meshkov, S.Shibuya, Ion lifetime at cooling stacking injection in HIMAC, HIMAC-087, May 2004.

Attainment of a High-Quality Electron Beam for Fermilab's 4.3 MeV Cooler[1]

A. Shemyakin[*,2], A. Burov[*], K. Carlson[*], M. Hu[*], G. Kazakevich[$],
B. Kramper[*], T. Kroc[*], J. Leibfritz[*], S. Nagaitsev[*], L. Prost[*], S. Pruss[*],
G. Saewert[*], C.W. Schmidt[*], S. Seletskiy[&], M. Sutherland[*], V. Tupikov[*],
A. Warner[*]

FNAL, Batavia, IL 60510, U.S.A.;
[$]*Budker INP, Novosibirsk, 63090 Russia;*
[&]*University of Rochester, Rochester, U.S.A*

Abstract. The recent demonstration of electron cooling of antiprotons in the Recycler ring required a stable 4.3 MeV electron beam with a DC current of hundreds of mA and an angular spread in the cooling section of a fraction of a mrad. This paper describes the achieved parameters of the Fermilab cooler's electron beam and details of operation.

Keywords: Electron cooling, electrostatic accelerator, electron beam
PACS: 29.17.+w, 29.25.Bx, 29.27.Eg, 41.75.Ht

INTRODUCTION

In 2004-2005, an electron cooling device was installed in the Fermilab's Recycler ring to assist in accumulating and cooling a large number of antiprotons. While employing the same energy recovery (or recirculation) scheme as existing low-energy coolers, the Recycler electron cooler (REC) differs from them dramatically in several aspects. First, the REC electron energy of 4.32 MeV is higher by an order of magnitude; second, the gun cathode is immersed in a longitudinal magnetic field, but most of the beam transport line is field-free; third, the electrons are not heavily magnetized in the cooling section, making only two Larmor rotations over the section's 20 meter length; and finally, the cooler shares the tunnel with a 150-GeV synchrotron, the Main Injector (MI). These features make the machine commissioning and operation very specific. Below we describe the setup and the beam properties that allowed the first demonstration of electron cooling in the MeV electron energy range [1].

---

1 Fermilab is operated by Universities Research Association Inc. under Contract No. DE-AC02-76CH03000 with the United States Department of Energy

2 E-mail: Shemyakin@fnal.gov

SETUP DESCRIPTION

The mechanical schematic of the setup is shown in Fig.1. The electron beam is generated by an electrostatic accelerator, the Pelletron [2], and transported through a beam "supply" line to the cooling section where it interacts with antiprotons circulating in the Recycler. After separation of the beams by a 180 degree bend, electrons move through the "return" beam line back to the Pelletron, are decelerated in the second Pelletron tube, and the beam is absorbed in a collector at the kinetic energy of 3.2 keV. The cooler optics is described in Ref. [3]. Some of the cooler parameters are listed in Table 1.

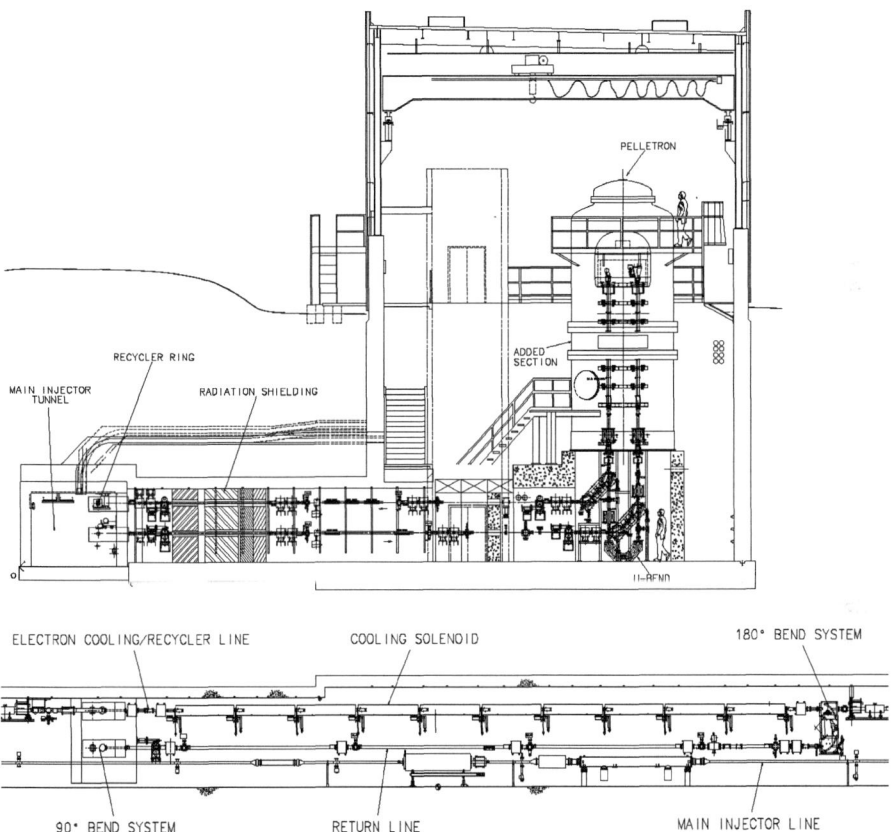

FIGURE 1. The top insert is an elevation view showing the Pelletron, the acceleration and deceleration beam lines, the transfer lines passing through connecting enclosure to Recycler ring, and the cross-section of the Main Injector tunnel which houses the Recycler ring. The bottom insert is an elevation view of the Main Injector tunnel showing the 90°-bend system which injects the electron beam from the transfer line into the Recycler ring, the cooling section of Recycler, the 180°-bend system which extracts the electron beam from the Recycler, and the return line.

The vacuum chamber is pumped down by ion and titanium sublimation pumps. The typical diameter of the beam line vacuum chamber is 75 mm, but the aperture is limited by the BPM's inner diameter of 47 mm. In the cooling section, the beams see a

47 mm round chamber except for the narrow diagnostics/pumping gaps between modules.

When both main bending magnets under the Pelletron are turned off, the beam can be recirculated through a short beam line, denoted as U-bend in Fig.1. This so-called U- bend mode was used for commissioning purposes. For instance, in this mode we were able to reach DC beam currents up to 1 A at the nominal energy.

TABLE 1. Some parameters of the cooler.

Parameter	Unit	Value
Electron energy	MeV	4.338
Beam current used for cooling	A	0.2
Maximum DC beam current	A	0.6
Magnetic field in the cooling section	G	105
Beam radius in the cooling section	mm	4.2
Pressure	nTorr	0.2 - 1
Total length of the beam line	m	90

A simplified electrical schematic of the accelerator is shown in Fig.2. Typically, the Pelletron chain supplies 115 µA, from which 95 µA go to the resistive strings and 20 µA are used to provide the voltage stabilization and compensation of the beam loss.

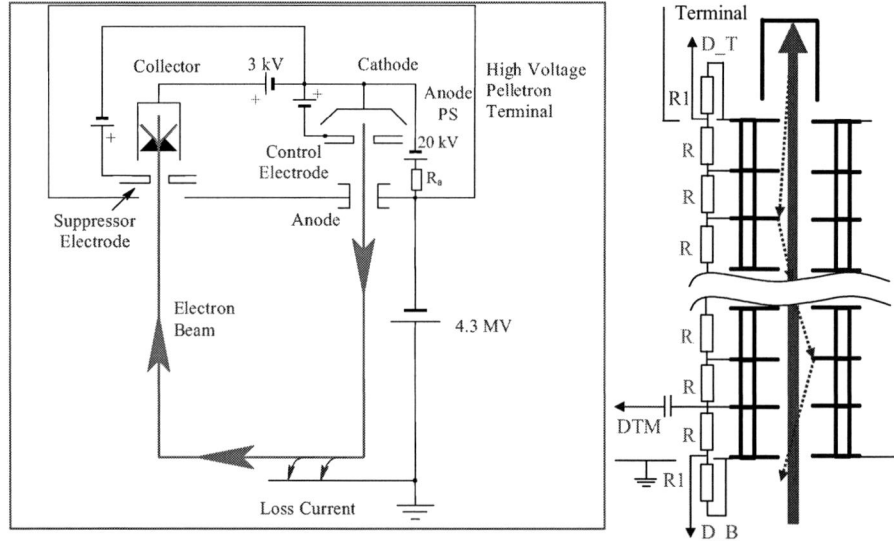

Figure 2. A simplified electrical schematic of the accelerator. The left insert shows the recirculation scheme. The sketch on the right represents the deceleration tube with its resistive divider. The divider current is measured at the top (D_T) and at the bottom (D_B). Fast changes in the potential distribution along the tube can be analyzed by measuring the AC component of the signal from the last tube electrode (Tube Monitor, DTM). Diagnostics on the acceleration side are identical.

Diagnostics

The beam line is equipped with several types of beam diagnostics. The beam trajectory is measured by 31 pairs of capacitive pickups, referred to further as BPMs

[4]. The BPMs can work in four modes: the pulse mode (2 μsec pulses at 1 Hz); negative pulsing, when a DC beam is interrupted for 2 μsec at 1 Hz rate; a sinusoidal modulation of a DC beam current at the frequency of 32 kHz; and the antiproton beam position measuring mode, when the signal is processed at the Recycler ring revolution frequency of 89 kHz.

Eleven scrapers, installed in the gaps between the modules of the cooling section, are used to measure the beam envelope. Each scraper is a retractable copper plate with a 15 mm round orifice. Also, in the pulse mode the beam size can be estimated by a multi-wire harp (fifty tungsten, 25-μm-diameter wires in each of two planes separated by 0.5 mm).

Several optical transition radiation (OTR) monitors have been installed in the beam line. However, unexpectedly high radiation from the Main Injector destroyed all cameras in the MI tunnel, and only two such monitors, located under the Pelletron, were used in measurements.

Protection System

During operation of the cooler prototype, twice its vacuum chamber was drilled through by the electron beam. To avoid such accidents at the cooler, several layers of protection have been implemented.

The most important one is a system that closes the gun if the Pelletron terminal voltage drops due to large beam losses. A capacitive pickup plate, positioned at the Pelletron tank opposite to the terminal shell, measures the AC component of the terminal voltage, which is compared to a threshold. If the voltage drop is too large (usually > 5 kV), the gun is closed in ~ 1 μs.

In addition, a dedicated controller monitors radiation loss monitors placed around the cooler's beam line. Radiation above a preset threshold results in closing of the gun in ~ 1 ms. Finally, the control system prevents manipulating critical power supplies in an unsafe manner.

RECIRCULATION STABILITY

To provide cooling, the electron beam should stay for hours at the nominal energy and the current of hundreds of mA. One of the necessary conditions for the stable operation is low current losses. At the beam current of 0.4 A, the relative value of the lost current is $1.2 \cdot 10^{-5}$ (Fig.3). Comparison of this number with the relative loss in the U-bend configuration, $6 \cdot 10^{-6}$, and with losses observed at a test bench, $\sim 2 \cdot 10^{-6}$ with the same collector, hints at a mechanism of losses dependent on the beam line length. The radiation monitors indicate that practically all losses occur under the deceleration tube; therefore, they are related to the collector efficiency. Simulations suggest that the mechanism is the reflection from the collector entrance of electrons that lost part of their energy due to intra-beam scattering [5].

At the level of 10 μA or below, this additional load for the Pelletron does not cause any problems for the voltage regulation. However, the part of losses coming to the tube electrodes redistributes the potential along the tube and may lead to full discharges. Ref. [6] describes our work to decrease the frequency of these discharges.

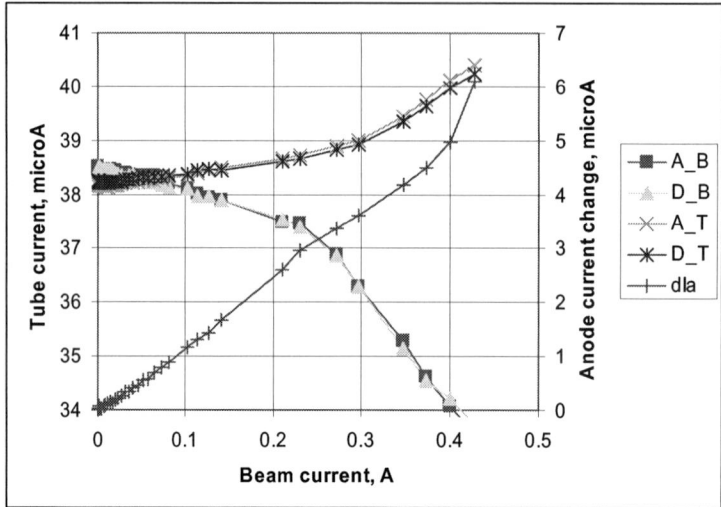

Figure 3. Current losses vs the beam current in the full beam line. dIa is the change in the anode current. The other four curves are currents of the resistive divider strings. 'A' and 'D' label the acceleration and deceleration tube strings, while 'B' and 'T' indicate readings at the bottom and top ends of the string (see Fig.2).

Significant difficulties in achieving a stable recirculation were caused by being in the neighborhood of the Main Injector (Fig. 4). When it ramps, stray magnetic fields from magnets and busses moved the electron beam, even though significant efforts had been made for proper shielding [7]. The beam motion in the return line was initially up to 6 mm and frequently resulted in the beam scraping either at a so-called crash scraper [6] or in the collector anode. Typically, losses were high enough to cause a recirculation interruption. The main reason for the extensive motion was found in an inadequate compensation of the MI buss currents. Magnetic shielding around the busses does not suppress the magnetic field generated by a non-zero total current summed over all busses and a dedicated compensation wire, running near the busses around the entire 3.3 km ring. A proper adjustment of the current in the wire as well as a modification of the return line shielding decreased the electron beam motion amplitude to 2 mm in the worst location. In addition, shortening of the protection system's response time from 0.8 ms to ~1 μs allowed operation with the crash scraper removed [5], which dramatically freed the beam line aperture. Presently, the beam motion in the return line does not significantly affect the recirculation stability.

Another difficulty is the losses of the MI proton beam that sometimes produce the radiation high enough to trip the electron beam protection system. Eventually we set the trip level above possible MI- initiated jumps of radiation. This value corresponds to an electron beam loss of about 20 μA, which is considered to be an upper boundary from the point of view of damaging the vacuum chamber.

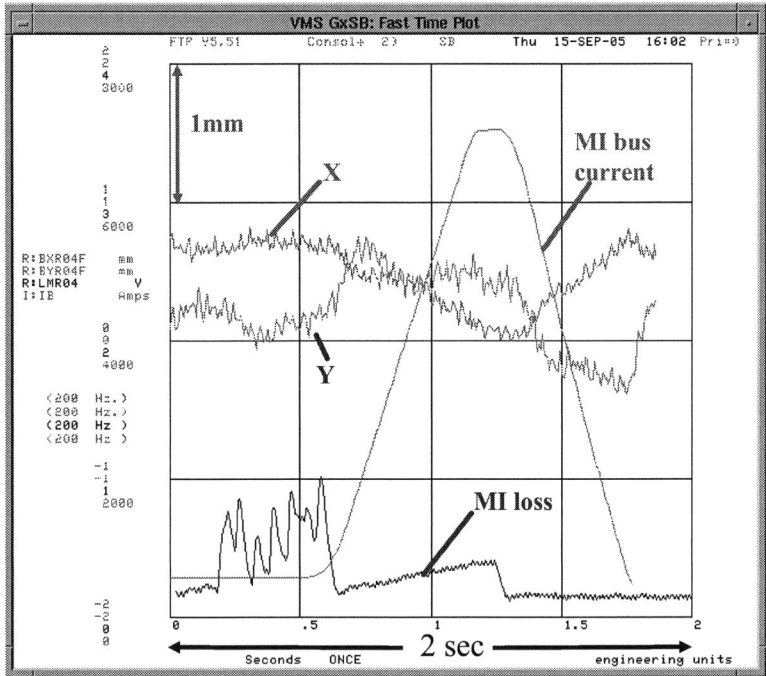

Figure 4. MI bus current, electron beam positions (X and Y), and radiation monitor signal in the return line in a typical MI ramp. A case of a low MI loss is shown. The 0.55 Hz oscillations outside the ramp are due to a 250 V r.m.s. ripple of the Pelletron voltage, because the dispersion in this location is 3 m.

At the nominal beam currents of 0.2 A, recirculation interruptions are observed once in several hours, and full discharges occur once in several weeks until the deceleration tube looses its electric strength. Several parameters in a 24 hours span including a full discharge are shown in Fig. 5.

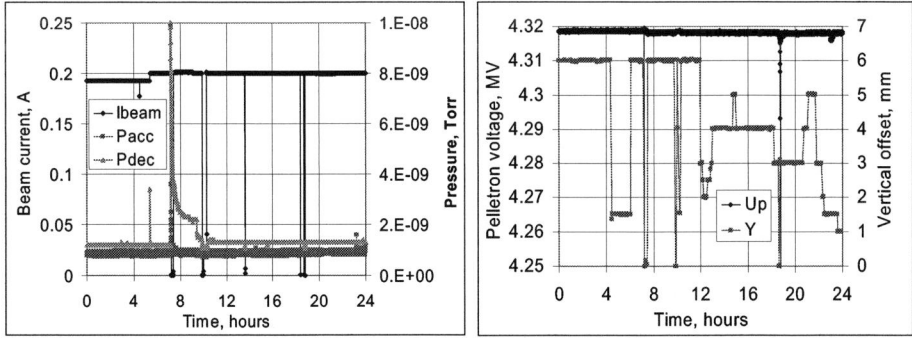

Figure 5. 24 hours of running the electron beam. Curves are labeled as follows: Ibeam is the electron beam current, Pacc is the pressure measured by an ion pump under the acceleration tube, Pdec is the pressure measured by an ion gauge under the deceleration tube, Up is the Pelletron voltage, and Y is the vertical offset in the cooling section. A large perturbation of Pdec corresponds to a full discharge. The vertical offset adjustments reflect regulation of the cooling strength.

ELECTRON BEAM PROPERTIES

For an effective cooling, the electron energy spread and the electron angles should be lower than the corresponding antiproton values. Also, the beams have to be well aligned in space and in momentum.

Electron Energy

Preliminary calibration of the electron energy was made by measuring of the length of a Larmor spiral pitch in the cooling section [8]. The precision, determined by the calibration of the Hall probe used for the longitudinal field measurements and by errors of the beam position measurements, was estimated to be ±0.2%. Comparison of the nominal Recycler energy with the energy of a low intensity, electron-cooled coasting antiproton beam shown that the actual discrepancy between both beams' energy calibration was $5 \cdot 10^{-4}$ [1]. While continuously running, the relative drift of the electron energy is about $3 \cdot 10^{-5}$; shutting the Pelletron off for a shift may result in changes of $1 \cdot 10^{-4}$; and opening the Pelletron and removing the terminal shell might modify the calibration by up to $3 \cdot 10^{-4}$.

The effective energy spread is determined by the electron energy distribution and fluctuations of the Pelletron voltage. The first component is calculated in Ref. [5] to be ~100 eV. The voltage fluctuations can be estimated by the signal from capacitive pickups in the tank and by beam motion in a high-dispersion region (Fig. 4). Both methods give the r.m.s. value of 250 V. Therefore, the effective energy spread is about 270 eV.

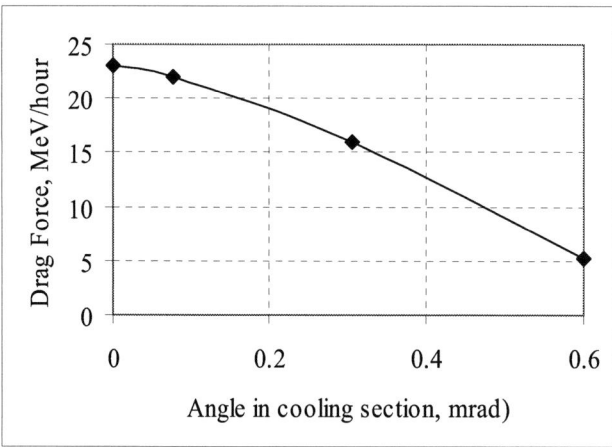

Figure 6. Drag force as a function of the dipole angle in the cooling section. The angle is applied with a corrector at the entrance of the cooling section and measured by BPMs. The drag force is measured by tracing the change of the average pbar energy after a 1 kV shift of the Pelletron voltage. The data were acquired with coasting $4 \cdot 10^{10}$ pbars, transverse emittance of 1.4 π mm mrad (95%, normalized), and the initial pbar momentum spread of 0.8 MeV/c (90%).

Electron Angles

Analysis of the electron angles in the cooling section, presented in Ref. [3], shows that presently the angles across the beam are determined by envelope oscillations. The original design assumed operation of the gun in the Pierce mode, with a nearly uniform current density distribution. However, stable operation at 0.7 A, which corresponds to this regime, has not been achieved, and the gun runs at 0.2 A with both current and angle distributions far from being homogeneous. The only diagnostics available for envelope angle measurements in the cooling section is a set of scrapers [9], and it allows aligning trajectories of outside electrons. While the gun aberrations do not contribute significantly into the angle spread in the beam core, the angles of the boundary particles may be far from a simple linear dependence on the radial offset. As a result, the procedure has provided angles low enough to make electron cooling operational, but still much higher than the 0.1 mrad initially foreseen for this angle component.

The large envelope oscillations of the beam core can be the reason for the observed sharp dependence of the drag force on the beam offset. The measurements were made with a low intensity, cold antiproton beam (see details in Ref. [1]) and shown that the drag force drops from ~ 20 MeV/h in the center to ~10 MeV/h with the 1 mm offset to ~5 MeV/h with 2 mm offset. Another indication of large envelope angles is a smooth dependence of the drag force on the amplitude of a dipole kick given to the beam at the entrance of the cooling section (Fig. 6). We estimate that while the electron angles near the axis are ~ 0.1 mrad, the maximum angles in the beam can be as high as 0.4 mrad.

OPERATIONAL ASPECTS

Electron cooling is now routinely used for storing and cooling of antiprotons in the Recycler ring. It was observed that overcooling may result in beam instability, a growth of the transverse emittance, and shortening of the beam life time. Two methods are used to regulate the cooling strength: changing the beam current or the electron beam offset.

Increasing the beam current from zero to 0.2 A decreases monotonically the equilibrium antiproton longitudinal emittance. Note that a further rise of the current does not lower this equilibrium. It may be explained by changes in the beam envelope, because for optimum cooling the focusing settings should be adjusted for every beam current.

More often, the cooling strength is regulated by a parallel shift of the electron beam in the cooling section. Because of the high energy, the electron beam space charge does not affect the pbar dynamics. Also, the combination of a long cooling time and large coupling in the Recycler ring allows cooling of all degrees of freedom even if the electron beam stays at a fix position off axis. As a result, cooling with at a constant 0.2 A current and the offset varying between 1 and 5 mm is at least as good operationally as on-axis cooling at lower currents. Although the life time deterioration has not been studied to the level of making a reliable conclusion about benefits of this mode, one can speculate that off-axis cooling may alleviate instabilities related to an overcooled antiproton beam core. In addition to cooling off axis, we plan to try cooling slightly off energy. Combined with containment of the beam in a barrier bucket, off energy cooling might be beneficial due to a flattened equilibrium momentum distribution.

SUMMARY

Electron cooling of 8 GeV antiprotons has been demonstrated and is presently in a routine operation.
- Recirculation is stable enough for the electron beam currents up to 0.2 A but drops for currents above 0.3 A because of increasing probability of full discharges.
- Electron angles in the cooling section are determined by envelope scalloping.
- In operation, the cooling strength is regulated by parallel shifts of the electron beam with respect to the antiproton beam.

ACKNOWLEDGMENTS

The authors are thankful to Fermilab's technical staff for installing and maintaining the cooling system and to V. Lebedev for many useful discussions about the beam optics. We acknowledge contributions made by R. Goodwin, J. Crisp, and L. Carmichael to the protection system; by L. Nobrega and S. Wesseln to the vacuum system design; and by B. Chase and P. Joireman to the design of the BPM system and to beam oscillation measurements.

REFERENCES

1. S. Nagaitsev et al., "Antiproton cooling in the Fermilab Recycler", *these Proceedings*.
2. Pelletrons are manufactured by the National Electrostatics Corporation, www.pelletron.com.
3. A. Burov et al., "Optics of Electron Beam in the Recycler", *these Proceedings*
4. P. W. Joireman, et al., "BPM System for Electron Cooling in the Fermilab Recycler Ring" in *Beam Instrumentation Workshop 2004*, edited by T. Shea and R. Coles Sibley, AIP Conference Proceedings 732, New York: American Insitute of Physics, 2004, pp. 319--326.
5. A. Burov et al., "IBS in CAM-Dominated Electron Beam", *these Proceedings*
6. L. R. Prost and A. Shemyakin., "Full Discharges In Fermilab's Electron Cooler", *these Proceedings*
7. T. Kroc et al., "Magnetic Shielding of an electron beamline in a hadron accelerator enclosure", to be published in Proc. of PAC'05, Knoxville, USA, May 16-20, 2005
8. S. M. Seletskiy and A. Shemyakin, Beam-based calibration of the electron energy in the Fermilab electron cooler, to be published in Proc. of PAC'05, Knoxville, USA, May 16-20, 2005
9. T. Kroc et al., "Electron Beam Size Measurements in the Fermilab Electron Cooling System", *these Proceedings*

The HESR Electron Cooling Proposal

D. Reistad

The Svedberg Laboratory
Uppsala University

Abstract. The maximum electron energy of the HESR electron cooling system has been decided to be 4.5 MeV, with a possible future upgrade to 8 MeV. Calculations with BETACOOL have been carried out; these include effects of imperfections of the electron beam as well as the internal hydrogen pellet target and intra-beam scattering. Design work is going on. This makes use of the experience gained at the FNAL Recycler and aims to result in a practical solution, which includes considerations for robustness as well as ease of assembly, bake-out, and to make use of proven solutions as much as possible.

Keywords: High energy electron cooling
PACS: 41.75.Lx

INTRODUCTION

The purpose of the electron cooling system at HESR [1] is to work together with the stochastic cooling system [2] in order to prepare the antiproton beam to provide the required momentum resolution for the experiments. Longitudinal and transverse blow-up of the beam due to the internal target and to intrabeam scattering is to be compensated by either of these systems or of them working together. The exact splitting of the tasks between the electron cooling system and the stochastic cooling system is not yet clear. However it is clear that stochastic cooling will be available for all degrees of freedom from 3 GeV to the maximum energy of HESR, and electron cooling will be available from the minimum energy 831 MeV (1.5 GeV/c) up to 8 GeV but (at least initially) not be available for energies higher than 8 GeV.

PANDA [3] requires an internal hydrogen target with effective thickness 4×10^{15} hydrogen atoms per cm^2 and $10^{10} - 10^{11}$ antiprotons circulating in HESR in order to produce luminosities ranging from 2×10^{31} to 2×10^{32} cm^{-2}s^{-1}. The only known internal target, which meets this specification, is the hydrogen pellet target [4]. The wish of the experimenters is to combine this with a momentum resolution of about $10^{-5} - 10^{-4}$.

The pellet target consists of a stream of frozen droplets of hydrogen produced in a vibrating nozzle at a rate of 60,000 s^{-1}. These are injected into vacuum in a vacuum injection capillary in which they also get accelerated by gas drag to a velocity of 60 m/s. In order to limit the transverse spread of the hydrogen pellets they are subsequently made to pass a skimmer, which gives the pellet stream a diameter of about 1.5 mm at the interaction point. The rate of pellets that make it through the skimmer is about 20,000 s^{-1}.

The ratio between the maximum instantaneous luminosity $\hat{L}$ (when the antiproton beam hits a pellet head-on) to the average luminosity $\langle L \rangle$ (for an antiproton beam of small emittance, which travels through the cylindrical pellet stream along a diameter) is

$$\frac{\hat{L}}{\langle L \rangle} = \frac{hR}{4\sigma^2},$$

where the pellet stream has a constant density of pellets inside a cylinder of radius R, the average vertical separation between the pellets is h and the antiproton beam has rms size σ in both planes. If we assume that $2R = 1.5$ mm, $h = 3$ mm, and allow $\hat{L}/\langle L \rangle \leq 5$ then we find that $\sigma \geq 0.36$ mm. Thus, we conclude that the rms. antiproton beam size at the target should not be smaller than 0.36 mm.

A thorough study of the electron cooler for HESR has been performed by the Budker Institute of Nuclear Physics in Novosibirsk, Russia [5, 6].

COOLING SCENARIO

The beam will always be injected at 3 GeV. We assume that the beam will be stochastically pre-cooled to rms un-normalised transverse emittances (both planes) of 0.1 μm and relative momentum spread of 1.5×10^{-4}. The beam will then be accelerated or decelerated to the experimental energy, where it will be exposed to the internal target and cooled with electron or stochastic cooling, or both.

In order for the electron energy to reach the desired stability the electron cooler should be kept at a constant voltage during the cycle, thus there is only electron cooling at the injection energy when this coincides with the experimental energy.

CHOICE OF BETA-VALUE AT THE INTERNAL TARGET

In order to make the beam not become too small at the pellet target, there is a preference for choosing a not too small beta-value at the target. The argument against too large beta-values at the target is that the cross section for single scattering out of the acceptance becomes large. This cross section is

$$\sigma_{ss} = \pi \left(\frac{2 r_e m_e c^2}{\beta^2 \gamma m_0 c^2} \right)^2 \times \frac{\beta_T^*}{A_0}$$

where r_e is the classical electron radius, m_e and m_0 is the electron and antiproton mass respectively, β and γ are the relativistic factors, β_T^* is the beta-value at the target, and A_0 is the acceptance.

The nuclear reaction cross section in the target is about 100 mbarn at 1.5 GeV/c, 70 mbarn at 3.8 GeV/c, 55 mbarn at 8.9 GeV/c and 52 mbarn at 15 GeV/c [7]. If we allow the single-scattering cross section σ_{ss} to be 40 % of the nuclear cross-section, and consider

$$A_0 = \begin{cases} 2.25 \times 10^{-6} \text{ m for } T \geq 3 \text{GeV (stochastic cooling pickups)} \\ 20 \times 10^{-6} \text{ m for } T < 3 \text{ GeV} \end{cases}$$

we find that the maximum allowed beta values at the target are 5.0 m at 1.5 GeV/c, 3.3 m at 3.8 GeV/c, 13.5 m at 8.9 GeV/c and 40 m at 15 GeV/c.

CHOICE OF MAGNETIC FIELD IN THE ELECTRON COOLING SECTION AND ELECTRON BEAM RADIUS

Busch's theorem implies that the flux, which is contained in the electron beam in the cooling section, determines parameters of the electron beam transport. There is a technological upper limit to this flux, because the accelerating tubes cannot be made with arbitrary apertures. In fact, in [5] an aperture of 20 mm (diameter) is assumed, and the High-Gradient tubes from National Electrostatic Corporation have an aperture of 1 inch (diameter), and there is also a technological limit to the magnetic field strength which can be produced by solenoids placed at high voltage because of limitations in the power that can be transmitted to high voltage with rotating shafts or other means, and there is a limit to the power, that it is possible to cool away with the insulating SF$_6$ gas. (If the magnetic field in the accelerating section will be produced with solenoids, which are placed at ground potential, then they will be of big diameter, and therefore also can only reasonably produce a limited magnetic field strength). Therefore, it is necessary to trade-off electron beam diameter in the cooling section against the magnetic field in the cooling section.

We conclude that the electron beam diameter in the acceleration sections shall be 17 mm in order to have a sufficient margin to a 25 mm aperture, and that the magnetic field in the accelerating section can be up to 0.07 T. Thus, the magnetic flux, which is contained in the electron beam in the cooling section, is going to be $0.07 \times \pi \times 0.085^2$ Tm^2.

In [5], it is shown that good magnetization of the electron beam at the highest momentum (15 GeV/c) requires a magnetic field strength in the interaction region of at least 0.2 T. As a safety margin, a magnetic field strength of 0.5 T is chosen in [5].

This higher magnetic field helps to ensure magnetized electron cooling in the case of magnetic field imperfections. It is also useful for keeping rotation of electron beam due to **E**×**B** drift small, also when some neutralization of the electron beam is present.

On the other hand, a lower value of the magnetic field in the drift tube allows a bigger diameter of the electron beam; this reduces worries for effects of resonances induced by the non-linear tune shift induced by the electron beam on the antiproton beam. We therefore choose a magnetic field strength of 0.2 T in the cooling section. The electron beam radius in the cooling section then becomes, $r_0 = 5$ mm.

Given this relatively low value of the longitudinal magnetic field, we believe that it would be reasonable to make the solenoid normal conducting and not superconducting, as was proposed in [5]. Detailed work on the solenoid has however not yet been done.

CHOICE OF THE BETA-VALUE AT THE COOLING SECTION

Large beta-values at the cooling section will speed up the cooling-down process and increase the longitudinal cooling force. On the other hand, if the beta values are too large, then a big fraction of the antiprotons will be outside of the electron beam in the beginning of the cooling process. We choose the horizontal and vertical beta values at the cooler to be $50\,\text{m} \times \beta\gamma/\beta_0\gamma_0$ (where β and γ are the relativistic parameters, and β_0 and γ_0 are those parameters at 3 GeV) in order to have more than 90 % of the antiprotons to be inside of the electron beam (with radius 5 mm) in the beginning of the electron cooling process (i.e. with $\varepsilon_{rms} = (\beta_0\gamma_0/\beta\gamma) \times 0.1$ µm).

CHOICE OF LENGTH OF ELECTRON COOLER AND ELECTRON CURRENT, AND CALCULATIONS WITH BETACOOL.

The cooling rate is essentially proportional to the product of electron cooler length and electron current. The cooling rate to be applied is limited by the required minimum beam size on the hydrogen pellet target. Calculations with BETACOOL [8], which assume that electron cooling is applied simultaneously with longitudinal stochastic cooling, give rms. beam size on target of about 0.36 mm for the combinations of electron current and electron cooler length given in the table 1.

TABLE 1. Summary of BETACOOL computations.

p GeV/c	RF?	N	tr. acc.. µm	long. acc. 10^{-3}	l_C m	I_e A	β_C m	β_T m	σ_x mm	σ_y mm	$\Delta p/p$ 10^{-5}	τ_N s	τ_L s
1.5	yes	10^{10}	20	1	15	0.2	20	5.0	0.56	0.53	12	780	890
1.5	yes	10^{10}	20	5	15	0.2	20	5.0	0.55	0.55	14	1730	1790
1.5	yes	10^{11}	20	1	15	0.2	20	5.0	0.71	0.76	19	570	610
1.5	yes	10^{11}	20	5	15	0.2	20	5.0	0.83	0.83	23	920	940
3.8	no	10^{10}	2.25	1	15	0.04	50	3.3	0.33	0.32	4.2	1340	1210
3.8	no	10^{10}	2.25	5	15	0.04	50	3.3	0.34	0.30	4.5	1700	1540
3.8	no	10^{11}	2.25	1	15	0.08	50	3.3	0.36	0.32	22	1260	1210
3.8	no	10^{11}	2.25	5	15	0.08	50	3.3	0.56	0.51	39	1760	1480
8.9	no	10^{10}	2.25	1	15	0.8	116	15	0.36	0.37	2.1	2230	1750
8.9	no	10^{11}	2.25	1	15	1.0	116	15	0.39	0.33	6.8	2250	1750
15	no	10^{10}	2.25	1	30	1.0	196	20	0.59	0.57	3.1	3440	2110
15	no	10^{11}	2.25	1	30	1.0	196	20	0.64	0.57	6.3	3490	1810

The calculations were carried out using the so-called Parkhomchuk phenomenological expression for the electron cooling force [9], and assuming that the solenoid can be made with a straightness of 10^{-5} radians. The magnetic field in the electron cooling section was 0.2 T and the electron beam radius 5 mm. The electron transverse and longitudinal temperatures were set to 1.0 eV (in the centre of the electron beam) and 0.001 eV respectively. Furthermore, unavoidable envelope oscillations in the electron beam were taken into account by assuming that the transverse electron velocity grows linearly with radius inside the electron beam, with a gradient of 7×10^8 s^{-1}. Thus, the electron velocity at the edge of the electron beam ($r = 5$ mm) becomes 3.5×10^6 m/s, corresponding to a cyclotron radius of 0.1 mm (amplitude of the envelope oscillations of 0.2 mm, measured on the radius of the electron beam, or $kT = 70$ eV). This feature has recently been introduced into BETACOOL [10] For antiproton energies from 3 GeV the transverse acceptance was set to 2.25 µm for both planes, given by the stochastic cooling pickups [2]. Below 3 GeV the transverse acceptance was assumed to be 20 µm. The longitudinal acceptance was set to 0.1 % [11]. This limits the beam lifetime, therefore some computations were also carried out assuming a longitudinal

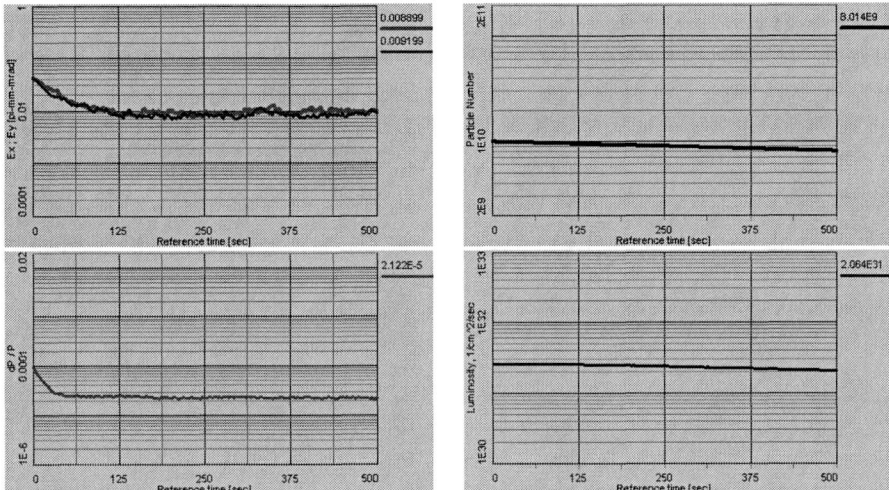

FIGURE 1. Calculated evolution of horizontal (red) and vertical (blue) emittance, momentum spread, particle number and luminosity with 10^{10} antiprotons of 8.9 GeV/c on hydrogen pellet target. The beta values at the electron cooler and the internal target are 116 m and 15 m respectively. The cooler length is 15 m and the electron current is 800 mA. Longitudinal stochastic cooling with bandwidth from 2 GHz to 4 GHz is assumed as well as electron cooling.

acceptance of 0.5 %. The energy loss in the target was calculated according to the Urban model, a realistic target model including large energy losses, recently implemented into BETACOOL [10]. The pellet target is represented as a cylindrical hydrogen fibre of diameter 1.5 mm and the same density, as of 20,000 s^{-1} hydrogen pellets of 30 µm diameter and velocity 60 m/s (i.e. vertical spacing 3 mm). This fibre target option is also newly implemented into BETACOOL [10]. The result of the calculations is summarized in table 1, where τ_N and τ_L are the e^{-1} lifetimes of the beam intensity and luminosity. These are calculated as $(250\,\text{s}/\ln(N(250\,\text{s})/N(500\,\text{s})))$ and $(250\,\text{s}/\ln(L(250\,\text{s})/L(500\,\text{s})))$ respectively.

The calculations assume intrabeam scattering according to the Martini model [12]. Longitudinal stochastic cooling, with parameters as indicated by Stockhorst [2], is included for the calculations from 3.8 GeV/c and up. At 1.5 GeV/c the energy loss in the target needs to be compensated; we assume a low-voltage (5 kV) first-harmonic rf at this energy. An example of the output of the BETACOOL computations is shown in figure 1.

We conclude that the electron beam radius will be 5 mm, and that the longitudinal magnetic field will be 0.2 T. We conclude that the beta values at the target and at the electron cooling section should be variable, as indicated above. We conclude that the electron cooler length can be 15 meters as long as the electron energy is limited to 4.5 MeV, but should be extended to 30 meters if the electron cooler is to be upgraded to 8 MeV. We conclude that we will allow the amplitude of envelope oscillations of the radius of the electron beam to be 0.2 mm, corresponding to a transverse velocity of the electrons of 3.5×10^6 m/s or $kT_\perp (r = 5\,\text{mm}) = 70\,\text{eV}$.

HIGH VOLTAGE

We identify two different techniques as serious options for the generation of the high voltage. These are the Dynamitron [13, 14] and the Pelletron [15]

Dynamitron

A Dynamitron uses capacitive coupling from dynode electrodes to corona rings at individual stages of the acceleration column. The transfer is done at 30-100 kHz. The induced signals are rectified and the stages are cascaded. The coupling to the individual stages leads to low impedance in the feeding of the acceleration electrodes, which is an advantage for electron beam stability. A power transfer of up to at least 100 kW is possible [16, 17].

This is the only technique for which a stability and ripple $\leq 10^{-5}$ has been demonstrated [18]. It is commercially available [16, 17].

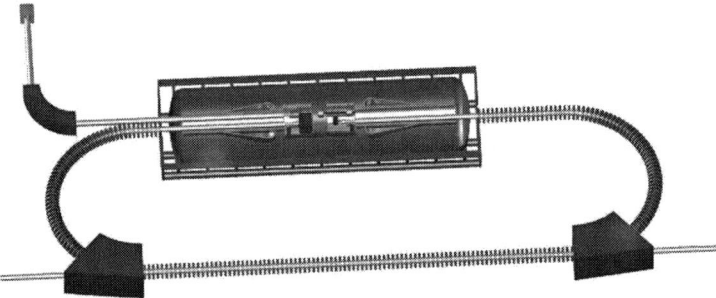

FIGURE 2. Tentative horizontal layout of the electron cooler with Dynamitrons for the acceleration and deceleration of the electrons. Coils generating longitudinal magnetic field surround the Dynamitrons. An H⁻ beam line for measurement and stabilization of the electron energy is also shown.

Pelletron

A Pelletron uses chain transport of charge to the high voltage terminal. The current is limited to about 150 µA/chain. The terminal voltage is regulated by shunting current to ground through corona needles. Pelletrons are commercially available with a modular design, which facilitates possible future increase of the high voltage to 8 MV [19]. At Fermilab a Pelletron is used for electron cooling at 4.5 MV, although with requirements on stability and ripple which are not as strict as for HESR and without continuous longitudinal magnetic field [20].

LAYOUTS

Horizontal

A horizontal layout avoids the need for a high tower and minimizes the number of bends in the electron beam transport. However a horizontal layout implies gravitational shear stress and bending of the acceleration columns. This rules out placing the coils for the longitudinal magnetic field on high voltage. Coils can instead be placed outside the pressure tank.

Vertical

A vertical layout permits putting heavy equipment on each voltage level, thus making it possible to place coils and current supplies for the generation of the longitudinal magnetic field on high voltage, close to the accelerating columns. On the other hand, the electron beam path will be more complicated, and a tower is needed. The longitudinal magnetic fields in the acceleration and deceleration columns must have opposite directions and therefore need separate coil systems.

Figure 3 shows a tentative design of this type with a Pelletron for the generation of the high voltage. The acceleration column, the deceleration column, the H⁻ voltage

FIGURE 3. Tentative vertical layout of the electron cooler with a Pelletron for generation of the high voltage. An H⁻ beam line for measurement and stabilization of the electron energy is also shown.

stabilizing column, the charging system, the power transfer axes, and the longitudinal magnetic field systems have to be placed in the same tank.

CHOICE OF TECHNIQUE AND LAYOUT

As discussed above, there are two techniques for the generation of the high voltage that we believe are close to fulfill the requirements for the HESR electron cooler, namely the Pelletron and the Dynamitron. Therefore we feel that it is reasonable to disregard from other proposed techniques although some of them might have some advantages. Substantial R&D is needed for proving of their feasibility.

Together with the manufacturers of Pelletrons and Dynamitrons more detailed studies of electron accelerators based on these techniques have been made. The layout for the Pelletron alternative has to be vertical whereas the Dynamitron alternative needs a horizontal layout.

The most important advantages of the Pelletron solution come from the experience gained at FNAL, thus possibilities to copy parts of the design from FNAL, and perhaps getting help from FNAL. The Pelletron solution also therefore has no need for very extensive R&D. Another advantage of the Pelletron alternative is its proven UHV performance (the Pelletron accelerating tube is ceramic, with no organic materials). Furthermore, the Pelletron is modular in its design, making it possible to increase the energy to 8 MeV at a later stage, by prolonging the high-voltage tank and adding sections to the accelerating column.

A main advantage of the Dynamitron alternative is that the accelerating electrodes in the Dynamitron are connected directly to the high-voltage supply. Therefore, the impedance on the electrodes is much smaller than in the case of the resistive high-voltage divider of the Pelletron (about 150 MΩ instead of 10 GΩ). Other important advantage of the Dynamitron is that it has shown performance with 10^{-5} voltage stability and ripple. Regulation is fast and is achieved without corona spikes. The high-voltage generation is achieved without moving parts.

The final choice between Pelletron and Dynamitron is not yet taken, however, most of the detailed design work that has been made is based on the Pelletron alternative.

ACCELERATION COLUMN AND HIGH VOLTAGE TANK

We identified the following changes, which we wanted to achieve, in a further development of the design proposed in [5]:
- to decrease number of HV levels with solenoids
- to combine a continuous 0.07 T magnetic field with possibility for assembly and bake-out
- incorporation of pumping at an intermediate level
- using commercially available elements

We made a design based on the overall geometry of the electron beam transport as in the Novosibirsk reports and NEC's Pelletron and the Fermilab experience. Because of the Fermilab experience the NEC tubes are the best-known tubes for this application; they combine a proven operation at voltages of 4.5 MV and above with excellent vacuum properties. The design is shown in figure 4.

The high voltage tank is equipped with an intermediate flange to be able to elongate the tubes if necessary. The high voltage column consists of 6 sections. For 4.5 MV terminal voltage the voltage over each section will be 750 kV.

The design has been made with consideration of the electric fields at the gas surfaces, the electron beam transport, how to mount, how to bake, cooling of solenoids, how to pump, and how to make magnetic alignment. Further work will address questions of mechanical vibrations and magnetic stray fields from motors and other equipment.

The design is based on the standard NEC structure with 1' length tube modules in pairs, separation boxes and posts. There are solenoids protruding from each side of separation boxes.

Each solenoid needs about 750 W to produce the desired field of 0.07 T. The power is provided by two 2 kVA generators in each separation box. Two rotating shafts drive the generators. Space for the power supplies of the solenoids is provided inside "piece-of-cake" shaped boxes mounted on both sides of the separation boxes. Cooling is provided by the SF_6 gas flow.

There are almost equidistant solenoid positions in the regular sections. These do not allow space for sputter-ion pumping. Therefore a so-called "big separation box" is in-

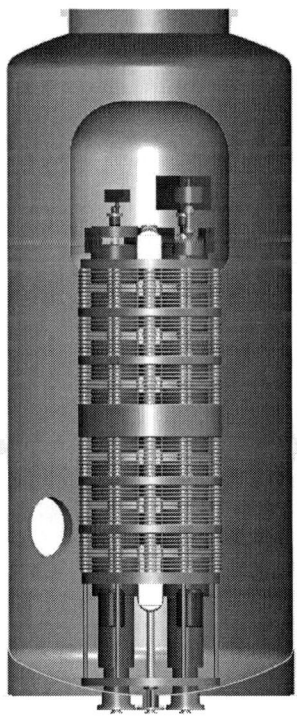

FIGURE 4 High voltage tank and column based on a 6 MV NEC Pelletron. An intermediate flange in the high-voltage tank, which allows extension of the high voltage, is not shown.

cluded. Inside the box an extra solenoid is mounted to maintain the equidistance of the solenoid positions. It cannot be mounted exactly halfway between the surrounding solenoids but has to be shifted 20 mm downwards to allow space for the pumping.

Electric field simulation for 750 kV/section shows maximum electric field strength of 99 kV/cm, well below the maximum electric field strength in SF_6. The tube modules are connected directly to the separation boxes, not through the solenoids. This lowers risks for damaging or changing magnetization of the flux return. Tube modules are mounted in the column in pairs. In the sections with a large separation box there is a dead section in between. Baking is done by installing heating jackets through 120 mm gaps between solenoids. Solenoid cooling is done by providing a good thermal contact between the solenoids and separation boxes. Magnetic alignment is done with dipole correctors based on magnetic measurements made prior the tube installation. The solenoid design will foresee the possibility to move solenoids away from the separation box by the amount of the gap (120 mm) to service the resistors.

ACKNOWLEDGMENTS

The work on the HESR electron cooler is teamwork. The contributions of all the members of the HESR electron cooling group, as well as other participants in the HESR design study, are acknowledged. The author also especially appreciates kind help from Alexander Shemyakin, Vladimir Reva, Anatoly Sidorin and Andrey Ivanov. This is gratefully acknowledged. Important discussions with industry, including Mark Sundquist, Dirk Mous, Henri van Oosterhout and Marshall Cleland are also acknowledged.

REFERENCES

1. An International Accelerator Facility for Beams of Ions and Antiprotons, Conceptual Design Report, GSI, 2001
2. H. Stockhorst, these proceedings
3. FAIR-ESAC/Pbar/Technical Progress Report
4. C. Ekström et al., NIMA 371 (1996) 572
5. O. Bazhenov et al., Electron Cooling for HESR, Final Report, Budker Institute of Nuclear Physics, Novosibirsk 2003
6. M. Steck et al., EPAC 2004, 1969
7. F. Hinterberger, "Study of the Beam-Target Interaction in the HESR Ring", unpublished, 2005
8. I. Meshkov et al., RuPAC XIX, Dubna 2004, 18
9. V.V. Parkhomchuk, NIMA 441 (2000) 9
10. A. Sidorin, private communication 2005
11. D. Prasuhn, private communication
12. M. Martini, CERN PS/84-9
13. http://www.highvolteng.com/
14. http://www.e-beam-rdi.com/Index.htm
15. http://www.pelletron.com/
16. D.J.W. Mous and H.A.P. van Oosterhout, private communication 2004
17. M. Cleland, private communication 2005
18. D.J.W. Mous, H.A.P. Oosterhout, "Feasibility Study on an 8 MV Electron Recuperator for HESR at GSI", unpublished, 2004
19. M. Sundquist, private communication 2005
20. S. Nagaitsev, these proceedings

The Proposed 2 MeV Electron Cooler for COSY

Juergen Dietrich[1], Vasily V. Parkhomchuk[2], Vladimir B. Reva[2], and Maxim A. Vedenev[2]

[1] *Forschungszentrum Juelich GmbH, Germany*
[2] *BINP Novosibirsk, Russia*

Abstract. The design, construction and installation of a 2 MeV electron cooling system for COSY is proposed to further boost the luminosity even with strong heating effects of high-density internal targets. In addition the design of the 2 MeV electron cooler for COSY is intended to test some new features of the high energy electron cooler for HESR at GSI. The design of the 2 MeV electron cooler will be accomplished in cooperation with the Budker Institute of Nuclear Physics in Novosibirsk, Russia. Starting with the boundary conditions of the existing electron cooler at COSY the requirements and a first general scheme of the 2 MeV electron cooler are described.

Keywords: Electron Cooling.
PACS: 29.20.Dh

INTRODUCTION

The COSY synchrotron accelerator and storage ring provides unpolarized and polarized proton or deuteron beams for internal or external hadron physics experiments in the momentum range from 300 MeV/c to 3.7 GeV/c [1]. Electron cooling is applied at low energies, at present mainly at injection energy, to prepare low-emittance beams to be used after acceleration and extraction for internal and external experiments. Stochastic cooling, covering the momentum range from 1.5 GeV/c up to the maximum momentum, is used to compensate energy loss and emittance growth at internal experiments.

Requests for future COSY experiments as WASA – a detection system from CELSIUS accelerator of The Svedberg Laboratory (TSL) at Uppsala with a pellet target [2] - are higher luminosities (> 10^{32} cm^{-2} s^{-1}).

There are two possible ways i) increasing the band width of the stochastic cooling system and/or ii) electron cooling up to maximum momentum. For operations with thick internal targets, fast (magnetized) electron cooling is the only technically feasible solution.

For electron cooling up to maximum momentum of COSY an electron cooler up to 2 MeV electron energy has to be developed together with the Budker Institute in Novosibirsk [3,4].

EXISTING ELECTRON COOLER

The design of the existing COSY electron cooler represents the state-of-the-art in the eighties [5], see Table 1. The capability to produce a 3 A, 100 keV electron beam was demonstrated during various tests of the electron cooler. At present, only 25 keV beam energy is necessary for the proton injection energy of 45 MeV. Electron beam currents in the range from 50 to 440 mA were used for cooling tests. Higher electron currents are not useful because the advantage of shorter cooling times is foiled by drastically increasing proton beam losses. Currents of 170 to 250 mA have turned out to be appropriate for the physics experiments. The typical cooling time of about 10 s can be tolerated in view of the duty cycle.

TABLE 1. Relevant Electron Cooler and COSY Ring Parameters

COSY Electron Cooler	Design Parameters	Used up to now	
Mechanical Length (Drift Solenoid)	2.00		m
Effective Cooling Length	≈ 1.5		m
Beam Tube Diameter throughout the Cooler	0.15		m
Potential Tube Diameter in Toroids	0.065		m
Electron Beam Diameter	0.0254		m
Electron Beam Radius in Toroids	0.60		m
Magnetic Field Range	80 ... 165	80	mT
Maximum Electron Energy	100	24.5	keV
Gun Perveance	0.84		µP
Design Electron Beam Current at 100 keV	4		A
Design Electron Beam Current at 25 keV	1.8	0.05 ... 0.5	A
Collector Loss Factor	$\leq 5 \times 10^{-4}$	$1 ... 4 \times 10^{-4}$	
Vacuum Pressure in the Cooling Region	$5 ... 10 \times 10^{-9}$	5×10^{-9}	hP
COSY Ring			
Particles	Protons and Deuterons (unpolarized and polarized)		
Type of Injection	H⁻, D⁻ Stripping Injection, 20 ...25 µg/cm² Carbon Foil		
Injection Energy	45 MeV for Protons, 76 MeV for Deuterons		
Shape of the Ring	Racetrack Type, Two Straight Sections and Two Arcs		
Nominal Circumference	183.473 m		
Dimensions of the Beam Tube	Round in Straight Sections, $d = 0.15$ m; Rectangular in Arcs, 0.15 m Horizontal (x), 0.06 m Vertical (y)		
Working Point Range	Variable Between 3.55 and 3.7 in Both Planes		
Optical Functions at the Electron Cooler	$\beta_x = 8$ m, $\beta_y = 16$ m, $D = -6$ m		

PROPOSED 2 MEV ELECTRON COOLER

Basic Parameters and Requirements

The basic parameters and requirements are listed in Table 2. The most important restrictions are given by the available space at the COSY ring itself. The height is limited by the building up to 7 m, the length of the cooler in beam direction by the existing electron cooler and the ring itself to 3 m. The acceleration of polarized beams at COSY must to be taken into account. Space for compensating magnets must be foreseen to achieve conservation of polarisation.

TABLE 2. Basic Parameters and Requirements.

COSY 2 MeV Electron Cooler	Parameter
Energy Range	0.025 ... 2 MeV
High Voltage Stability	$< 10^{-4}$
Electron Current	0.1 ... 3 A
Electron Beam Diameter	10 ... 30 mm
Cooling Length	3 m
Toroid Radius	1.5 m
Variable magnetic field (cooling section solenoid)	0.5 ... 2 kG
Vacuum at Cooler	10^{-8} ... 10^{-9} mbar
Available Overall Length	7 m
Maximum Height	7 m
COSY beam Axis above Ground	1.8 m

Cooling of 2 GeV Proton Beam at COSY

Calculations are performed with the trubs.exe code [3], in which the cooling force is approximated by the well known Parkhomchuk formula [6]. The effect of intra beam scattering is included by the simple model of relaxation distribution velocity. The increase of the angle spread due to scattering of an internal target is also taken into account. The simulation was made with following parameters: cooler length 3 m, beta function in the cooling section 13 m, electron beam radius 0.5 cm, electron beam current 2 A, magnetic field 2 kG, initial normalized emittance 10^{-6} m rad, 2 GeV proton beam energy and number of protons $2 \cdot 10^{10}$ (5 mA).

As it is seen in Fig. 1, the ion beam emittance is effectively decreased during 10 s. The reached equilibrium emittance is a result of balance between intra-beam scattering and electron cooling.

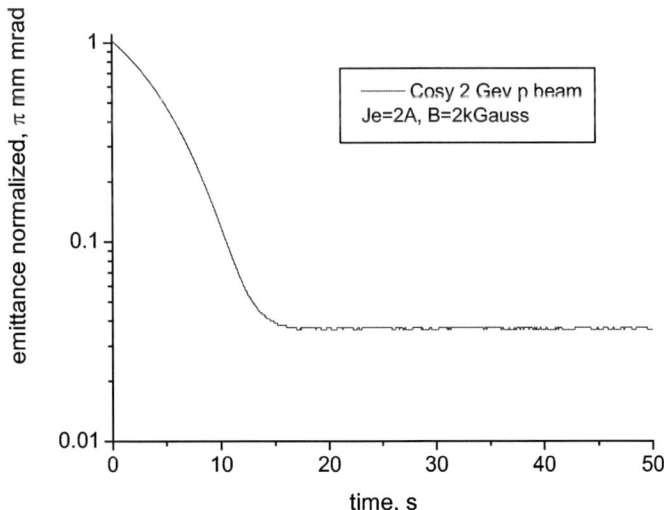

FIGURE 1. Normalized beam emittance versus time at electron cooling of 2 GeV proton beam (without target, parameters see text).

The presence of a target introduces multiple scattering which will be suppressed by the cooling and single scattering at large angles. The single scattering on the aperture limit leads to losses of proton current and can be described as life time of the proton beam. Figure 2 shows the proton beam current versus time with and without target.

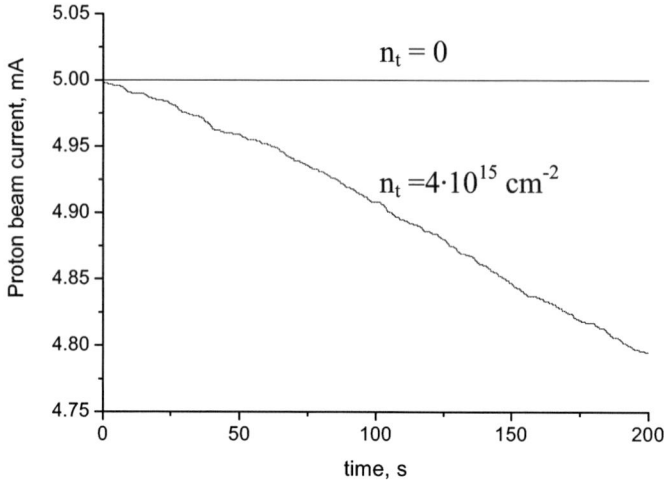

FIGURE 2. Proton beam current versus time with and without target (n_t – target density, parameters see text).

Without target we do not observe any losses of proton beam but with target the life time of the proton beam inside an aperture at the cooling section of 1 cm is about 5000 s. The single scattering at angles larger than the equilibrium but less than the aperture limit leads to formation of beam tails in the distribution.

Preliminary Cooler Technical Design

The proposed electron cooler consists of a high voltage vessel with electrostatic acceleration and deceleration columns, two bending toroids and cooling drift section. The preliminary scheme of the cooler is shown in Fig. 3 [3]. The basic features of the design are i) the longitudinal magnet field from the electron gun to the collector, in which the electron beam is embedded, ii) the collector and electron gun placed at the common high voltage terminal and iii) the power for magnet field coils at accelerating and decelerating column is generated by turbines operated on SF_6 gas under pressure. The gas flux which drives the turbines is also used for cooling the magnetic coils and for keeping the temperature inside the vessel constant. The cathode of the electron gun is immersed in the magnetic field. The electron beam is accelerated to an energy up to 2 MeV. After that the electron beam is bent in the toroid and is guided to the cooling section. After the main solenoid the beam is returned to the electrostatic column. Here it is decelerated and is absorbed in the collector located in the head of the electrostatic column. Each toroid consists of two parts. The first one bends the magnetized electron beam in the vertical plane on 90^0. The second one bends the electron beam on 180^0 in a plane, which is inclined on 45^0 to the vertical plane. Such a complicated 3-D geometry provides compactness of the system. The dipole kick for protons in the bending toroids near the cooling section will be compensate by dipole magnets which will be installed near the large toroid coils as close as possible. The Electrons receive dipole kicks due to the inhomogeneity of the magnetic field. These kicks must be compensate by electrostatic kickers which will be inserted in front of the cooling section. Electrostatic bending for better recuperation efficiency will be used [7,8].

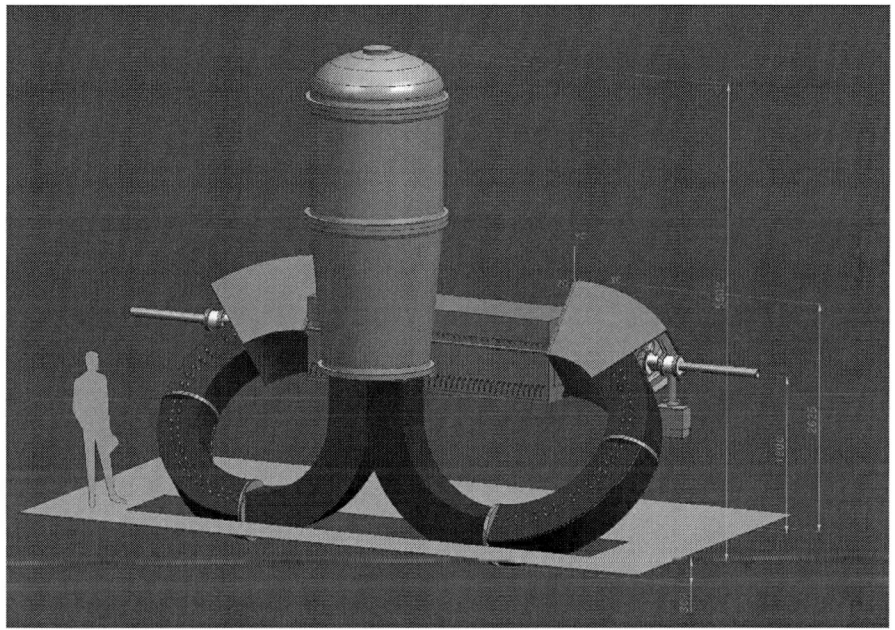

FIGURE 3. Layout of the proposed 2 MeV electron cooler for COSY.

High Voltage System

The high voltage system consists of the vessel, the accelerating and decelerating column, high voltage sections and high voltage head with gun and collector. For the vessel and column the main parameters of the industrial accelerator ELV-8 which works on 2.5 MV are taken [9]. The vessel geometry is identical to the vessel of the ELV-8. The vessel withstands pressures up to 10 bars. The diameter of the high voltage sections amounts to 80 cm. At the ELV-8 accelerator SF_6 gas is used as insulation gas. The ELV-8 has no magnetic coils inside. The electron current is equal to 0.05 A. The beam power is 100 kW. Budker Institute has experience in recuperation of high voltage beams with an energy of 1 MeV and a current of 1A [10].

The high voltage sections (Fig. 4) contains: high voltage power supply, coils for the magnet field along acceleration and deceleration columns, power source and control units for measurement and control of parameters for each section.

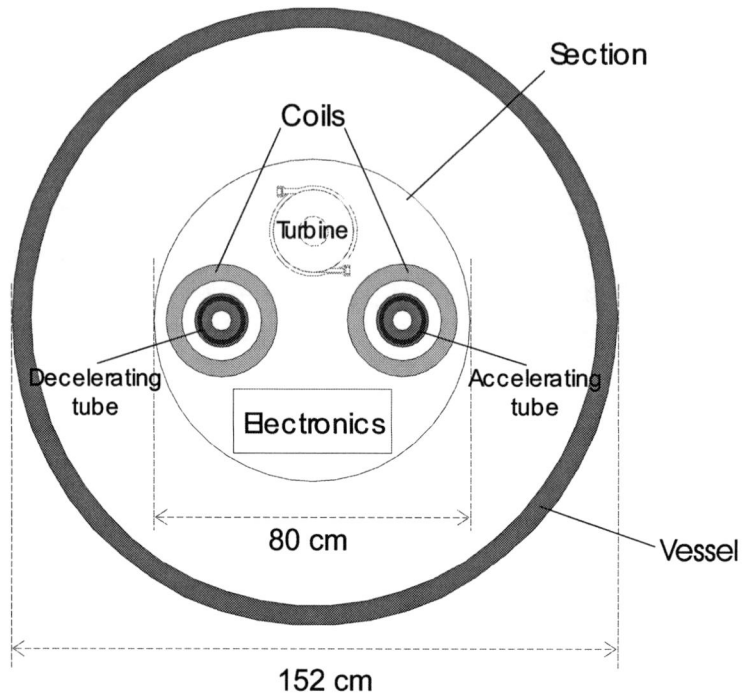

FIGURE 4. High voltage section

The experience from the Budker Institute in the design of high voltage sections with ± 30 kV is used. Each section has two high voltage power units on 30 kV. Using of two power units allow to decrease the voltage for insulation from 60 kV to 30 kV.

Using 60 kV for the sections (±30 kV) means that the whole 2 MV column consists of 34 sections. The electric field between the sections will be 30 kV/cm. The pressured SF_6 gas can be used for protection from sparking. At a pressure of about 10 bars an electric field strength of up to 500 kV/cm can be stable operated [11].

Computer simulation of the electric field distribution at the high voltage column showed that the maximum electric field strength occurs at the edge of the upper section and amounts 132 kV/cm. To suppress sparking a SF_6 gas pressure of about two bars is sufficient in this case [11].

Special measures must be taken to prevent destructions from sparks. Accelerating rings are surrounded by collar rings. The width of the gaps between the collar rings are chosen in a manner that in case of a discharge, it occurs between the collar rings.

Magnetic Field at the Acceleration Tube

The magnetic field at the cathode of the electron gun and at the drift section define the electron beam radius in the cooling section. The electron beam size should be close to the proton beam size for effective cooling. The proton beam size at low energies is larger than at higher energies. For a cathode radius of 1.5 cm in the low energy case the magnetic field in the electron gun should be the same as the magnetic field at the cooling section. But in the high energy case it is difficult to obtain an optimal ratio of magnetic fields. A maximum magnetic field in the cooling section of 2 kG corresponds in the gun region to a magnetic field of 150 G. The minimum necessary magnetic field in the electron gun is limited by the space charge of the electron beam inside the anode region which requires a higher field. With an electron current of 3 A the minimum value of the magnet field is 120 G. But the experiments with electron coolers built by the Budker Institute shows that for magnetic fields a few times higher (about 250-300 G) the normal regime of recuperation of the beam became very sensitive and very often crashed. For safe operation the magnetic field value in the electron gun should be about 1 kG. Along the acceleration tube it is not easy to realize the electric power for the 1 kG magnetic field coils. But in the acceleration tube the electron beam has an energy higher than in the cathode region and therefore a magnetic field of 0.5 kG is sufficient. The diameter of the acceleration tube ceramic rings is about 120 mm and the pancake coils for the magnetic field can be made with inner diameter 240 mm, external diameter 320 and thickness 40 mm. These coils are installed at each HV section along acceleration and deceleration tubes with an gap of 20 mm. The consumption of electric power for one coil at a section with a value of magnetic field of 500 G is equal 130 W (for filling efficiency of copper 0.7, weight of a single coil 6.7 kg). The current density at each coil amounts to 2.1 A/mm^2.

For maximal voltage of 60 kV the high voltage power at each section is about 60 W. For powering of the high voltage sections a mechanical generator with a maximum electric power of 0.5 kW with enough power reserve will be used.

The electron gun and the electron beam collector are placed very close to each other. In the electron gun a magnetic concentrator (magnetic steel) is used to increase the magnetic field by a factor of two from 500 G at the column solenoids to 1000 G at the surface of the cathode. The gun concentrator also improves the field homogeneity at the cathode surface. The collector magnetic shielding is used to decrease the magnetic field to spread the electron beam inside the collector. The fast decreasing magnetic field at the collector entrance produces a magnetic mirror for the soft secondary electrons emitted from the collector due to bombardment from the primary electron beam. The magnetic mirror together with the electrostatic suppression electrode suppress secondary electron emission.

To adjust the electron beam radius in the cooling section the magnetic field at the cathode of the electron gun will be changed with additional coils. Increasing the magnetic field up to 1000 G is possible. Decreasing magnetic field is achieved by reversing the current direction in these coils.

The power consumption of the complete magnetic system including the coils of the high voltage column, toroids and cooling section amounts to 280 kW. The total mass of copper is about 2.7 tons.

Electric Generator at the High Voltage Section

The simplest system of powering the high voltage sections and power supply for the magnetic field is a mechanical electric generator. The most popular system consists of a electric engine on ground potential and an insulation shaft (plastic) which transfers power to an electric generator on high voltage potential. In the present case too many generators (>35) along the acceleration column and to the high voltage terminal would be necessary. The twisting moment of the shaft for the first generator would be 35 times larger than for the last one. Vibrations of the whole system could be an other disadvantage. Therefore turbo engines with integrated electric generators at each section are proposed. A compressor at ground potential will pump SF_6 gas from the vessel, compress it to 4-5 bar and feed it to a thermo exchange chamber and gas filter. After this the pressurized gas is directed with plastic tubes along the high voltage column. At each section the pressurized gas is used to drive a turbo generator for production of the electric power and after this the gas is used for cooling and regulating the temperature constant.

The electron cooling requires very low level of high voltage ripple, less than $\Delta U/U < 10^{-5}$ over the whole dynamic voltage range (0.025 – 2 MV). For simplification of the high voltage power supply it is possible to use for low voltage only 1 section. In this case, instead of 60 kV, this section will operate on a voltage of 25 kV.

At each section an optical communication block, few DAC for control of magnetic coil current and high voltage power supply (0-60 kV) and few ADC for measuring parameters of section operation should be installed.

Matching Section

After acceleration in the longitudinal magnetic field of 500 G electrons enter into the toroid with a field of 2000 G. This transition excites transverse motion to a big temperature which is unacceptable for electron cooling. Therefore a special matching section is foreseen to smooth the magnetic field in the transition.

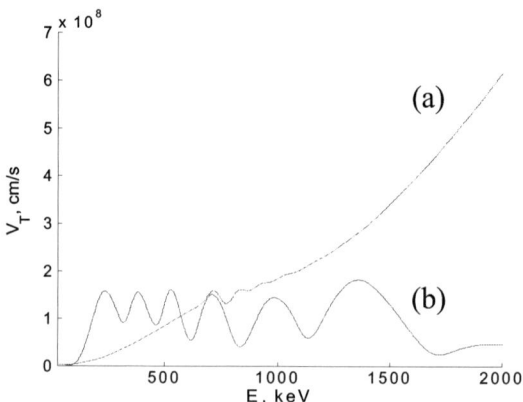

FIGURE 5. Transverse electron velocity (rest frame) in the toroid without (a) and with (b) matching section.

SUMMARY

The development of a 2 MeV electron cooling system for COSY is essential for the future COSY physics program, it delivers higher beam quality and higher luminosity.

The operation of a 2 MeV cooling system at COSY together with a high-density internal target of WASA detector would uniquely allow to optimize the cooling performance for "tail" particles. This involves both the electron cooling system alone and a combination of the electron and stochastic cooling systems. Realization of such a cooling system will be an important step toward creation of a novel experimental technique aiming to reduce significantly parasitic effects related to halo in accelerated beams – a step to "backgroundless" detection systems.

For operations with thick internal targets, fast (also known as *magnetized*) cooling is the only technically feasible solution. Engineering design of a magnetized cooling system would be sufficiently different from the Fermilab 4.3-MV system [12] to warrant a dedicated effort to design a 2-4 kG warm or superconducting solenoid of a high field quality. The 2 MeV COSY electron cooler would be an intermediate energy step to future high-energy magnetized cooler projects like the HESR high energy electron cooler in the FAIR project [13] and would be extremely useful for finding optimal technical solutions and prototyping many elements.

ACKNOWLEDGMENTS

The authors like to thank R. Maier, D. Prasuhn, H. J. Stein and H. Stockhorst for the permanent support of this work and helpful discussions.

REFERENCES

1. R. Maier, Nucl. Instr. Meth A 390, 1-8 (1997).
2. J. Zabierowski et al., The CELCIUS/WASA Detector Facility, Physica Scripta T99, 159-168 (2002).
3. V. V. Parkhomchuk, Electron Cooling for COSY, Internal Report (2005).
4. V. B. Reva et al., Budker INP proposals for HESR and COSY electron cooling system, these proceedings.
5. H. J. Stein et al., R Present Performance of Electron Cooling at COSY-Jülich. Proceedings of the Russian Particle Accelerator Conf. (RUPAC 2002) ,October 1-4, 2002, Obninsk, and in Journal Atomnaya Energiya, Russia.
6. V. V. Parkhomchuk, Nucl. Instr. Meth. A 441, 9-17 (2000).
7. V. V. Parkhomchuk, Recuperation of Electron Beam in the Coolers with Electrostatic Bending, Poster, these proceedings.
8. V. V. Parkhomchuk, Development of a generation of coolers with a hollow electron beam and electrostatic bending, these proceedings.
9. Y. I. Golubenko et al., Accelerators of ELV-Type: Status, Development, Applications, BudkerINP 97-7, Novosibirsk (1997).
10. R. Salimov et al., Nucl. Instr. Meth A 391,138-141 (1997).
11. I. M. Bortnik, Physical properties and electric strength of sulfur hexafluoride (SF_6), Energoatomizdat, Russia
    (1988).
12. S. Nagaitsev, Antiproton Cooling in the Fermilab Recycler, these proceedings.
13. D. Reistad, HESR Electron Cooling System Proposal, these proceedings.

Budker INP Proposals for HESR and COSY Electron Cooler Systems

V.Bocharov, M.Bryzgunov, A.Bubley, V.Gosteev, I.Kazarezov,
A.Kryuchkov, V.Panasyuk, V.Parkhomchuk, V.Pavlov, D.Pestrikov,
V.Reva, V.Shamovskij, A.Skrinsky, B.Sukhina, M.Vedenev, V.Vostrikov

Budker Institute of Nuclear Physics, Novosibirsk, Russia

Abstract. The subject of the report is the problem of the technical feasibility of fast electron cooling in the energy range between 0.8 and 14.5 GeV. It is very useful for one of the major objectives of the GSI and COSY future plans. For the realization of the cooler device BINP team proposes the design that is like the conventional and elaborated for the low energy cooling (up to 300 keV). The main features of this design are the accelerating tube immersed in the magnetic field along the whole length and the strong magnetic field in the cooling section. The physics of electron cooling is based on the idea of the fast magnetized cooling. The cooling force at strong magnet field was measured at many experiments and can be surely estimated. The magnetized cooling rate enables to obtain the required beam parameters, eliminate the beam heating due to intrabeam scattering, fluctuations of ionization energy losses and multiple scattering in the internal target

Keywords: high energy cooler, electron cooling.
PACS: 29.27:

DIFFERENT SOLUTION FOR ELECTRON COOLER DESIGN

The technical realization of the medium energy electron cooling can be done with different variants. The first method is an electrostatic machine. This method is conventional and elaborated for the low energy cooling (up to 300 keV) [1-2]. The research program of 4.3 MeV electron cooling device is carried out at FNAL/USA [3-4]. Electrostatic machine has the advantage of small spread of the electron beam energy and it provides the continuous electron beam without any time structure that is a good for the cooling of the coasting antiproton beam. Moreover it enables to vary the electron energy in wide range that is necessary for HESR research program.

The second method is an RF linear accelerator. A problem in using of the RF accelerator for the electron cooling device is the requirement to maintain extremely low values of the spreads of longitudinal and transverse momentum ($\Delta p_\parallel / p_\parallel$ and $\Delta p_\perp / p_\perp \approx 10^{-4} - 10^{-5}$) at the maximum length of the electron bunch. At the electron energy about 8 MeV the energy spread is few units of 10^{-4} may be attainable. A further improvement of the stability of the electron beam parameters by regular methods like thermostabilisation of the cavities, using feed backs and etc seems problematic. So, the special efforts need for the cooling of antiproton beam to the extremely low value of the longitudinal momentum spread. Another serious problem is the variation of the cooling energy in the wide range. It is probably impossible to do it in the whole energy range from 0.44 to 7.9 MeV because of the large variation of the electron flight – time through an RF resonator. The more realistic variant is to restrict oneself by the more narrow energy range 5–8 MeV. Thus, RF linac is not reasonable choice for this project, but it is most likely next milestone in the manufacture of electron coolers. The electron beam energy about 8–10 MeV required for this project seems to be close to the maximum attainable value for electrostatic type of accelerator. The project of the high relativistic electron cooler (electron energy 55 MeV) is designed at BNL/USA [5].

To avoid the problems of the traditional electron cooling system the cooler based on the "modified betatron" scheme was proposed in [6-7]. The electron beam circulates in longitudinal (quasitoroidal) magnetic field. The long-term stability of the beam is provided with additional spiral coils. The magnetic field of such device is similar to the "stellarator" one. The acceleration of the electron beam is produced by using induction (betatron) acceleration. This type of the acceleration is very cheap, enables to obtain coasting (continuous) electron beam and does not induce coupling between the longitudinal position of an electron and its energy in contrast to RF linac system. During few msec it is possible to accelerate the electron from 20 kV to 10 MeV at inductor 1x1 m and the magnetic field 20 kG. But the longitudinal magnet field cannot be changed during of few msec. Thus, it leads to crossing many linear resonances during acceleration process. The possibility to pass theirs without large loss of electron beam should be studied. It is one of the research problems of Meshkov's team at Dubna [8]. For low heating rate at time of acceleration the magnet system should have the smooth changing magnet field along the electron orbit. The low momentum of electron beam (at 1836 times les antiprotons) results to extremely high sensitivity of a costing electron beam to development of coherent instabilities.

The certain thermal capacity of the electron single ring restricted the rate of electron beam refresh. The new cold electron beam should replace the old heated electron beam in the betatron frequently enough for the effective cooling process. For parameters of the Fermilab Recycler ring the repetition rate of the electron injection is about 2 kHz [8].

PHYSICS OF ELECTRON COOLING FOR HESR AND COSY

The process in the target is the source of the energy spread and loss of the antiprotons. The typical method of this effect suppression is use of magnetized cooling [9]. This mechanism was detail investigated in [10-12].

In the case of magnetized cooling the cooling force is enough strong to suppress the antiproton scattering and the average energy loss in the internal target. According the expression for the cooling force supposed in [9]

$$\Delta \vec{p} = \vec{F} \cdot \tau = -\frac{4e^4 n_e \vec{V} \tau}{m_e (\sqrt{V^2 + V_{eff}^2})^3} \ln\left(1 + \frac{\rho_{max}}{\rho_L + \rho_{min}}\right) \quad (1)$$

the strong magnetic field in the cooling region is essential on two reasons. The first is reducing the role of the transverse electron velocity $v_{e\perp}$. If the impact parameter of ion

$$\rho_{max} \approx V \cdot \tau \tag{2}$$

is larger than the Larmour radius of the electron $\rho_L = m_e c v_{e\perp}/eB$ then the effective electron velocity

$$V_{eff}^2 = V_{\Delta\Theta}^2 + V_{E\times B}^2 + V_{e\parallel}^2 \tag{3}$$

doesn't contains the term with $v_{e\perp}$. The other terms are the effective velocity induced by the curve of the magnetic field lines $V_{\Delta\Theta}$ and the electron drift velocity in the crossed the space charge fields of the beams and the guiding magnetic field of the cooling device $V_{E\times B}$.

At the small value of the magnetic field and large transverse velocity of the electron can be realized the condition $\rho_L > \rho_{max}$. In this case the Coulomb logarithm can be written as

$$\ln(1+\rho_{max}/\rho_L) \approx \tau \omega_L V/v_{e\perp} \tag{4}$$

where τ is flight time ion through the cooling section in the bam reference system and ω_L is the electron Larmour frequency. Thus, the transverse electron velocity becomes the essential factor and the cooling force drops as $F \propto 1/(V v_{e\perp})$. This situation was observed in NAP-M experiments [13]. The friction force dropped down only as $F \propto 1/V$ and not as $F \propto 1/V^2$. The measurements made in these experiments showed that the cooling force decreased with increasing the transverse Larmour velocities of the electrons as $F \propto 1/v_{e\perp}$.

Another role of the strong magnetic field is reducing the drift velocity $V_{E\times B}^2$ induced by the space charge. At the electron current

$$J_{opt} = \frac{1}{2\sqrt{2}} \gamma \beta B a_e \sqrt{V_{\Delta\Theta}^2 + V_{e\parallel}^2 + V^2} \tag{5}$$

the maximum cooling force is achieved. In Figure 1 one can see the cooling rate versus antiproton/proton energy (Eq.1). Two variants are considered. The first variant deals with the injection parameters of the antiproton beam when the emittances and momentum spread are large. The emittance of pbar beam is taken as $\varepsilon_n=1$ mm·mrad (normalized, 1 σ) and the momentum spread is $\sigma_p=10^{-3}$. The second variant describes situation after some cooling procedure. The pbar parameters are taken as $\varepsilon_n=0.05$ mm·mrad and $\sigma_p=10^{-4}$. The dash lines are calculated at absence any restriction on the electron current and the magnetic field in the electron gun. The last parameter enables to change the electron beam radius in the cooling section and regulate the electron density in the cooling section. The solid lines are calculated at the following restriction. The minimal and maximal magnetic fields on the cathode are 200 G and 1000 G. The maximum electron beam current is 3 A. The other parameters of the calculation are the cooling length 30 m, the electron temperature on the cathode 0.3

eV, beta function in the cooling section 100 m. The radius of the electron beam is equal to the radius of the ion beam if it allows by the magnetic field on the cathode. In a different case it is nearest maximum or minimum value. The magnetic fields in the cooling section are taken 500 G, 2 kG and 5 kG. The cathode radius is taken 0.33 cm, 0.63 and 1 cm correspondingly.

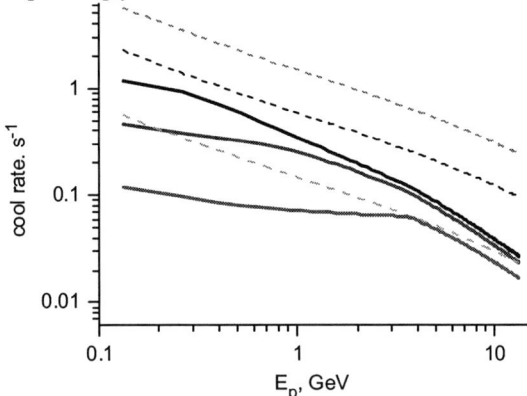

FIGURE 1. Cooling rate versus proton/antiproton energy. The normalized emittance (1 σ, r.m.s) is 1 π·mm·mrad, the momentum spread is 10^{-3}. The dotted lines is highest possible value of the cooling rate without taking technical restriction into consideration. The solid lines are cooling rates calculated with the following technical restrictions. The minimal and maximal magnetic field on the cathode are 200 and 1000 G. The maximum electron current is 3 A. The magnetic filed in the cooling section are 500 G, 2 kG and 5 kG from bottom to top.

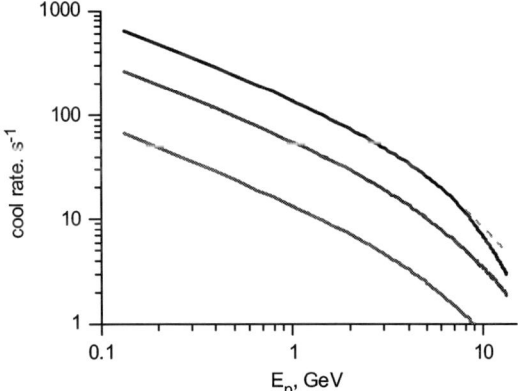

FIGURE 2. Cooling rate versus proton/antiproton energy. The normalized emittance (1 σ, r.m.s) is 0.05 π·mm·mrad, the momentum spread is 10^{-4}. The dotted lines is highest possible value of the cooling rate without taking technical restriction into consideration. The solid lines are cooling rates calculated with the following technical restrictions. The minimal and maximal magnetic field on the cathode are 200 and 1000 G. The maximum electron current is 3 A. The magnetic filed in the cooling section are 500 G, 2 kG and 5 kG from bottom to top.

One can see that the cooling rate isn't sensitive to the magnetic field value at the high pbar energy and injection parameters (Figure 1). At low energy the high value of the magnetic field gives preference for obtaining maximal cooling rate. After the cooling the high magnetic field enables to obtain large cooling rate (Figure 2). The space charge effect isn't viewed essential but the high ratio between the magnetic fields in the electron gun and the cooling section enables to have a very large density of the electron beam in the cooler.

KEY SOLUTIONS.

The technical solution of the BINP team is based on the standard low-energy design for the electron coolers.

The longitudinal magnetic filed in the kilogauss range is used for the transportation of the electron beam. The magnetic filed in the cooling section is strong enough for guarantee magnetizing collision of the ions and electrons. The acceleration tube is also located in the magnetic field with value about 500 G. The equal value of the magnetic field in the cooling and transportation section enables to close the magnetic flux without the large iron circuit (the flux closed in the toroidal device as plasma tokamak).

The cooling solenoid is constructed as the set pancake section assembled in series. The design of solenoid enables to incline and rotate each coil so the quality of the magnetic field can be obtained $\Delta B_\perp/B = 10^{-5}$ [14].

The bending of the electron beam is realized with help of the electrostatic fields. In this case the dynamic of the primary electrons and secondary electrons reflected from the collector is similar. Such optics of the electron beam is capable to pass the beam both the forward and reverse directions. In this case the electron, which was not absorbed by the collector, has several attempts to get in the collector. There exists a friction force between the bulk electrons and the scattered electrons. Thus, the velocities of the scattered electrons and others are equalized and the scattering electrons are absorbed by the collector. The decrease of the leakage current leads to an improvement of the vacuum condition. These effects were observed in the cooler manufactured for IMP (Lanzhou, China)[15]. The point of full compensation of the centrifugal force by the electrical force is characterized by the minimum of leakage current and good vacuum condition.

DESIGN OF ELECTRON COOLER FOR HESR

A layout of the electron cooling device is shown in Figure 3. The electron beam starts its path in the gun located in the head of the electrostatic column (Fig.3, 2). The cathode of the gun is immersed in the magnetic field. The electron beam is accelerated to energy up 8 MeV. Immediately after the electrostatic accelerator the electron beam transferred from magnetic field 500 G to 5 kG. After that the electron beam is bent in the vertical and horizontal planes and is moved to the cooling section (Fig.3, 4). After the main solenoid the beam is returned (Fig.3, 5) to the electrostatic column. Here it is decelerated and is absorbed in the collector located in the head of the electrostatic column (Fig.3, 2).

The bending of the electron beam is realized with the electrostatic field for creating a centrifugal force. In this case there is no drift of the electron across the driving magnetic field. The value of the electrical field is 21 kV/cm at the electron energy 8 MeV and the bending radius 400 cm.

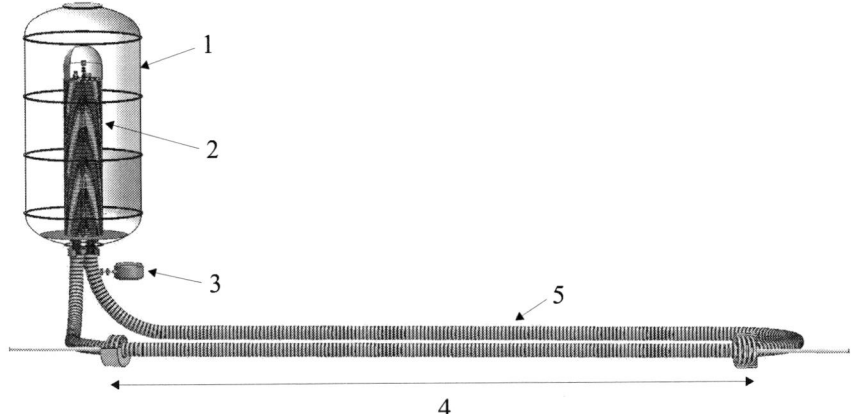

FIGURE 3. Layout of the high voltage cooler for HESR .1 – high voltage tank; 2 – electrostatic column; 3 – cyclotron for charging of the head of electrostatic column; 4 – coolling section; 5 – reversal track.

The main cooling solenoid will have length 30 m and maximum magnetic field up to 5 kG. This solenoid may be constructed as the set pancake section assembled in series. The diameter of one pancake solenoid is 80 cm. The step of the regular structure is 20 cm. The flexjoint should enable the incline 0.6° in the longitudinal direction for the field correction in the cooling section. The total length of the coil with isolation and their construction should enable the incline 3° in the arc section of the cooler.

The obtaining of the parallelism of the magnetic field 10^{-5} is realized by tuning of each pan-cake solenoid. This method was used in the cooler for the IMP. The non-parallelity of the magnetic field lines 8×10^{-6} at a length of 300 cm was obtained. This tuning is hard to be done "on-the- fly" because for fast and operative regulation of the magnetic filed the set of the dipole correction coil is planned. The correction coil may be low-power as the main tuning will be done with the pan-cake coils.

The important part of this project should be a system, which automatically measures and corrects the straightness of magnetic field. It should operate inside the vacuum chamber at ultra-high vacuum condition. For this purpose the laser beam will be sent through a vacuum window and system of mirrors at the center of vacuum chamber. The magnetic compass with movable mirror reflects the laser beam. The deflection of the laser beam is used as the signal for the correction magnetic lines. This system was used at BINP at the single pass measuring of the magnetized cooling force and the accuracy was achieved near 10^{-5}. These experiments made on this system show that for keep the magnet line straightness on level 10^{-5} rad we need to did correction each week.

The electrostatic column contains of three optic channels. Two optic channels are used for the beam acceleration and deceleration on its way from the electron gun to the collector. Third optic channel is used for the charging of the head of electrostatic column by the H⁻ beam. At the same time it used for the transport test H⁻ beam generated in the head of electrostatic column. The voltage of the head is measured with the energy of the H⁻ beam.

There are some possible methods for the charging of the head of electrostatic column.

The most popular high voltage generator system for a voltage over 5 MV are either a mechanical charging device like PELETRON or Van De Graff. In this system a small pellet charged at a ground potential to a voltage near of 50 kV mechanically moved to a high voltage terminal. The charging current of this system is near 100 mkA. For reaching a current near 1 mA used a many chain systems should be used. The ion accelerators with such type of a high voltage system work stably for many days without problems. But the system of moving pellet is very complicate mechanical system in any case.

For the electrostatic cooler we propose to use a small cyclotron at an energy 10 MeV H⁻ ions as a charging system. The commercially available cyclotrons with similar parameters are widely used for isotope production. The ion beam charging eliminates the mechanical vibration of electrostatic columns and gives the high charging current for more stable operation of a cooler at present of some sort sparking effects in the high voltage system.

Table 1. Parameters list of the HESR cooler.

Length of cooler section, m	30 m
Magnetic field in the cooler section, G	5 kG
Magnetic field in the transport section, G	5 kG
Magnetic field in the acceleration section, G	0.5 kG
Electron current, A_y	1 A
Electron energy	8 MeV
Radius of the electron beam	0.3-0.5 cm
Divergence of the magnetic force line	10^{-5}

DESIGN OF ELECTRON COOLER FOR COSY

Electron design for COSY is similar to the HESR design. It is shown in Figure 4. The electron beam is immersed to the magnetic field from the gun to the collector. The electron beam is accelerated to energy up to 2 MeV. After that the electron beam moves in the matched section for the smooth transition from the magnetic field 500 G to the magnetic field 2000 G. After that it is bent in the toroid and it is guided to the cooling section. After the main solenoid the beam is returned to the electrostatic column. Here it is decelerated and is absorbed in the collector located in the head of the electrostatic column. Each toroid consists of two part. The first one bend magnetized electron beam in vertical plane on 90^0. The second one bends the electron beam on 180^0 in plane that canted on 45^0 from vertical. Such complicated 3-D geometry provides compactness of system.

FIGURE 4. Layout of COSY cooler.

The power supply of the acceleration section is planned to be done with gas turbine. The turbine is located in each section and it is provided by SF_6 gas under pressure. The expanding gas rotates the turbine blade and the motor generator shaft. This generator supplies electrical power to the solenoid coil. The flux of gas after turbine is used for cooling coils and keep the temperature inside vessel constant. The preliminary test of such turbine was made in BINP.

The high voltage power supply is located in each section too. Thus there are the set of the high voltage power supplies in series connection. The voltage per section is about ±30 kV. This solution enables easy regulation from the high to low energy. However it is convenient from optic of electron beam point view. It is possible to have the voltage on the limits number section. This may be useful at operation in low energy of the electron beam.

CONCLUSION

The magnetized cooling enables to obtain high cooling rate. The convenient technical decisions for the low energy coolers (up to 300 keV) can be extrapolated to the region of 2 MeV electron cooler (COSY project) or even of 8 MeV (HESR project). The projected based on the quality-checked solutions is reliable with phyics point of view. The technical problem related to this way looks solvable as it is shown in this report.

REFERENCES

1. V.V. Parkhomchuk, A.N. Skrinskiî. *Usp. Fiz. Nauk,* **170**, issue 5, p. 473 (2000) [*Phys. Usp.* **43**, p.433 (2000)].
2. K.Beckert, P.Beller, B. Franzke, F.Nolden, M.Steck.. *Beam Cooling and Related Topics*, Japan, May 19 – 23, 2003.
3. S.Nagaitsev et al. *Nucl. Instr. Meth.*, **A 441** p.241-245, (2000).
4. S.Nagaitsev. *Beam Cooling and Related Topics*, Japan, May 19 – 23, 2003.
5. V.Parkhomchuk, I.Ben-Zvi. C-A/AP/47, April 2001.
6. G. Jackson. *Proceedings of the International Workshop on Medium Energy Electron Cooling*, Novosibirsk, 1997, p.171-182.
7. I.Meshkov, A.Sidorin. *Proceedings of the International Workshop on Medium Energy Electron Cooling*, Novosibirsk, 1997, p.183-188.
8. I.Meshkov, A.Sidorin et al. *Nucl. Instr. Meth.* **A 441**, p.267-270, (2000).
9. V.V.Parkhomchuk. *Nucl. Instr. Meth.,* **A 441**, pp.9-17, (2000).
10. V.Parkhomchuk. *Proc. of Workshop on Electron Cooling and Related Applications*, 1984, Karlsruhe.
11. N.Dikansky, V.Kononov, V.Kudelainen. *Proc. VII All-Union Meeting on Accelerators of Charged Partcles*, Dubna, 1978.
12. Ya. Derbenev, A.Skrinsky. *Fizika Plazmy*, **4**, N3, p.492-500, (1978).
13. V.V.Parkhomchuk. *Workshop on the Medium Energy Electron Cooling*, Novosibirsk, Russia, 26-28 February 1997, pp.11-18
14. V. Bocharov, A. Bubley, S. Konstantinov et al. In proceeding of COOL-05 conference.
15. V.Bocharov, A.Bubley et all BINP, Yang X D, Zhao H W et all IMP. Beam *Cooling and Related Topics*, Japan, May 19 – 23, (2003).

Summary Report: Working Group on COSY 2 MV Cooler

Sergei Nagaitsev[*] and Igor Meshkov[1&]

FNAL, Batavia, IL 60510, U.S.A.;
&JINR, Dubna, 141980, Russian Federation

The working group has discussed several technical issues associated with the proposed 2MV electron cooler for the COSY ring. The topics discussed can be divided into two categories: (1) physics and technical aspects of the project that would advance the state-of-the-art in electron cooling, thus potentially benefiting the COSY science program and (2) technical aspects that could prove beneficial as prototypes for future electron cooling projects.

The working group discussion topics included:
1. Relevance of magnetized cooling.
2. Technical differences between magnetized and non-magnetized electron cooling.
3. Could we study physics of magnetized cooling at existing electron coolers?
4. Importance of an intermediate energy electron cooler (2 MeV) between existing and the future HESR cooler.
5. Importance of interaction studies between ion beams and high density targets (WASA at COSY, pellet target) in the presence of electron cooling.
6. Importance of combination of high energy electron cooling and stochastic cooling against target heating.

The working group has concluded that:
1. For operations with thick internal targets, fast (also known as magnetized) cooling is the only technically feasible solution. Engineering design of a magnetized cooling system would be sufficiently different from the Fermilab 4.3-MV system to warrant a dedicated effort to design a 2-4 kG warm or SC solenoid of a high field quality. An intermediate energy step between the existing COSY cooler and the future high-energy magnetized cooler projects would be extremely useful for finding optimal technical solutions and prototyping many elements. Thus, it would be a natural extension of COSY program to develop a 2MV cooling system on a 3-4 year time scale.

2. The operation of a 2MV cooling system at COSY together with a high-density internal target of WASA detector would uniquely allow to optimize the cooling

---

[1] Working group convener

performance for "tail" particles. This involves both the electron cooling system alone and a combination of the electron and stochastic cooling systems. Realization of such a cooling system will be an important step toward creation of a novel experimental technique aiming to reduce significantly parasitic effects related to halo in accelerated beams – a step to "backgroundless" detection systems.

3. The realization of a 2-MV electron cooler at FZ Juelich (if accomplished on 3-4 year time scale) is invaluable for development of nearest future projects, specifically for the FAIR project, which relies on very efficient electron cooling for its high-luminosity experiments.

Detailed Studies of Electron Cooling Friction Force

A.V. Fedotov[1], D. L. Bruhwiler[2], D.T. Abell[2], A.O. Sidorin[3]

[1]*Brookhaven National Lab, Upton, NY 11973*
[2]*Tech-X Corp., Boulder, CO 80303*
[3]*JINR, Dubna, Russia*

Abstract. High-energy electron cooling for RHIC presents many unique features and challenges. An accurate estimate of the cooling times requires detailed simulation of the electron cooling process. The first step towards such calculations is to have an accurate description of the cooling force. Numerical simulations are being used to explore various features of the friction force which appear due to several effects, including the anisotropy of the electron distribution in velocity space and the effect of a strong solenoidal magnetic field. These aspects are being studied in detail using the VORPAL code, which explicitly resolves close binary collisions. Results are compared with available asymptotic and empirical formulas and also, using the BETACOOL code, with direct numerical integration of less approximate expressions over the specified electron distribution function.

Keywords: electron cooling, beam dynamics, friction force
PACS: 29.28.Bd.,41.75.Lx

INTRODUCTION

The first step towards accurate calculation of cooling times is to use an accurate description of the cooling force. The achievable Coulomb logarithm in the analytic expression for the magnetized cooling force is not large. In addition, in some regimes there is a significant discrepancy between available formulas. For this reason, the VORPAL code [1] is being used to simulate from first principles the friction force and diffusion coefficients for RHIC parameters [2]. VORPAL uses molecular dynamics techniques (i.e. simulating every particle in the problem) and explicitly resolves close binary collisions to obtain the friction force and diffusion coefficient with a minimum of physical assumptions [3].

Only a few topics are addressed in this paper, but the goals of this work include the following: 1) to resolve differences in analytic formulae and semi-analytic calculations, which make assumptions such as uniform electron density, no space charge forces, infinite magnetic field, etc.; 2) to determine the validity of Z^2 scaling for friction forces, which could be broken by non-linear plasma effects in the Debye shielding for a magnetized plasma; 3) to understand the effects of bulk space charge forces; 4) to understand the effects of magnetization, from strong to weak, including magnetic field errors; 5) to accurately simulate the friction force due to magnetized collisions, in the regime of small Coulomb logarithm. In addition, for the non-magnetized concept of the RHIC cooler [4], which uses a long wiggler to focus the electron beam and suppress recombination, the goals of this work include the following: 1) to explore the effects of the wiggler on the friction force, and 2) to study the effects of magnetic field imperfections. In both the magnetized and non-magnetized approaches, if the friction force for RHIC parameters deviates significantly from the description based on available formulas, it will be necessary to generate with VORPAL a table of friction coefficients for use in other codes.

The BETACOOL code has been recently enhanced [5], in order to provide benchmarking with VORPAL. In addition to the asymptotic formulas, typically used for the magnetized and non-magnetized friction force estimates, BETACOOL now includes direct numerical integration over the electron velocity distribution. This numerical evaluation of the force enables an accurate comparison with VORPAL results, both for the magnetized and non-magnetized friction force with an anisotropic velocity distribution of electrons. The results of such benchmarking are summarized in this paper.

NON-MAGNETIZED FRICTION FORCE

The friction force on an ion inside an electron beam with velocity distribution function $f(v_e)$ is described by the formula:

$$\vec{F} = -\frac{4\pi n_e e^4 Z^2 L}{m} \int \frac{\vec{v}_i - \vec{v}_e}{|\vec{v}_i - \vec{v}_e|^3} f(v_e) d^3 v_e, \qquad (1)$$

where v_e and v_i are the electron and ion velocity, while L is the Coulomb logarithm:

$$L = \ln \frac{\rho_{max}}{\rho_{min}}, \qquad (2)$$

where ρ_{max} and ρ_{min} are the maximum and minimum impact parameters, respectively. For an isotropic Maxwellian distribution of electrons, described by the function

$$f(v)d^3v = \left(\frac{\mu}{2\pi T_e}\right)^{3/2} \exp(-\mu v_e^2 / 2T) v_e^2 dv_e d\Omega, \qquad (3)$$

the friction force $\vec{F}_{NM}$ is described by the well known formula [6, 7]:

$$\vec{F}_{NM}(\vec{v}_i) = -\frac{\vec{v}_i}{v_i^3} \frac{4\pi n_e e^4 Z^2 L}{m} \varphi\left(\frac{v_i}{\Delta_e}\right), \text{ where } \varphi(x) = \sqrt{\frac{2}{\pi}} \int_0^x e^{-y^2/2} dy - \sqrt{\frac{2}{\pi}} x e^{-x^2/2}, \qquad (4)$$

and Δ_e is the rms electron velocity in the rest frame of the beams (PRF), corresponding to an isotropic electron temperature $T_e = m\Delta_e^2$. Budker first suggested that this friction could be used to cool heavy ions in a storage ring [8].

The more general case of an anisotropic velocity distribution, which is the typical situation for electron cooling, can be approximated by a Maxwellian distribution with different temperatures for the longitudinal and transverse degrees of freedom:

$$f(v)d^3v = \left(\frac{m}{2\pi}\right)^{3/2} \frac{1}{T_\perp \sqrt{T_\parallel}} e^{-mv_\perp^2/2T_\perp - mv_\parallel^2/2T_\parallel} 2\pi v_\perp dv_\perp dv_\parallel. \qquad (5)$$

The friction force components can be accurately calculated using numerical evaluation of the integral in Eq. (1). However, asymptotic expressions based on the Coulomb analogy [9, 10] are frequently used instead, due to their simplicity. If the transverse rms velocity of the electrons $\Delta_\perp$ is substantially larger than the longitudinal velocity spread $\Delta_\parallel$, the friction force can be approximated in three ranges of the ion velocity, for example, as follows [9].

For high ion velocities, $v_i \gg \Delta_\perp$, the longitudinal and transverse components of the friction force are equal:

$$\vec{F} = -\frac{4\pi Z^2 e^4 n_e L}{m} \cdot \frac{\vec{v}_i}{v^3}, \tag{6}$$

and, in this range, the friction force shape agrees asymptotically with Eq. (4). For ion velocities in the range $\Delta_\parallel \ll v_i \ll \Delta_\perp$, the friction force components are given by:

$$\vec{F}_\perp = -\frac{4\pi Z^2 e^4 n_e L}{m} \cdot \frac{v_{i,\perp}}{\Delta_\perp^3}, \qquad F_\parallel = -\frac{4\pi Z^2 e^4 n_e L}{m} \frac{v_\parallel}{|v_\parallel|\Delta_\perp^2}. \tag{7}$$

In the limit of very low ion velocities, $v_i \ll \Delta_\parallel$, the transverse component of the friction force is zero, and the longitudinal component is given by:

$$F_\parallel = -\frac{4\pi Z^2 e^4 n_e L}{m} \frac{v_\parallel}{\Delta_\parallel \Delta_\perp^2}. \tag{8}$$

The friction force based on the asymptotic expressions has been compared with direct numerical integration, based on Eq. (1), using the BETACOOL code. It was found that these asymptotic formulas can overestimate the friction force by a factor of two or higher, even for relatively large anisotropy of the electron distribution. Since our primary goal is to have an accurate description of the friction force, for comparison with VORPAL simulations, we prefer to use numerical integrals rather than the asymptotic expressions given above.

Figure 1 compares VORPAL data (dots with error bars) with the force described by Eq. (4) (blue solid line) and the result of numerical integrations based on Eq. (1) (red dashed line), for the case of an isotropic Maxwellian distribution of electrons, where $\Delta_\perp = \Delta_\parallel = 1.0e5$ m/s ($Z=79$, $n_e=2e15$ m^{-3}). For all VORPAL simulations presented in this paper, the electron/electron interactions are neglected, which is valid for sufficiently short interaction times. The interaction between ions and electrons is limited by the time of flight through the cooling section, which is 4e-10 s in the PRF, for a cooling section length L=13 m and relativistic factor $\gamma=108$ for the Au^{+79} ions in RHIC. For typical parameters of the RHIC cooler, the interaction time is smaller than or comparable to the plasma period.

Since only very short interaction times are simulated with VORPAL, the rms spread in the ion velocity changes, due to diffusion, is significant and can be larger than the velocity reduction due to friction for a single pass. This makes it difficult to accurately extract the friction force from simulated ion velocities. Some special tricks are being employed to suppress the diffusive aspect of the ion dynamics [3], because our primary interest here is the friction force. For each VORPAL data point shown in the figures, corresponding to a single initial ion velocity, we have generated 100's of ion trajectories, $N_{traj.}$, and plotted the mean velocity drag (or, equivalently, the mean friction force). According to the Central Limit Theorem, the uncertainty in these mean values is +/- 1 rms/$(N_{traj.})^{1/2}$, which is what we've used for the error bars.

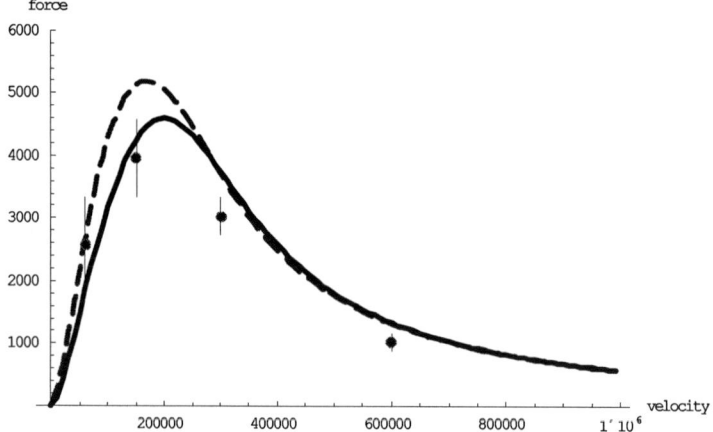

FIGURE 1. Non-magnetized friction force for an isotropic Maxwellian distribution of electron velocities with $\Delta_\perp = \Delta_\| = 1.0\text{e}5$ m/s. Force [eV/m] vs. ion velocity [m/s]: solid blue line – Eq. (4); dashed red line – numeric integration using BETACOOL; points with error bars –VORPAL simulations.

Figure 2 compares VORPAL data (dots with error bars) with the result of numerical integration based on Eq. (1) (solid red line), for the case of an anisotropic Maxwellian distribution of electrons, where $\Delta_\perp = 4.2\text{e}5$ and $\Delta_\| = 1.0\text{e}5$ m/s ($Z=79$, $n_e = 2\text{e}15$ m^{-3}).

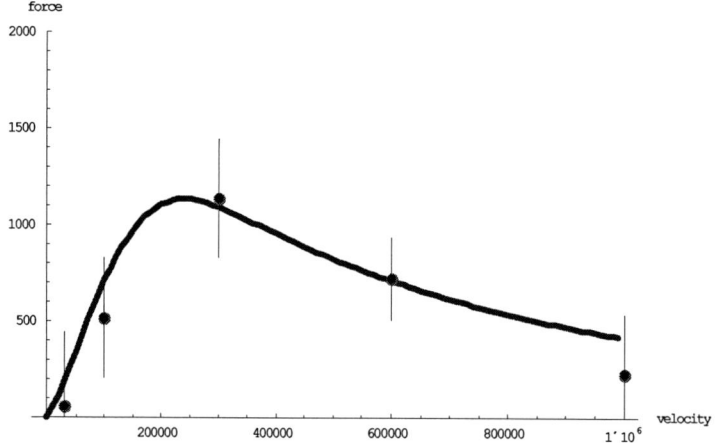

FIGURE 2. Non-magnetized friction force for an anisotropic electron distribution. Force [eV/m] vs. ion velocity [m/s]: solid line (red) – numeric integration using BETACOOL; points with errors bar – simulations using VORPAL.

Simulations were done for other degrees of anisotropy of electron velocity as well. We find agreement between VORPAL simulations and numeric integration satisfactory, and thus use the non-magnetized friction force in BETACOOL based on numerical integration in our dynamics studies of the non-magnetized cooling [4].

MAGNETIZED FRICTION FORCE

The presence of a strong longitudinal magnetic field changes the collision kinetics. The magnetic field limits transverse motion of the electrons. In the limit of very strong magnetic field, the transverse degree of freedom does not take part in the energy exchange, because collisions are adiabatically slow relative to the Larmor oscillations. As a result, the efficiency of electron cooling is determined only by the longitudinal velocity spread of electrons. In typical low-energy electrostatic coolers, the longitudinal velocity spread of electrons is much smaller than the transverse one. Thus an effect of strong velocity anisotropy together with the magnetic field leads to very fast magnetized cooling. In applications to higher-energy cooling, this advantage is somewhat limited due to the significant contribution to the effective velocity spread of electrons from the imperfections of the cooling solenoid. For example, for the RHIC cooler, assuming solenoid imperfections with an rms angle at the level of 1e-5, the effective angular spread in the beam frame is larger by the factor $\gamma=108$, or ~1e-3 m/s.

Since the friction force cannot be written analytically in closed form for arbitrary strength of the magnetic field, numerical simulations are required to explore in detail collisions between magnetized electrons and ions. In recent years, a lot of studies in this area were done by the Erlangen group [11]. However, to the best of our knowledge, a systematic comparison with the friction force formulas used by the electron cooling community have not been reported. In this paper, we attempt such a comparison with available formulas using the VORPAL code.

In the limit of a very strong magnetic field, where transverse motion of electrons is completely suppressed, the result for the magnetized friction force was written by Derbenev and Skrinksy in the following form [10]:

$$\vec{F} = \frac{2\pi Z^2 e^4 n_e}{m} \frac{\partial}{\partial \vec{V}} \int \left[\frac{V_\perp^2}{U^3} L_M + \frac{2}{U} \right] f(v_e) dv_e, \qquad (9)$$

where Z is the ion charge number, e is the electron charge, n_e is the electron density, m is the electron mass, $\vec{V} = (V_\perp, V_\parallel)$ is the ion velocity, $U = \sqrt{V_\perp^2 + (V_\parallel - v_e)^2}$ is the relative velocity of the ion and an electron "Larmor circle," with transverse electron velocities being completely suppressed (i.e. approximation of infinite magnetic field). The actual values of the magnetic field and transverse rms electron velocity spread enter only via the cutoff parameters under the Coulomb logarithm, which is defined as

$$L_M = \ln\left(\frac{\rho_{max}}{\rho_L}\right) \qquad\qquad \rho_L = \frac{cm\Delta_\perp}{eB}. \qquad (10)$$

The function in Eq. (9) has asymptotes in the region of small and large ion velocities. When $V \gg \Delta_\parallel$, the electron distribution can be approximated by the delta-function $f(v_e) = \delta(v_e)$, and integration of (9) gives [10]:

$$F_\parallel = -V_\parallel \frac{2\pi Z^2 e^4 n_e}{mV^3} \left(\frac{3V_\perp^2}{V^2} L_M + 2 \right), \qquad (11)$$

$$F_\perp = -V_\perp \frac{2\pi Z^2 e^4 n_e L_M}{mV^3} \frac{V_\perp^2 - 2V_\parallel^2}{V^2}. \qquad (12)$$

When the ion velocity is sufficiently less then electron velocity spread V << $\Delta_\parallel$, the friction force can be expressed as [10]:

$$F_\parallel \approx -2\sqrt{2\pi}\frac{Z^2 e^4 L_M}{m\Delta_\parallel^3} V_\parallel, \qquad (13)$$

$$F_\perp \approx -2\sqrt{2\pi}\frac{Z^2 e^4 L_M}{m\Delta_\parallel^3}\ln\left(\frac{\Delta_\parallel}{V_\perp}\right) V_\perp. \qquad (14)$$

We note that Eq. (9) and the asymptotic expressions in Eq.'s (11) and (12) were originally derived based on a perturbative treatment of the collective plasma response. It was later suggested by Parkhomchuk [12] that one gets slightly different asymptotic expressions using the binary collisions approach. Because the difference between these two approaches only slightly affects the final expression for the asymptotics given in Eq.'s (11) and (12), we leave this issue to future work.

As for the non-magnetized case, one can compare the accuracy of asymptotic expressions with numerical integration over the electron velocity distribution. Such numerical integration was recently implemented in the BETACOOL code, using Eq, (9), in the following form:

$$F_\perp(V_\perp, V_\parallel) = -\frac{2\pi Z^2 e^4 n_e L_M}{m}\int \frac{V_\perp\left(V_\perp^2 - 2(V_\parallel - v_e)^2\right)}{\left(V_\perp^2 + (V_\parallel - v_e)^2\right)^{5/2}} f(v_e) dv_e, \qquad (15)$$

$$F_\parallel(V_\perp, V_\parallel) = -\frac{2\pi Z^2 e^4 n_e}{m}\int\left(L_M\frac{3V_\perp^2(V_\parallel - v_e)}{\left(V_\perp^2 + (V_\parallel - v_e)^2\right)^{5/2}} + 2\frac{V_\parallel - v_e}{\left(V_\perp^2 + (V_\parallel - v_e)^2\right)^{3/2}}\right) f(v_e) dv_e. \qquad (16)$$

Recently, to account for finite values of the magnetic field, an empirical expression for the magnetized friction force was suggested by Parkhomchuk [13]:

$$\vec{F} = -\vec{v}\frac{4Z^2 e^4 n_e L_P}{m}\frac{1}{\left(v^2 + \Delta_{e,eff}^2\right)^{3/2}}, \qquad (17)$$

where $\Delta_{e,eff}$ is the effective electron velocity spread. In the absence of space-charge effects, $\Delta_{e,eff}$ is the longitudinal velocity spread of electrons, with an additional contribution from the effective angles of variations in the magnetic field lines. The Coulomb logarithm in Eq. (17) is given by

$$L_P = \ln\left(\frac{\rho_{max} + \rho_{min} + \langle\rho_\perp\rangle}{\rho_{min} + \langle\rho_\perp\rangle}\right). \qquad (18)$$

Figure 3 shows the longitudinal friction force as a function of the longitudinal ion velocity, for the case of zero transverse ion velocity. The VORPAL simulations are done for the following parameters: B=5T, time of interaction in beam frame 0.4 ns, $\Delta_\perp$=1.1e7 and $\Delta_\parallel$=1.0e5 m/s, Z=79 and n_e=2e15 m^{-3}. One can see that the asymptotic expressions overestimate the friction force by a significant factor. The numerical integrations from BETACOOL show friction force values similar to the expression in Eq. (17), which compare well with VORPAL results but with the maximum shifted towards higher ion velocities. For the effective velocity in Eq. (17) we did not assume any contribution except for $\Delta_\parallel$=1.0e5 m/s. For ion motion along the magnetic field lines, the expression in Eq. (17) is very close to the results of VORPAL, for the parameters mentioned above.

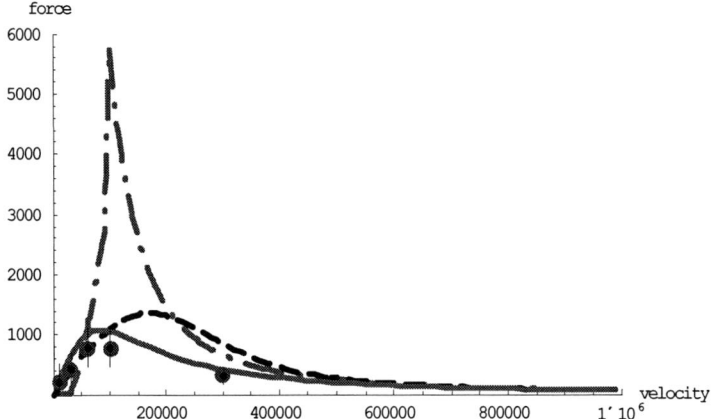

FIGURE 3. Magnetized friction force [eV/m] vs. longitudinal ion velocity [m/s]: solid line (green) – empiric formula by Parkhomchuk (Eq. 17); dot-dash line (gray) – Derbenev-Skrinksy-Meshkov asymptotic expressions [9]; dash line (blue) – Derbenev's expression with numerical integration over electron distribution (Eq. 16); dots with error bars – VORPAL results.

Figure 4 shows simulations using the same parameters as in Fig. 3, but with two different sets of transverse rms electron velocities: $\Delta_\perp$=1.1e7 m/s (pink, lower data set) and 4.2e5 m/s (red, upper data set), with good magnetized cooling seen in both cases. This difference in the transverse electron velocities results in a factor of 2.2 change in the magnetized Coulomb logarithm for fixed ion velocity. The corresponding change in the cooling force is reproduced by the VORPAL simulations and also by Eq. (17), which is plotted as green lines in Fig 4.

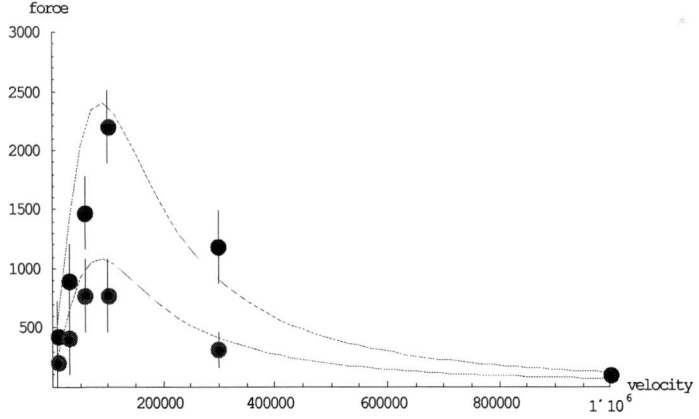

FIGURE 4. Magnetized friction force for ion velocities parallel to magnetic field lines. Force [eV/m] vs. longitudinal ion velocity [m/s]: upper curve (Eq. 17) and red dots (VORPAL data) for $\Delta_\perp$=4.2e5 m/s; lower curve (Eq. 17) and pink dots (VORPAL data) for $\Delta_\perp$=1.1e7 m/s.

Angular Dependence for Large Relative Velocities

An important feature of rigorous description in a strong magnetic field is that for relative velocities much higher than the longitudinal spread of electrons, the component of the friction force in the longitudinal and transverse directions have very different forms, as can be seen from Eq. (11) and (12). The first important feature is a specific dependence of the transverse angle between an ion velocity and the direction of the magnetic field line, which is plotted in Fig. 5 for the longitudinal component of the friction force. The solid blue line corresponds to Eq. (11).

In order to simulate the same behavior, we need to reproduce conditions similar to those assumed in the derivation of first Eq. (9) and then Eq. (11). This corresponds to zero rms electron velocities in both the transverse and longitudinal degrees of freedom. The lower cutoff parameter in the Coulomb logarithm of Eq. (9) is modeled, for such idealized "cold" case, with a 'cloud' or softening parameter introduced in the direct numerical simulations of the drag force. Figure 5 shows that, when such approximations are specifically introduced in the simulations, the agreement between simulations and Eq. (11) is very good.

However, this "cold-beam" approximation is far from valid in real situations. For example, the blue dots without error bars correspond to finite rms velocity spreads of the electron beam, as is expected for the case of magnetized cooling in RHIC, with $\Delta_\perp = 1.1e7$ and $\Delta_\parallel = 1.0e5$ m/s. The dependence on the angle between the ion velocity and the magnetic field lines, which is dramatic in the idealized cold case, is still seen, but it is now much less pronounced. In fact, the finite temperature results are not much different from the empirical formula in Eq. (17), which completely ignores any angular dependence with respect to the strong magnetic field. Similar arguments were made previously by Parkhomchuk [13].

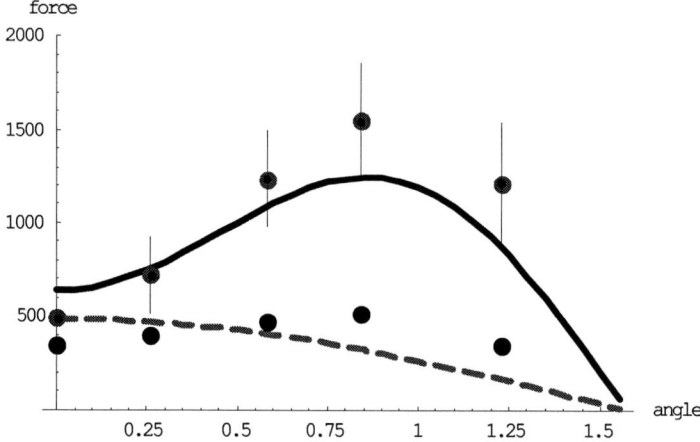

FIGURE 5. Dependence of the longitudinal component of the friction force on transverse angle with respect to magnetic field lines. Force [eV/m] vs. angle [rad]: solid blue line – Derbenev-Skrinsky asymptotics; dashed green line – VP empirical formula (Eq. 17); pink dots with errors bars – VORPAL simulations of ultra-cold electron beam; blue dots without errors bars – VORPAL simulations of finite temperature electrons.

For the transverse component of the friction force, VORPAL simulations show 'anti-friction' as predicted by the asymptotic formulas. This is shown for the case of cold electron beam simulations in Fig. 6. The detailed discussions of the amount of observed anti-friction, comparison with different approaches [10,12], and discussion for the finite temperature electron beam will be reported elsewhere.

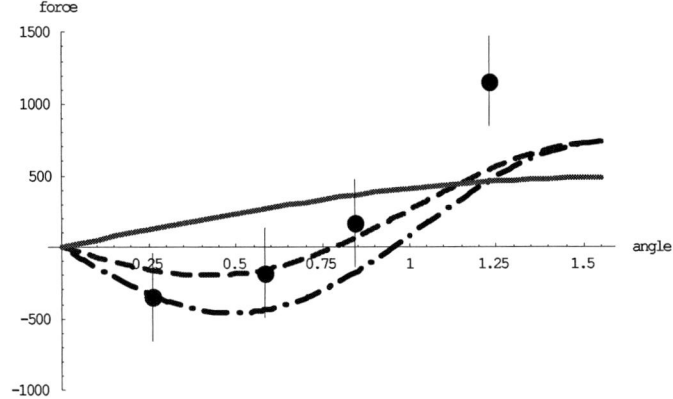

FIGURE 6. Dependence of the transverse component of the friction force on transverse angle. Force [eV/m] vs. angle [rad]: blue dot-dash line – Eq. (12) asymptotics based on the dielectric approach; red dashed line – asymptotics based on the binary collisions approach [12]; green solid line – empirical formula in Eq. (17); dots with errors bars – VORPAL simulation for ultra cold electron beam.

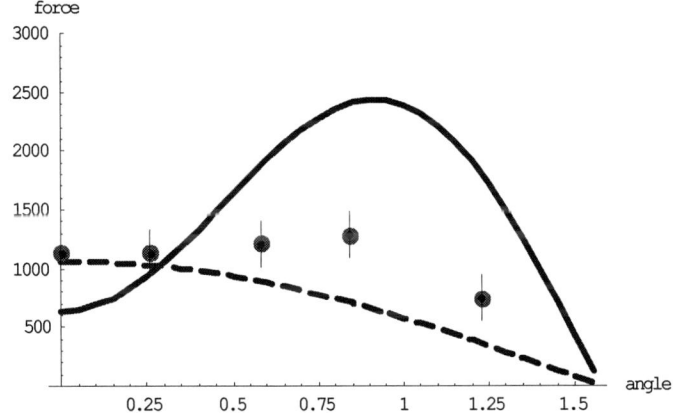

FIGURE 7. Dependence of the longitudinal component of the friction force on transverse angle with respect to magnetic field lines. Force [eV/m] vs. angle [rad]: red solid line – asymptotics in Eq. (11); red dashed line – Eq. (17); dots with errors bars – VORPAL simulations for $\Delta_\perp$=4.2e5 m/s.

Another important feature of Eq. (11) is the behavior at zero transverse angle (i.e. zero transverse ion velocities), which would lead to zero longitudinal friction force in the absence of a non-logarithmic term. However, the origin of this constant term in Eq. (11) has a completely different nature, and is due to phenomena which are not included in the present simulations. (Detailed discussion of these issues is beyond the scope of this paper.) As a result, this constant term has no dependence on the magnetized Coulomb logarithm and, in fact, should be zero in our simulations.

The appearance of finite values for the longitudinal friction force at zero transverse angle in our simulations is due to the fact that we allow for all possible exchanges of electron energy with ions during the finite time of interaction. Our simulations show scaling with the magnetized logarithm at zero transverse angle. This is shown in Fig. 7 for $\Delta_\perp$=4.2e5 m/s. The friction force is increased by the factor expected based on the magnetized logarithm, as compared to the case with $\Delta_\perp$=1.1e7 m/s, shown in Fig. 5. Such scaling with the magnetized logarithm at zero transverse angle is also in good agreement with Eq. (17), although that formula does not include the full dependence on the transverse angle, as can be seen for other values of angles plotted in Fig. 7.

SUMMARY

Direct numerical simulations of binary collisions with the VORPAL code have allowed us to explore the accuracy of various formulas and approximations, both for the non-magnetized and magnetized friction force. Using these simulations, we are in a position to explore detailed aspects of magnetized cooling, some of which are reported in this paper. For some simple cases, a comparison with numerical integration of the friction force expressions using the BETACOOL code have also been shown. These studies enable us to use appropriate friction force formulas with a known degree of accuracy, which is very important for future facilities that will depend on high-energy magnetized cooling.

ACKNOWLEDGMENTS

We are grateful for Ilan Ben-Zvi and Vladimir Litvinenko for many useful discussions and suggestions during these studies. We also thank A. Burov, Ya. Derbenev, and the Accelerator Physics Group of Electron Cooling Project of RHIC for constant support and stimulating discussions. The support and constant help of Tech-X Corp. with the VORPAL code and the Dubna Cooling group with the BETACOOL code is greatly appreciated. Work supported by the US Department of Energy.

REFERENCES

1. C. Nieter, J. Cary, J. Comp. Phys. **196** (2004), p. 448.
2. RHIC E-cooler Design Report, http://www.agsrhichome.bnl.gov/eCool
3. D.L. Bruhwiler et al., AIP Conf. Proceed. **773** (Bensheim, Germany, 2004), p. 394.
4. A.V. Fedotov et al., Proceedings of PAC'05 (Knoxville, TN), TPAT089.
5. BETACOOL program , http://lepta.jinr.ru
6. S. Chandrasekhar, *Principles of Stellar Dynamics* (U. Chicago Press, 1942).
7. NRL Plasma Formulary, ed. J.D. Huba (2000).
8. G.I. Budker. At. Energy **22** (1967), p. 346.
9. I. Meshkov, Phys. Part. Nucl. **25** (1994) , p.631. (and references therein)
10. Ya. Derbenev, A. Skrinsky, Part. Acc. **8** (1978) , p. 235; Sov. Phys. Rev **1** (1981), p. 165.
11. C. Toepffer, Phys. Rev. A **66**, 022714 (2002); H.B. Nersisyan, G. Zwicknagel ands C. Toepffer, Phys. Rev. E **67**, 026411 (2003). (and references therein)
12. V. Parkhomchuk, "Physics of fast electron cooling," Electron Cooling Workshop (Karlsruhe, 1984).
13. V. Parkhomchuk, Nucl. Inst. Meth. A **441**, p. 9 (2000).

Simulations of Dynamical Friction Including Spatially-Varying Magnetic Fields

G.I. Bell[1], D.L. Bruhwiler[1], V.N. Litvinenko[2], R. Busby[1], D.T Abell[1], P. Messmer[1], S. Veitzer[1] and J.R. Cary[1,3]

[1]*Tech-X Corp., Boulder, CO 80303*
[2]*Brookhaven National Lab, Upton, NY 11973*
[3]*University of Colorado, Boulder, CO 80309*

Abstract. A proposed luminosity upgrade to the Relativistic Heavy Ion Collider (RHIC) includes a novel electron cooling section [1], which would use ~55 MeV electrons to cool fully-ionized 100 GeV/nucleon gold ions. We consider the dynamical friction force exerted on individual ions due to a relevant electron distribution. The electrons may be focussed by a strong solenoid field, with sensitive dependence on errors [2], or by a wiggler field [3]. In the rest frame of the relativistic co-propagating electron and ion beams, where the friction force can be simulated for nonrelativistic motion and electrostatic fields, the Lorentz transform of these spatially-varying magnetic fields includes strong, rapidly-varying electric fields. Previous friction force simulations for unmagnetized electrons or error-free solenoids used a 4th-order Hermite algorithm [2,3,4], which is not well-suited for the inclusion of strong, rapidly-varying external fields. We present here a new algorithm for friction force simulations, using an exact two-body collision model to accurately resolve close interactions between electron/ion pairs. This field-free binary-collision model is combined with a modified Boris push [5,6], using an operator-splitting approach, to include the effects of external fields. The algorithm has been implemented in the VORPAL code [7] and successfully benchmarked.

Keywords: electron cooling, beam dynamics, friction force
PACS: 29.28.Bd.,41.75.Lx

INTRODUCTION

We present a new algorithm for nonrelativistic, electrostatic simulations of the dynamical friction force exerted by an electron distribution on an individual ion. In particular, we consider the 'high-energy cooler' for the proposed luminosity upgrade of RHIC [1], where the spatial variations of the lab-frame magnetic fields must be Lorentz-transformed into the beam frame. This situation results in rapidly-varying electric and magnetic fields, which are not easily incorporated in the 4th-order variable time step algorithm used previously for friction force simulations [2,3,4]. Especially for the wiggler-based approach to electron focussing and suppression of electron-ion recombination, these beam-frame electric fields can be very strong. The new algorithm, using an operator-splitting approach, combines a semi-analytic model for binary electron-ion collisions in the absence external fields, with a slightly modified Boris push [5,6] to incorporate all external fields. We discuss the binary collision model in detail and present some comparisons with the Hermite algorithm.

THE BINARY COLLISION MODEL

We consider the interaction of a large number of electrons with a small number of ions (often a single ion), neglecting the electron-electron interactions. Let $\vec{e}_i$ be the coordinate vector of electron i, and $\vec{g}_j$ the coordinate vector of ion j, which we take below to be fully-ionized gold. To advance particle positions and velocities in time, we solve the two-body problem between electron i and ion j exactly, and then add the contributions from all electrons and ions. Because we model close collisions exactly, we can take a constant time step Δt. In the binary collision model, we neglect three-body or higher n-body collisions, which is a good approximation for the low electron densities found in electron coolers. One must choose Δt small enough so that there is at most one 'strong' collision occurring per time step; however, the requirement to adequately resolve time-varying external fields (through operator splitting) is typically the more stringent constraint.

This differs from an existing Hermite algorithm [4] in that we resolve two-body collisions using an exact model. The Hermite algorithm, adapted from astrophysical applications [8], uses a 4th order predictor-corrector algorithm to advance particle positions and velocities, and very small time steps are required to resolve close collisions. In order to obtain good overall time performance, particles are binned in time step levels that differ by factors of two, and particles are moved up and down in levels using an estimate of the single-step error. The algorithm retains 4th order accuracy in the case of a constant magnetic field, but in the case of a more complex magnetic field, 4th order accuracy is not guaranteed.

In the binary collision model we solve the classical two-body, central-force problem. This is a classical problem in celestial mechanics [9], and involves moving into the center-of-mass reference frame. In this frame, the problem reduces to central force motion of a single particle of reduced mass $\mu = m_g m_e / (m_g + m_e)$. If we define the "reduced coordinate" $\vec{x}_{ij} = \vec{e}_i - \vec{g}_j$, then the Lagrangian of the reduced one-particle system is

$$L = \frac{1}{2}\mu |\dot{\vec{x}}_{ij}|^2 - V(|\vec{x}_{ij}|), \quad \text{where} \quad V(r) = -\frac{k}{r} \quad \text{and} \quad k = \frac{Ze^2}{4\pi\varepsilon_0}. \tag{1}$$

By conservation of angular momentum, the motion lies in the plane defined by $\vec{x}_{ij}$ and $\dot{\vec{x}}_{ij}$. To solve exactly, we perform an 'Euler 123' sequence of three rotations into the plane of motion [9].

In the gravitational case, which is typically treated by standard texts, the force between two bodies is always attractive; however, we allow for like-signed particles, for which $k < 0$. In the majority of cases, the orbit of the reduced particle will be unbounded and hyperbolic. As shown in classical mechanics texts [9], the solution to the reduced two body problem involves solving the classical hyperbolic Kepler equation of the form:

$$\varepsilon \sinh\psi - \text{sgn}(k)\psi = C, \tag{2}$$

where $\varepsilon > 1$ is the eccentricity, ψ is the "eccentric anomaly", and C is a constant which depends on initial conditions and other parameters of the problem. Note the sign difference for the opposite-sign ($k > 0$) versus like-sign ($k < 0$) particles. The Kepler equation cannot be solved analytically, but ψ can be calculated using root finding methods.

Elliptical (bound) orbits are rare, but possible for opposite sign particles. These interactions are not of great importance because they (presumably) do not affect the velocity of the ion strongly, and can either be ignored, or else solved using the same methodology and standard elliptic Kepler equation (currently implemented).

We advance the particle coordinates and velocities one time step by the formulas:

$$\vec{e}_i = \vec{e}_i + \Delta t \dot{\vec{e}}_i + \frac{\mu}{m_e} \sum_j \delta \vec{x}_{ij} \quad \dot{\vec{e}}_i = \dot{\vec{e}}_i + \frac{\mu}{m_e} \sum_j \delta \dot{\vec{x}}_{ij}$$
$$\vec{g}_j = \vec{g}_j + \Delta t \dot{\vec{g}}_j - \frac{\mu}{m_g} \sum_i \delta \vec{x}_{ij} \quad \dot{\vec{g}}_j = \dot{\vec{g}}_j - \frac{\mu}{m_g} \sum_i \delta \dot{\vec{x}}_{ij} \quad (3)$$

These specify that the particles drift in straight lines, except for the sum of "correction terms" $\delta \vec{x}_{ij}$ and $\delta \dot{\vec{x}}_{ij}$. These corrections are not assumed to be small, but can be arbitrarily large. The term $\delta \vec{x}_{ij}$ comes from the exact solution to the two-body problem between electron i and ion j; it is the difference between the exact solution and the "drift solution" where the two particles drift freely without interaction. The sums represent equal and opposite "impulses" in coordinate and velocity that must be added to the drift solution to take into the account all the two-body interactions:

$$\delta \vec{x}_{ij} = \vec{x}_{ij}(\Delta t) - \left[\vec{x}_{ij}(0) + \Delta t \dot{\vec{x}}_{ij}(0) \right]$$
$$\delta \dot{\vec{x}}_{ij} = \dot{\vec{x}}_{ij}(\Delta t) - \left[\dot{\vec{x}}_{ij}(0) \right] \quad (4)$$

Here $\vec{x}_{ij}(t)$ and $\dot{\vec{x}}_{ij}(t)$ are the exact solution at time t of the reduced problem for the interaction of electron i with ion j, and $t = 0$ corresponds to the initial condition at the beginning of the time step. Calculating these involves solving the reduced 2D problem in polar coordinates, and then converting back first to 2D Cartesian and then back to 3D via the inverse (transpose) of the rotation matrix.

Solving the two-body problem exactly for all electron-ion pairs can be computationally expensive, and for many such pairs the interaction is very weak. We speed up the algorithm by using a simpler technique for ion-electron pairs for which the ratio of potential to kinetic energy is smaller than some threshold (.001, for example) during the entire time interval Δt. The coordinate and velocity impulses can be approximated by solving the two-particle system using Euler's method, or some other fast technique. For example, for Euler's method we would approximate by using

$$\delta \vec{x}_{ij} \cong 0$$
$$\delta \dot{\vec{x}}_{ij} \cong \frac{\Delta t}{\mu} \vec{F}(\vec{x}_{ij}(0)) \quad (5)$$

where $\vec{F}(\vec{x}) = -k\vec{x}/|\vec{x}|^3$ is the central force. Euler's method is only first order accurate in time, but for these weak interactions the force varies slowly and the overall error is small. In the current VORPAL [7] implementation, we use Heun's method, an explicit 2nd-order predictor-corrector algorithm.

HERMITE MODEL COMPARISON

We compare the results from the new binary collision algorithm with an existing Hermite predictor-corrector algorithm [4]. Figure 1 presents one such comparison, for parameters relevant to the proposed RHIC cooler. These simulations assume a periodic 3D domain, 0.4 mm on a side, with an electron density of 2e15 m^{-3}, for a total of 128,000 electrons.

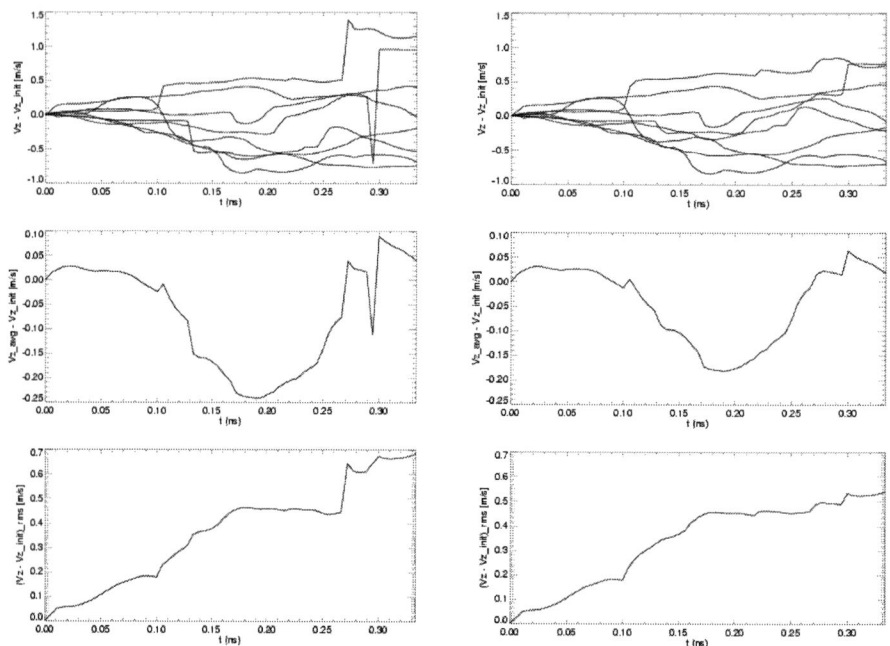

FIGURE 1: The change in longitudinal velocity of the gold ions versus time, the average and rms spread for the Hermite algorithm (left) and binary collision model (right).

Figure 1 shows the trajectories of eight gold ions, for the Hermite (left) and binary collision (right) algorithms. The two sets of trajectories are virtually indistinguishable until the middle of the time period. Close collisions are very sensitive to the initial conditions, and small differences in the algorithms are expected to cause trajectories to diverge eventually. On this time scale, the simulation is noise-dominated and no overall friction is observed. Many such simulations must be run over a longer time period to obtain reasonable statistics and show friction.

SUMMARY

Previous simulations of the dynamical friction force on heavy ions have proven useful [2,3,4] for exploring the new regime of high-energy electron coolers, such as the one proposed for a luminosity upgrade at RHIC [1]. We have presented here a new algorithm, which is superior to the previous approach in that it will allow for accurate inclusion of strong, oscillating external fields. Without external fields, the new algorithm can in some cases tolerate much larger time steps. This new binary-collision algorithm, implemented in the VORPAL code [7], is currently being used to explore the effect of wiggler magnetic fields on the friction force.

ACKNOWLEDGMENTS

We thank A. Fedotov, A. Burov and Ya. Derbenev for helpful discussions. This work is supported by the U.S. DOE Office of Science, Office of Nuclear Physics under grants DE-FG03-01ER83313 and DE-FG02-04ER84094.

REFERENCES

1. RHIC E-cooler Design Report, http://www.agsrhichome.bnl.gov/eCool
2. D.L. Bruhwiler, R. Busby, D.T. Abell, S. Veitzer, A.V. Fedotov and V.N. Litvinenko, "The Effect of Magnetic Field Errors on Dynamical Friction in Electron Coolers," Proc. Part. Accel. Conf. (IEEE, 2005), p. 4206.
3. A.V. Fedotov, D.L. Bruhwiler, D.T. Abell and A.O. Sidorin, "Detailed Studies of Electron Cooling Friction Force," these proceedings.
4. D.L. Bruhwiler *et al.*, AIP Conf. Proceed. **773** (Bensheim, Germany, 2004), p. 394.
5. J. Boris, in Proc. Fourth Conf. on Numerical Simulation of Plasmas, eds. J.P. Boris and R.A. Shanny (Naval Research Laboratory, Washington, D.C., 1970), p. 367.
6. P.H. Stoltz, J.R. Cary, G. Penn and J. Wurtele, Phys. Rev. Special Topics - Accel. & Beams **5**, 094001 (2002).
7. C. Nieter and J.R. Cary, J. Comp. Phys. **196** (2004), p. 448.
8. J. Makino & S.J. Aarseth, Publ. Astron. Soc. Japan **44** (1992), p. 141.
9. H. Goldstein, *Classical Mechanics*, 2nd Ed. (Addison-Wesley, 1980), Chapter 3.

Comission of Electron Cooler EC-300 for HIRFL-CSR

E.Behtenev[*], V.Bocharov[*], V.Bubley[*], M.Vedenev[*], R.Voskoboinikov[*], A.Goncharov[*], Yu.Evtushenko[*], N.Zapiatkin[*], M.Zakhvatkin[*], A.Ivanov[*], V.Kokoulin[*], V.Kolmogorov[*], M.Kondaurov[*], S.Konstantinov[*], G.Krainov[*], V.Kozak[*], A.Kruchkov[*], E.Kuper[*], A.Medvedko[*], L.Mironenko[*], V.Panasiuk[*], V.Parkhomchuk[*], V.Reva[*], A.Skrinsky[*], B.Smirnov[*], B.Skarbo[*], B.Sukhina[*], K.Shrainer[*], Yang X.D[†], Zhao H.W[†], Li J[†], Lu W[†], Mao L J[†], Wang Z X[†], Yan H B[†], Zhang W[†], Zhang J H[†]

[*]*BINP, Novosibirsk, Russia*
[†]*IMP, Lanzhou, China*

Abstract. HIRFL-CSR, a new ion accelerator complex, is under construction at IMP, Lanzhou, China. It is equipped with two electron cooling devices. This article describes the commissioning of cooler at electron energy 300 keV. The cooler is one of the new coolers with some unique manufactured in BINP, Russia. It has a new electron gun producing a hollow electron beam, electrostatic bending and a new structure of solenoid coils at the cooling section. The test results of cooler obtained in Novosibirskand Lanzhou are reported.

INTRODUCTION

HIRFL-CSR is a multi-purpose system that consists of HIRFL (injector), main ring (CSRm) and experimental ring (CSRe) [1]. It is planned, that the heavy ion beams with the energy range of 8-50 MeV/u from the HIRFL is accumulated, cooled and accelerated to 100-450 Mev/u in CSRm. After that the ions is extracted fast to produce RIB or highly charged heavy ions. The secondary ions are stored by the experimental ring for many internal-target experiment or high-precision spectroscopy. Two electron coolers are manufactured for CSR complex. In CSR-m, ecooling will be used to increase the beam intensity at injection energy. It was commissioned in Lanzhou in Spring, 2003 [2]. In CSRe, e-cooling will be used to compensate the growth of beam emittance induced by internal target processes and to provide high-quality beams for the high-resolution mass spectroscopy experiment. The main parameters of the CSRe cooler are shown in Table1.

Table 1. Parameters CSRe

Operation Energy: Ion [MeV/u] Electron [keV]	10-450 5-300
Max. electron current [A]	3
Cathode diametr [cm]	3
Magnetic field of cooling section [kG]	0.5-1.5
Length of cooling section [m]	4
Beta-functions at cooling section [m]	12,16

CSRE ELECTRON COOLER

The cooler EC-300 consists of cooling section solenoid, two bending toroids, electron gun and collector with solenoids forming magnetic field near cathode and collector body (see Fig. 1). Pumping system produces vacuum value 3.10_{-11} mbar. Dipole magnets and set of special coils are included to provide correction of both electron and ion trajectories. The location of the electron beam is measured by system of pick-up electrodes.

ELECTRON GUN WITH VARIABLE ELECTRON BEAM PROFILE

The electron gun with a control electrode was designed to produce hollow electron beams[4]. The electron gun under consideration is shown on Fig. 1. By digits on the figure are marked: 1 – cathode, 2 –forming electrode, 3 – control electrode, 4 – anode. The gun is immersed into the longitudinal magnetic field of 700-1000 Gs. Convex oxide cathode Ø29 mm is used.

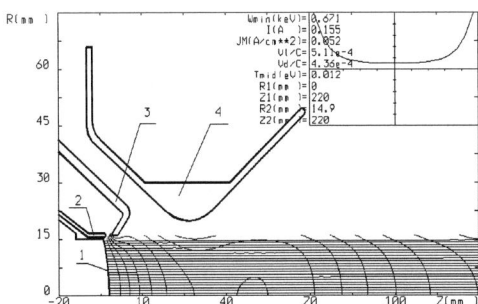

Figure 1. Electron gun calculation.

The control electrode is situated near the cathode edge, so its potential strictly influences on the emission from this area. By varying the potential of this electrode it is possible to obtain on the gun output the beam with parabolic, flat or hollow profile. The potential of the forming electrode is equal to the cathode potential; the purpose of this electrode is to dump exceeding emission from the cathode edge.

ELECTROSTATIC BENDING

The electron and the ion beams convergence before entrance at cooling section base on bending at electric field. Advantage of using electric field instead of usually used transverse magnet field is compensation of drift shift at bending for both direction of moving electrons from cathode to collector and reflected electrons moves from collector to gun. First experimental testing of this idea was made Tim Ellison at Indiana University cooler.

The electrostatic field affects equally on both primary electrons of beam and secondary electron reflected from collector. Thus, the secondary electron from collector has a number of attempts for absorption in collector. The leakage current is very small in this case. The figure 3 shows the dependence of the leakage current versus voltage of the electrostatic bending plates for different value of suppressor voltage. In this experiment the voltage of the electrostatic plates and the current in the correction coil was chosen in such a way that the electron beam doesn't shift in the collector and cooling section. So, the integral of drift motion of electron in the toroid section was constant. From figure 3 one can see that the optimum of the electron loss is changed with suppressor voltage. The energy of the electron reflected from the collector is slightly changed. So, the electrostatic bending should be adjusted with the energy of secondary electrons.

The obtainable value of the recuperation efficiency is better than 10_{-6} in optimum. It leads to high vacuum condition of cooler, the pressure of residual gas was better than 10_{-11} at commissioning in BINP. There are no electrons falling on the vacuum chamber and inducing the degassing process. Really, after some training procedure the switching on of electron beam improves the vacuum pressure.

The small leakage current decreases the problem with the radiation condition. The energy 300 keV is enough for producing radiation level 25 $\propto$Rem/sec/mA at not shielded place of cooler device. But this effect was observed at magnetic bending only. At electrostatic bending the radiation level was less than noise level.

Pan-Cake Structure of the Cooling Solenoid

Main solenoid of the cooling section consists of 68 pancake coils connected in series. All of them are adjustable because of tree points of support. After several iterations of measurement and adjustment the sufficiently high level of magnetic field homogeneity was achieved. A Fig.2 show practical solution of this idea at BINP coolers.

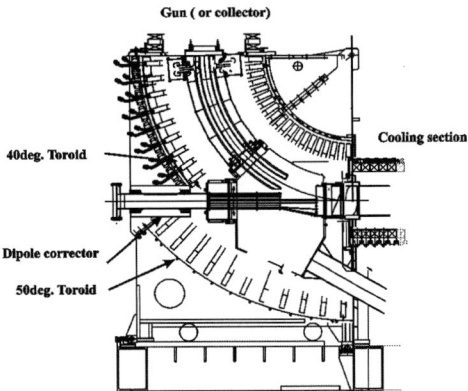

Figure 2. Design of electrostatic bending plates for electron and ion beams convergence.

Absent drift motion for electrons oscillated between electron gun and collector results to very high efficiency capture its at collector after few attempts. The force on bending orbit is:

$$F = \frac{\gamma \beta^2 m_e c^2}{R} = \frac{q[B \times V]}{c} + qE, \quad (14)$$

where B,E magnet and electric field for moving electron at trajectory with radius R. It is possible to have pure magnet bending B=Bmax, E=0 and pure electrostatic bending B=0, E=Emax. At case magnet bending electrons reflected from collector have twise large drift from geometric trajectory by action cenrifugial and magnet force at the same direction instead compensation. The figure 3 show losses current versus voltage on electrostatic bending plates (0→2kV). The magnet field at this experiment simultaneously change (Bmax→0) so that orbit of the primary electron beam is not change.

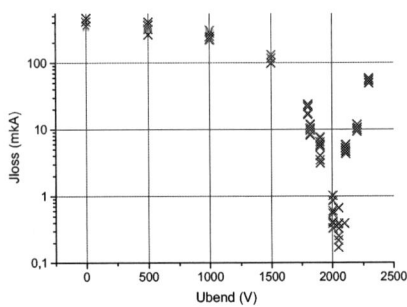

Figure 3. Losses current at EC-35 versus voltage on electrostatic bending plates. (The electron beam 15 keV*1A.).

From this figure clear see that the recuperation efficiency change from 4×10^{-4} for pure magnet bending to $6-7 \times 10^{-7}$ for pure electrostatic bending.

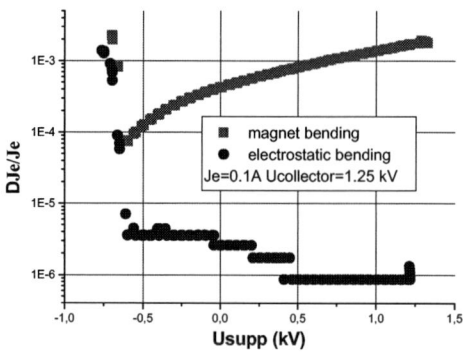

Figure 4. Relative losses electron current versus suppresser voltage for magnet and electrostatic bending.

The minimisation of the electron beam losses current differ for electrostatic and magnet bending (fig. 4). At case the magnet bending using the suppresser electrode for more effective capture electron help decrease relative losses from 2×10^{-3} to 8×10^{-5} when voltage on the suppresser change from collector voltage to -0.7 kV. At the voltage less them -0.7 kV the tails of the primary beam reflected from collector and losses very sharp increased. But at case the electrostatic bending more impotent to have free enter at collector then suppress reflection from the surfer collector . The multiply coming the electrons at the collector made requarement on capture efficiency not so significant. This phenomena open new possibility for operation with the low voltage on collector without strong suppression at the entrance.

Decreasing bombarding the vacuum chamber the electrons with high energy on few order magnitude decrease outgasing and radiation problems for high energy cooler. For example if we need vacuum 10^{-12} Torr and have losses current 400 mkA the pumping power of the cooler vacuum system should be 50000 l/s but for electrostatic bending and loss current 1 mkA we need only 130 l/s pumping power that at many times easy. Figure 5a and 5b show experiments with measuring vacuum pressure at LEIR cooler with switch of ion pumps. At this case gas components with weak pumping by NEG produced fast increasing pressure at cooler with rising time near 0.6E-9 Torr/(175-75 s).

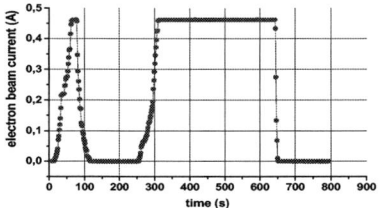 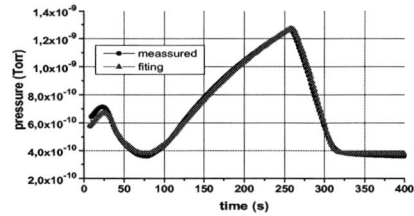

Figure 5a. Electron current versus time at LEIR cooler.

Figure 5b. Pressure versus time with switch off ion pump.

The combination low voltage on collector and low losses the electron current open perspective to have good vacuum at the electron cooler. After relatively short time operation with the electron beam the out gassing inside cooler becomes so low that electron beam switch on improve vacuum at cooler.

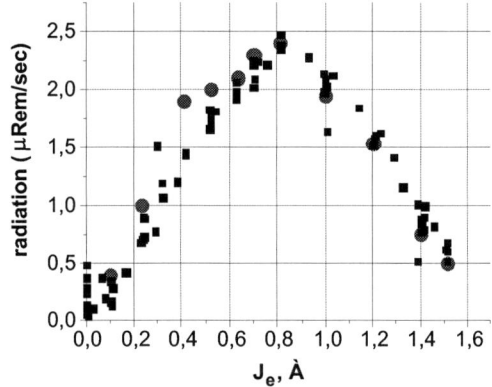

Figure 6. Radiation level near cooler with beam energy 260 keV and losses current (at units 1=40 mkA) versus electron beam current.

Figure 6 show results of measuring radiation at EC300 cooler before final tuning capture efficiency. The radiation level is proportional of loss current and equal to 25 µRem/mkA (for 260 keV). Fast increasing the radiation with increasing cooler voltage made this problem important for the next generation high voltage coolers.

CONCLUSION

The commissioning of cooler with electron beam demonstrated high performance. The electron current up to 3 A was obtain that few times high that can used for the cooling ion beams. The nearest future these coolers will tested at experiments with cooling ion beam and it can open many interesting physics pheromones. The development new generetion of the electron coolers are the result of effort high intellectual team of sciences, engineers, workers at BINP with using world experience technique of the electron cooling.

REFERENCES

[1] Parkhomchuk V.V., Skrinsky A.N. Report on progress in physics 1991, v. 54, n.7, p.919-947.
[2] Parkhomchuk V.V., Skrinsky A.N., Physics-Uspekhi 43(5) 433-452 (2000)
[3] Reistad D. et al., in Proc. Workshop on Beam cooling and Related Topics (Montreux, Switzerland, 4-8 Oct. 1993) (CERN (Series), 94-03, Ed. J.Bosser) (Geneva: European Aorganization for Nuclear Research, 1994) p.183.
[4] Hermanssson L., Reistad D., NIM in Physics Research A 441 (2000) 140-144.
[5] http://accelconf.web.cern.ch/AccelConf/e02/PAPERS/WEPRI049.pdf
[6] http://accelconf.web.cern.ch/AccelConf/r04/papers/TUAI02.PDF

Recuperation of Electron Beam in the Coolers with Electrostatic Bending

M.Bryzgunov, V.Panasyuk, V.Parkhomchuk, V.Reva, M.Vedenev.

Budker Institute of Nuclear Physics, Novosibirsk, Russia

Abstract. An important aspect of the cooler operation is the collector efficiency. The low loss current improves the vacuum condition, the radiation condition and makes the easy design of the power supply system. This article deals with the reuse of collector at the electrostatic bending in the toroidal section. It's possible to compensate the centrifugal force for the secondary electrons by the electric field because the effect of both forces is independent from the direction of the electron longitudinal velocity. As result the secondary electrons may return into the collector after reflection from the gun. Nonzero loss current is defined mainly by the energy spectrum of the secondary electrons. The review of the experimental data for the different cooler is described in this article. The range of the electron energy is 2 – 300 keV. Some physical model is proposed for the experiment explanations.

Keywords: electron cooler, collector efficiency, recuperation and electrostatic bend.
PACS: 29.27

INTRODUCTION

Two electron-cooling devices EC-35 and EC-300 were built at BINP and commissioned at IMP (Lanzhou in China) in 2003 – 2004 [1-2]. One electron-cooling device EC-40 [3] was built at BINP and commissioned at LEIR (CERN) in 2005. Recuperation of electron beam was investigated on these devices in energy range from *2.5 keV* to *200 keV*. The coolers are equipped by the electrostatic plates for bending in the toroidal sections.

These devices are equipped the same electron gun and collector. Ratio of magnetic field between the magnetic fields of the cathode (B_{GUN}) and the collector input (B_{COL}) is equal 2. Maximal values of B_{GUN} are next: EC-300 – 5 kG, EC-35 and EC-40 – 2.5 kG. The bending toroids of these devices are like. The bending radius is *R = 100 cm*, the bending angle is π/2, the electrostatic plates are occupied near π/3 of bending arch. The maximal value of toroid magnetic field B_{TOR} is *1.5 kG*. Magnitude of field in cooling section is usual such as in toroid. Thus, the described above devices are identical in respect of recuperation and obtained results are comparable. One may see layout of devices in [1, 2].

Condition of electron passage through bending along centerline without drift displacement is

$$U(T) = \frac{F_C}{e} \cdot \frac{\pi}{2} R \cdot \frac{d}{L}, \quad F_C = \frac{\gamma m c^2 \beta^2}{R} \qquad (1)$$

Here, $\pm U$ – required electrostatic plate voltages, d – half-gap between plates, $L = \pi \cdot R/3$ – middle length of plates, F_C – centrifugal force, γ and β calculates at electron kinetic energy T_b.

PROFILE OF SECONDARY ELECTRON FLOW

The radial profiles of the flux of the secondary electrons were estimated in the experiments when the centrifugal force is compensated by combination of the magnetic and electric fields. Combining magnetic and electrostatic bending the primary electron beam is aimed to the collector. The secondary electron is shifted in the horizontal direction depending on ratio between the electrical field on electrostatic plates and magnetic field in the correction coils. At pure magnetic bending the secondary electrons has the maximal displacement so the large loss current is observed. At electrostatic bending the both primary and secondary beams don't drift and the leakage current is minimal (Figure 1, left picture).

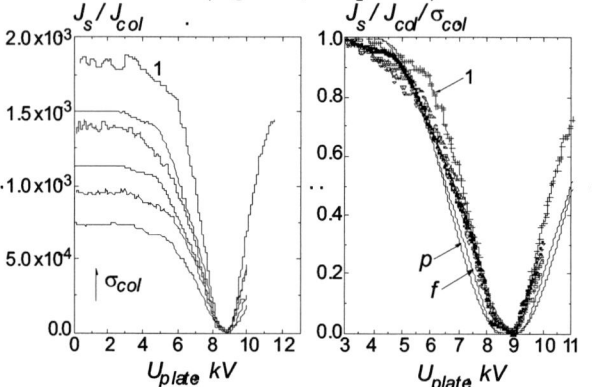

FIGURE 1. Loss current versus the electrostatic plates voltage for the different regimes. The left picture is the same but it is normalized on the collector efficiency (the loss current at U_{plate}=0). The curves p and f are constructed with a parabolic and flat profiles of the secondary electrons. The model of "moon-phase" is used for loss current estimation.

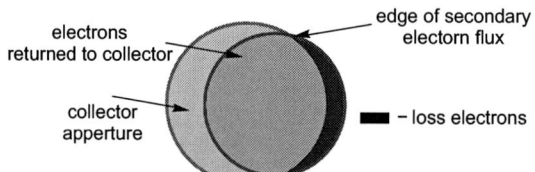

FIGURE 2. Model of the secondary model losses.

The theoretical model of the loss current is based on the picture of the phases of the moon (Figure 2). The flow of the secondary electrons leaves the effective collector aperture and is lost. Not only the geometrical size but also the shape of the electrostatic equipotential surface and the distribution of the magnetic field in the collector define the collector aperture.

Measurement accuracy is enough for compare measured dependence $J_S(U_{plate})$ with estimated curve $J_{APPR}(U_{plate})$ calculated with some axial-symmetric density profiles.

The typical results of such comparison are shown in Figure 1 (right picture). Measured normalized dependences $J_S(U_{plate}/(J_{col}\sigma_{col}))$ are shown by set of points. It is seen that curve shapes are similar except for curve 7 where other value B_{COL}. This is narrowed as $\sqrt{B_{COLL\_7}/B_{COLL}}$. Constructed dependences $J_{APPR}(U_{plate})$ for some selected profiles are shown by solid lines ($B_{COL} = 0.8$ kG). The curve p is constructed with parabolic profile and curve f is constructed with flat profile. It is seen that measured curves are narrower in minimum. Non-zero secondary current density outside of beam radius results from comparison.

Fitting results are shown for some variant in details (Figure 3). The corresponding beam density profile is shown too ($U_{grid}/U_{anode}=0.2$).

OPTIMAL PLATE VOLTAGES

The accuracy of experiments is enough for determination of optimal plate voltage. The optimal plate voltages are dependent as $\gamma\beta^2$ (see Eq.1) but measured values U_{opt} correspond to electron energy $T_b - (2\pm0.5)$ keV. Probably such difference results from energy spread of the secondary electrons.

Due to the electron collision with metal surface the secondary electrons have roughly flat spectrum. The width of the energy spread relates with the suppressor voltage $\delta T_s \approx e \cdot U_{sup}$. The effective value of width defines by suppressor potential sagging and beam volume charge into collector. If the ground surface is located on a larger radius than the suppressor or the anode electrode then there is an additional way to form energy spread of the secondary electron beam. The secondary electrons impact with anode and collector and generate the flow of the new emission electrons. The energy spread defines by maximal value from $e \cdot U_{anode}$ and $e \cdot U_{coll}$. Such situation was observed on EC-300.

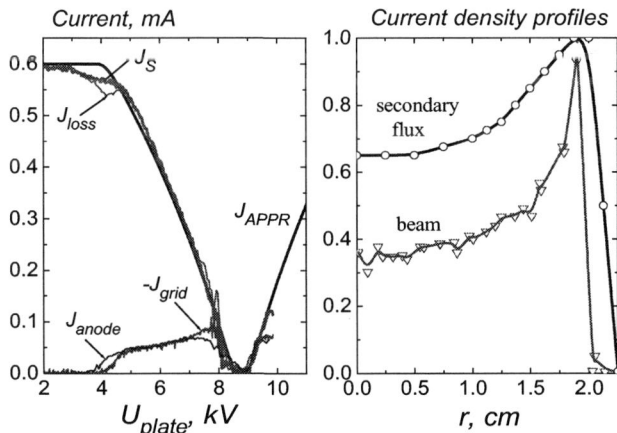

FIGURE 3. Loss current versus the voltage of electrostatic plates (left picture). The estimation of the current density profile of the secondary electrons leaving collector (right picture). The electron beam parameters: $U_{an}=3$ kV, $U_{grid}=0.54$ kV, $U_{cath}= -100$ kV, $U_{sup}=0.4$ kV, $U_{col}=1.9$ kV, $J_{col}=0.5$ A. All voltages are counted off from the cathode.

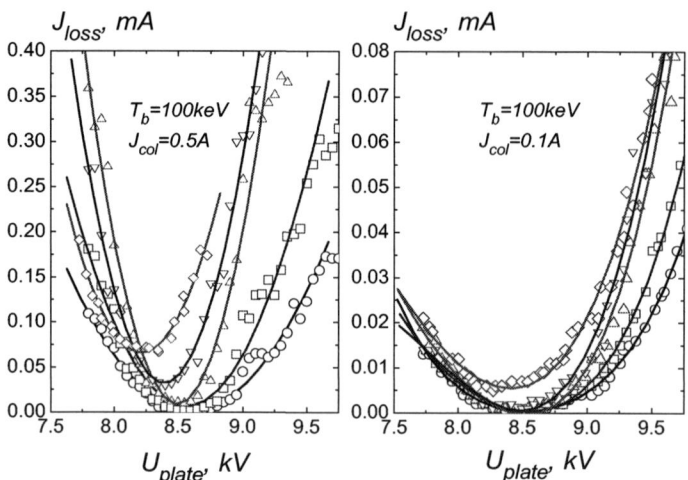

FIGURE 4. Loss current versus the voltage of electrostatic plates for the different value of the suppressor voltage. The electron beam parameters: $U_{an}=3$ kV, $U_{grid}=0.5$ kV, $U_{cath}= -100$ kV, $U_{sup}=0.42$, 1.0, 2.1, 3.0 and 4.5 kV (from right to left), $U_{col}=4.1$ kV, $J_{col}=0.5$ A (left picture); $U_{an}=3$ kV, $U_{grid}=-0.2$ kV, $U_{cath}= -100$ kV, $U_{sup}=0.42$, 1.0, 2.1, 3.0 and 4.5 kV (from right to left), $U_{col}=4.1$ kV, $J_{col}=0.1$ A (right picture). All voltages are counted off from the cathode.

The scanning with combined electric and magnetic bending was carried out on EC-300 in order to investigate dependence of optimal voltages U_{opt} from suppressor voltages U_{sup}. (Figure 4). The scanning regions were selected near optimum for the primary beam. The experiments was done at $B_{TOR} = 0.95$ kG, $B_{COL} = 0.8$ kG, $U_{cath} = -99.4$ kV, $U_{coll} = 4.1$ kV, $U_{anode} = 3$ kV. One series of measurements was done at circle profile of primary beam (left picture) and parabolic beam (right picture).

The optimum voltage of the electrostatic plates is shifted to the low value at the increasing of the suppressor voltage. In this case the collector is more "open" and the electrons with large energy spread leave the collector. The minimum leakage current is observed at tuning of the electrostatic bend on the typical energy of the secondary electrons. The primary beam falls into the collector in any case.

The effect of the electron current value on the recuperation level is shown in Figure 5. The large current in the collector locks the secondary electrons into the collector due to the space charge effect. So, the best recuperation results was observed at the electron current about *0.5-1.0 A* (left picture).

Moreover, the potential sagging in the beam at low energy leads to the shift of the optimum voltage of the electrostatic plates to the low voltage. The optimal voltages U_{opt} weakly dependents from beam current J_{coll} at $T_b \geq 24$ keV (right picture).

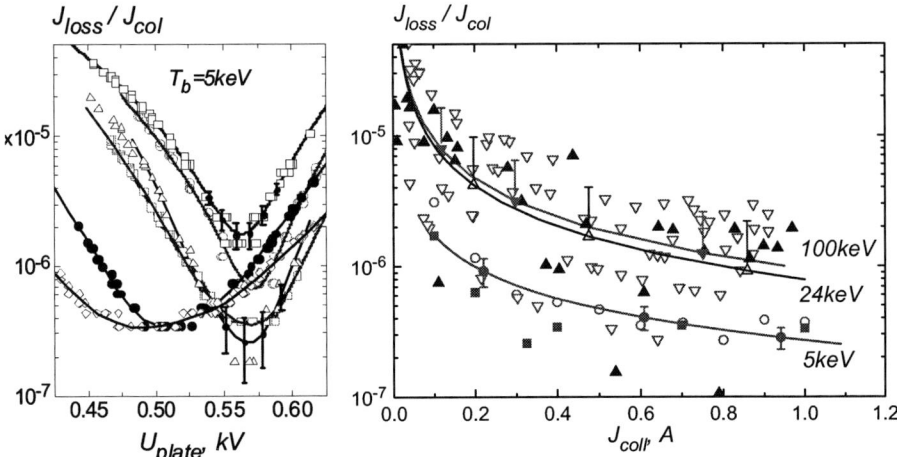

FIGURE 5. Leakage current versus at the different electron current J_{coll} = 0.2, 0.33, 0.4, 0.7 and 1 A (left picture). Recuperation level as function beam current at different beam energies (right picture).

SUMMARY

The collector can be reused at some optimal plate voltages U_{opt}. The main part of the secondary electrons is returning into the collector and the recuperation level can be done ($\sim 10^{-6}$) that is much less than degree of collector efficiency ($\sim 10^{-3}$). The density profile of secondary electron flux depends on beam density profile weakly. The flux density in outer layer is equal or some larger than inside. The important detail is presence of outputting secondary electrons in small gap (0.25 cm) between the collector aperture and the beam.

The optimal plate voltages U_{opt} for leakage current minimization are proportional to $\gamma\beta^2$ but the measured values U_{opt} correspond to energy of leaving electrons which some less than the beam energy T_b. The observed energy shift may be explained by energy spectrum of secondary electrons.

Measured recuperation levels at $U_{plate} \approx U_{opt}$ change inversely proportional to J_{coll}, i.e. inversely proportional to space charge potential into collector. Recuperation levels appreciably depend on suppressor voltage. Optimal voltages U_{opt} is independent of beam current J_{coll} at $T_b \geq$ 24 keV.

If the ground surface is outside than the anode or suppressor electrodes then the secondary electrons can form the population of the trapping electrons here. The continuous process of the absorbing and emitting electrons in this domain produces the large current to the anode or suppressor electrodes.

REFERENCES

1. E.A. Bekhtenev, V.N. Bocharov, A.V. Bubley, *Proceeding of EPAC 2004, Lucerne, Switzerland*, pp. 1419.
2. V.Bocharov, A.Bubley, Yu.Boimelstein et al. *Nucl. Instr. Meths* **A 532**, 144-149 (2004)
3. G.Tranquille, *Nucl. Instr. Meths* **A 532**, 399-402 (2004).

Low Energy Electron Cooling and Accelerator Physics for the Heidelberg CSR

H. Fadil*, M. Grieser*, R. von Hahn*, D. Orlov*, D. Schwalm*, A. Wolf* and D. Zajfman*,†

*Max-Planck-Institute für Kernphysik, Saupfercheckweg 1 D-69117 Heidelberg, Germany
†Department of Particle Physics, Weizmann Institute of Science, Rehovot, 76100 Israel

Abstract. The Cryogenic Storage Ring (CSR) is currently under construction at MPI-K in Heidelberg. The CSR is an electrostatic ring with a total circumference of about 34 m, straight section length of 2.5 m and will store ions in the $20 \sim 300$ keV energy range (E/Q). The cryogenic system in the CSR is expected to cool the inner vacuum chamber down to 2 K. The CSR will be equipped with an electron cooler which has also to serve as an electron target for high resolution recombination experiments. In this paper we present the results of numerical investigations of the CSR lattice with finite element calculations of the deflection and focusing elements of the ring. We also present a layout of the CSR electron cooler which will have to operate in low energy mode to cool 20 keV protons in the CSR, as well as numerical estimations of the cooling times to be expected with this device.

Keywords: Heavy Ion Storage Ring, Lattice, Finite Elements, Electron Cooling

INTRODUCTION

At the Max-Planck Institut für Kernphysik a new electrostatic storage ring, operating at cryogenic temperature is under construction. The temperature down to approximately 2 K of the **C**ryogenic **S**torage **R**ing (CSR) will provide ultrahigh vacuum by reducing the residual gas density to extremely low values ($\approx 10^3$ cm^{-3}), corresponding to a pressure of the order of a few 10^{-16} $mbar$. In addition, the thermal blackbody radiation will be eliminated, opening up the possibility of controlling the rotational quantum state in complex molecular systems [1]. By the use of electrostatic deflectors and focusing elements ions with kinetic energies in the range $20 - 300$ keV can be stored without any mass limitation. An electron cooler, based on photo-cathode technology developed for the TSR [2], will be used to cool the ions and to study electron ion interactions.

THE LATTICE

The ring lattice consists of electrostatic cylindrical deflectors and quadrupole doublets (Fig. 1, left). In each corner section, the overall 90° bend is divided into two 6° deflectors with a bending radius of 2 m and two 39° deflectors with a radius of 1 m. This separation has many advantages. In particular, the injection of the ion beam to be stored, but also the merging of neutral atom and laser beams can be realised along a straight line. Moreover, not only neutral, but also charged reaction fragments, created in the straight section can be detected in this arrangement of the deflectors. The standard lattice functions of

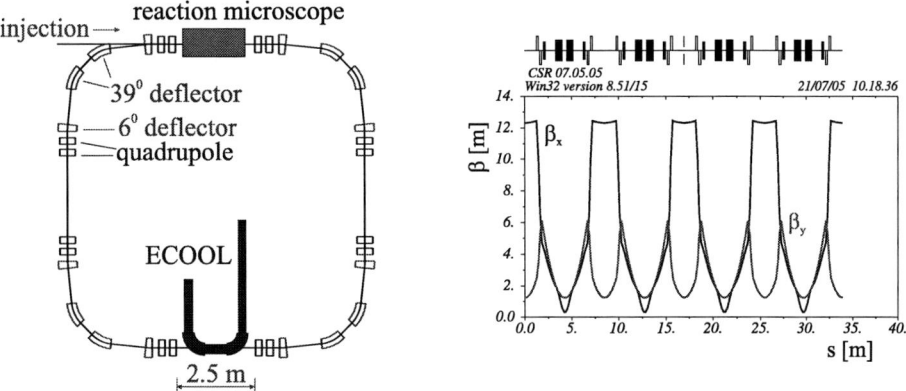

FIGURE 1. Left; schematic layout of the CSR showing the lattice configuration. Right; the horizontal and vertical beta functions of the CSR lattice.

CSR are shown on the right of Fig. 1. The tunes are $Q_x = 2.59$ and $Q_y = 2.60$ and the dispersion in the centre of the straight section is $2.1\ m$. The optics elements will exhibit deviations from the ideal field distributions, as well as fringing fields. Hence, three dimensional numerical calculations of realistic CSR elements were performed using the finite elements electrostatic code TOSCA3D. The calculations have shown that the cylindrical deflectors have fringing fields which can affect the particle dynamics. We have therefore installed at each end of the deflectors grounded electrodes which serve as field clamps and have trimmed (shortened) the deflectors to match the design deflection angles exactly. For the focusing elements we use hyperbolically shaped electrodes in order to achieve a quadrupole field as free from higher orders as possible. Due to the proximity of the two doublet quadrupoles, we have installed a grounded shield between them. which effectively decoupled them. The horizontal distribution of the deviations of

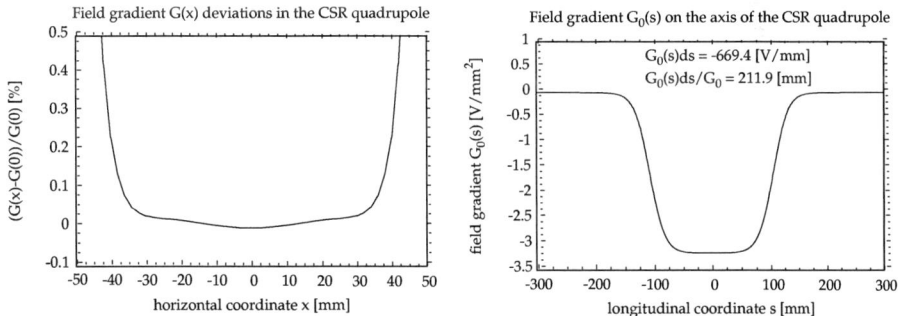

FIGURE 2. Left; horizontal dependence of the field gradient deviations in the CSR quadrupole. Right; the longitudinal dependence of the field gradient gives the effective length of $21.2\ cm$, which is about 6% larger than the electrode length of $20\ cm$.

the field gradient $G(x) = dE(x)/dx$, where E denotes the electric field, is shown on the left of Fig. 2. We observe that outside the region $x = \pm 40$ mm there is large deviation in the field gradient which will cause tune shifts and limit the horizontal acceptance. The longitudinal distribution of $G_0(s)$ plotted on the right of Fig. 2, shows that the effective length of the quadrupole is about 6% larger than the electrode length of 20 cm. The quadrupole strenghts will therefore be scaled down accordingly in order to achieve the same integrated field strength of the quadrupoles as the value used in the linear optics calculations. Table 1 summarises the parameters of the CSR elements. After the investigation of the individual elements, a model of the entire ring was simulated. In this model we have made use of all the symmetry conditions present in the CSR. After the field solution was obtained, tracking of a 300 keV proton was performed and we could store the particle for a maximum of 670 turns which was limited only by the program run time. In order to compare the lattice parameters with those found with the transfer

TABLE 1. Design parameters and simulation results for the CSR optics.

Deflectors		39°	6°	
	Curvature radius	1	2	m
	Electrode gap	6	12	cm
	Electrode height	16	24	cm
	Applied voltage (300 keV p)	+17.74/-18.28	+17.85/-18.40	kV
Quadrupoles				
	Bore radius	5		cm
	Pole length	20		cm
	QF voltage	±3.95		kV
	QD voltage	±4.98		kV
Lattice		Linear	Realistic	
	Horizontal function	12.3	12.1	m
	Vertical function	1.2	1.3	m
	Dispersion	2.1	2.1	m
	Acceptance horizontal	–	120	$mm \cdot mrad$
	Acceptance vertical	–	180	$mm \cdot mrad$

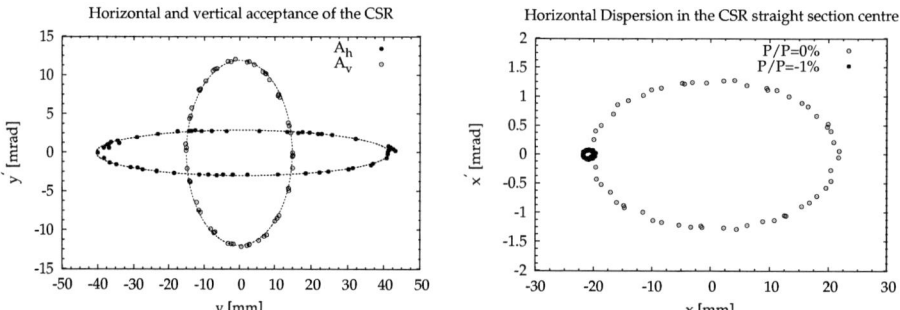

FIGURE 3. Left; the horizontal acceptance of the CSR is about 120 $mm \cdot mrad$ and the maximum horizontal beam size at the centre of the straight section is ±4 cm. Vertically these figures are 180 $mm \cdot mrad$ and ±1.4 cm respectively. Right; tracking an off-momentum particle ($p/p = -1\%$) we find that the dispersion in the straight section is about $D_x \approx 2.1$ m.

matrix formalism, we have calculated the phase space ellipse of the stored particle at the centre of the straight section, then fitted the phase space ellipse equation to the data. The results are consistent as shown in table 1. The dispersion at the centre of the straight section as well as the ring acceptance were also calculated, and the results are shown in Fig. 3. We think that the horizontal beam size of ± 4 cm is limited by the good field region in the quadrupoles. The vertical acceptance is larger because of the smaller vertical beta function.

CSR ELECTRON COOLER

The CSR will be equipped with a compact electron cooler, which will serve the double purpose of phase space compression of the stored ion beam as well as an electron target for recombination experiments. The entire device will be installed inside the cryostat, and therefore no iron shielding will be used, and the entire cooler will consist of air magnets only. Also due to space restrictions as well as heat dump considerations all coils of the cooler will constructed from a high temperature superconducting material and operated at about liquid nitrogen temperature. The entire electron cooler should fit in the 2.5 m straight section of the CSR. The preliminary design status of the CSR electron cooler/target is discussed in more detail in the following sections. The layout of the CSR

TABLE 2. Design parameters of the CSR electron cooler/target.

Electron Energy	1-100	eV
Cooling solenoid coil length	1.47	m
Cooling solenoid coil radius	0.115	m
Toroid bending radius	0.1	m
Maximum magnetic field (cooling)	100	G
Field uniformity in cooling solenoid	1×10^{-4}	
Correction coils length	0.1	m

electron cooler is shown in Fig. 4 and the main parameters are summarised in table 2. The device will be installed horizontally in the CSR, so that the stored ions move in the X-Z plane. The present design consists of three solenoid coils for the gun, cooling and collector sections respectively. Ultra cold electrons will be generated in the gun section

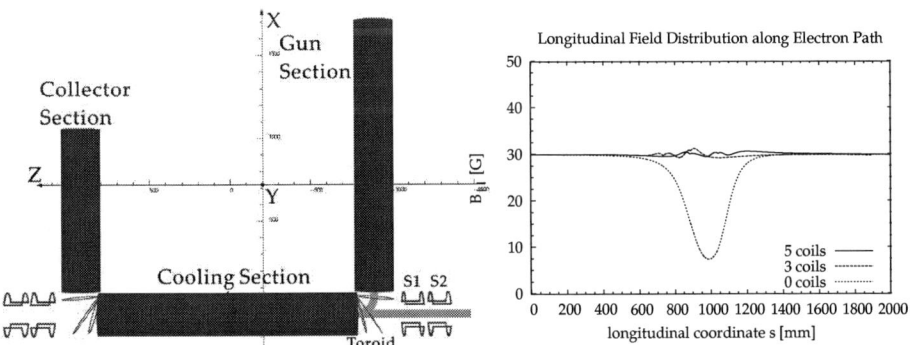

FIGURE 4. Left; layout of the CSR electron cooler/target. Right; calculated longitudinal magnetic field along the electron path for different toroid coil arrangements.

with the photo-cathode developed at MPI-K[2], and then transported through a bending toroid section towards the cooling section where they are merged with the stored ions.

The toroid consists of three coils only in order to fit inside the cryostat as well as provide space for the inner vacuum chamber. The angular position as well as the current of each single coil of the toroid was optimised by numerical calculations in order to achieve the best field homogeneity along the electron beam orbit. The toroid field however, perturbs the stored ions especially for low energy light particles. This effect can be corrected with a pair of saddle coils (Fig. 4 S1 and S2) located at each end of the straight section. The calculated longitudinal magnetic field along the electron orbit is shown on the right of Fig. 4.

We have performed numerical estimations of the cooling rates, assuming a cooler setup with a photocatode with a cathode temperature $T_c = 10$ meV, delivering a 10 eV electron beam of current 0.1 mA and radius 6.7 mm. After expansion the electron transverse temperature is assumed to be $T_\perp = 1$ meV. The longitudinal temperature after the 10 V

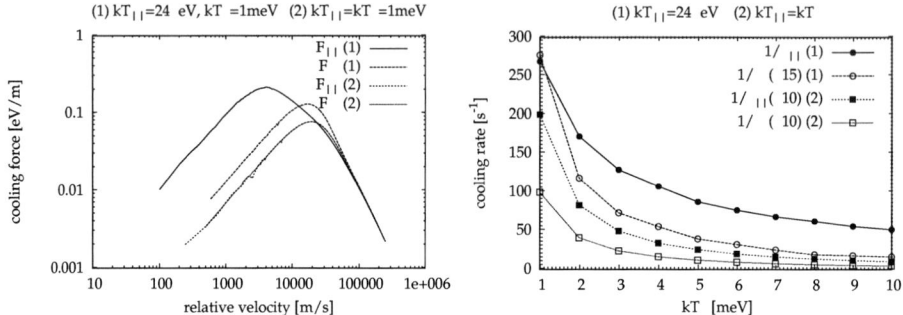

FIGURE 5. Left; calculated electron cooling forces for protons for the two cases of (1) a flattened distribution ($T_\parallel \ll T_\perp$) and (2) an isotropic distribution ($T_\parallel \approx T_\perp$). Right; the dependence of the longitudinal and transverse cooling rates on the transverse electron temperature for these two cases. ($n_e = 2.2 \times 10^6 cm^{-3}$)

acceleration is $T_\parallel = 24$ eV for electron density $n_e = 2.2 \times 10^6$ cm^{-3}. The electron cooling force is estimated using the well-known binary collision model formula neglecting the magnetic field [3], since standard operation will be with a magnetic field of 30 G or below. The result is shown on the left of Fig. 5. In the linear region of the cooling force (small relative velocities) we can define the cooling rates, and on the right of Fig. 5 we show their dependence on $T_\perp$. We observe that cooling rates increase for decreasing $T_\perp$, and that efficient cooling is possible with the photo-cathode for our conditions. However, when the electrons are confined in a low magnetic field it is possible that a relaxation process takes place, which can change the flattened velocity distribution ($T_\parallel \ll T_\perp$) to an isotropic one ($T_\parallel \approx T_\perp$), thus reducing the cooling rates. A reduction of about 1 order of magnitude of the longitudinal cooling rate was calculated. The transverse cooling rate is not so severely affected with a decrease of only about factor 2.

REFERENCES

1. D. Zajfman et al., J. Physics: Conference Series 4 (2005) 296.
2. D.A. Orlov et al., J. Physics: Conference Series 4 (2005) 290.
3. YA.S. Derbenev and A.N. Skrinsky, Part. Acc. 8 (1977) pp.135-297.

Electron Cooling of Bunched Beams

T. Uesugi, K. Noda*, E. Syresin, I. Meshkov† and S. Shibuya**

*National Institute of Radiological Sciences (NIRS), 4-9-1, Anagawa, Inage-ku, Chiba-shi, Chiba-ken, 263-8555, Japan.
†Joint Institute for Nuclear Research (JINR), Dubna, Moscow-region, 141980, Russia
**Accelerator Engineering Corporation (AEC), 4-9-1 Anagawa, Inage-ku, Chiba-shi, Chiba-ken, 263-0043, Japan

Abstract. Experiments of electron cooling have been done with the HIMAC synchrotron in NIRS. Limitation on cooled beam-sizes in longitudianl and transverse spaces were measured. The effect of space-charge field and intra-beam scattering are investigated.

Keywords: Electron cooling

INTRODUCTION

The heavy ion medical accelerator in Chiba (HIMAC) [1] is the accelerator complex constructed in national institute of radiological sciences (NIRS) for cancer therapy and other researches. Since 2000, electron cooling (EC) experiments have been carried out at the synchrotron in order to develop new technologies in heavy-ion therapy and related fields. One of the objectives of the HIMAC cooler is to increase the beam intensity of heavier ions, such as Fe, for risk estimations under low-dose exposure in space [2].

Such a high density beam includes strong space-charge effects. In the past experiments with the HIMAC synchrotron, the bunch length of cooled beams was measured for different ion-intensities and RF voltages to discuss the longitudinal cooling force, the heating force by intra-beam scattering (IBS) and space-charge effect [3]. The transverse beam-size was measured at cool equilbrium of a coasting beam in Ref. [4]. Analyzing the beam-size measurements during slow beam-loss, it was found that the beam-size was limited by the cross-section density at 9×10^8 ions/cm^2. Such a restriction on real-space density is related to some resonance of betatron oscillation. At the intensity of less than $\sim 10^8$ ion/ring, there was a stronger limitation related to the phase-space density, which can be understood with the IBS [5].

In this paper, the limitation of the transverse beam-size is generalized to a bunched beam. The longitudinal and transverse beam-sizes were simultaneously measured for weakly bunched beams, and volume density was compared with that of a coasting beam. The experimental result showed that the peak volume density was constant indendent of the bunching factor.

EXPERIMENTAL METHOD

Experiments were done with bunched beams of Ar^{18+} ions at the injection energy of the HIMAC synchrotron. Experimental conditions of the HIMAC synchrotron are listed in Table 1.

The transverse ion distributions were non-destructively mesured with gas-sheet beam profile monitor (SBPM) [6]. The SBPM, at which the betatron amplitude function is (β_x,β_y)=(9 m,7 m), can measure the transverse ion-distribution with 0.8 mm resolution at full-width of half maximum (FWHM). On the other hand, the longitudinal ion distributions are measured with sum-pickup of beam position montior (PON). The time resolution of the PON is estimated around 8 ns.

The longitudinal and transverse beam sizes were measured after cooling of 5.9 s from the injection. This cooling-time was long enough to reach the cool-equilibrium. The number of injected ions were controled by changing the length of the injected beam. The beam-size was measured for different RF amplitudes in order to investigate the dependence on the bunching factor. The momentum spread was measured for a coasting beam by the Schottky diagnostic in order to evaluate the emittance.

EXPERIMENTAL RESULTS

Line-Density Control by RF Amplitude

First, we show in Fig. 1 the FWHM bunch length (T_{fw}) as a function of RF voltage at a fixed ion-intensity. The ion intensity was kept at 1.4×10^8 ions/ring. The bunch length was varied twice by changing the RF voltage from 4.5 V to 130 V.

It is worth noting that the bunch length behaves as $V_c^{-1/4}$, as shown by the solid line in Fig. 1. If we neglect the space-charge force, the longitudinal emittance is proportional

TABLE 1. Experimental conditions

Parameter	Value
Synchrotron	
Ring circumference	$2\pi R$=129.6 m
Particle, Energy	$^{40}Ar^{18+}$, 6 MeV/u (β=0.113,γ= 1.006)
Betatron tune	(3.69 / 3.13)
RF frequency, Harmonic no.	1042.1 kHz, h=4
RF amplitude	5 V$\leq V_c \leq$130 V
Phase-slip factor	η=−0.91
Total number of ions	N=1.4 $\times 10^8$ ions/ring
EC	
Current	100 mA
Cathode diameter, Magnetic expansion factor	35 mm , R=3.3
Cathode temperature	100 meV
Solenoid field	0.05 T, 1.2 m

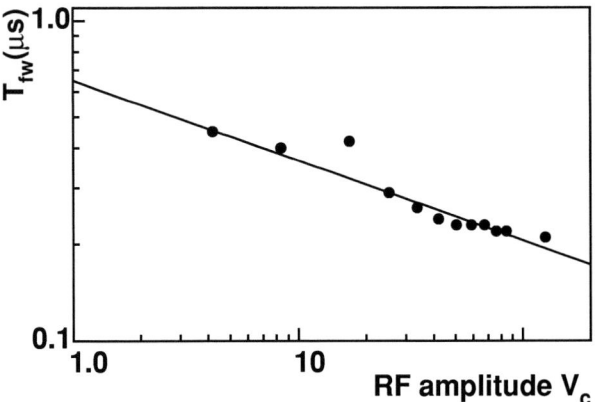

FIGURE 1. FWHM bunch length as a function of RF amplitude. Fitting function, shown by the solid line, corresponds to $T_{fw} = 0.65 V_c^{-1/4}$.

to $\sqrt{V_c} T_{fw}^2$. Therefore, the longitudinal emittance at its equilibrium was independent of the RF amplitude.

Transverse Cooling

Now we investigate the limitation on the transverse size of a bunched beam at cool equilibrium. In the case of a coasting beam, the transverse size was limited by the cross-section density at 9.0×10^8 ions/cm² [4]. Therefore, we first calculate the peak cross-section density of a bunched beam, taking into account the bunching factor.

On the assumption of Gaussian distribution in each spaces, the peak volume density (n_3) of a bunched beam is given by

$$n_3 = \left(\frac{dN}{dxdydz} \right)_{peak} = \left(\sqrt{\frac{8\ln 2}{2\pi}} \right)^3 \frac{N/h}{X_{fw} Y_{fw} Z_{fw}}, \quad (1)$$

where X_{fw} and Y_{fw} are the horizontal and the vertical FWHM beam-sizes, and $Z_{fw} = \beta c T_{fw}$ is the longitudinal FWHM size, respectively. The peak volume density corresponding to the coasting beam with 9.0×10^8 ions/cm² is 2.3×10^8 ions/cm²µs.

In Fig. ?? the peak volume-density is plotted as a function of FWHM bunch length. The peak volume-density was about 1.75 µs independent of the bunch length, as shown by the solid line in Fig. 2. This value is near to the limit for a coasting beam (dashed line). Thus, the limitation on cross-section density of for a coasting beam was generalized to a bunched beam in terms of the volume-density.

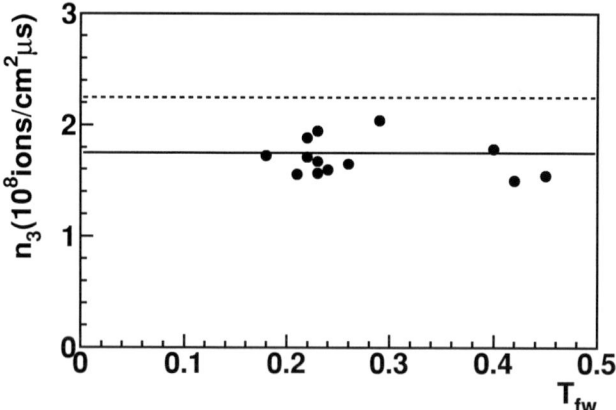

FIGURE 2. Peak volume-density of a bunched beam as a function of FWHM bunch length. Dashed line corresponds to the limit for a coasting beam, which is measured in Ref. [4]. The averaged value of the data points corresponds to 1.75×10^8 ions/cm$^2\mu$s (solid line).

CONCLUSION

The longitudinal and transverse sizes of a bunched beam at cool equilibrium was measured for different RF amplitude. Bunch length was varied by the RF amplitude, while the longitudinal emittance was kept constant.

The transverse size was limited by the volume density. The peak volume density did not exceed the limit of a coasting beam. The effect of the IBS, which is related to the phase-space density was not seen in the experimental condition.

ACKNOWLEDGMENTS

The authors would like to express their thanks to members of the Department of Accelerator Physics and Engineering at NIRS for continuous encouragement, and to the crew of Accelerator Engineering Corporation for their skillful operation of the HIMAC accelerator complex. This work was carried out as a part of the Research Project with Heavy Ions at NIRS-HIMAC.

REFERENCES

1. Y. Hirao *et al.*, Nucl. Phys., A 538(1992) 541c.
2. K. Noda *et al.*, NIM A441(2000), 159.
3. K. Noda *et al.*, Nucl. Instr. and Meth in Phys. Res. A 532(2004) 129-136.
4. T. Uesugi *et al.*, Nucl. Instr. and Meth in Phys. Res. A 545(2005) 45-56.
5. T. Uesugi *et al.*, Proc. of PAC'05, Tenessee, (2005) *to be published*.
6. Y. Hashimoto *et al.*, NIM A527 (2004), 289.

First Tests of LEIR – Cooler at BINP

Valentin Bocharov, Maxim Brizgunov, Alexander Bubley, Viacheslav Ershov, Anatoly Goncharov, Sergey Konstantinov, Alexey Lomakin, Vitaly Panasyuk, Vasily Parkhomchuk, Valery Polukhin, Vladimir Reva, Boris Skarbo, Boris Sukhina, Maxim Vedenev, Mikhail Zakhvatkin and Nikolay Zapiatkin

Budker Institute of Nuclear Physics
Lavrentieva ave. 11, 630090, Novosibirsk, Russia

Abstract. New electron cooling device was constructed for LEIR accumulator ring according to ILHC project at CERN. The cooler was designed, manufactured and completely tested with electron beam at BINP (Novosibirsk, Russia). Special features of the device and the results obtained are presented in the paper.

Keywords: Electron cooling, electron gun, electron collector, magnetic measurements.

INTRODUCTION

New electron cooling device was designed and constructed at BINP in collaboration with CERN. It is intended to be a significant part of newly upgraded low energy ion ring (LEIR) under ILHC project. The cooler is equipped with electron gun with variable beam profile, high perveance collector and electrostatic bending. Other important features are precise cooling section solenoid made of adjustable pancake coils and ultra high vacuum system. All systems were completely tested at BINP before shipping to CERN.

MAGNETIC MEASUREMENTS

One of the main parts of the cooling device is magnet system. It consists of drift solenoid as a cooling section, toroidal magnet bends and solenoids for electron beam expansion at a gun and compression at a collector region. Also it includes various correction coils for both electron and ion beams adjustments. Special dipole correctors are built in the toroid magnets for ion beam deflection compensation. Results of Hall probe measurements are sown in Fig.1.

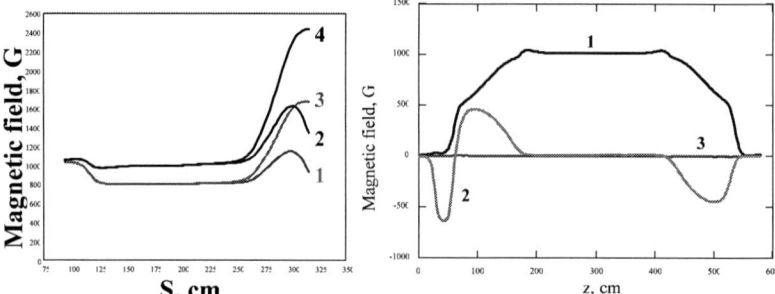

FIGURE 1. Curvilinear Hall probe measurements along electron orbit (in the left) and 3D straight measurements along ion trajectory. Magnetic field distribution along straight section (ion orbit) of the cooler. Right picture: 1-longitudinal, 2-vertical, 3- horizontal components of magnetic fiels. Left picture: magnetic field distribution along electron orbit of the cooler for different combinations of currents applied to the power supplies

Vertical component of magnetic field in drift solenoid is dependent on its longitudinal value because of toroidal field penetration from bending magnets. This fact is rather important for those cases when cooler is intended for use at different field values. Linear and cubic correction wound on drift solenoid were foreseen to eliminate vertical field slope dependent on longitudinal value. Measurements of those correction replies (Fig.2 right) are very important further operation of the magnet system and drift solenoid in particular.

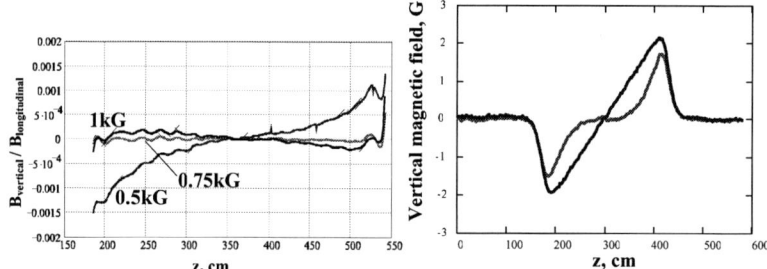

FIGURE 2. Vertical field slope in the cooling section dependence on longitudinal field value (on the left). Vertical magnetic field produced by linear (blue curve) and cubic (red curve) 2.5m long correctors installed on cooling section (on the right).

Magnetic field in the drift solenoid was measured and tuned to meet designed requirements. For more complete description of the methods applied for measurements and adjustments refer to [1].

ELECTRON GUN AND COLLECTOR OPERATIONAL CHARACTERISTICS

One of the main tests were done was putting into operation the electron gun and collector. While backing out of the vacuum system was being carried out cathode activation was performed step by step with applying required voltages to the gun electrodes. Electron current of 1.5A emitted by the gun was obtained (Fig. 3 on the left).

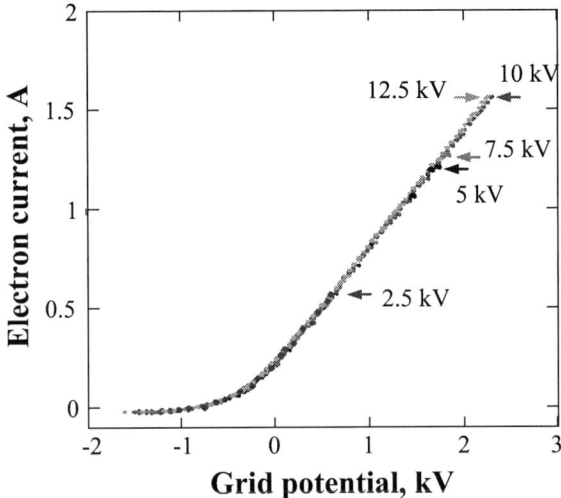

FIGURE 3. Electron gun current. Different curves show measurements for different potentials applied to the collector when arrows point them.

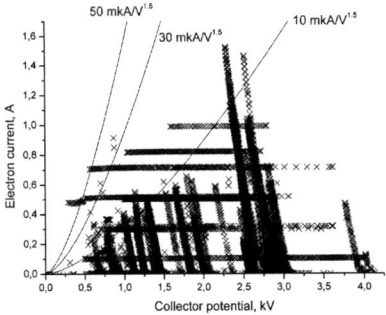

FIGURE 4. Electron collector current versus collector potential. Rows – series of measurements when collector potential was decreased at constant suppressor voltage and electron current. Solid curve is asymptote of the collector perveance for each series.

Much attention was paid for collector properties study. Rows on right part of Figure 3 show how varying collector potential al constant suppressor voltage extremely high value of collector perveance (solid curves) was achieved. Recuperation efficiency in those measurements was very high that meet all requirements. This was obtained thanks to using electrostatic bending. Recuperation efficiency arise up to 10^{-6} or even better in comparison with magnet bending (left picture in Figure 5). This fact opens results in opening new possibilities in cooling device techniques [2], [3].

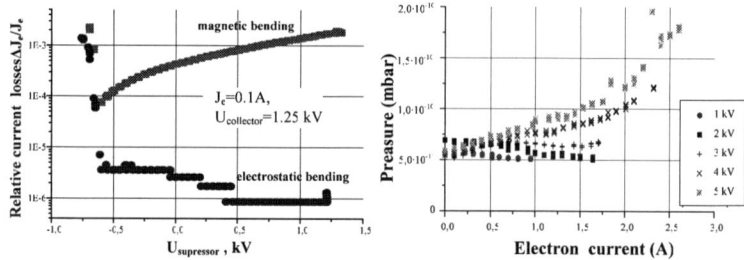

FIGURE 5. Left picture: relative current losses versus suppressor potential at constant electron current and collector potential. Left picture: vacuum pressure versus electron current.

Lower curve in left part of Fig.5 shows that current loss is almost independent on suppressor potential if electrostatic bending is used, and it even gets worse at some regimes of operation. It means that collector should be opened as much as possible for better capturing secondary electrons [4].

One of the advantages of electrostatic bending use is low desorption because of low current loss. Pressure in vacuum system was observed to get even better under the electron beam action (lower curves in right part of Fig.5). Intensive electron current ionizes residual gas atoms and then they are captured by ion pumps belonged to the vacuum system. Using of electrostatic bending seemed to be the only way out to meet very strict requirement to the vacuum condition in LEIR machine (10^{-12} mbar).

Space Charge Effects in the Collector

Collector potential was being decreased while collector current was kept constant by value of 0.5A and suppressor potential by 0.47kV. When collector potential went down below suppressor potential, current losses increased significantly, because of potential barrier disappearance. Collector current changed a little with losses increased, the much beam losses the less collector current. It may be explained as an additional space charge of secondary electrons in the gun influenced on electron current.

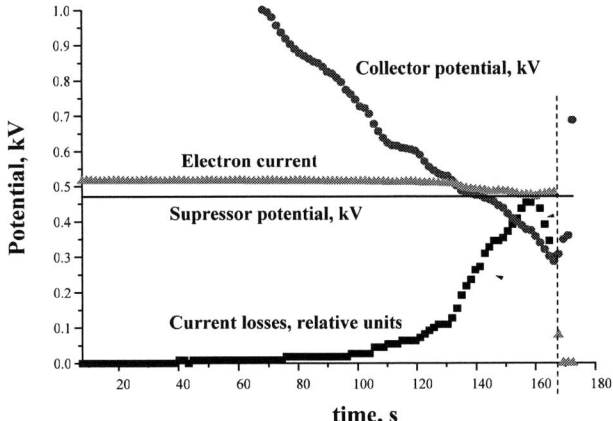

FIGURE 6. Beam current losses, collector potential, collector current and suppressor potential depending on time.

Break-down occurred when collector potential was below suppressor potential so that virtual cathode supposed to arise inside the collector cavity rather than suppressor. Beam current losses decreased just before break-down, because secondary electrons were suppressed by space charge inside the collector cavity. Collector perveance was observed to be about $100\mu A/V^{3/2}$ in this experiment. This interesting result gives us an opportunity to use space charge inside collector as additional suppressor.

CONCLUSION

All results obtained during tests will serve for operation conditions setting and are important for cooling process tooling. Measurements of the space charge in the gun and collector look very interesting and after complete study may result in further technique improvement. High recuperation efficiency gives a possibility to use high voltage power supplies with low power consumption. On the other hand it allows achieving ultra high vacuum condition in the cooler.

REFERENCES

1. V. Bocharov, A. Bubley, S. Konstantinov, V. Panasyuk, V. Parkhomchuk, Precise Measurements of a Magnetic Field at the Solenoids for Low Energy Coolers, proceedings of this conference.
2. Bocharov V, Bubley A, Boimelstein Yu, et. al, HIRFL-CSR Electron Cooler Commissioning, NIM. 2004. A 532, p.144.
3. V.V. Parkhomchuk, Development of a New Generation of Coolers with a Hollow Electron Beam and Electrostatic Bending, proceedings of this conference.
4. M.Bryzgunov, V.Panasyuk, V.Parkhomchuk, V.Reva, M.Vedenev, Recuperation of electron beam in the coolers with electrostatic bending, proceedings of this conference.

Precise Measurements of a Magnetic Field at the Solenoids for Low Energy Coolers

V. Bocharov, A. Bubley, S. Konstantinov, V. Panasyuk, V. Parkhomchuk

Budker Institute of Nuclear Physics
Lavrentieva ave. 11, 630090, Novosibirsk, Russia

Abstract. Description of equipment developed at BINP SB RAS for precision solenoid magnetic field measurement is presented in the paper. Transversal field components are measured by small compass-based sensor during its motion along the field line. The sensor sensitivity is a few tenth parts of mG and is limited in this range by external noise sources only. Scope of the device application is illustrated by results obtained at BINP during tests of cooling solenoids for electron coolers built at the Institute recently.

Keywords: Electron cooling, magnetic measurements, precise solenoid.

The idea of cooling of ions with an electron beam, moving with the same average speed, is based on coulomb interactions increase between particles at small relative speeds [1]. Efficiency of this process in a strongly depends on quality of the leading magnetic field in a cooling section solenoid. Acceptable cooling rate can be reached if non-parallelism of magnetic force lines $B_\perp/B_0$ in a vicinity of ions trajectories does not exceed angular dispersion of the ion beam. The striving to receive extremely high speeds of cooling makes rigid demands to straightforwardness of force lines - from 10^{-4} for low electron energies up to 10^{-5} and even less - for high energies.

Electron cooling is applied now on many accelerators all over the world. Last years some electron cooling devices were designed and constructed for foreign laboratories in Budker Institute of Nuclear Physics SB RAS. All of them contain solenoid with high quality magnetic field for ion and electron beams interaction.

SOLENOID FOR LOW ENERGY COOLERS

Central (drift) solenoids for low energy cooling devices are 2÷4m long and consist of so-called pancake coils with the thickness of about 6 cm distributed with the step of 6.5 cm. Each coil has three points of support relative to the magnet yoke so that it can be inclined or turned around vertical axis (Fig.1). Due to these movements transversal field components can be eliminated and required value of magnetic force lines parallelism achieved.

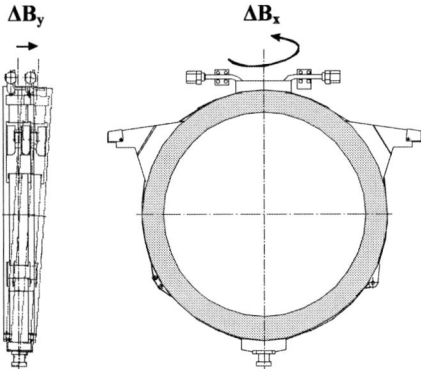

FIGURE 1. Movable coils for drift (cooling) section. Inclination or rotation of the coil in a solenoid produce vertical or horizontal field component correspondingly.

Below are steps to archive necessary high homogeneity of magnetic field in drift solenoid based on BINP experience [2]. All coils for cooling section are tested separately namely magnetic axis declination and its azimuth are measured on the special test bench. Next they are arranged inside magnet yoke according to data obtained in order to minimize initial field inhomogeneous. This step is quite important because precise system has relatively narrow dynamic range and might be overloaded during first measurement. Hall probe measurement is performed as a preliminary for rough adjustment. When transversal field value is within volume range of the precise system two calibration passages are done, one before and one after single coil inclination (rotation) by definite space (say 1mm) (Fig.2). Using calibration data shifts for all coils in the solenoid are calculated so that they can be moved with the help of special tuning screws.

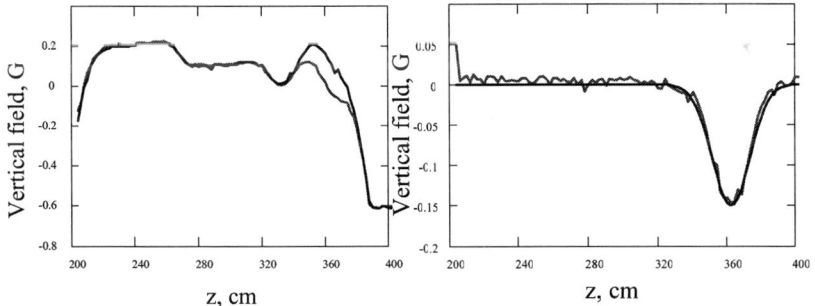

FIGURE 2. Vertical field measurements before and after one coil inclination by 1mm (in the left) and correspondent reply (in the right).

Required field quality is obtained after several iterations of measurement and coils adjustment (Fig.3). Certainly it depends on mechanical properties of coils and magnet yoke.

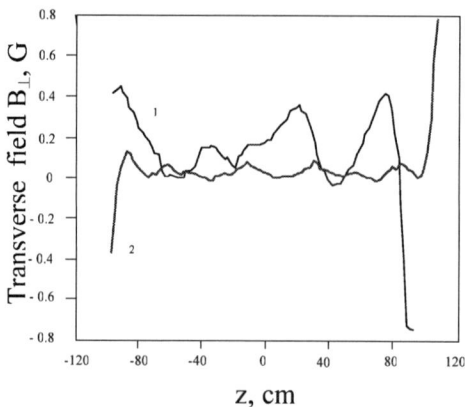

FIGURE 3. Transversal magnetic field in the solenoid measured initially (curve 1) and after few iteration of the coils adjustment (curve 2).

COMPASS-BASED MEASURING SYSTEM

Prototype of the compass-based measuring system was designed and constructed at BINP and then shipped to Fermilab according to mutual collaboration in 2000 [3]. Since that time it has being modified and improved but the basic ideas are the same. A head of the system consists of two permanent magnet cylinders and a mirror attached to one of them. This assembly is suspended with thin thread fixed on aluminum holder (Fig.4) which also is used for damping mechanical oscillations.

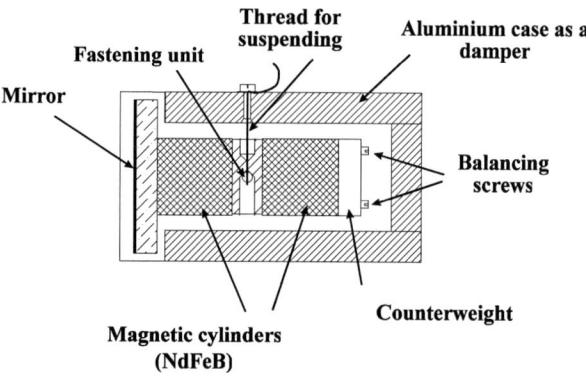

FIGURE 4. Scheme of the compass head.

Principle of Operation

The schematic layout of the measuring system is shown in Figure 5. Light beam, generated in a laser goes to sensor head placed inside measured solenoid. The mirror, attached to the compass reflects the beam to a 4-segmented photodiode. Using a pair of differential signals (for horizontal X and vertical Y directions) from photodiode segments, electronic feedback system generate currents in X and Y compensation dipole coils. The value of these currents is measured by digitizing of voltage drops on shunt resistors. When solenoid transverse fields are compensated by the current in the coils, the reflected laser beam comes to the center of the photodiode and the system comes to the equilibrium. More completely the principle of operation is described in [6].

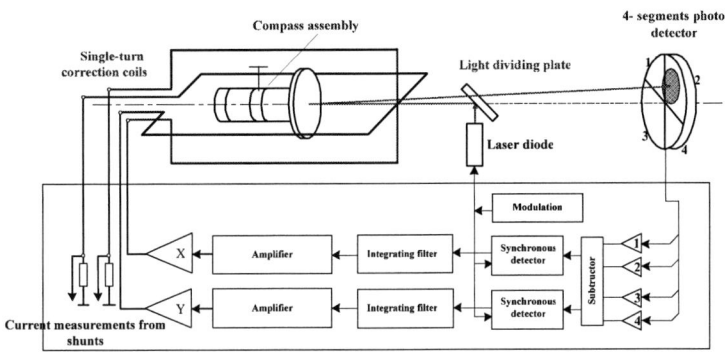

FIGURE 5. Scheme of the measuring system.

The transport system is made in such a way that single turn dipole correction coils serve as guiding wires for the movable cart with the compass assembly installed inside. Ordinary fishing-line attached to the cart and wounded around two pulleys may be used for moving. Driving system should be equipped with step motor and device for cart position (coordinate along drift solenoid axis) measurement.

Calibration and Equilibration

The sensor has angular errors because of several reasons [4], [5]. One them is discrepancy between magnetic and geometrical axes of the compass assembly resulted from non-homogeneous of permanent magnet material and imperfection of the mirror attaching. Equilibration of the sensor is made with the help of rotatable solenoid (Fig.6) which is shielded with µ-metal layer to avoid external field influence. Having two measurements of transversal field before and after rotation by 180° with motionless sensor imperfectness of the field can be taken into account. Sins magnetic component of the discrepancy is linear with longitudinal field and mechanical component is independent on it series of measurements is performed at different current applied to the solenoid winding. It helps to distinguish those components in order to eliminate them. One of the ways to make equilibration is using four screws incut to the counterweight (Fig.4), two of those are made from magnetic material and two from non-magnetic. Coincidence of magnetic and geometrical axes can be established while tightening magnetic screws. After that mechanical balance is restored screwing (unscrewing) non-magnetic ones.

FIGURE 6. Special test bench for sensor equilibration.

Calibration is made by applying definite current to single wire stretched inside solenoid in a well measured distance from the sensor.

CONCLUSION

Mentioned above technique of precise solenoid adjustment was successfully applied for low and middle energy coolers production at BINP. System for precise magnet measurement is being improved permanently and supposed to be used for future projects where requirements for field quality are significantly higher (up to 10^{-6}).

ACKNOWLEDGMENTS

Collaboration with Electron Cooling group (Fermilab) group was very useful and important at first stages of the measuring system development.

REFERENCES

1. Parkhomchuk V.V., Skrinsky A.N., UFN, 2000, T. 170. №5. p. 473, (in Russian).
2. A.V.Bublei et al. New Technology for Production of Precision Solenoid for Electron Cooling
3. Systems. Proc. of MEEC-98, Dubna, 1998, Russia.
4. V.N.Bocharov et al., Sensor for a precise measurement of the magnetic field direction, XVIIth Workshop on Particle Accelerators, Protvino, Russia, Oct. 17- 20 (in Russian).
5. C. Crawford, E. McCrory, S. Nagaitsev, A. Shemyakin*,FNAL,.V.Bocharov, A. Bubley, V. Parkhomchuk, V. Tupikov Budker INP, S. Seletsky, Univ. of Rochester, Fermilab Electron Cooling Project: Field Measurements In The Cooling Section Solenoid, Proceedings of PAC, 2001, Chicago.

Electron Cooling for Cold Beam Synchrotron for Cancer Therapy

B. Grishanov, M. Kumada[*], V. Parkhomchuk, S. Rastigeev, V. Reva, V.Vostrikov

*BINP, Novosibirsk, Russia: * NIRS, Chiba, Japan.*

Abstract. A wide usage of carbon ions for cancer therapy is limited mostly due to technical difficulties, resulting in higher cost. This cost problem can be solved by our CBS (Cold Beam Synchrotron) proposal. In this paper a conceptual design of the facility for the carbon beam cancer therapy using a high precise active beam scanning system of synchronizing with respiration. The main feature of the CBS facility is an application of electron cooling device. The use of cold ion beam allows to decrease the aperture of synchrotron and components of high energy beam transport lines, significantly. The precise ion beam energy variation and two unique schemes of beam extraction ("pellet" extraction and extraction on recombination) enclose the list of possibilities appearing with EC applying.

Keywords: Electron Cooling; Ion Cancer Therapy.

INTRODUCTION

A carbon ion beam is a superior tool to x-rays or a proton beam in both physical and biological doses in a cancer treatment. Carbon ions compared with protons, have a better relative biological effect (RBE), an effective Linear Energy Transfer (LET) to a cancer tumor are more effective to treat radiation resistant and deep-seated tumors, a mitotic independence of the breakdown of a DNA double strands and so forth. Carbon ions have the property of lower scattering along their trajectory. It has been shown that a superior QOL (Quality Of Life) therapy is made possible by the carbon beam. A major focus of the hadron therapy community is transformation of particle therapy into a practical and affordable treatment option. A wide use of carbon ions cancer therapy is limited mostly due to technical difficulties, resulting in higher cost. This cost problem can be solved by our CBS (Cold Beam Synchrotron) proposal [1]. The main feature of the CBS facility is an application of electron cooling device.

The electron cooling technology applied to the heavy ion medical accelerator will open a world of high performance yet with considerably lower cost. Novel extraction techniques, a new approach to a high intensity beam and a new scanning method of low emittance beam is possible. It also enables high energy economic beam lines less power consumption. The highlight of the cold beam accelerator is a two axis rotating carbon gantry.

COLD BEAM SYNCHROTRON

The general specifications for the CBS facility are based on the following premises for the clinical requirements;
Clinical spec: 2 fixed port (horizontal and 45 degree), 1 gantry;
Type of particles: Carbon, proton (option);
Ion energy: 100-400 MeV/u, a variable beam energy from spill to spill;
Average dose rate: 5 Gy/min;
Field size: 15 cm x 15 cm;
Dose uniformity: ± 4% of the prescribed dose over treatment field;
Delivered dose accuracy: 2%;
Irradiation method: revised spot scanning system with synchronization of respiration.

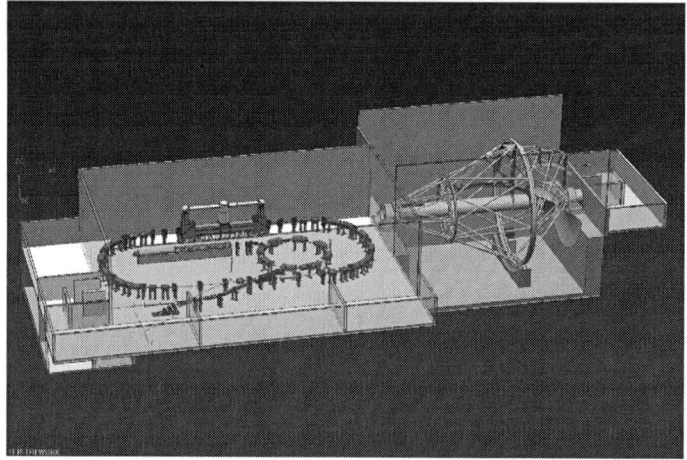

FIGURE 1. The layout of carbon beam facility for cancer treatment.

The accelerator part of CBS complex consists of linac (C^{+4}, 6 MeV/u), fast cycling booster (C^{+4}, 30 MeV/u), main synchrotron with electron cooling and HEBT (High Energy Beam Transport). In Fig. 1 the layout of CBS with the low aperture gantry is shown. In Tab. 1, the basic parameters of the main ring are listed.

TABLE 1. Basic Parameters of Main Synchrotron.

Parameter	Units	Value
Type of Particle		$^{12}C^{+6}$
Injection Energy	MeV/u	30
Extraction Energy	MeV/u	100 - 400
Magnet Rigidity	Tm	6.4
Circumference	m	80.6
Betatron Tunes (hor/vert)		3.42/ 2.43
Magnet Gap	mm	36

ELECTRON COOLING APPLICATION

Storage & Cooling of Intense Carbon Beam

The electron cooling process will be used both on the injection energy and on the final energy. In Fig.2. the evolution of the normalized transverse emittance and longitudinal momentum spread at injection are presented. The simulation was made for follow parameters: the beta-function in cooling section is 12.5 m, the ion energy is $E_i = 30$ MeV/u, the electron beam radius is 1.2 cm, the electron current is $I_e = 0.6$ A, the length of ion bunch is 0.33 from the synchrotron perimeter. As it is seen, the ion beam emittance and momentum spread are effectively decreased during less than 100 ms. The limiting factor for cooling is tune shift induced by the ion beam space charge.

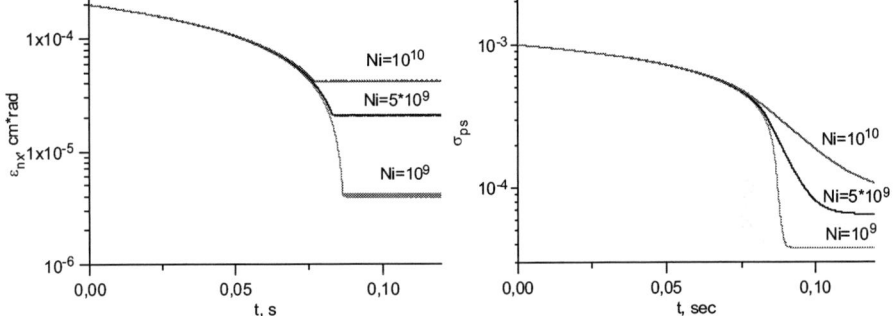

FIGURE 2. Time evolution of normalized transverse emittance (left) and longitudinal momentum spread (right), for different beam intensity, $E_i = 30$ MeV/u.

The main task for the cooler is the cooling of ion beam at the top energy (100 – 400 MeV/u). In Fig. 3 the evolution of normalized transverse emittance and longitudinal momentum spread for $E_i = 400$ MeV/u are presented. The simulation was made for follow parameters: the electron beam radius is 0.3 cm, the electron current is $I_e = 0.6$ A, the length of ion bunch is equal to the synchrotron perimeter. As it is seen, the ion beam emittance and momentum spread are effectively decreased during about 200 ms.

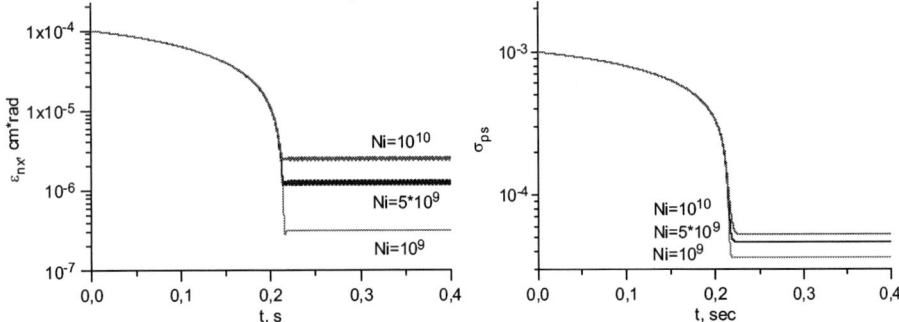

FIGURE 3. Time evolution of normalized transverse emittance (left) and longitudinal momentum spread (right), from top to bottom $N_i = 10^{10}$, $5 \cdot 10^9$, 10^9 accordingly, $E_i = 400$ MeV/u.

Precise Ion Beam Energy Scanning

For the active 3D scanning, the changing of the extracted beam energy with high accuracy is necessary. The possibility of the accelerating or decelerating the beam by means of the friction force of the electron beam has been demonstrated in the set of experiments. The electron cooling device allows operation with energy of the extracted beam by varying the electron beam energy simultaneously with the synchrotron magnet field. Dependence of the accelerating rate on the stored ion beam intensity at different ion energy is shown in Fig. 4.

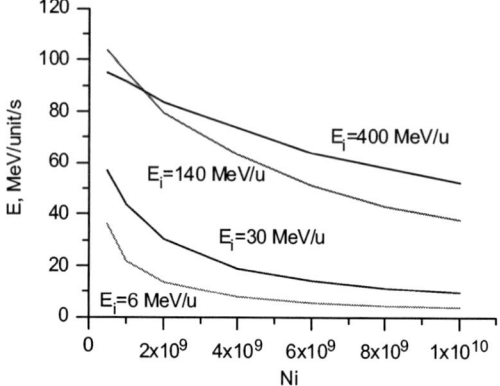

FIGURE 4. Dependence of the accelerating rate on the stored ion beam intensity. The parameters of electron beam are the following: I_e=0.4 A, a_e=1.5 cm (E_i=6 MeV/u), I_e=0.6 A, a_e=1.2 cm (E_i=30 MeV/u), I_e=0.8 A, a_e=0.5 cm (E_i=140 MeV/u), I_e=1.0 A, a_e=0.3 cm (E_i=400 MeV/u).

Extraction

Two different schemes of the extraction are possible to use. At condition of electron cooled ion beam, it is possible to used kicker extraction with very high repeating rate of extraction. The total ion beam can be divided up to 10 000 portions with controllable intensity of portion, by such method named as "pellet" extraction.

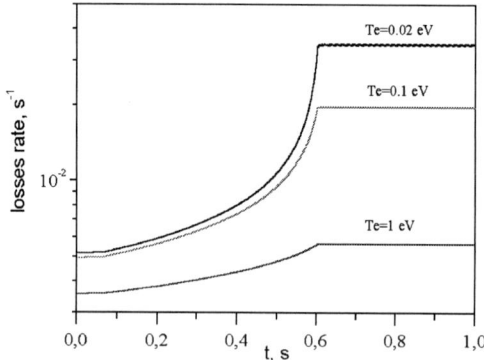

FIGURE 5. Recombination losses rate for different electron beam temperature.

Ion charge exchange by the electron beam recombination is utilized for a slow extraction scheme. The very small relative velocity between ions and electrons leads to the certain probability for a recombination process. Operating with the electron beam intensity and transverse profile gives the possibility for precise varying of the intensity and emittance of $^{12}C^{5+}$ beam extracted from the electron cooler.

In Fig.5. the recombination rates for different values of electron beam temperature are presented. The simulation was made for follow parameters: $N_i=10^{10}$, $I_e=3$ A, $E=400$ MeV/u.

ELECTRON COOLER DESIGN

As base for out project we take the design of the cooler EC-300 for CSRe storage ring, IMEP, China, designed and produced by BINP [2]. The main parameters of electron cooler are presented in Tab. 2. The sketch of the electron cooler design is shown in Fig 6.

FIGURE 6. The layout of the electron cooler device for CBS.

TABLE 2. Electron Cooler Parameters.

Parameter	Units	Value
Electron Energy	keV	Up to 250
Total Length	m	10.2
Cooling Length	m	7
Magnet Field	kG	1.5
Quality of the magnetic field		Better 10^{-4}

REFERENCES

1. V.V.Parkhomchuk et al., Proc. of PAC 05.
2. V.V.Parkhomchuk et al., "Commission of electron cooler EC-300" in Proc of this Workshop.

Electron Beam Size Measurements in the Fermilab Electron Cooling System

T.K. Kroc[*], A.V. Burov[*], T.B. Bolshakov[*], A. Shemyakin[*], and S. M. Seletskiy[†]

[*]*FNAL¶, PO Box 500, Batavia, Il 60510 USA*
[†]*University of Rochester, Rochester, NY 14627 USA*

Abstract. The Fermilab Electron Cooling Project requires a straight trajectory and constant beam size to provide maximum cooling of the antiprotons in the Recycler. A measurement system was developed using movable apertures and steering bumps to measure the beam size in a 20m long, nearly continuous, solenoid. This paper will focus on results of these measurements of the beam size and the difficulties in making those measurements

Keywords: Electron cooling, diagnostics, beam measurement
PACS: 29.27.Eg, 29.27.Fh

INTRODUCTION

This paper reports on experimental results of measurements of the size of the electron beam in Fermilab's electron cooling system. The general understanding of the system and results of preliminary measurements performed in a prototype system have recently been reported[1]. Since then measurements have been made in the final production configuration.

The Fermilab electron cooling system [2] has recently provided cooling of the complex's antiproton beam in the Recycler. Typical electron currents have ranged between 50 - 200 mA although the system is capable of supplying 500 – 600 mA as antiproton intensities increase. This beam of cold electrons moves along a trajectory coincident with the trajectory of the beam of antiprotons. When the velocities of the two beams are matched, Coulomb interactions transfer energy ("temperature") from the antiprotons to the electrons.

For electron cooling to be effective, the electrons need to have a total transverse motion of no more than 200 μrads[3]. Of this, a maximum of 100 μrads is allocated for variations of the beam envelope. The beam's effective temperature due to trajectory perturbations and remnant variations of its envelope due to the optics of the beam line can be measured by the technique described here.

The cooling section consists of 10 two meter long solenoids [4]. Before and after the cooling section and between each solenoid is a movable circular aperture, a "scraper", which consists of a copper bar, .125" thick with a 15 mm hole through which the beam can pass. There are a total of 11 scrapers.

¶Operated by Universities Research Association Inc. under Contract No. DE-AC02-76CH03000 with the United States Department of Energy.

The scrapers are inserted so that the DC beam passes through the aperture and then the beam is moved within the scraper to determine the shape of the beam at that location. Doing this at all 11 locations throughout the cooling section determines the evolution of the beam envelope as a function of the longitudinal position. The beam has very distinct edges which are used for determining its size.

TECHNIQUE

The procedure involves creating an offset trajectory through the cooling section using "4-bumps". A scraper is inserted so that the hole is centered on the central trajectory. Then the beam is moved until it touches the aperture of the scraper. The same 4-bump is used with each of the 11 scrapers.

The 4-bumps are created using combinations of corrector coils with multiplicative coefficients so that they can be varied in unison to create the offset trajectory. The transfer function of the correctors, measured using the response of the beam in downstream Beam Position Monitors (BPMs), is used to compute these coefficients.

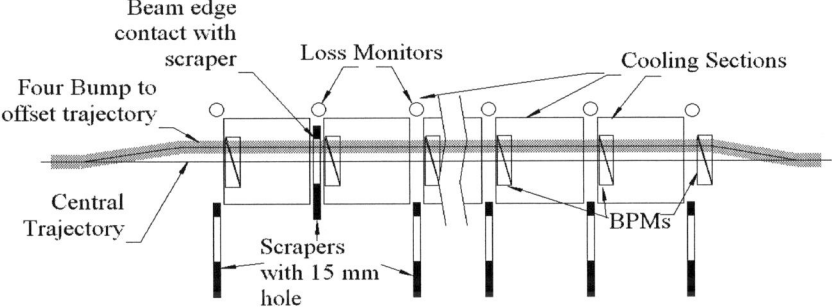

FIGURE 1. Schematic of the beam size measurement system in the cooling section.

The scrapers are inserted individually so that the aperture of the scraper is centered on the beam trajectory (Figure 1) and the beam is turned on. The beam is then moved up/down, left/right, and in 45° diagonals until it touches the aperture of the scraper. The touch is determined by: the response of a nearby loss monitor, a reduction in intensity in a BPM immediately downstream of the scraper, or loss of electron current.

The eight points of contact provide the information for determining the parameters of the beam envelope. These parameters include: the axes of the ellipse, its eccentricity, and tilt. The eleven ellipses along the length of the cooling section show the evolution of the beam envelope. This information about the variations of the envelope along the length of the solenoid can then be used to adjust the focusing (and trajectory if necessary) into the cooling section to reduce these variations below the allowed tolerances using a procedure described in reference [5].

RESULTS

Figure 2 shows the calculated ellipse for a recent measurement, displaced to the eight positions where a touch was indicated. The solid line shows the limit of motion in the X direction and the solid ellipses show its position at the edge. The dotted line indicates Y motion. The dot-dash shows the –XY motion and dot-dot-dash +XY motion. While the mults that produce the motion are not at 45° to each other, they are sufficiently distinct to allow the calculation of the ellipse.

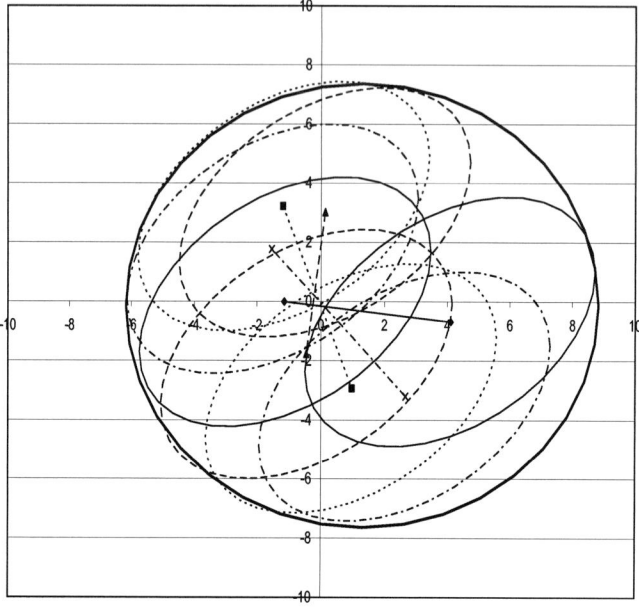

FIGURE 2. The positions of the beam ellipse in scraper SCQ01 at the point of interaction with the scraper aperture at each of the eight points of contact.

Table 1 shows the resulting ellipse parameters for five of the scrapers measured during a recent study period. The calculated ellipse parameters correspond to the following equation:

$$x = A_x \cos \Psi + x_0$$
$$y = A_y \cos(\Psi - \upsilon) + y_0 \tag{1}$$

TABLE 1. Ellipse Parameters at Measured Scrapers.

	SCQ01	SCC90	SCC80	SCC60	SCC00
A_x	4.60 ± .22	4.56 ± 1.09	4.35 ± .50	4.11 ± .29	4.91 ± .34
A_y	4.22 ± .19	4.17 ± .91	4.20 ± .18	4.45 ± .19	3.73 ± .44
Angle υ	1.16 ± .14	1.38 ± .43	1.34 ± .16	1.29 ± .12	1.17 ± .19
Major Half-axis	5.24	4.81	4.75	4.85	5.27
Minor Half-axis	3.39	3.87	3.74	3.62	3.20
Area (mm^2)	55.9	58.5	55.8	55.2	53.0
Average Area			55.7 ± 2.0		

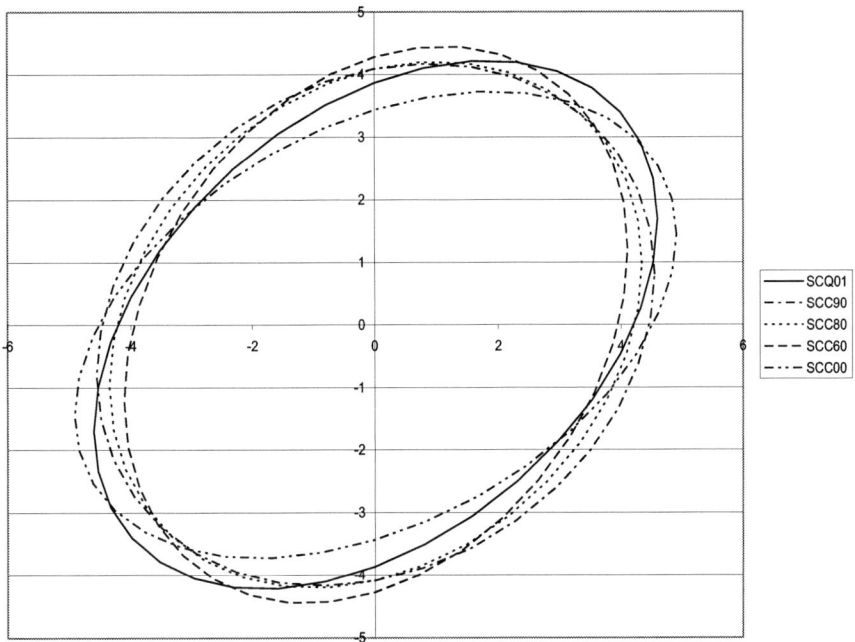

FIGURE 3. Ellipses for 5 scraper positions plotted together to show variation of measured ellipse.

Figure 3 shows all of these ellipses plotted together. As noted in the table, the physical area of the ellipse is 55.7 mm^2 with a statistical error of 2 mm^2. The position of the centroid of each ellipse is determined to about .4 mm. The error of the tilt angle v is about 16% and the A_x and A_y are determined to within about 10%. These results indicate that the envelope oscillations are ~220 μrads[6]. The beam shape appears to be fairly uniform through the cooling section although it is somewhat elliptical. As stated in the introduction, the quality of the beam is sufficient to provide cooling for the present high energy physics program. Further analysis of the measurements is continuing.

USER INTERFACE

As operational experience has been accumulated, it has become apparent that a single automated process may not be possible. As noted above, three different signals can be use to determine the beam edge. The response of these signals is not necessarily consistent. Therefore, the judgment of the user is needed to best determine the beam edges. A new program interface has been developed, an image of which is shown in figure 4. It incorporates an X-Y plot of the limits of beam motion, which gives the user an indication of the beam shape as the measurement is being made. Surrounding this plot are 4 smaller plot (one for each of the four motion directions) showing the response of the signals as the beam is moved. Using this, the user can analyze the measurement just taken and determine if the thresholds are appropriate and if a measurement should be repeated.

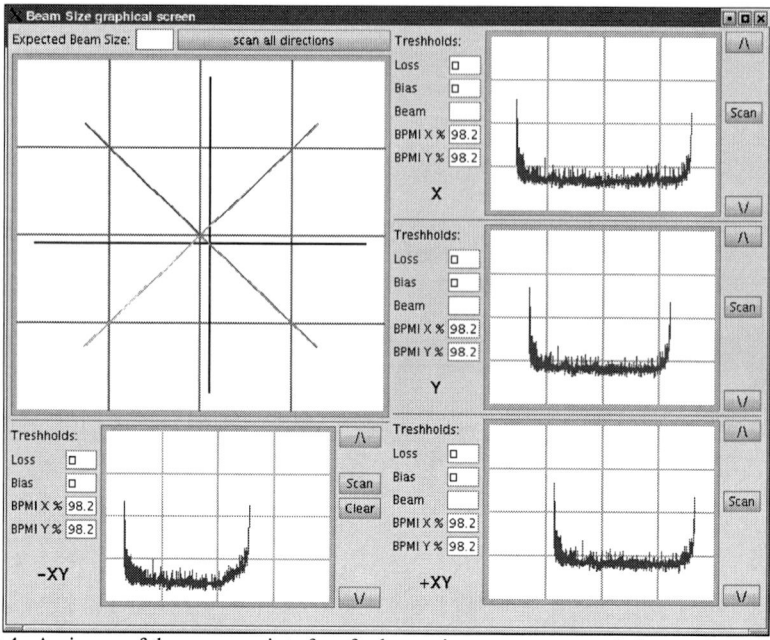

FIGURE 4. An image of the new user interface for beam size measurements. Only one signal for each of the four signal response plots is shown.

CONCLUSIONS

A technique has been developed for measuring the size, shape, and uniformity of a DC electron beam through the Fermilab Electron Cooling cooling section. Measurements indicate that the technique provides the data necessary to tune the optics of the line and that the present transverse velocity parameters of the beam are on the order of our tolerances. Initial results show r.m.s. angles of 220 μrads. Cooling has been observed and is now part of the regular operational program of the laboratory.

REFERENCES

1. T. K. Kroc, et. al., "Electron Beam Size Measurements in a Cooling Solenoid", Proc. 2005 Part. Acc. Conf.
2. S. Nataitsev, et. al., "Commissioning of Fermilab's Electron Cooling System for 8-GeV Antiprotons", Proc. 2005 Part. Acc. Conf.
3. A. Burov, "Electron Cooling Scenarios at Fermilab", NIM-A 532(2004) p. 291-297.
4. J. Leibfritz, et. al., "Fermilab Electron Cooling Project : Engineering Aspects of Cooling Section", Proc. 2001 Part. Acc. Conf., pg. 1414.
5. A. Burov and V. Lebedev, "Cylindric Electron Envelope for Relativistic Electron Cooling", FERMILAB-TM-2303-AD (2005).
6. A. Burov, et. al., "Optics of Electron Beam in the Recycler: Analysis of First Results", these proceedings.

Magnetic Field Measurement and Compensation in the Recycler Electron Cooler

V.Tupikov[#,1], G..Kazakevich[2], T.K.Kroc[1], S.Nagaitsev[1], L.Prost[1],
A.Shemyakin[1], C.W.Schmidt[1], M.Sutherland[1], A.Warner[1]

[1] *FNAL, Batavia, IL 60510, USA*
[2] *BINP, Novosibirsk, 630090, Russia*

Abstract. Cooling of 8.9-GeV/c antiprotons in the Recycler Electron Cooler requires a round 4.34-MeV electron beam with a small angular spread propagating through a 20-m long cooling section. To confine the electron beam tightly and to keep its total transverse angles below 0.2 mrad the cooling section is immersed in a solenoidal field of 50-200 G. The field was measured with a compass-based sensor (transversal) and a hall-probe (longitudinal) after installation of the solenoids into the Recycler tunnel. For the field strength of 105 G, the transverse field components were compensated to the level that provided corresponding dipole beam oscillations below 0.1 mrad, which in turn allowed the first cooling of antiprotons in the GeV energy range. This paper discusses the field measurements and compensation scheme including the results of dipole oscillation measurements.

Keywords: Electron Cooling.

INTRODUCTION

The Electron Cooling project at Fermilab is characterized by a low 100-G magnetic field in the cooling section (CS), despite conventional coolers with kilogauss range fields. As a result, the helical length, λ, of an electron in the CS for a closed Larmor radius becomes much longer ($\lambda \gg d_{SOLENOID}$), which makes a low value of $\int B_\perp dz$ more important then an absolute low value of $B_\perp$. The system used for magnetic measurements has been discussed in previous publications [1-2]. The cooling section solenoid design is discussed in [3].

FIGURE 1. (a) New wheeled cart design with compass. Compass is replaced by hall-probe during longitudinal measurements; (b) Four copper rods compensator.

MAGNETIC MEASUREMENTS

Preliminary Test Bench Run

Test bench measurements for each of 10 of the cooling section solenoids preceded the final measurements in the Recycler tunnel. They resulted in a field map for each 2-m long solenoid modules. The map consists of longitudinal and transverse field components. The longitudinal component is a product of the "main" and two "trim" solenoids, measured separately. The transverse component represents the field generated by compensation coils, which are ten X and Y pairs per module. The field shape generated by the compensation coil is well described by a Gaussian distribution with the center given with respect to a "zeroed" longitudinal lead trim-solenoid peak.

Following the idea of the BINP (Novosibirsk) design, the transverse field measurement scheme had been revised in comparison to the one described in [1-2]. We omitted the cylindrical cart design with wrapped compensation coils because of its deficiencies: unpredictable X and Y component coupling caused by cart rotation; occasional cart sticking in the chamber; and jerky movement. The new design, based on a wheeled cart (FIGURE 1a) and 4-rod compensator (FIGURE 1b) was a significant improvement and–removed the sticks, jerks, and rotations. The cart movement became smoother giving precise transverse measurements due to elimination of the cable connection.

Longitudinal Field

All longitudinal field measurements were done with use of a BH204 hall-probe from "F.W.Bell Co". The initial probe calibration was done against a calibrated Teslameter DTM-141 at the Fermilab Technical Department's magnetic measurement test bench. The achieved 0.1% calibration accuracy was crucial not only for magnetic field mapping but possibility for matching the energies of electron and antiproton beams as well. For procedure details refer to [4] and to [5] for how the matching was successfully achieved.

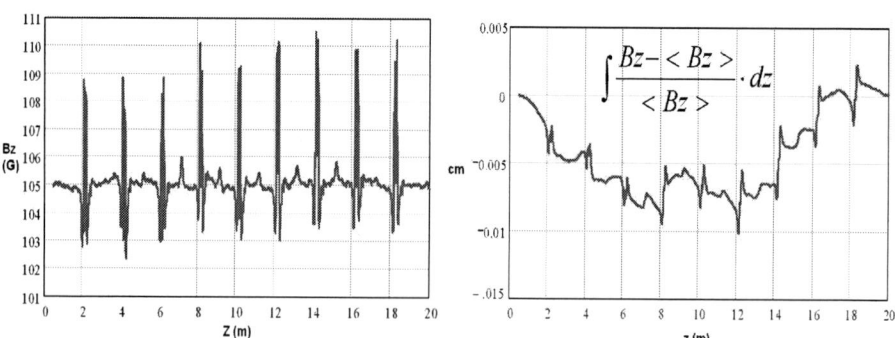

FIGURE 2. Bz measurement after solenoid current correction (left plot). Integral of Bz field deviation from the average with <Bz> = 105.1 G (right plot).

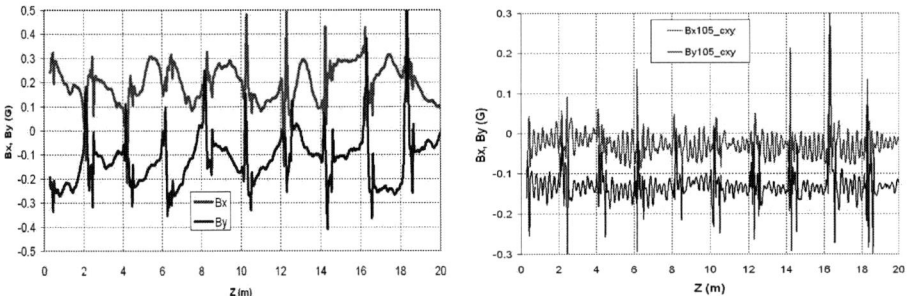

FIGURE 3. *Left plot* - First measurement of transverse field taken at <Bz> = 105.1 G. *Right plot* - Second measurements after field correction by dipoles. There is a large By-offset, caused by a broken wire between initial and second measurements.

The first measurements were made with the lead trim-solenoids on and the return trim- and main solenoids off. It permitted defining dipole elements to the absolute Z-coordinate in the Recycler tunnel, based on dipole peaks, with coordinates for each solenoid found during preliminary measurements.

Then, longitudinal field was measured along the full cooling section (CS) at the expected nominal settings for solenoid currents. Due to varying winding numbers for each solenoid during manufacturing, the fields generated by the solenoid differed slightly. The field was divided into the 'main solenoid' and 'gap' intervals, where the average field values were calculated. Each 'main solenoid' is powered by an individual power supply, but all trim-solenoids share just two power supplies feeding lead- and return- solenoids connected in series respectively. The equalization of the 'main solenoid' averages were done by adjusting DAC settings for their individual power supplies. In the case of averages for trim-solenoids, - the equalizations were achieved by resistive shunts connected in parallel to the coils to bypass some current.

The final longitudinal field corresponding to 105 G average is shown in FIGURE 2. The plot of Bz integral deviation along the CS is less than 0.1 cm which is one order of magnitude better than cooling requirements.

For possibly working at different Bz values, a set of measurements of Bz(z) versus trim-solenoid currents in more problematic regions (gaps), were made. Then, calculated integrals of bell-like fields as a function of induced current were plotted, and were perfectly linear. Having measured the longitudinal field in the CS at 190 G, we are now able to approximate field settings for any field from 50 G through 190 G..

Transverse Field

The first transverse measurements started at longitudinal field Bz=105 G and all dipole correctors set to zero current (FIGURE 3, left plot). The choice of 105-G field was defined by the period of the electron's Larmore radius. In a case of 100 G longitudinal field it is very close to an integer number of BPM placements in the CS, which makes difficult to make a field correction by zeroing the electron beam in BPMs as it will be described in next chapter.

The measured field map combined from longitudinal and transverse components was entered into a simulation program, which calculated the dipole corrector values necessary to compensate the transverse field in the CS. Corrected currents were applied to the correctors and a second run of transverse measurement were taken. Again, the discrepancy in the transverse field between measured (during the second run) and expected field (superposition of fields measured during the first run and calculation of field generated by dipole corrector settings found after first run) was simulated. Corrector changes were found for this discrepancy and applied to the settings of the first run. The results of the composite measurements is shown in FIGURE 3 on the right plot, where dark color curves present measured transverse components. The calculated electron angles in the CS were equal to 50 μrad, disregarding offsets, mostly determined by residual tension of the compass suspension and unresolved during the measurements.

Transverse fields at Bz=190 G was measured once, and gave a reduction of field compensation accuracy. The estimated electron angles in the CS at this field were 130 μrad.

Further Field Correction By Electron Beam Position (BPM)

After the Pelletron was commissioned and first beam in the CS was obtained, additional field improvement was achieved. In this case, corrections made to the dipole settings relied on beam position monitor (BPM) readings. All CS BPMs had been very precisely calibrated against the pbar beam.

First attempt was to get rid of unresolved transverse field offsets. Monitoring the e-beam trajectory in the CS and minimizing the trajectory oscillation required tuning the current, by the same value, of all of horizontal and then vertical dipole correctors. The current changes of 170 mA for horizontal and -215 mA for vertical planes led to an electron trajectory with 1 mm peak-to-peak oscillation in the CS. Then, changing the current by the same value in all of the correctors (separately for X and Y planes) of any 2m-long section gave a zeroed beam displacement in the closest downstream BPM (**FIGURE 4**). Due to a ground fault in CXC80 and CYC80 correctors 20% of the CS was lost from effective cooling (see electron trajectory after 16 m on the plot), but we are going to get it back by replacing the regular power supplies with isolated outputs.

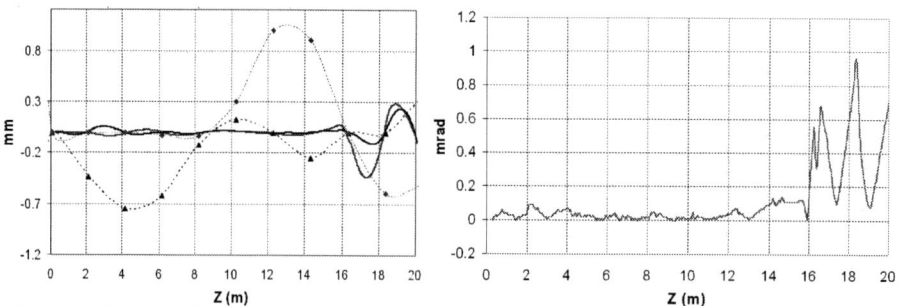

FIGURE 4. Left plot - Electron trajectories in CS before (dash curves) and after (solid curves) section by section adjustment Right plot – Estimated electron angles in CS after compensation.

Transverse field correction was done at a longitudinal field of 190 G as well. Now it is possible to compensate for any longitudinal field in the range of 105 to 190 G. New corrector settings were calculated by formula:

$$C_{X\{Y\}}^{150G} = C_{X\{Y\}}^{105G} + \frac{(150-105)G}{(190-105)G}(C_{X\{Y\}}^{190G} - C_{X\{Y\}}^{105G}) \quad (1)$$

where C is the corrector current.

With these settings, and no additional tuning, the beam angle in the CS was estimated to be 130 μrad. This is the same number found for the compensated field at Bz=190 G, thus it is sort of an error translation.

SUMMARY

The magnetic field in the cooling section was measured for both transverse and longitudinal components. Then the transverse filed was compensated with dipole correctors. Good cooling requires electron angles in the CS to be less than 100 μrad. The first estimation of electron angles in the CS with use of a simulation program and the compensated field map gave 50 μrad. The second estimation, by drag force measurements, gave 100 μrad [6].

ACKNOWLEDGEMENTS

The authors would like to express appreciations to V.Parkhomchuk, whose ideas made the measurement system possible; J.Leibfritz, V.Sidorov and L.Nobrega for their engineering support; K.Carlson – for electrical engineering help; and the teams of technicians and electricians working with us, especially: R.Kellet, A.Germain, J.Nelson, W.Johnson.

REFERENCES

1. The Precise Magnetic Field Sensor for Solenoids of Cooling Section, V.N.Bocharov et al, *Proc. Of XVII Workshop on Particle Accelerators, Protvino, Russia, October 17-20, 2000 (in Russian).*

2. Field Measurements In The Fermilab Electron Cooling Solenoid Prototype, V.Tupikov, et.al., *Fermilab-TM-2224, October, 2003.*

3. Fermilab Electron Cooling Project: Estimates for the Cooling Section Solenoid, Fermilab-FN-689, S.Nagaitsev, A.Shemyakin and V.Vostrikov, April 2000.

4. Beam-Based Calibration of the Electron Energy in the Fermilab Electron Cooler, S. M. Seletskiy , A. Shemyakin; Proceedings of PAC2005, Knoxville, TN, USA, May 16-20, 2005.

5. Attainment of high-quality electron beam for Fermilab 4.3-MV cooler, A.Shemyakin, et al., This Workshop Proceedings.

6. Antiproton cooling in the Fermilab Recycler, S.Nagaitsev, This Workshop Proceedings.

OTR Measurements and Modeling of the Electron Beam Optics at the E-Cooling Facility

A. Warner[1], A. Burov[1], K. Carlson[1], G. Kazakevich[2], S. Nagaitsev[1], L.Prost[1], M. Sutherland[1], and M. Tiunov[2]

[1] *Fermi National Accelerator Laboratory, P.O.Box 500, Batavia IL 60543*
[2] *Budker Institute of Nuclear Physics, 630090 Novosibirsk, Russia*

Abstract. Optics of the electron beam accelerated in the Pelletron, intended for the electron cooling of 8.9 GeV antiprotons in the Fermilab recycler storage ring, has been studied. The beam profile parameters were measured under the accelerating section using Optical Transition Radiation (OTR) monitor. The monitor employs a highly-reflective 2 inch-diameter aluminum OTR-screen with a thickness of 5 μm and a digital CCD camera. The measurements were done in a pulse-signal mode in the beam current range of 0.03-0.8 A and at pulse durations ranging from 1 μs to 4 μs. Differential profiles measured in pulsed mode are compared with results obtained by modeling of the DC beam dynamics from the Pelletron cathode to the OTR monitor. The modeling was done with SAM, ULTRASAM and BEAM programs. An adjustment of the magnetic fields in the lenses of the accelerating section was done in the simulations. The simulated electron beam optics downstream of the accelerating section was in good agreement with the measurements made with pulsed beam.

Keywords: Optical transition radiation, CCD camera, electron beam, accelerator, bremsstrahlung
PACS: 29.27.Bd, 29.27.Eg, 29.27.Fh

INTRODUCTION

Optical transition radiation monitors are being used with pulsed beam to image and model the charge distributions of the 4.3 MeV electron beam used in the electron cooler at Fermilab. The transition radiation is produced by the charged particles as they traverse the boundary between media with different dielectric constants [1]. The OTR monitors have several advantages over more traditional imaging devices; they provide an image of the 2-D beam distribution in a plane, they have a linear response to beam charge with good spatial resolution and they have a wide possibility in data processing. Pulsed operation of the electron beam is achieved by modulating the voltage of the gun control electrode and was developed primarily for diagnostics and tuning purposes. To compensate for pulsing and optic variations during the pulse, differential processing of profiles taken at different pulse durations were done. Comparison of the measured beam differential profile dynamics with that of simulated DC beam dynamics show good agreement.

OTR Beam Profile Monitor Set-Up

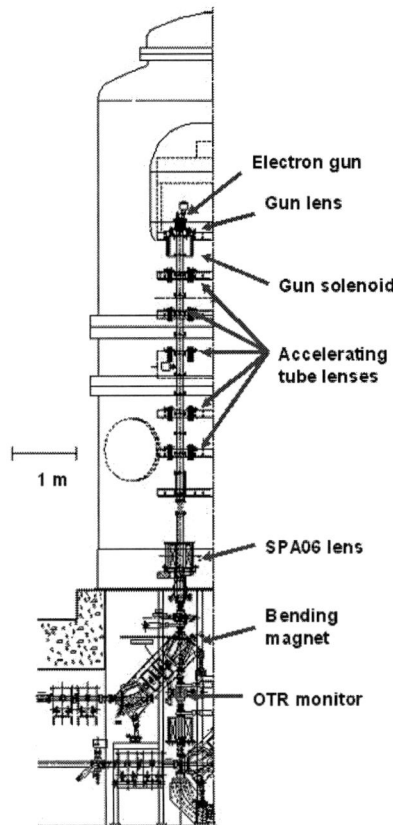

FIGURE 1. Layout of the experimental setup.

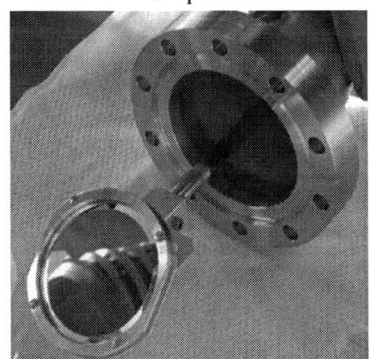

FIGURE 2. OTR screen mount.

The OTR beam profile monitor is mounted in a standard 6-way cross downstream of the accelerating section of the cooling facility, FIG.1, and is composed of a 2"-in diameter OTR screen made from a mirror-surface 5 μm aluminum foil tensioned with high flatness onto a ring frame movable with a stepper-motor actuator. Such a thin foil is essential to reduce the background caused by beam scattering and bremsstrahlung. The linear drive provides insertion of the OTR screen with accuracy of 0.1 mm and the screen is inserted into the beam at an angle of 38.9 degrees to the beam direction. This corresponds to the angle $\theta \approx 1/\beta\gamma$ [2], where the intensity of the light has a maximum. This allows to significantly increase the monitor sensitivity and to operate with low beam current. A photo of the OTR screen mounted on the actuator is shown in FIG. 2.

The backward OTR passes through the vacuum glass window to the digital CCD camera that provides an image the beam. To transform images to real-world coordinates, four 75 μm tungsten wires are mounted over the radiation screens to form a rectangular 10mm x 10mm grid. The CCD camera has an objective with: focal length f =25 mm; relative aperture F =1.4. The distance between the OTR screen and the camera lens is of 320 mm. The CCD camera and associated optics are incased in a light tight housing to omit external sources of light.

The digital camera is connected to computer via a IEEE 1394 fire-wire interface. A Labview based data acquisition and processing systems have been developed to obtain with real-time the beam spot images and to store them. The processing system provides the operator with tools for image analysis. The OTR diagnostic system employed at the cooler is designed to be routinely used to optimize the beam transport and to measure the transverse beam profiles and the respective beam sizes.

Data Acquisition

The images taken from the optical monitors are digitized in the CCD camera and saved in an uncompressed format. As a result external electronic and environmental noise sources have little effect on the image quality. In addition, the IEEE 1394 firewire interface allows control of the camera's gain and other features remotely. The images are then displayed and analyzed with application software that was developed using LabView and IMAQ vision utility tools. The program allows the user to display live (real-time) images of the beam and to simultaneously analyze either current (live) data or stored data from files. Two profiles taken with different pulse durations or optical settings can therefore be compared. This type of development differs from the standard application in that it incorporates image digitization, image display, as well as image analysis and system calibration in a real-time module. These image analysis tools are used to combine techniques that compute statistics and measurements based on the gray-level intensities of the image pixels; linear (convolution) filters can be applied to remove unwanted background at any stage of the analysis if necessary.

Results of the Measurements

The system so far described has been used to measure the properties of the electron beam while the Pelletron was operating in a pulse mode with the bending magnet turned off. This allows accelerated beam to be passed straight onto the OTR monitor. The residual field in the bending magnet was compensated with a coil mounted in the bending magnet. Images of 1μs beam pulses were subtracted from 2 μs beam pulses to compensate for pulser and gun optics variations during the pulse. The electron beam was aligned with BPM monitors located upstream of lens SPA06 and downstream of the OTR monitor before all measurements to avoid passing the beam off axis through the OTR.

During the measurements we checked the response of the monitor versus beam current which was varied by changing the pulse voltage of the gun control electrode. FIG. 3 shows the dependence of the integral of detected light in the beam spot versus the beam current and demonstrates good linearity of the OTR monitor. Moreover we measured dependence of the monitor sensitivity versus position of the beam along both the X and Y axes of the screen at a fixed beam current. Motion of the beam along the axes was done using correctors inside the focusing lens SPA06, FIG. 1.

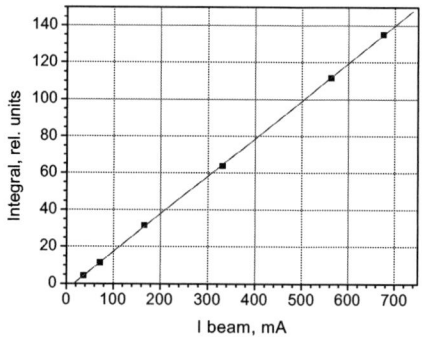

FIGURE 3. Beam spot integrals versus the beam current.

The measured dependence of the sensitivity on the beam position over the surface area of the screen is relatively weak; as a result this allows the use of the full area of the screen for the beam monitoring.

The beam profiles were measured as functions of the current of the lens SPA06 with a beam current of 0.56 A. FIG. 4 shows the variation of the profiles versus the lens current. One can see that the beam profiles are dissimilar with various lens current.

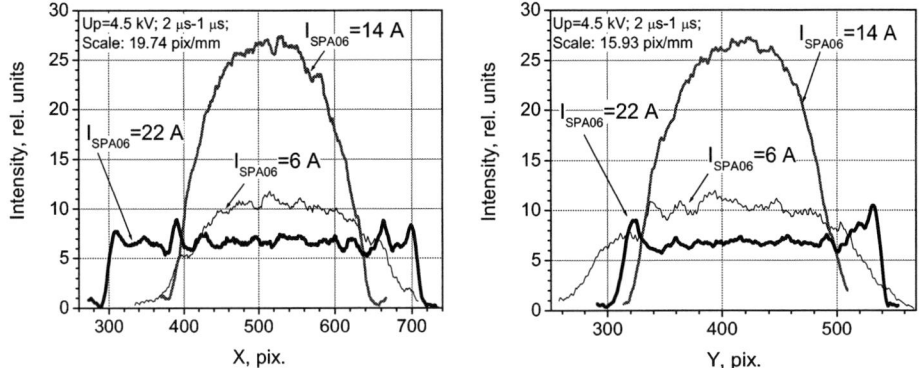

FIGURE 4. Beam profiles versus current in lens SPA06.

Simulation Results

The beam profiles plotted in FIG. 4 can not be explained as a trivial effect of focusing with SPA06 lens. We therefore studied this phenomenon with simulations of the beam optics in DC mode from the cathode to the OTR monitor with space charge effects included. The accuracy of the simulations was improved by using fitted magnetic fields in the lenses of the accelerating section. The fitting of the fields were done with SAM code package [3] and the measured distributions of the fields.

Simulation of the beam dynamics and optics in the electron gun was done with the 2D ULTRASAM code [4]. The results of the simulation, including the trajectories of the electrons, the contours of electrodes, the equi-potentials, and the distribution of the electric and magnetic fields are shown in FIG. 5.

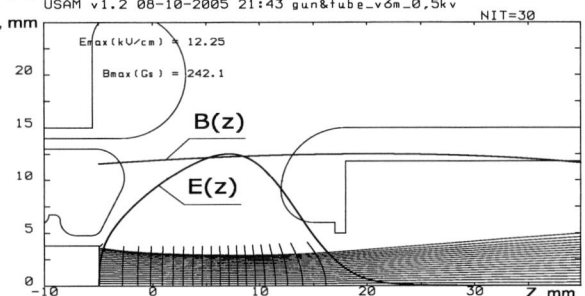

FIGURE 5. Results of the electron gun optics calculation.

The beam optics from the exit of the gun to the OTR-monitor was calculated with the BEAM program [5] using the gun exit beam parameters from the ULTRASAM code as input. The calculated electron beam envelopes versus SPA06 current for full beam current and 25% of the beam current, respectively, are shown in FIG. 6 (left) together with the corresponding distribution of the axial magnetic fields. The figure demonstrates that the similarity in the beam profiles along the beam axis is not conserved. Such phenomenon can be explained as a nonlinear dependence of the transverse velocities in the beam versus the radial coordinate for the existing gun optics; an inhomogeneous current density distribution in the gun leads to similar results. Note that with a linear profile of transverse velocities and a homogeneous current density the ratio of the radii for the 25% and 100% beam envelopes is a constant along the Z axis.

Figure 6 (right) shows measured (dots) and calculated (solid lines) beam X-profiles for the SPA06 lens current values of 6A, 14A, and 22A. The measured Y-profiles also approximately coincide with calculated profiles considering the slightly elliptic shape of the real beam. All measured and calculated profiles demonstrate good agreement.

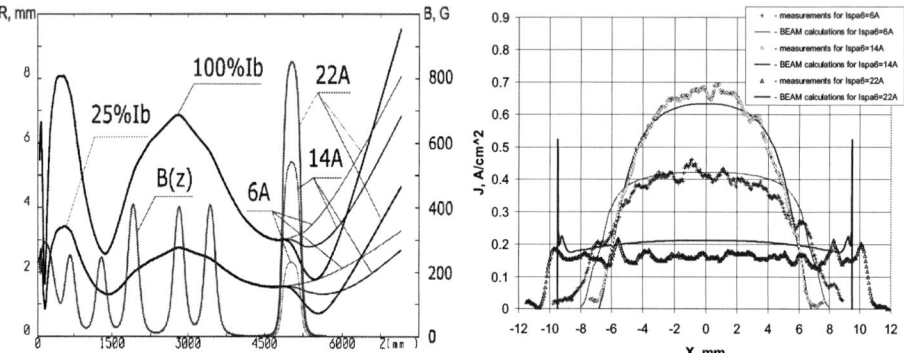

FIGURE 6. Beam envelopes and beam profiles versus SPA06 current.

SUMMARY

OTR monitoring, data acquisition and processing have been developed for analysis of beam optics with the Pelletron in the pulse mode. Using those systems measurements were done of the beam profiles downstream of the Pelletron accelerating section at the OTR screen location. Modeling of the beam optics in DC mode from the cathode to the OTR screen was done with SAM, ULTRASAM and BEAM programs. The measured result show good agreement with the modeling and also explains the variation of the beam profiles versus the SPA06 lens current.

ACKNOWLEDGMENTS

The authors would like to express appreciations to the team of expert technicians that worked with us to assemble and install all the systems, especially: R.Kellet, A.Germain, J.Nelson, W.Johnson, F.Juarez and M.Frett.

REFERENCES

1. V.L. Ginzburg and I.M. Frank, "Radiation from a Uniformly Moving Electron passing from One Medium to Another", Journ. of Experimental and Theoretical Physics (JETP) V.16, pp. 15-26 (1946).
2. A. Warner, G. Kazakevich, S. Nagaitsev, G. Tassotto, W. Gai and R. Konecny "Beam Profile Diagnostics for the Fermilab Medium Energy Electron Cooler." IEEE Trans.Nucl.Sci, V. 52, No 5, 2005
3. B.M. Fomel, M.A. Tiunov, V.P. Yakovlev. SAM – an Interactive Code for Evaluation of Electron Guns. Preprint BINP 96-11, Novosibirsk, 1996.
4. A.V. Ivanov, M.A. Tiunov. ULTRASAM - 2D Code for Simulation of Electron Guns with Ultra High Precision. Proceeding of EPAC-2002, Paris, 2002.
5. M.A. Tiunov. BEAM – 2D-code package for simulation of high perveance beam dynamics in long systems. Proceedings of International Symposium "SPACE CHARGE EFFECTS IN FORMATION OF INTENSE LOW ENERGY BEAMS". February 15-17, 1999, JINR, Dubna, Russia

Beam-Based Alignment of Magnetic Field in the Fermilab Electron Cooler Cooling Section*

S.M. Seletskiy[#], V. Tupikov

University of Rochester, USA, Fermilab, USA

Abstract. The Fermilab Electron Cooling Project requires low effective anglular spread of electrons in the cooling section. One of the main components of the effective electron angles is an angle of electron beam centroid with respect to antiproton beam. This angle is caused by the poor quality of magnetic field in the 20 m long cooling section solenoid and by the mismatch of the beam centroid to the entrance of the cooling section. This paper focuses on the beam-based procedure of the alignment of the cooling section field and beam centroid matching. The discussed procedure allows to suppress the beam centroid angles below the critical value of 0.1 mrad.

INTRODUCTION

The Recycler Electron Cooling (REC) [1] at Fermilab has been applied to 8 GeV antiprotons [2] and requires a high quality DC beam of 4.3 MeV electrons. A general layout of the REC system is shown in Fig. 1. The Pelletron [3] accelerates an electron beam. Then the beam is bent in two planes to bring it into the cooling section (CS). The cooling section is immersed into the 100 G solenoidal field. After the CS, the electrons make a U-bend down the cooler, and finally come back to the Pelletron where they are decelerated and dumped into a collector.

The electron beam suitable for the REC has to have effective rms angular spread in the CS less than 0.2 mrad. One of the main sources of growth of electron angular spread is the misalignment of electron beam centroid with respect to the axis of the cooling section (or with respect to the antiproton beam). The conventional requirement on the upper limit for this misalignment is 0.1 mrad.

THE ALGORITHM OF THE ALIGNMENT

General Idea

There are two different parts to the problem of the angle, that the electron beam centroid has with respect to the antiproton beam, that have to be considered.

---

* Operated by Universities Research Association Inc. under Contract No. DE-AC02-76CH03000 with the United States Department of Energy.
[#] smsvm@pas.rochester.edu

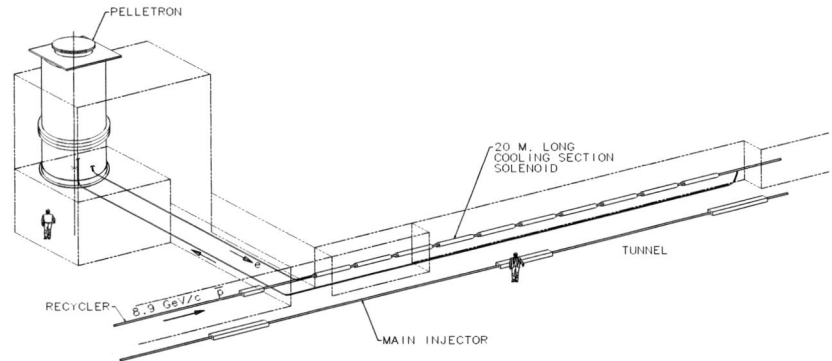

FIGURE 1. The schematic layout of the Fermilab electron cooler.

First, the beam has to be matched to the cooling section. With respect to the beam centroid angle that means that the beam has to have zero angle θ_0 and zero displacement ξ_0 at the entrance of the cooling section. Here $\xi = x + i \cdot y$, $\theta = \theta_x + i \cdot \theta_y$, i is the imaginary unit, θ_x and θ_y are the x and y components of the transverse angle of an electron, and we introduce Cartesian coordinates x, y, z in the laboratory frame such that the beginning of the coordinates coincides with the entrance of the cooling section, the z axis coincides with the axis of the cooling section, the y axis is directed upward; $\vec{x} \times \vec{y} = \vec{z}$. When the beam is matched to the cooling section its angle in the CS stays zero if the magnetic field is properly compensated. Whence we immediately come to the second part of the problem: a technique for fine alignment of the magnetic field in the cooling section has to be found.

The requirements to the transverse component of the magnetic field in the cooling section [4] are given by the following formula:

$$\int B_\perp(z)dz \leq 1.5\, G \cdot cm \qquad (1)$$

where $B_\perp = B_x + i \cdot B_y$, and B_x and B_y are the x and y components of the transverse field in the cooling section. The preliminary compensation of magnetic field in the CS was done with the aid of magnetic compass technique [4], [5]. The cooling section consists of ten 2 m long solenoidal modules; every module is equiped with 20 dipole coils that are used for field compensation. Preliminary compensation allows to suppress the transverse component of the magnetic field in each solenoidal module with an error only:

$$B_{\perp j} = \alpha_j \cdot B_z + \beta_j \qquad (2)$$

Here B_z is the longitudinal component of magnetic field in the cooling section, j is the number of the module, $\alpha_j = \alpha_{xj} + i \cdot \alpha_{yj}$, $\beta_j = \beta_{xj} + i \cdot \beta_{yj}$, and α_{xj}, α_{yj}, β_{xj}, β_{yj}

are the x and y components of the errors in compensation of the solenoid #j inclination angle and residual field respectively. The analisys of preliminary compensated field shows that the error in transverse field compensation stays constant in every module and may vary from solenoid to solenoid, and differences between $B_{\perp j}$ in two modules can be as high as 80 mG. These errors in compensation are intolerable and result in large beam centroid angle even for a well-matched electron beam.

The CS is equipped with 10 beam position monitors (BPM). The BPMs have 10 µm absolute precision and are longitudinally positioned with a precision better than 1 mm. For such configuration, probably the only feasible way to both match the beam to the CS and to align the magnetic field in the CS is to analyze how the beam's trajectory changes depending on the change of B_z. For the perfectly matched beam and ideally compensated $B_\perp$ the change of B_z will not affect the electron's trajectory in the cooling section; while in case of mismatched beam and (or) uncompensated transverse fields the trajectory of the beam will change with changing B_z. In principle, measuring the beam positions in the BPMs for different values of longitudinal field in different solenoids one can find θ_0 (ξ_0 is known from the reading of the very first BPM that is positioned at the entrance of the CS) as well as α and β for every solenoidal module.

Theoretical Consideration

The motion of an electron in a pure magnetic field [6] is described by (3):

$$\begin{cases} \xi' = \theta \\ \theta' = i\dfrac{e}{pc}(B_\perp - B_z \theta) \end{cases} \quad (3)$$

where p is the momentum of the beam, c is the speed of light and e is the electron charge. It follows from (3), that in a hard-edge approximation of solenoidal field, at the location of BPM #j [6]:

$$\begin{pmatrix} \xi_j \\ \theta_j \end{pmatrix} = M^j \cdot \begin{pmatrix} \xi_0 \\ \theta_0 \end{pmatrix} + D^j \quad (4)$$

where

$$M^n = \prod_{m=n}^{0} H^m, \quad D^n = \sum_{m=0}^{n-1}\left[\prod_{l=n}^{m+1} H^l\right] \cdot F^m + F^n, \quad D^0 \equiv 0 \quad (5)$$

$$H^j = H_{trans}^{j-1,j} \cdot H_{sol}^{j-1}, \quad H^0 \equiv \begin{pmatrix} 1 & 0 \\ -\dfrac{i \cdot k_0}{2} & 1 \end{pmatrix}, \quad F^j = H_{trans}^{j-1,j} \cdot d^{j-1}, \quad F^0 \equiv 0 \quad (6)$$

$$H_{trans}^{n-1,n} = \begin{pmatrix} 1 & 0 \\ \frac{i \cdot (k_{n-1} - k_n)}{2} & 1 \end{pmatrix}, \quad H_{sol}^{n} = \begin{pmatrix} 1 & \frac{i}{k_n}\left(e^{-ik_n L} - 1\right) \\ 0 & e^{-ik_n L} \end{pmatrix}, \quad d^n = \frac{1}{B_{zn}}\begin{pmatrix} I_{\perp n} - e^{-ik_n L} I_n \\ ik_n e^{-ik_n L} I_n \end{pmatrix} \quad (7)$$

Here $k_n = \frac{eB_{zn}}{pc}$, $I_{\perp n} = \int_L B_{\perp n} dz$, $I_n = \int_L B_{\perp n} e^{ik_n z} dz$, and L is the length of solenoidal module. We use a superscript index for the number of solenoid when we deal with matrices and vectors, so that matrix # (n,m) can not be confused with the $(n,m)^{th}$ element of the matrix. For scalar values we use subscript index.

Substituting (2) into (7), we obtain from (4):

$$\xi_j = \left(\{M^j\}_{0,0} \quad \{M^j\}_{0,1}\right) \cdot \begin{pmatrix}\xi_0 \\ \theta_0\end{pmatrix} + \sum_{n=1}^{j-1}\left[\left(\alpha_{n-1} + \frac{\beta_{n-1}}{B_{z\,n-1}}\right) \cdot \left\{\left(\prod_{l=j}^{n+1} H^l\right) F^n\right\}_0\right] + \\ + \left(\alpha_{j-1} + \frac{\beta_{j-1}}{B_{z\,j-1}}\right) \cdot F^j \quad (8)$$

where $\{a\}_{m,n}$ means $(m,n)^{th}$ element of matrix a. Equation (8) solves our problem. It relates θ_0, αs and βs with displacements of beam's trajectory in the CS BPMs measured at different B_zs.

Alignment Algorithm

To study the feasibility of the alignment algorithm the longitudinal filed was changed in the very first solenoidal module only, with the rest of the cooling section field undisturbed. This measurement allows to distinguish between θ_0, α and β in the first solenoid only (solenoid # 0), thus reducing the total number of unknowns to 11. Every trajectory (taken for each value of B_z) gives 10 measurables. Four trajectories were taken for B_{z0} =40 G, 70 G, 105 G and 140G. The field in the rest of the cooling section was 105 G. The obtained overdetermined system of equations (8) must be solved in a least-squares sense [7], and its solution can be found as:

$$r = (A^{*T} \cdot A)^{-1} \cdot A^{*T} \cdot \Xi \quad (9)$$

where $r = (\theta_0 \quad \alpha_0 \quad \beta_0 \quad ...)$ is the vector of unknowns, $\Xi = (\xi_0 \quad \xi_1 \quad \xi_2 \quad ...)$ is the vector of measurables and A is the matrix determined by equation (8); A^{*T} is the transposed complex conjugate of A. The precision of the solution (9) is given by [8]:

$$\sigma_{r_n} = \sqrt{\{(A_\sigma^{*T} \cdot A_\sigma)^{-1}\}_{n,n}}, \quad \{A_\sigma\}_{n,m} = \frac{\{A\}_{n,m}}{\sigma_{\Xi_n}} \quad (9)$$

THE RESULTS OF THE MEASUREMENTS

The result of the measurement and respective fit are presented in Fig. 2. The squares stand for the measured trajectories and the fit of measured data is given by the solid lines.

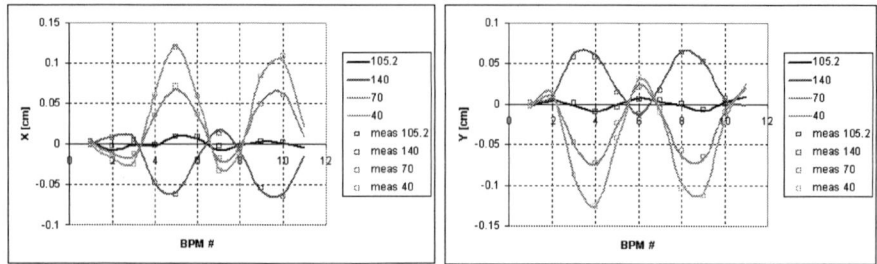

FIGURE 2. The response of beam's trajectory to the change of longitudinal filed in the first solenoidal module (September 13, 2005; the beam trajectories were taken for 4 different longitudinal fields).

The precision of the measurement of the angles (θ_0, α_0 and β_0 / B_{z0}) found from the shown fit is 60 μrad.

CONCLUSION

We have devised and tested the algorithm of beam-based alignment of magnetic field in the REC cooling section. The measurements show that in principle this algorithm allows one to align the beam centroid in the CS with the precision of 60 μrad.

ACKNOWLEDGMENTS

Authors acknowledge the critical contribution of A. Shemyakin to the development of the idea of the beam-based alignment of magnetic field in the CS and are thankful to the entire Electron Cooling group for help and fruitful discussions.

REFERENCES

1. J. A. MacLachlan, ed., "Prospectus for an Electron Cooling System for the Recycler", Fermilab-TM-2061, November 1998.
2. S. Nagaitsev et al., "Antiproton Cooling in the Fermilab Recycler", COOL'05 workshop, Galena, USA, 2005.
3. http://www.Pelletron.com.
4. A. C. Crawford et al., "Field Measurements in the Fermilab Electron Cooling Solenoid Prototype", Fermilab-TM-2224, September 2003.
5. V. Tupikov, "Magnetic Field Measurement and Compensation in Recycler Electron Cooler", COOL'05 workshop, Galena, USA, 2005.
6. S. M. Seletskiy, "Attainment of Electron Beam Suitable for Medium Energy Electron Cooling", Ph.D. Thesis, University of Rochester, 2005.
7. W. H. Press et al., "Numerical Recipes", Cambridge University Press, 1988.
8. L. Lyons, "Statistics for Nuclear and Particle Physicists", Cambridge University Press, 1986.

Full Discharges in Fermilab's Electron Cooler

L. R. Prost and A. Shemyakin

Fermi National Accelerator Laboratory, P.O. Box 500, Batavia, IL 60510, USA

Abstract. Fermilab's 4.3 MeV electron cooler is based on an electrostatic accelerator, which generates a DC electron beam in an energy recovery mode. Effective cooling of the antiprotons in the Recycler requires that the beam remains stable for hours. While short beam interruptions do not deteriorate the performance of the Recycler ring, the beam may provoke full discharges in the accelerator, which significantly affect the duty factor of the machine as well as the reliability of various components. Although cooling of 8 GeV antiprotons has been successfully achieved, full discharges still occur in the current setup. The paper describes factors leading to full discharges and ways to prevent them.

Keywords: Discharge, electron, cooling, Pelletron, conditioning, protection, beam loss.
PACS: 29.17.+w; 41.75.Fr; 52.59.Mv; 52.80.Vp

INTRODUCTION

Recent demonstration of electron cooling of 8 GeV antiprotons in the Recycler ring [1] became possible, in part, due to the stable performance of the electron beam generator [2]. The beam is generated in a Van-De-Graaff type electrostatic accelerator, Pelletron [3], which operates in a so-called energy recovery (or recirculation) scheme [4]. In this case the term means that the electron beam returns to the high voltage terminal, where it is dumped in the collector at the electron energy of 3 keV. The relative value of the current lost in the beam line is normally low, $\sim 10^{-5}$, and the protection circuitry closes the gun if the terminal voltage decreases because of higher losses. After these so-called "recirculation interruptions", the beam current is restored in 20 s. Because the typical cooling time of antiprotons is 30-40 mn, the interruptions only result in a slight decrease of the average electron beam current. However, sometimes the Pelletron terminal discharges in a microsecond to a nearly zero voltage. These full discharges dramatically increase the pressure in the tubes, decrease the tube electric strength, and may damage the Pelletron electronics. The recovery time may take from 5 mn for a simple check of the electronics and a modest pressure improvement, to a several hours long reconditioning of the tubes, to days to repair the electronics. A significant amount of efforts was put into understanding the mechanisms triggering the full discharges and into decreasing their frequency.

TUBE ELECTRIC STRENGTH AND CONDITIONING

Without beam, the discharge frequency depends on the magnitude of the electric field applied to a tube and on the tube history. If the high voltage applied to an acceleration tube is increased, at some level the tube pressure starts jumping. The jumps indicate discharges of single gaps of the tube and do not necessarily result in an immediate full discharge. However, the probability of a full discharge increases very rapidly above this point Below in the text we will refer to this level of the potential gradient along the tube as the tube's electric strength, E_{ts}.

Freshly installed acceleration tubes had tube strengths of 9 kV/cm, which is well below the nominal gradient, 12 kV/cm. The strength is significantly increased by conditioning. In our case, conditioning means a slow increase of the voltage applied only to one of six 60-cm sections forming the Pelletron tubes. The voltage increase results in partial discharges, which improve the tube strength. Full discharges still occur, but because the stored energy at a given voltage gradient is much lower than for all sections together, they are much less harmful. Namely, when conditioning a single 60-cm section, only ~5% of all full discharges deteriorate the tube strength; for two sections, it increases to ~10%; and, in attempts to condition all sections at once, the electric strength dropped for ~50% of all full discharges.

Typically, conditioning of a single section is easy up to an electric field E of 18 kV/cm but further increases are much slower. Usually, we stop the process when the electric field stays above 19 kV/cm for 5 minutes. After such conditioning the terminal voltage U_t can be raised to 5.2 MV ($E = 14.4$ kV/cm) before the first vacuum bursts are observed.

FULL DISCHARGES WITH A BEAM

The simplest explanation of why the frequency of the full discharges increases with the beam current is in the re-distribution of the electric field along the tube caused by beam losses. All tube electrodes are connected to a string of 0.48 GOhm resistors. At the nominal voltage $U_t = 4.32$ MV, the string current is 39 µA, and a beam loss of only several µA can significantly change the electrode potentials. As a result, the electric field in some locations may approach E_{ts}, provoking a full discharge. Hence, to decrease the frequency of the full discharges, one needs to increase the tube strength beyond the working point as much as possible and minimize the beam loss to the tube electrodes.

The first recommendation was followed when the tube length was extended in the time of relocating the Pelletron from a test building [5] to its final location at the Recycler ring, thus decreasing the nominal electric field by 20%. We feel that it made a dramatic effect on the reliability of operation.

There are several scenarios of how electrons can reach the tube electrodes. The most obvious one is when the primary beam touches an electrode because of errors in steering or focusing. Because the beam has sharp edges, scraping it off of any electrode by ~0.1 mm results in a beam loss of 10 µA. Thus, every error of this type induces a full discharge. However, after the very first shifts, the beam position in the acceleration and deceleration tubes is well controlled by use of beam position monitors at the bottom of the tubes and by beam centering in the collector, such that these events are eliminated. Much more serious difficulties arise from primary beam loss in the time of a recirculation interruption and re-distribution of the potential in the deceleration tube caused by secondary electrons escaping the collector.

Beam in the Acceleration Tube

If for some reason the beam is experiencing losses to the ground more than tens of µA, both the gun voltage and the Pelletron potential decrease until the protection circuitry closes the gun. As a result, the beam trajectory and envelope are changing, and the primary electrons can reach the tube electrodes. The resulting full discharges in the acceleration tube were the major problem in the recirculation test [5], where it was overcome by the cut-and-try method.

The capacitance of the terminal to the ground is 300 pF, and the capacitance between the cathode and anode potential surfaces is 800 pF. A beam loss to the ground discharges both these capacitances, but the relative changes in the anode voltage U_a are much stronger than in the Pelletron voltage. As a result, in the acceleration tube, where the beam trajectory is straight, the main effect is the changes of the beam envelope caused by the dropping anode voltage.

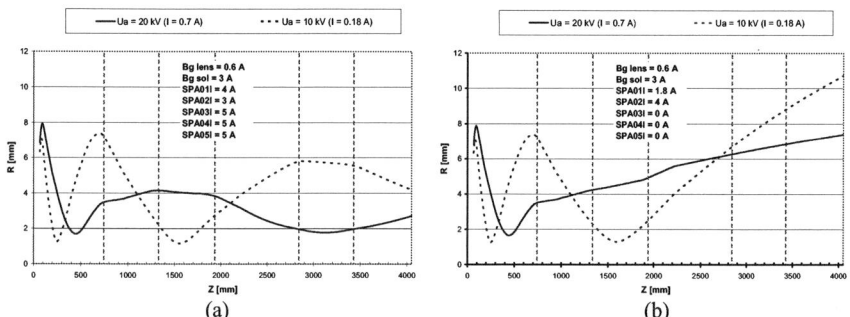

FIGURE 1. Simulations of the beam envelope in the accelerating column at the nominal anode and terminal voltages (solid curve) and at the time of recirculation interruption (dotted curve). (a) Nominal focusing settings; (b) alternate focusing settings. The limiting aperture is 12 mm. Vertical dotted lines represent the approximate locations of the magnetic lenses.

To prevent this type of discharges, the beam envelope in the acceleration tube was simulated with the SAM/SuperSAM/BEAM software package [6,7,8] before beginning the work in the final setup. First, a set of focusing lens currents which allowed the beam transport at nominal parameters (U_t =4.32 MV, beam current I_b = 0.7 A, U_a = 20 kV). was found. Then, the simulation was repeated for lower anode voltages with all other parameters fixed, modeling the beam behavior in the time of interruptions. The maximum beam envelope radii were typically found at U_a ~10 kV (I_b = 0.18 A). Finally focusing settings were adjusted to pass the beam far from the aperture at this intermediate voltage, and the process was repeated.

The final result and one of the intermediate steps are shown in Fig. 1. This work allowed operating with no full discharges in the acceleration tube.

Protection of the Deceleration Tube

We still do not know the optical properties of the entire beam line accurately enough to reliably predict the beam envelope in the deceleration tube. Hence, the procedure described in the previous section is not applicable there. In addition, at present settings, the dispersion in the deceleration tube is comparatively large, ~ 0.5 m, and the beam shifts significantly when the Pelletron voltage decreases.

The first solution, which worked successfully in the recirculation test [5], is the use of a so called crash scraper. The scraper was kept close to the beam so that in the time of an interruption the disturbed electrons are lost at the scraper and not at the tube electrodes. However, for the concept to work, the beam size modifications in the time of an interruption and the dispersion at the location of the crash scraper should be significantly larger than the corresponding values in the deceleration tube, which imposes inconvenient restrictions to the beam line settings. In addition, the cooler is mounted in the tunnel of a synchrotron, the Main Injector, which magnetic fields move the electron beam by several mm in the time of ramping. This motion forces to keep the electron beam far from the crash scraper, decreasing its efficiency.

A dramatic reduction in the full discharge frequency came from an improvement of the protection system, when the time to close the gun was decreased from 1 ms down to 1 µs. The signal to close the gun is sent if the terminal voltage drops by more than 5 kV (Fig. 2b), and neither the beam position nor the beam envelope change significantly. This improvement allows operating with no crash scrapers.

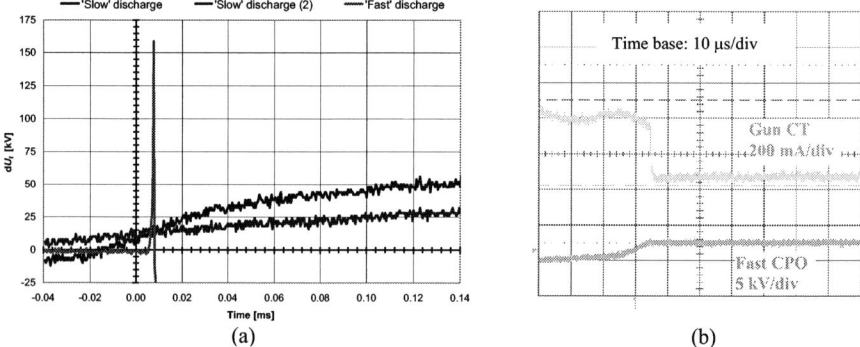

FIGURE 2. (a) Oscillograms of the terminal voltage in the time of a 'fast' (red) and 'slow' (blue and dark blue) full discharges. (b) Oscillograms of a recirculation interruption. The yellow trace is the terminal voltage, and the grey trace is the cathode current.

Fast Discharges

Full discharges provoked by the primary beam are "slow", i.e. have a tens of µs - long pre-history (Fig. 2a). The terminal is discharged first at the rate of 3- 300 V/µs, which corresponds to a current loss of 1 – 100 mA, and only later, when the beam touches the tube electrodes, the main drop of voltage occurs. In contrary, all recently observed full discharges are "fast" (the red curve in Fig. 2a). The voltage derivative is large from the very beginning and corresponds to the current discharging the terminal. At ~10 A, this current is much larger than the beam current. It can only be interpreted as a developing discharge on the vacuum side of the deceleration tube. Typically, after several microseconds the derivative increases by orders of magnitude. It means that the vacuum discharge has already shorted a large portion of the tube, while the terminal potential is almost unchanged. As a result, the voltage between electrodes of the unaffected part reaches the threshold at which the spark gaps mounted at each electrode fire. From this point on, the terminal discharges through the gaps and the electric strength of the remaining part of the tube does not deteriorate.

Frequency of these discharges correlates with changes in the potential distribution along the tubes caused by secondary electrons escaping the collector. The redistribution of potential is estimated by measuring the currents of the tube resistive divider strings at the tube bottom and top (Fig. 3).

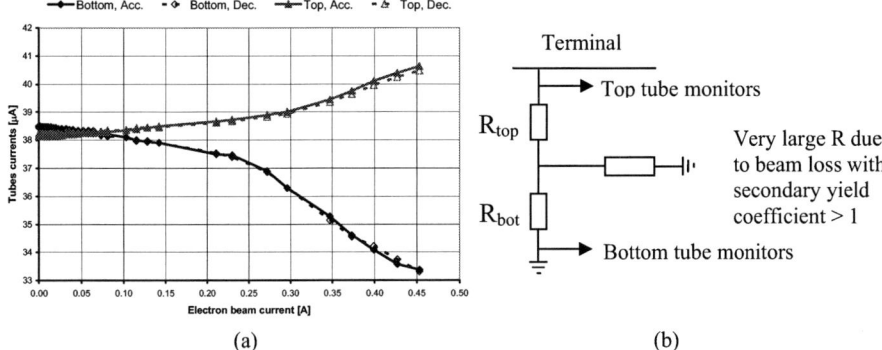

FIGURE 3. (a) Acceleration and deceleration DC tube monitor currents as functions of the electron beam current. (b) Simplified electrical schematic of the divider strings of the accelerating tubes assuming localized beam loss to the tubes.

The bottom current is always decreasing with the beam current, which means that the coefficient of secondary emission from the tube electrodes is above unity. Hence, the electric field between electrodes is increasing at the top portion of the tube and may eventually lead to a full discharge. Assuming electrons are lost to a single electrode, one can estimate from Fig. 3a the position of the electrode in the resistive string using the simple model of Fig. 3b. This position is found to be 1/3 from the bottom and do not change with the beam current.

Though the losses occur low in the tube, the balance of currents still causes the maximum overvoltage to appear at the top of the tube. Moreover, another indication that full discharges originate from the top sections is that the largest degradation of the tube's electric strength is always observed in the first two top sections of the deceleration tube, while its lowest sections and the acceleration tube do not deteriorate.

This model of a direct relation between the discharge frequency and the tube overvoltage gives several guidelines that have proven to be useful:
- keep the tube strength well above the nominal electric field;
- minimize the amount of the current lost at the tube rather than the total loss;
- move the location of the loss as close to the tube bottom as possible.

On the other hand, the model can't numerically predict the tolerable level of the resistive string current change. At $I_b = 0.45$ A, Fig. 3a predicts an overvoltage of 6.5%. Attempts to stay an hour at this current resulted in full discharges, while the Pelletron stayed reliably with no beam with the terminal voltage increased by the same 6.5%. Also, in Fig. 3a the currents recorded on the acceleration and deceleration sides coincide. Because the resistive strings are electrically connected at 5 intermediate voltage levels, the equality of currents indicates that the loss occurs in the middle portion of the tube. Therefore, the overvoltage is identical on both sides, but only the deceleration tube is affected by the discharges. One can speculate that while the frequency of the discharges is determined by overvoltage, they are triggered by secondary particles created by the lost electrons.

Note that recently, we did not observe full discharges while operating at $I_b = 0.1 - 0.2$ A. However, attempts to work at $0.4 - 0.6$ A still resulted in full discharges, and after ~10 discharges we had to repeat conditioning of the top sections (~ 6 hours).

SUMMARY

Decreasing the frequency of the full discharges was an important prerequisite for the electron cooling of 8 GeV antiprotons in the Recycler ring. The decrease was achieved by careful adjustment of the beam envelope in the acceleration tube, by implementation of a fast gun shut-off system, and by working at settings where changes of the resistive strings currents are below 1 µA. Presently, the operation at the beam currents of $0.1 - 0.2$ A does not produce any full discharges in weeks.

ACKNOWLEDGMENTS

The system of the fast gun shut-off was developed by G. Saewert with contributions by J. Crisp and support from J. Simmons. Commissioning and operation of the cooler was performed by the entire ECool group. Authors acknowledge participating in the data taking and analysis presented in this report by our summer student D. Artamonov. Finally, we greatly appreciate the help provided by M. Tiunov and D. Myakishev for the SAM/SuperSAM/BEAM codes.

REFERENCES

1. S. Nagaitsev, *these proceedings*
2. A. Shemyakin, *et al.*, *these proceedings*
3. Pelletrons are manufactured by the National Electrostatics Corporation, www.pelletron.com
4. A. Burov, *et al.*, in: Peter Lucas, Sara Weber (Eds.), Proceedings of the 2001 Particle Accelerator Conference, Chicago, 2001, IEEE, Piscataway, NJ, 2001, p. 2548.
5. A. Shemyakin, *et al.*, Nucl. Instr. and Meth. A, **532** (2004) 403-407
6. B. Fomel, M. Tiunov, V. Yakovlev. SAM - an interactive code for evaluation of electron guns, Preprint Budker INP 96-11, 1996
7. D.G. Myakishev, M.A. Tiunov, V.P.Yakovlev, SUPERSAM - the Code for Calculation of High-Perveance Guns, Program and Abstracts of 6th International Computational Accelerator Physics Conference, Germany, 2000
8. M. Tiunov, BEAM - 2D Code For Simulation Of High Perveance Beam Dynamics In Long Systems, Proceedings of SCHEF'99, Dubna, Russia, 1999, pp. 202-208

Cooling Rates of the USR as Calculated with BETACOOL

C.P. Welsch[1],*, A. Smirnov[2]

[1]*Max-Planck Institute for Nuclear Physics, Heidelberg, Germany*
**Present address: CERN, Geneva, Switzerland*
[2]*JINR, Dubna, Moscow Region, Russia*

Abstract. The ultra-low energy storage ring (USR) will be a multi-purpose facility providing electron-cooled antiprotons in the energy range between 20 keV and 300 keV for both in-ring experiments and effective injection into traps. The low beam energies and high beam quality to be provided by this accelerator will enable new studies of antimatter/matter interactions using in-ring experiments with an internal gas jet target as well as particle traps, which can be efficiently filled using the decelerated and cooled antiproton beam. High luminosity, low emittance and low momentum spread are some of the main characteristics of the electron-cooled antiproton beam that shall be achieved and that the various experiments may take advantage of. The layout of an electron cooler at such low energies is a great challenge and questions like the competition between multiple scattering and electron cooling, the needed cooling power with an installed internal target or the influence of the electron temperature on the cooling time have to be addressed for the first time. In this contribution, the layout of the USR is summarized and results from simulations with the BETACOOL code are presented.

Keywords: Electrostatic storage ring, electron cooling, antiproton physics, USR.
PACS: 29.17.+w , 29.20.Lq

INTRODUCTION

Within the Facility for Low-energy Antiproton and Ion Research (FLAIR) [1, 2] the ultra-low energy storage ring will be used to decelerated antiprotons and ions in a final step from 300 keV down to 20 keV, giving access to both in-ring experiments with the stored ions as well as external trap experiments with particles extracted via slow or fast extraction.

Using only electrostatic elements for the beam optics, i.e. electrostatic quadrupoles and cylinder deflectors, one avoids problems with remanence and hysteresis effects that would occur in a "standard" magnetic storage ring. In addition, costs of the ion optical elements can be reduced and experience gained from other low-energy electrostatic rings be used [3-5].

A clear advantage of the USR in comparison to alternative structures like decelerating RFQs is the availability of a cooled ion beam at all intermediate energies and the possibility to guide low-emittance extracted beams directly to external experimental installations. The basic design parameters of the USR are summarized in table 1 with the corresponding lattice functions at the mentioned working point shown in Fig. 1.

TABLE 1. Summary of the USR design parameters.

Machine Parameter	
Circumference	22.28 m
Gamma transition	3.43
Base pressure	$< 5.10^{-11}$ mbar
Betatron tunes (h/v)	2.29 / 1.08
Chromaticity	-2 / -1.5
# of pbars at 20 keV	10^7
Initial momentum spread	10^{-3}
Antiproton beam	
Base Energy	20 keV
# pbars @ 20 keV	10^7
$\varepsilon_{initial}$ (h/v) [mm mrad]	5 / 5
$\Delta p/p$	10^{-3}

The machine is given access to an extremely wide physics program, ranging from the study of antimatter–matter interactions at lowest energies in in-ring experiments at an internal gas jet target, as well as to a all the experiments employing traps as they can be efficiently filled using the decelerated and cooled antiproton beam. High luminosity, low emittance and low momentum spread are some of the main characteristics of the electron-cooled antiproton beam that shall be achieved and that the various experiments may take advantage of.

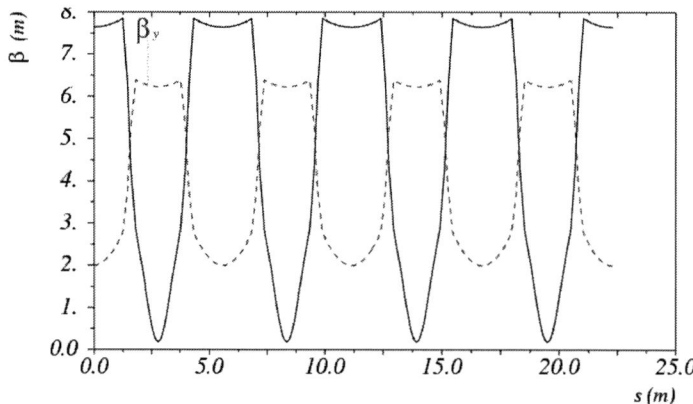

FIGURE 1. USR lattice functions at the working point $Q_x=2.29$, $Q_y=1.08$ as calculated with MAD.

Including beam growth during transport within FLAIR, the USR will be filled with cooled antiprotons from the LSR at 300 keV and an emittance of $\varepsilon_x=\varepsilon_y= 1.5$ mm mrad. The beam will then be decelerated in one step to the final energy of 20 keV and then cooled again.

RESULTS FROM THE SIMULATIONS

The following calculations were done using the *rms dynamics* function of the BETACOOL code. The goal of the algorithm is to calculate the growth rates of the beam's rms parameters. More detailed studies with the code are feasible and will be done in the future to help optimizing the design of the cooler. The general design parameters used in the calculations are shown in the following table 2.

TABLE 2. Overview of the electron cooler parameters.

Design parameter	
Length [m]	0.8
Magnetic field [kG]	0.1
Beta function [m], horizontal / vertical	7.5 / 2
Horizontal dispersion [m]	0.77
Electron beam radius [cm]	0.5
Electron beam current [mA]	0.05
Electron temperature [meV], transverse/longitudinal	4 / 0.5
Field homogeneity in cooler	1×10^{-3}

Neglecting the influence of the internal target, the overall cooling time is defined by the equilibrium between the intra beam scattering (IBS) heating rates and the cooling rates achieved by the electron cooler. BETACOOL allows a momentum spread dependent analysis of the IBS rates for both the horizontal and longitudinal component, the results are shown in Fig. 2.

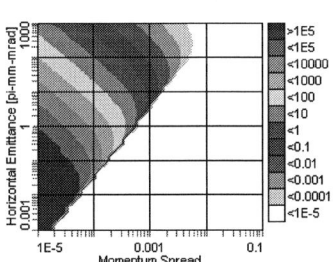

FIGURE 2. Intrabeam scattering heating rates (positive).

These results of IBS beam growth need to be compared with the achievable electron cooling rates under the given specifications. Again, a component split representation can be done and is shown in Fig. 3.

horizontal component   **longitudinal component**

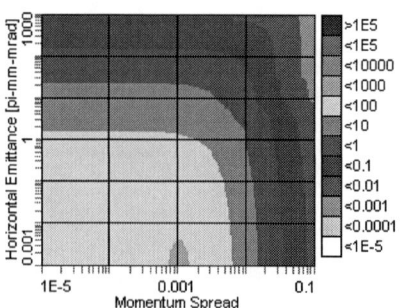 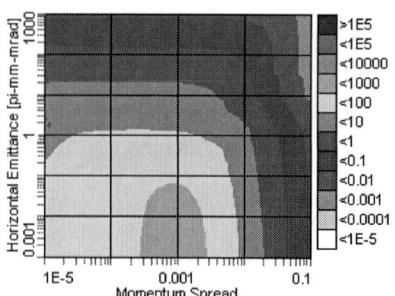

FIGURE 3. Electron cooling rates (negative).

By overlapping the results from the figures 1 and 2 with each of the two components, one gets a direct picture of the beam dynamics during the cooling process as indicated by the black points in Fig. 4.

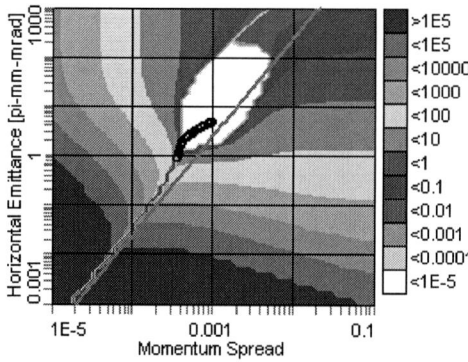

FIGURE 4. Overlapping components of figures 2 and 3. Beam dynamics shown by the black points.

Using these results, one can calculate the horizontal/vertical emittance and the momentum spread of the antiproton beam as a function of time. It can clearly be seen in Fig. 5 that even at lowest energies the cooling times are below one second.

A possible operating scheme for FLAIR could thus foresee the rf bunching in the LSR at e.g. the higher harmonics $h=4$ allowing to fill the USR up to its space charge limit with each of the bunches. Due to the short cooling times, each shot can be decelerated, cooled at 20 keV and be used either for the in-ring experiments or extracted from the machine. With a space charge limit of $\sim 2.5 \times 10^7$ pbars and a cycle time of 5 s, this gives an ideal number of 5×10^6 pbars/s or – assuming (conservative) 90% losses – 5×10^5 pbars/s.

horizontal and vertical emittances   momentum spread

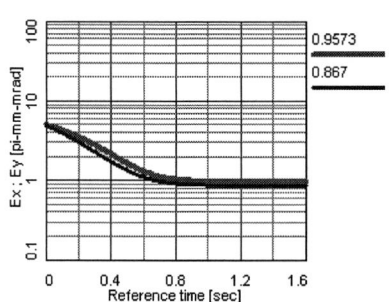
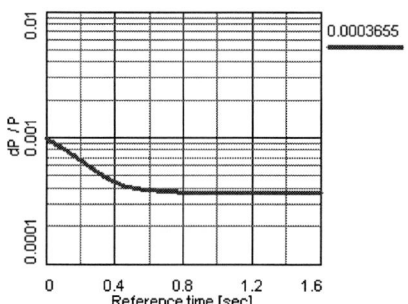

FIGURE 5. Beam dynamics in the USR during the cooling process.

A summary of the results obtained with BETACOOL is shown in the following table 3.

TABLE 3. Summary of the different results obtained with BETACOOL.

Parameters after the cooling process	
Equilibrium emittance [mm mrad] (h / v)	0.96 / 0.87
Equilibrium momentum spread	3.65×10^{-4}
IBS heating rates at equilibrium [s^{-1}] (h / v)	1.7 / 12.2
RestGas heating rates at equilibrium [s^{-1}] (h / v)	12.2 / 1.3
Electron cooling rates at equilibrium [s^{-1}] (h / v)	-13.9 / -13.5
Beam lifetime on RestGas [s]	100

REFERENCES

1. http://www.flair.eu.tt
2. C.P. Welsch, M. Grieser, J. Ullrich, A. Wolf, „FLAIR Project Proposal at GSI", these Proceedings
3. S.P. Møller, "ELISA – an Electrostatic Storage Ring for Atomic Physics", Proc. European Part. Acc. Conf., Stockholm, Schweden (1998)
4. T. Tanabe et al, "An Electrostatic Storage Ring for Atomic and Molecular Science", Nucl. Instr. and Meth. A 482 (2002) 595
5. C.P. Welsch, et al., "Electrostatic Ring as the Central Machine of the Frankfurt Ion Storage Experiments", PRST-AB, 7, 080101 (2004)

MUON COOLING

Recent Innovations in Muon Beam Cooling[*]

Rolland P. Johnson[ɛc], Mohammad Alsharo'a[c], Charles Ankenbrandt[a],
Emanuela Barzi[a], Kevin Beard[d], S. Alex Bogacz[d], Yaroslav Derbenev[d],
Licia Del Frate[a], Ivan Gonin[a], Pierrick M. Hanlet[c], Robert Hartline[c],
Daniel M. Kaplan[b], Moyses Kuchnir[c], Alfred Moretti[a], David Neuffer[a],
Kevin Paul[c], Milorad Popovic[a], Thomas J. Roberts[c], Gennady Romanov[a],
Daniele Turrioni[a], Victor Yarba[a], Katsuya Yonehara[bc],

[a]*Fermi National Accelerator Laboratory, Batavia, Illinois, U.S.A.*
[b]*Illinois Institute of Technology, Chicago, Illinois, U.S.A.*
[c]*Muons, Inc., Batavia, Illinois, U.S.A.*
[d]*Thomas Jefferson National Accelerator Facility, Newport News, Virginia, U.S.A.*

Abstract. Eight new ideas are being developed under SBIR/STTR grants to cool muon beams for colliders, neutrino factories, and muon experiments. Analytical and simulation studies have confirmed that a six-dimensional (6D) cooling channel based on helical magnets surrounding RF cavities filled with dense hydrogen gas can provide effective beam cooling. This helical cooling channel (HCC) has solenoidal, helical dipole, helical quadrupole, and helical sextupole magnetic fields to generate emittance exchange and achieve 6D emittance reduction of over 3 orders of magnitude in a 100 m segment. Four such sequential HCC segments, where the RF frequencies are increased and transverse physical dimensions reduced as the beams become cooler, implies a 6D emittance reduction of almost five orders of magnitude. Two new cooling ideas, Parametric-resonance Ionization Cooling and Reverse Emittance Exchange, then can be employed to reduce transverse emittances to a few mm-mr, which allows high luminosity with fewer muons than previously imagined. We describe these new ideas as well as a new precooling idea based on a HCC with z dependent fields that can be used as MANX, an exceptional 6D cooling demonstration experiment.

Keywords: Muon Beam Cooling, Helical Magnet, Muon Collider
PACS: 29.27.-a, 29.20.-c, 14.60.Ef, 41.85.Lc

INTRODUCTION

The eight ideas discussed below are to improve the cooling of intense muon beams, which could reduce costs of neutrino factories and facilitate designs of high-luminosity muon colliders. These ideas have initiated Small Business Innovation Research (SBIR) and Small business Technology Transfer (STTR) projects that represent a coherent program to develop high intensity, low emittance muon beams.

HIGH-PRESSURE RF CAVITIES

A gaseous energy absorber enables an entirely new technology to generate high accelerating gradients for muons by using the high-pressure region of the Paschen curve [1]. This idea of filling RF cavities with gas is new for particle accelerators and is only possible for muons because they do not scatter as do strongly interacting protons or shower as do less-massive electrons. Measurements by Muons, Inc. and IIT at Fermilab have demonstrated that hydrogen gas suppresses RF breakdown very well, about a factor six better than helium at the same temperature and pressure. Consequently, much more gradient is possible in a hydrogen-filled RF cavity than is needed to overcome the ionization energy loss, provided one can supply the required RF power. Hydrogen is also twice as good as helium in ionization cooling effectiveness, viscosity, and heat capacity. Future research efforts will include tests of pressurized RF Cavities in magnetic fields and high radiation environments and the use of new cavity construction materials [2], including beryllium RF windows for improved cavity performance [3].

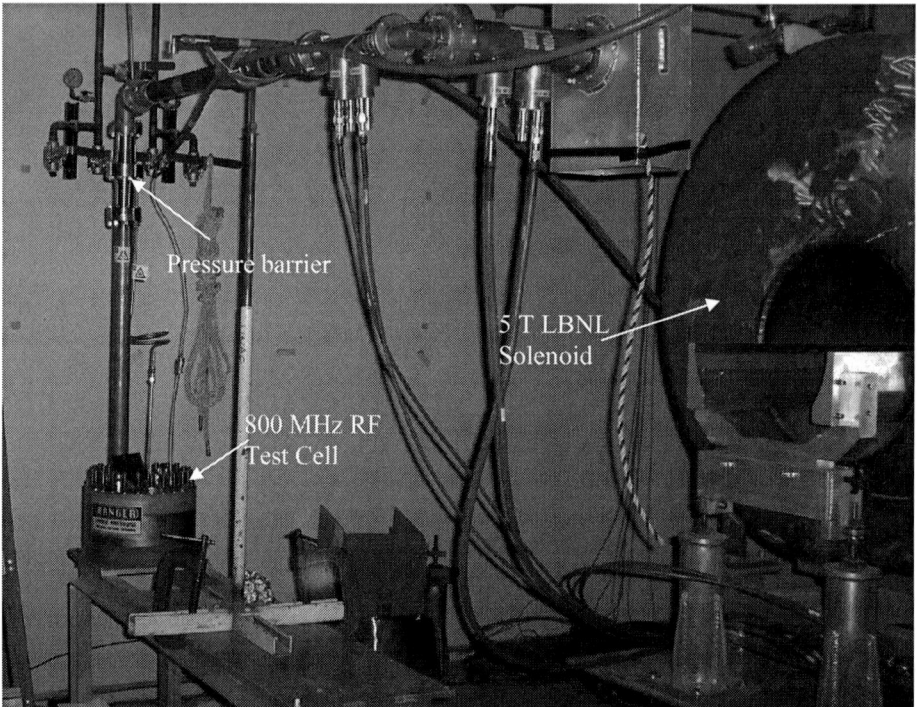

FIGURE 1. MuCool Test Area view of 800 MHz hydrogen-filled high-pressure test cell. The cell is capable of testing RF breakdown of gases up to 2000 PSI and down to 77 K. Future tests include operation of the test cell in the 5 T solenoidal magnet shown at the right of the picture and in the extracted 400 MeV H- Linac beam, which should be available in 2006.

COMBINED CAPTURE, PHASE ROTATION, AND COOLING

High-pressure RF cavities near the pion production target can be used to simultaneously capture, bunch rotate, and cool the muon beam as it emerges from the decaying pions [4]. We have started an R&D effort to develop RF cavities that will operate in the extreme conditions near a production target and an effort to simulate the simultaneous capture, phase rotation, and cooling of muons as they are created from pion decay.

EMITTANCE EXCHANGE IN A CONTINUOUS ABSORBER

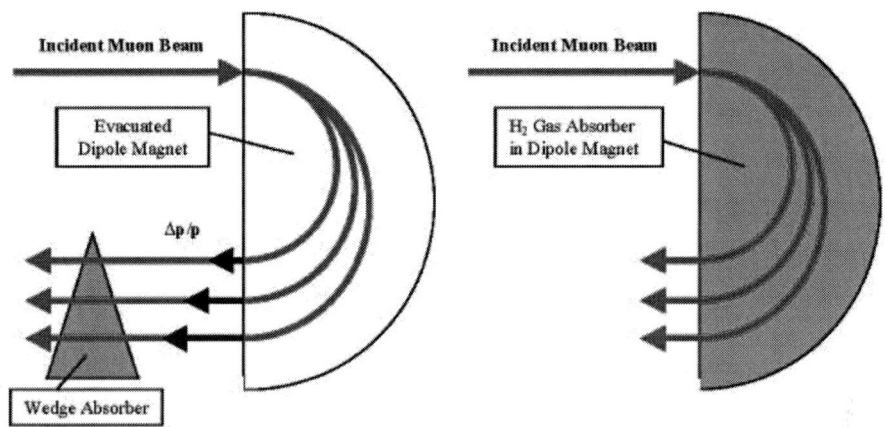

FIGURES 2 (LEFT) AND 3 (RIGHT). Comparison of the usual method of emittance exchange based on a wedge absorber (Left) with the method advocated in this paper using a continuous energy absorber as is possible with pressurized RF cavities (Right).

Ionization Cooling is only transverse. To get 6D cooling, emittance exchange between transverse and longitudinal coordinates is needed. Figure 2 shows the usual mechanism based on a wedge absorber to achieve reduction in momentum spread at the expense of increasing transverse emittance. In figure 3, positive dispersion gives higher energy muons larger energy loss due to their longer path length in a low-Z absorber.

HELICAL COOLING CHANNELS

The idea of pressurized RF cavities led to the concept of a cooling channel filled with a continuous homogeneous absorber. Such a cooling channel provides longitudinal ionization cooling by exploiting the path length (and therefore energy loss) correlation with momentum in a magnetic channel with positive dispersion. Using this approach in a helical cooling channel creates emittance exchange and excellent 6D muon beam cooling.

The mathematical treatment of the original helical cooling channel exploits the fact that the magnitudes of the fields are constant and only the directions of the helical fields change with a constant frequency. By transforming to the rotating frame, a z or time independent Hamiltonian can be formed, which leads to an elegant treatment of the properties of the HCC [5]. The two radial forces from the cross products of the muon longitudinal and transverse momenta with the helical dipole and solenoidal fields, respectively, oppose each other in the stable solution. The solution has unusually large acceptance and exactly the required dispersion characteristics. Figure 4 shows the simulation of a series of four such helical channels with a factor of 50,000 6D emittance reduction to demonstrate the idea that the dimensions of the channel and the RF frequency can be reduced as the beam is cooled to improve the cooling efficiency. A program to develop HCC magnets has started at Fermilab [6].

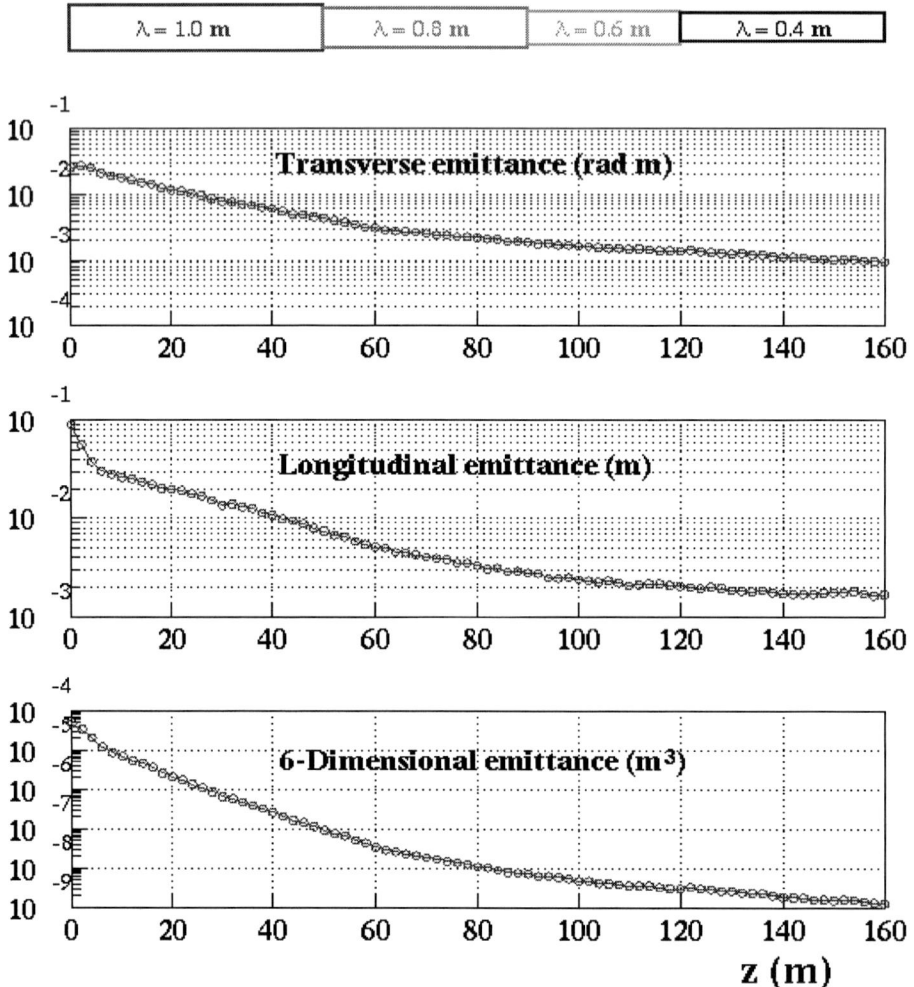

FIGURE 4. Emittance evolution for a series of four helical cooling channels with shorter and shorter helix periods and higher magnetic fields as calculated by the G4Beamline simulation program. Hydrogen-filled 200 MHz RF cavities embedded in solenoidal and helical dipole, quadrupole, and sextupole fields provide continuous 6-dimensional cooling. The simulation uses Maxwellian fields for the magnets. The practical design of devices that can provide these fields is under study.

MOMENTUM DEPENDENT HCC USES: PRECOOLER

The solution for a particle in a HCC with period $\lambda = 2\pi/k$ on an equilibrium orbit with radius a and momentum p can be written:

$$p(a) = \frac{\sqrt{1+\kappa^2}}{k}[B - \frac{1+\kappa^2}{\kappa}b(\kappa)], \quad (1)$$

where $\kappa = ka = p_\perp/p_z$ is the arctangent of the helix pitch angle at the periodic orbit, B is the magnitude of the solenoidal field and b is the magnitude of the helical dipole field.

The new idea is that this relationship can be exploited in cases where the beam momentum is not constant to provide the required dispersion and orbit for effective cooling. We can use equation (1) to change fields and helix parameters to maintain the orbit and dispersion properties. The HCC concept can thus be extended to have momentum dependent magnetic field strengths for several new applications.

This conceptual change that allowed it to become a z or momentum dependent device has made the HCC a potential work-horse for different components of the muon beam cooling channel. For example, by filling the HCC with hydrogen or helium, the beam can be decelerated and cooled by ionization energy loss over more than 100 MeV/c, then reaccelerated by a series of RF cavities (pressurized or conventional). The HCC magnet parameters must be varied to match the momentum of the beam as it slows down, according to equation (1). Filled with gas, the HCC with z dependent magnetic field parameters can be used as a transition between HCC sections with different RF frequencies and fields.

Filled with liquid, the HCC with momentum dependent field parameters followed by RF cavities can be a 6D precooler, a 6D demonstration experiment, or an alternative to the original gas-filled HCC (where the momentum is kept almost constant). Two examples discussed below are a 6D precooler and a 6D Muon Collider And Neutrino Factory muon beam cooling demonstration eXperiment, MANX [7], which is being designed to follow MICE [8,9].

Figure 5 shows a G4BL simulation of a 5 m HCC precooler that follows a 40 m HCC decay channel. Figure 6 shows the normalized transverse (the average radial and azimuthal), longitudinal, and 6D emittances plotted as a function of the distance down the channel. The settings of the helical dipole, helical quadrupole and the solenoidal magnets are chosen to give equal cooling decrements in all three planes. The combined 6D cooling factor is about 5.4, corresponding to 1.7 coming from each of the three planes. The improved performance of this precooler simulation relative to previous non-HCC cooling channel designs comes from the effectiveness of HCC, from the greater path length in the hydrogen absorber ($5m/\cos(45^0) = 7.07m$), and from less heating by high-Z windows. The previous designs have aluminum windows on each side of many short liquid hydrogen containment vessels for separation from evacuated RF cavities, while the two thin windows needed for the precooler have not yet been included in the simulation.

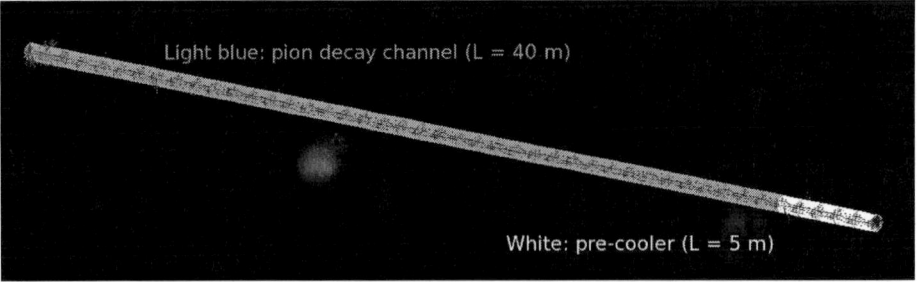
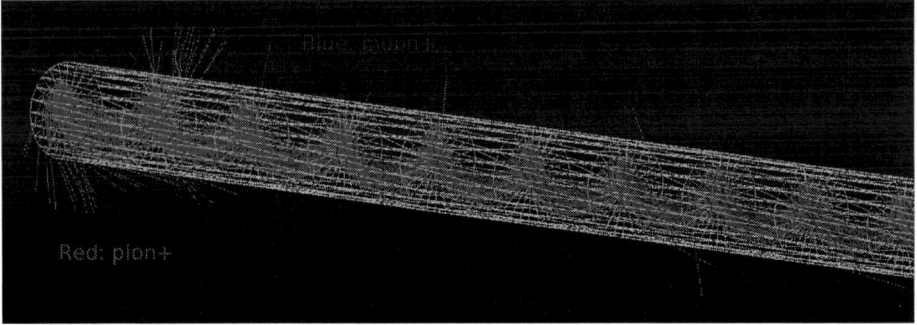
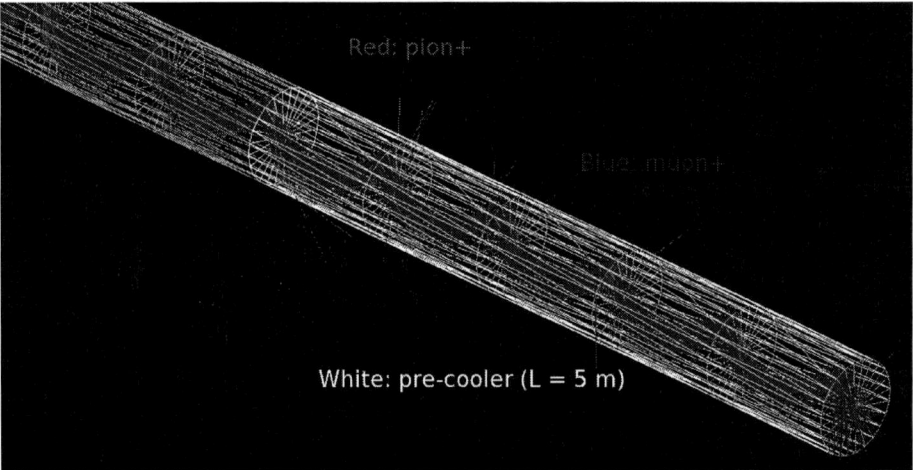

FIGURE 5. Examples of a momentum dependent HCC used as a decay region and as a precooler. The top view shows an overview of a 40 m HCC used as a pion decay channel followed by a 5 m liquid hydrogen filled HCC used to precool the muon beam. Pions are shown in red and muons in blue. The center view shows the pion production region. The bottom view shows the liquid hydrogen precooler region of the channel. The emittance evolution of the beam in the precooler is shown in the next figure.

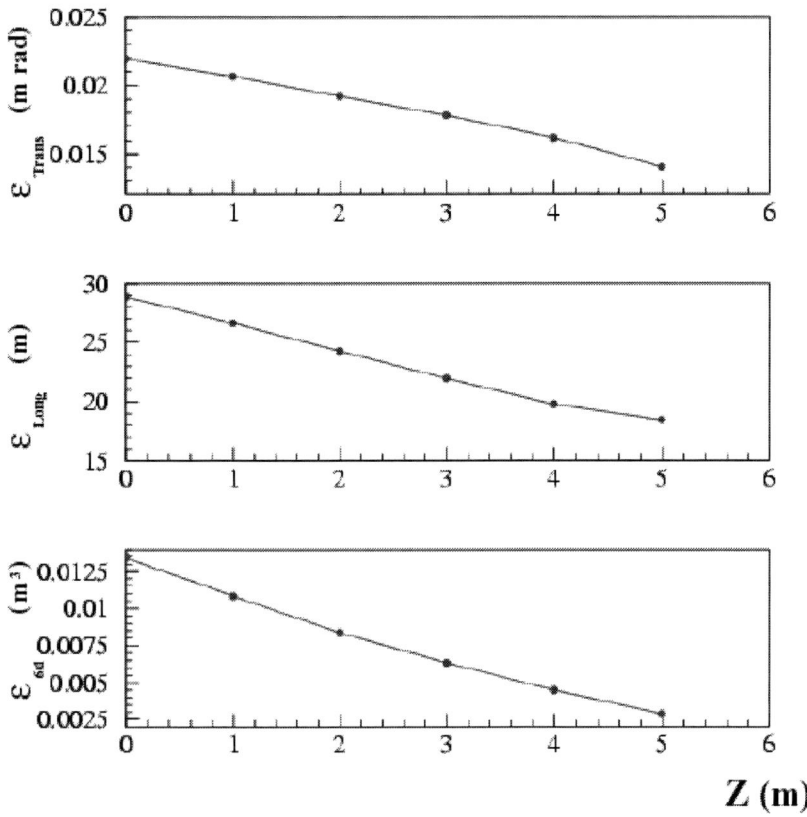

FIGURE 6. The transverse, longitudinal, and 6D emittance evolution of the beam in the precooler.

MANX 6-DIMENSIONAL COOLING DEMONSTRATION

The excellent cooling of the previous precooler design has inspired a new idea for a MANX 6D demonstration experiment. Namely, we can use a liquid-filled HCC without RF to demonstrate emittance exchange in a prototype precooler. Recent work reported at this workshop [10] has been to optimize the momentum dependent HCC design to work with liquid helium (LHe) instead of liquid hydrogen (LH_2) for safety reasons and to relax the beam momentum and HCC magnetic field parameters to allow easier construction using available technology. Figure 7 shows the emittance evolution in the new design based on LHe which has a maximum field of 5.5 T, 4.5 m length, and inner diameter of 70 cm.

The factor of 4.2 in 6D emittance reduction shown in figure 7 is only slightly worse than the factor of 5.5 in the LH_2 case. Although each transverse equilibrium emittance should be twice as large for LHe compared to LH_2 based on their energy loss and radiation lengths, the cooling factors are very similar when far from equilibrium.

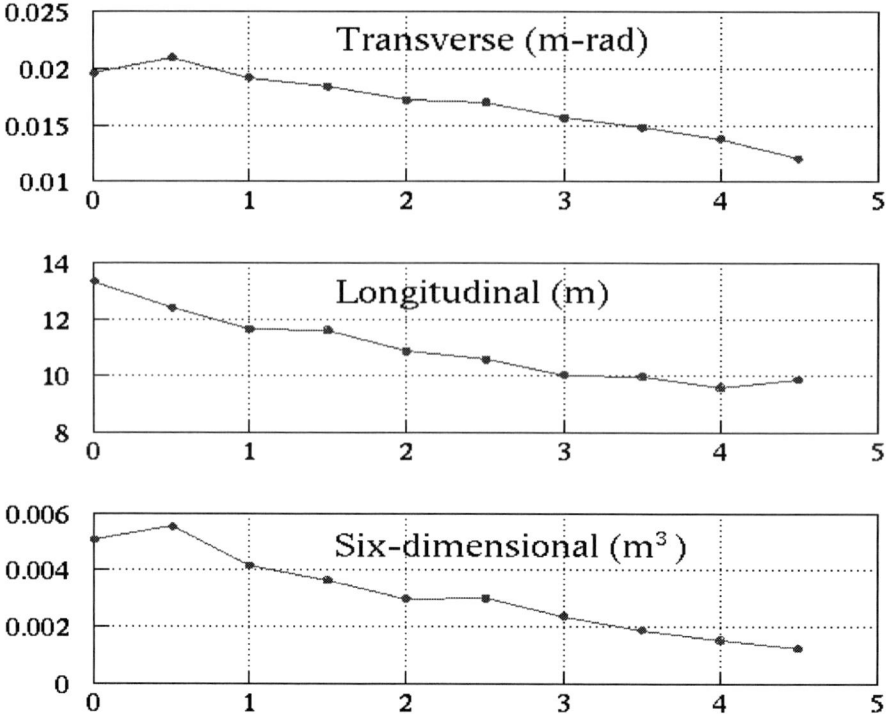

FIGURE 7. The emittance evolution of the beam in a liquid helium HCC designed to have reduced field requirements by using lower initial momentum, longer period, and smaller pitch angle.

PARAMETRIC-RESONANCE IONIZATION COOLING

To reduce the number of muons required for a high-luminosity muon collider, additional cooling is required beyond what can be accomplished with a HCC even using 20 T magnets. The large number of muons that have been assumed in previous muon collider plans have been an impediment to acceptance for several reasons: the proton driver and required targetry to produce the required number of muons is difficult, the decay of muons in the collider produces a large electron background in the detectors, and the neutrinos from the same decays can be a site-boundary radiation problem. Additionally, smaller muon beam emittances can mean cost savings in muon acceleration and can allow smaller β^* at the interaction regions.

Accelerator physicists are used to ½-integer parametric resonances as a way to extract beam from a synchrotron. The normal elliptical motion of particles in a beam as shown on the left of figure 8 is transformed into hyperbolic motion by operating on the ½-integer resonance as shown on the right of the figure. Thus particles move to larger and larger x as they circulate in a synchrotron until they pass an electrostatic septum and are extracted from the beam. In the scheme that is being developed now for muon cooling beyond the HCC, the perturbations that drive the ½-integer resonance have a phase that causes the particles to stream the other way such that they go to smaller and smaller x and larger and larger x'. Ionization cooling is then employed to constrain or shrink the x' dimension. Thus the beam size is reduced by the action of the resonance and the angular divergence is reduced by ionization cooling. An essential aspect of this method of cooling, which we anticipate can reduce each transverse dimension another factor of 10, is the control of the aberrations that cause detuning and loss of the resonance condition. Some clever techniques to control the chromatic and spherical aberrations have been developed analytically and are now being simulated.

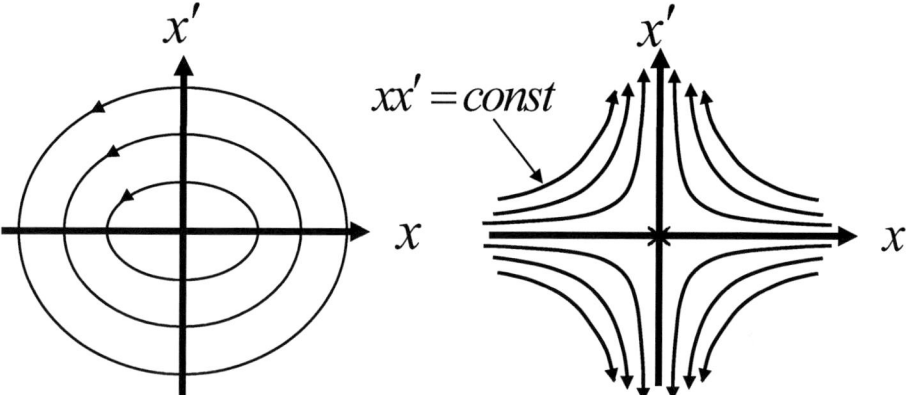

FIGURE 8. Comparison of particle motion at periodic locations along the beam trajectory in transverse phase space for: LEFT ordinary oscillations and RIGHT hyperbolic motion induced by perturbations at a harmonic of the betatron frequency.

REVERSE EMITTANCE EXCHANGE

Reverse emittance exchange is another technique to reduce the transverse emittances of a muon beam beyond the limits of a HCC in order to increase muon collider luminosity. For example, in the case of a collider of 2.5 TeV/c on 2.5 TeV/c, the momentum spread $\Delta p/p$ will decrease by a factor of over 10,000 from when it was cooled so that the bunch length will be much shorter than the β^* of the interaction region.

While longitudinal cooling is essential to reduce the physical dimensions and increase the RF frequency of the cooling and low energy accelerating systems, it will be more than is required for the collider. The new idea is to use wedge absorbers to exchange this unusable longitudinal emittance with the transverse emittance as shown in figure 9. The first analytical treatments of reverse emittance exchange and parametric-resonance ionization cooling have been presented at this conference [11] and have shown that it is possible and perhaps necessary that that these two techniques will be done in the same cooling channel segment.

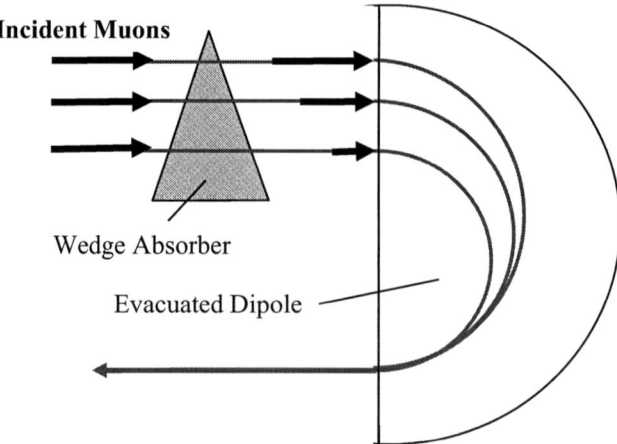

FIGURE 9. Conceptual picture of reverse emittance exchange. A wide beam with narrow momentum spread is transformed using a wedge absorber into a narrow beam with wider momentum spread. The corresponding decrease in transverse emittance leads to higher luminosity while the corresponding increase in longitudinal emittance has little effect on the luminosity as long as the bunch length remains smaller than the focal length of the interaction region.

CONCLUSIONS

The most exciting development in the last few years has been the realization that there are many new ideas for muon beam cooling. The eight that we have described above may be only the first of many to come that will make muon accelerators and storage rings realistic options for physics research in the near future.

REFERENCES

[*] Supported by DOE SBIR/STTR grants DE-FG02-02ER86145, 03ER83722, 04ER84015, 04ER86191, and 04ER84016
[ϵ] Corresponding author, rol@muonsinc.com
[1] R. P. Johnson et al. LINAC2004, http://www.muonsinc.com/TU203.pdf
[2] P. M. Hanlet et al., PAC05, http://snsapp1.sns.ornl.gov/pac05/TPPP054/TPPP054.PDF
[3] M. Alsharo'a et al., PAC05, http://snsapp1.sns.ornl.gov/pac05/TPPP053/TPPP053.PDF
[4] K. Paul et al., PAC05, http://snsapp1.sns.ornl.gov/pac05/TPPP055/TPPP055.PDF
[5] Y. Derbenev and R. P. Johnson, *Phys.Rev.STAB* **8**, 041002, (2005)
[6] L. Del Frate et al. PAC05, http://snsapp1.sns.ornl.gov/pac05/TPPP050/TPPP050.PDF
[7] T. J. Roberts et al., PAC05, http://snsapp1.sns.ornl.gov/pac05/TPPP056/TPPP056.PDF
[8] D. M. Kaplan, this workshop
[9] M. A. Cummings et al., this workshop
[10] K. Yonehara, et al., this workshop
[11] Y. Derbenev, this workshop

6D Cooling of a Circulating Muon Beam

A Garren[a], D. Cline[a], S. Kahn[b], H. Kirk[c], and F. Mills[d]

[a] University of California,, Los Angeles, California
[b] Muons Inc., Batavia, Illinois
[c], Brookhaven National Laboratory, Upton,, New York
[d] Fermilab, Batavia, Illinois

Abstract. We discuss the conceptual design of a system to reduce the 6D emittance of a circulating muon beam. This system utilizes ionization cooling to achieve 6D phase reduction of the beam. Our design is based on a hydrogen gas filled ring which incorporates optics consisting of weak-focusing dipoles and 200 MHz rf cavities which restore the ionization energy loss due to the muons traversing the hydrogen gas.

INTRODUCTION

Muon beams can be cooled by ionization cooling, in which the muons lose momentum traversing a material absorber, and regain the longitudinal momentum component traversing an RF cavity. Focusing magnets, such as solenoids or quadrupoles, contain the beam. The system can be closed into a ring by bending magnets or by tilting solenoids. A ring system provides 6D cooling, while a straight one only cools transversely. We have been investigating cooling rings with lattices composed of various combinations of dipoles and quadrupoles. The first rings designed included short liquid hydrogen (LiH$_2$) absorbers, but subsequently we have have investigated rings in which the energy absorbtion occurs in compressed hydrogen gas that fills the entire beam enclosure. We will briefly discuss some LiH$_2$ based rings, then turn to the gas-filled rings. and lastly discuss a small demonstration ring intended to validate the principle of 6D muon cooling.

DIPOLE-QUADRUPOLE RINGS WITH Li H$_2$ ABSORBERS

Before studying gas-filled rings, many rings were designed for cooling with short LiH$_2$ absorbers. Their performance was simulated with ICOOL The magnet lattices used various arrangements of dipoles and quadrupoles, or of dipoles alone with edge focusing. The main lattice objectives were to have low beta-function values in the absorbers to reduce heating, to minimize the maximum beta-function values elsewhere to obtain large acceptances and to reduce the cell lengths to increase cooling efficiency. Figure 1 shows an example of this type or ring cooler.

1 m drift available for rf
Low □ (25 cm) at absorber
Combined function dipole simulated
Dispersion only at absorber
(allows for matching straight sections)
Cell tune ~ 3/4
Beam momentum 250 MeV/c
25 cm LiH$_2$ wedges
Wedge angle 20^0
rf frequency 201.25 MHz
E_{max} = 16 MV/m
Transmission 50%
Total Merit = *Transmission* x $(e_x e_y e_z)_{initial}/(e_x e_y e_z)_{final}$ = 15

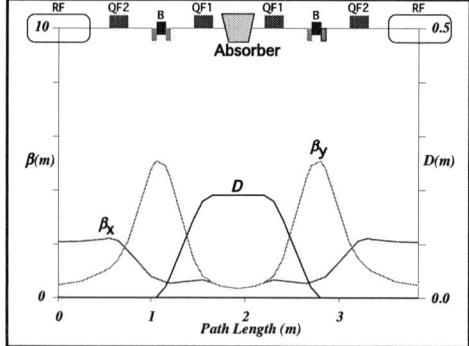

FIGURE 1. Cell of an 4-cell quadrupole-dipole ring with LiH$_2$ absorber in the center. The 'ears' on the dipoles (**B**) indicate edge focusing.

DIPOLE-ONLY GAS-FILLED RINGS

We have adopted the following approach to muon beam cooling in gas-filled rings:

Rings filled with high-pressure hydrogen gas for energy absorption.
 efficient cooling (absorber everywhere)
 RF breakdown voltage increased

Dipole-only, scaling lattices
 compact rings
 lower betamax values, high acceptances

Two types of scaling lattice rings have been investigated: Alternating Gradient Rings and Zero-gradient sector dipole **rings. Examples are shown in Figures 2 and 3.**

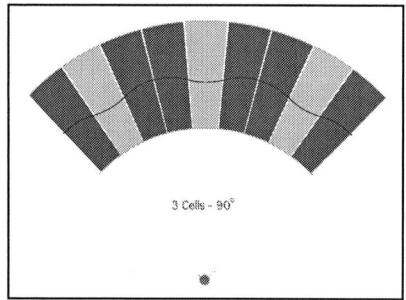

Lattice consists of alternating vertically defocusing and horizontally focusing magnets

No drift spaces between dipoles.

B_o = 2.6T and P_o = 250 MeV/c

Total merit with decay = 120

FIGURE 2. Three cells of FFAG Alternating Gradient 12 cell Ring.

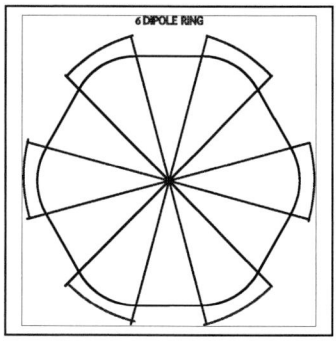

Key parameters at r = 60 cm
β_x = 53 to 72 cm ; β_y = 60 to 64 cm
Dispersion = 60 to 64 cm
Circumference = 3.91 m

FIGURE 3. Schematic diagram of 6 cell weak focusing high-field ring.

RINGS TO DEMONSTRATE COOLING

We have proceeded to make a design scenario to demonstrate 6D muon cooling with a small zero-gradient dipole ring, In order to make this demonstration economically feasible we have reduced the cooling goals to correspond to a merit factor of at least 10, and set the following design parameters:

1.8T conventional magnets, 200 MHz RF cavities, 40 Atmosphere compressed H_2
For each harmonic, the beam momentum that corresponds to the field is calculated and the cooling performance evaluated. Comparisons are made for different harmonics. It was found that 4 or 6 dipole rings and harmonic number 3 were optimum.

Harmonic 2
  Circumference = 1.76 m, P_0 = 77 MeV/c
Harmonic 3
  Circumference = 3.76 m, P_0 = 165 MeV/c
Harmonic 4
  Circumference = 5.45 m, P_0 = 240 MeV/c

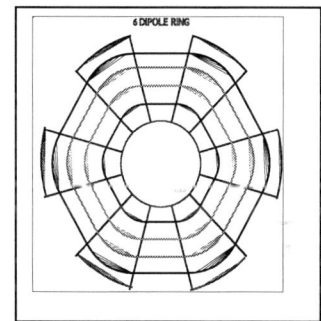

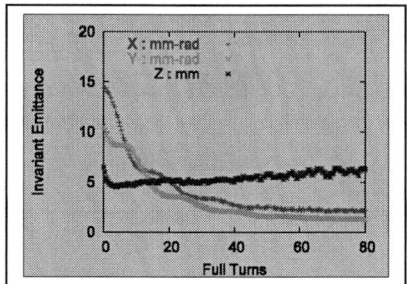

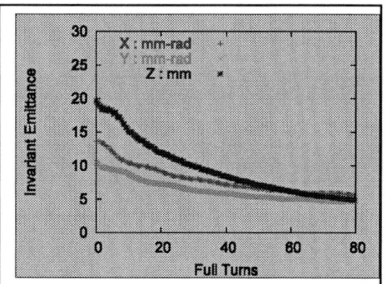

FIGURE 4. 6-dipole ring: schematic, parameters and performance for harmonic numbers 2 and 3.

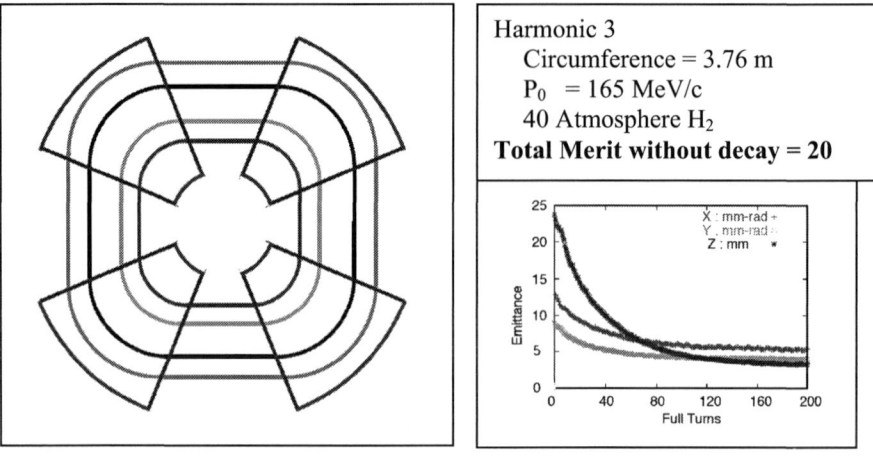

FIGURE 5. 4-dipole ring: Schematic, parameters and performance for harmonic number 3.

SMALL MUON RING FOR A COOLING DEMONSTRATION

A preliminary plan has been made to demonstrate muon cooling in a gas-filled ring. The conceptual design and parameters are shown in Figure 8.

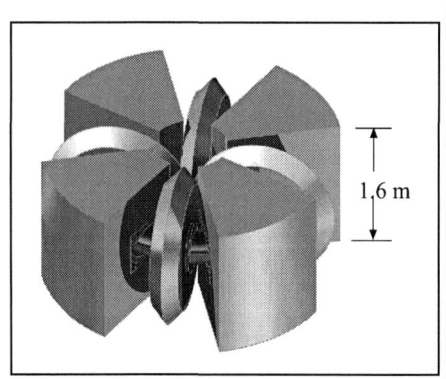

Parameter	Value
Dipole Field	1.8 T
Number of Cells	4
Reference Momentum	172.12 MeV/c
Ring Circumference	3.81 m
X Aperture	±20 cm
Y Aperture	±10 cm
P_z Acceptance	±10 MeV/c
Minimum β_X	38 cm
Maximum β_X	92 cm
Minimum β_Y	54 cm
Maximum β_Y	66 cm
Hydrogen Gas Pressure	40 Atm @ 300½K
RF Gradient	10 MV/m
RF Frequency	201.25 MHz
Total RF Length	1.2 m
Total Orbit Turns	100

FIGURE 6. Proposed gas-filled muon storage ring for cooling demonstration.

In order to model the fields more realistically, these have been calculated with the TOSCA code and tracking and performances obtained using the field maps generated.

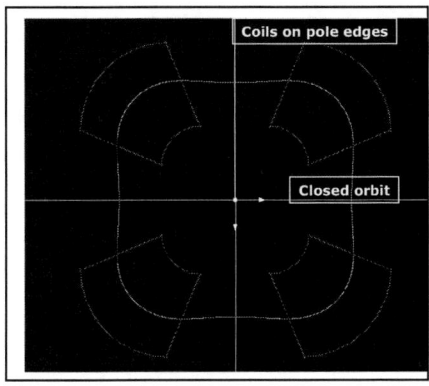 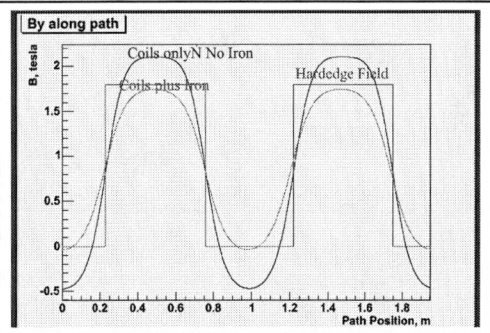

FIGURE 7. Azimuthal fields generated by TOSCA with three models.

Not surprisingly, cooling is best with the hardedge model. Studies have been made and will continue to shape the poles and coil configuration to maximize performance.

SUMMARY

- Quadrupole-dipole rings with LiH_2 absorbers require high fields for cooling.
- Dipole-only rings are more compact and thus have better cooling performance.
- Rings filled with compressed hydrogen gas with scaling lattices are promising.
- A small weak focusing ring system has been designed to demonstrate feasibility of 6-D muon ionization cooling.

ACKNOWLEDGEMENTS

The authors wish to thank S. Berg, R. Fernow, Y. Fukui, R. Palmer and D. Summers for their contributions. They also wish to acknowledge the work of R. Johnson et al on the effects of compressed H_2 gas on RF breakdown voltage. The lattices were designed using the SYNCH program, the cooling performances were simulated with ICOOL, and realistic fields generated with TOSCA.

This work was performed with the support of the US DOE under Contact No. DE-AC02-98CH10886 and Grant No. DE-FG02-92ER40695.

REFERENCES

[1] H.G. Kirk et. al., *A Compact Ring for the 6D Cooling of a Muon Beam*, Proc. of the 2005 PAC, TPPP048 (2005)
[2] A. Garren et al., *SYNCH: A Program for Design and Analysis of Synchrotrons and Beamlines*, Fermilab-Pub-94/013.
[3] R.C. Fernow, *ICOOL: A Simulation Code for Ionization Cooling of Muon Beam*, Muon Collider Note 290 (2003)
[4] The Tosca program is described in the Opera-3D Reference Manual VF-05-03-B2 from Vector Fields.

Parametric-Resonance Ionization Cooling and Reverse Emittance Exchange for Muon Colliders

Yaroslav Derbenev[a] and Rolland P. Johnson[b]

[a]*Thomas Jefferson National Accelerator Facility, Newport News, Virginia, U.S.A.*
[b]*Muons, Inc., Batavia, Illinois, U.S.A.*

Abstract. Two new ideas are being developed to reduce the transverse emittance of muon beams in order to increase the luminosity of muon colliders. The first idea involves driving a ½-integer parametric resonance in a beam line or ring such that particle motion becomes hyperbolic, where $xx'=constant$. With the proper phase of the resonance driving term, particles move to larger and larger x' and smaller and smaller x at the position of a thin wedge absorber. The usual mechanism of ionization cooling reduces or constrains the excursion in x' while the dynamics of the resonance reduces the spread of x. The second idea takes advantage of the large reduction of relative momentum spread with increasing momentum in going from a few hundred MeV/c where the beam is cooled to a few TeV/c for an energy frontier collider. In this case we can use thin wedge absorbers to exchange the transverse and longitudinal emittances to make the transverse emittance smaller. These two ideas depend on careful control of the lattice functions and corrections for chromatic and spherical aberrations. We discuss these ideas and their potential luminosity implications considering the limitations of aberration corrections and of space charge effects.

Keywords: Muon Beam Cooling, Muon Collider
PACS: 29.27.-a, 29.20.-c, 14.60.Ef, 41.85.Lc

INTRODUCTION

Muon collider luminosity depends on the number of muons in the storage ring and on the transverse size of the beams in collisions. Ionization cooling (IC) as it is presently envisioned [1] will not cool transverse beam sizes sufficiently well to provide adequate luminosity without large muon intensities. A new idea to combine ½-integer parametric resonances and IC (PIC) [2] in a linear focusing channel has been proposed that will lead to much smaller transverse beam emittances so that high luminosity in a muon collider can be achieved with fewer muons. A second new idea follows from the observation that the normalized longitudinal emittance after 6D IC is very small compared to what is needed for an advanced interaction region design. Thus, there is room for further reduction of transverse emittances by entropy exchange with the longitudinal degree of freedom. To realize this potential, an absorber-based method of fast reverse emittance exchange (REMEX) was recently proposed [3]. As will be discussed, REMEX follows PIC in order to minimize the heating effect of energy straggling on transverse emittance.

We would like to emphasize that the strong reduction of transverse emittance has at least eight very beneficial consequences for a muon collider. The reduction of the required muon current for a given luminosity diminishes several problems: 1) the radiation levels due to the high energy neutrinos from muon beams circulating and decaying in the collider that interact in the dirt at the site boundary; 2) the electrons from the same decays that cause background in the experimental detectors; 3) the difficulty in creating a proton driver that can produce enough protons to create the muons in the first place; 4) the proton target heat deposition and radiation levels; and 5) the beam loading and wake field effects in the accelerating RF cavities. Smaller emittance also: 6) allows smaller, higher-frequency RF cavities with higher gradient for acceleration; 7) makes beam transport easier; and 8) allows stronger focusing at the interaction point since that is limited by the beam extension in the quadrupole magnets of the low beta insertion.

In this report we describe basic principles and potentials of PIC and REMEX and address the main constraints associated with tune spreads and energy straggling in the muon beam.

BASIC PRINCIPLES OF PIC

Beam Focusing Using a Parametric Resonance

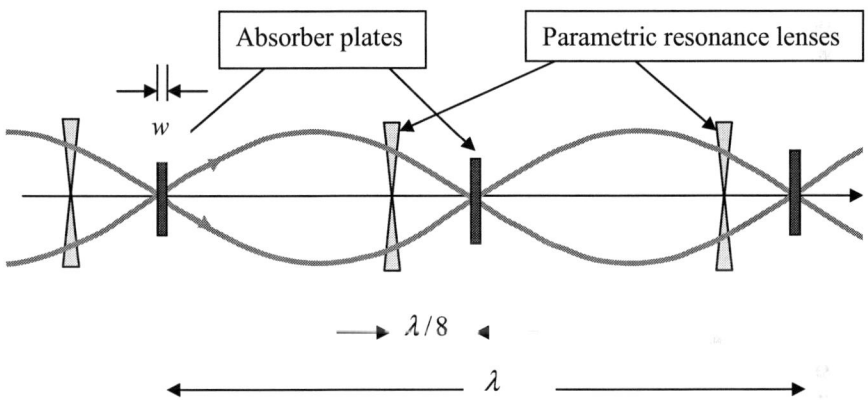

FIGURE 1. Conceptual diagram of a beam cooling channel in which hyperbolic trajectories are generated in transverse phase space by perturbing the beam at the betatron frequency, a parameter of the beam oscillatory behavior. Neither the focusing magnets that generate the betatron oscillations nor the RF cavities that replace the energy lost in the absorbers are shown in the diagram.

We assume initially that the tune spread for a beam in a focusing channel is zero, leaving tuning demands and compensation for aberrations to be discussed later. Figure 1 illustrates an arrangement for resonance beam focusing along a cooling channel. Weak lenses installed every half oscillation period drive a half-integer parametric resonance that creates a hyperbolic beam evolution at the absorber plates:

$$\begin{pmatrix} x \\ x' \end{pmatrix}_{n+1} = -\begin{pmatrix} k^{-1} & 0 \\ 0 & k \end{pmatrix}\begin{pmatrix} x \\ x' \end{pmatrix}_n ; \quad k = \exp(\Lambda_d \lambda / 2) ; \quad 0 < \Lambda_d \lambda \ll 1$$

When the resonance instability evolves, the amplitudes grow while the phase distribution shrinks, so the beam size σ and angle spread θ alternate with 90 degrees relative phase shift. At points where the size is minimal (focused at the absorber plates), the angle spread is maximal, and vice versa – in accordance with Liouville's theorem. There are no correlations in the (x, x') distribution at these points, so that the normalized emittance can be defined as the product $\beta\gamma\sigma\theta = \varepsilon_\perp$ at the plates, where $\beta\gamma = p/mc$ are Lorentz factors. By using IC to impose a damping force on transverse velocities in the absorber with decrement $\Lambda_c = 2\Lambda_d$, we obtain damping for both the angle spread and the beam size at the plate.

Due to scattering in the absorber plates, the angle spread at the plates and oscillation amplitudes evolve to a conventional equilibrium regardless of the thickness of the plates. However, the diffusion of phases and particle transverse positions at the plates is suppressed drastically due to the small width of the plates with respect to the focal parameter, $\lambda/2\pi$: $\delta x = -s\delta x'$, where $-w \leq 2s \leq w$.

PIC Compatibility with Emittance Exchange

PIC should be accompanied by emittance exchange in order to maintain the energy spread near the equilibrium value that was achieved after 6D cooling; this requires the use of wedge absorbers (with height $h >> \sigma$) and the introduction of dispersion D. (D is the sensitivity of particle orbit to energy: $x = D(\Delta p/p) + x_b$, where Δp is the momentum deviation from the reference particle, and x_b is associated with the particle oscillation around the energy dependent orbit.) The strength of emittance exchange is determined by the ratio D/h. But there is also a heating effect on transverse emittance associated with energy straggling in the absorber. Namely, energy jumps due to collisions with electrons cause excitation of muon transverse oscillations: $\delta x_b = -D\delta p/p$; $\delta x_b' = \delta x' - D'\delta p/p$. Emittance exchange for simultaneous longitudinal and transverse cooling requires the ratio D/h to be about a unit; since the beam size is initially reduced by basic cooling, large dispersion is not required for emittance exchange and the straggling impact on transverse emittance does not overwhelm the PIC. It is also important that the beam size at the plates decreases during the PIC process to allow further reduction of the necessary dispersion.

PIC AT A PERFECT PARAMETRIC RESONANCE

Balance Equation

Combining the average rates due to the parametric resonance, transverse cooling, emittance exchange, scattering and straggling, we obtain the following balance equations for θ, σ and longitudinal emittance $\varepsilon_z = \sigma_z \Delta p / m_\mu c$:

$$\beta^2 {\theta^2}' = -(\Lambda - 2\beta^2 \Lambda_d)\theta^2 + \Lambda \frac{m_e}{2\gamma m_\mu}(Z + 1 + \frac{\gamma^2 + 1}{4\log} D'^2) \quad (1)$$

$$\beta^2 {\sigma^2}' = (\Lambda \frac{D}{2h} - 2\beta^2 \Lambda_d)\sigma^2 + \Lambda \frac{m_e}{2\gamma m_\mu}(\frac{Z+1}{12} w^2 + \frac{\gamma^2 + 1}{4\log} D^2) \quad (2)$$

$$\beta^2 \varepsilon_z' = \Lambda [-\frac{D}{2h} + \frac{1}{\gamma^2} + \frac{m_e}{8\gamma m_\mu} \frac{\gamma^2+1}{\log} (\frac{p}{\Delta p})^2] \varepsilon_z \qquad (3)$$

where $\Lambda = 2<E_i'>/\gamma mc^2$ is the invariant 6D cooling decrement [1], E_i' is the ionization energy loss rate in the absorber, Z is absorber's atomic number, $\log \approx 12$ is the Coulomb logarithm of ionization energy loss by a fast particle, m_e/m_μ is the electron and muon mass ratio, and $(1/h) \equiv (\partial <E_i'>/\partial x)/<E_i'>$.

Optimum Cooling and PIC Equilibrium

We define optimum cooling by equating the three emittance cooling rates (thus, making each of them equal to $\Lambda/3$) to obtain the following relationships:

$$\frac{D}{h} = 2 - \frac{4}{3}\beta^2; \quad \frac{\Lambda_d}{\Lambda} = \frac{1}{2\beta^2} - \frac{1}{6}. \qquad (4)$$

Then the balance equations lead to equilibrium as follows:

$$\theta^2 = \frac{3m_e}{2\gamma\beta^2 m_\mu}(Z+1+\frac{\gamma^2+1}{4\log}D'^2); \quad \sigma^2 = \frac{3m_e}{2\gamma\beta^2 m_\mu}(\frac{Z+1}{12}w^2 + \frac{\gamma^2+1}{4\log}D^2);$$

$$(\frac{\Delta p}{p})^2 = \frac{3m_e}{8\gamma\beta^2 m_\mu} \cdot \frac{\gamma^2+1}{\log}.$$

For the conditions $D' << 2\left(\frac{Z+1}{\gamma^2+1}\log\right)^{1/2}$ and $D << w\left(\frac{Z+1}{3}\frac{\log}{\gamma^2+1}\right)^{1/2}$, the equilibrium normalized transverse emittance and beam size σ will be close to minimum values:

$$\varepsilon_\perp \Rightarrow \varepsilon_{\perp 0} = \frac{\sqrt{3}}{4\beta}(Z+1)\frac{m_e}{m_\mu}w, \quad \sigma = \frac{\theta w}{2\sqrt{3}}. \qquad (5)$$

PIC Potential

The equilibrium emittance (5) can be expressed as function of the intrinsic energy loss in the absorber, E_i' and the average energy loss or accelerating field using the relationships $(2w/\lambda)E_i' = <E_i'> = <E_{acc}'>$:

$$\varepsilon_{\perp 0} = \frac{\sqrt{3}}{8\beta}\frac{m_e}{m_\mu}\lambda\frac{<E_i'>}{E_i'}(Z+1).$$

The emittance that can be achieved after an "ordinary" (non-resonance) 6D cooling [1] in hydrogen absorber is $\varepsilon_{ord} = (3\lambda m_e/2\pi\beta m_\mu)$; so PIC results in the reduction of transverse emittance by a factor $\frac{\pi}{4\sqrt{3}}\frac{<E_i'>}{E_i'}(Z+1)$. As a function of Z, the factor $E_i'/(Z+1)$ is about $(60/\beta^2)$ MeV/m in case of beryllium (compare with $(15/\beta^2)$ MeV/m in case of liquid hydrogen) and does not change significantly with Z for heavier elements. Note that use of absorbers with large atomic number is disadvantageous because of the small thickness of the plates, which makes it difficult to tune to resonance. Note also that the equilibrium emittance can be decreased by decreasing the plate thickness and lowering the accelerating field, which makes the beam line longer. Thus the cooling rate and equilibrium emittance are limited by beam loss due to muon decay. For an optimal PIC design, the plate thickness w should diminish along the cooling channel, starting from a maximum determined by the available accelerating voltage. Table 1 illustrates the PIC effect.

TABLE 1. Potential PIC effect.

Parameter	Unit	Initial	Final
Beam momentum, p	MeV/c	100	100
Distance between plates, $\lambda/2$	cm	19	19
Plate thickness, w	mm	6.4	1.6
Intrinsic energy loss rate (Be)	MeV/m	600	600
Average energy loss	MeV/m	20	5
Transverse emittance, norm.	μm	600	25
Beam transverse size at plates	mm	6.0	0.15
Angle spread at plates, $\theta_x = \theta_y$	mrad	200	200
PIC channel length	m		100
Integrated energy loss	GeV		0.7
Beam loss due to muon decay	%		15
Number of particles/bunch*			10^{11}
Space charge tune spread	%		.2

*To overcome the space charge impact on tuning, one can implement a beam recombining scheme: generate a low charge/bunch beam and recombine bunches after cooling and acceleration to sufficiently relativistic energy (under investigation).

REVERSE EMITTANCE EXCHANGE

The normalized longitudinal emittance after the basic 6D and parametric resonance cooling is too small to be efficiently used (or maintained) in a collider. Therefore, reverse emittance exchange should be implemented after PIC to gain luminosity while preserving the achieved 6D emittance.

The absorber based concept under study is based on the following plan:
1) Continue the resonance regime to maintain minimum beam size at the plates.
2) Use the strongest reverse wedge: $h = \sigma/\xi$; ($\xi = 1/3$).
3) Make the dispersion at the plates maximal: $D(\Delta p/p) = \sigma$
4) Design the parametric resonance to equalize σ and θ decrements according to equations (1) and (2). This implies:

$$\Lambda_d = -\frac{\Lambda}{4\beta^2}(\frac{\xi}{2}\frac{p}{\Delta p}-1)(1-\chi), \qquad (6)$$

where the parameter χ is associated with straggling impact:

$$\chi \equiv \frac{\gamma}{4\xi}\frac{p}{\Delta p}\frac{m_e}{m_\mu \log} \quad (<1). \qquad (7)$$

Then, the relationship $\sigma = w\theta/2\sqrt{3}$ is maintained.

5) The relative momentum spread should also be maintained at a reasonable value implying bunch stretching (the first stage of REMEX) and then beam acceleration (second stage).

The optimum design according to (6) e leads to the REMEX balance equations:

$$\varepsilon'_z = \Lambda_z \varepsilon_z; \quad \varepsilon'_\perp = -\frac{\Lambda_z}{2}\frac{1-\chi}{1+\chi}\varepsilon_\perp + \frac{\Lambda}{3}\frac{w\beta_0}{w_0\beta}\varepsilon_{\perp 0}; \quad \Lambda_z \equiv \frac{\Lambda}{2\beta^2}\frac{p}{\Delta p}\frac{1+\chi}{\xi} \gg \Lambda.$$

The asymptotic solution of equation for $\varepsilon_\perp$ can be found as follows:

$$\frac{\varepsilon_\perp(z)}{\varepsilon_{\perp 0}} \Rightarrow \approx \sqrt{\frac{\varepsilon_{z0}}{\varepsilon_z} + \frac{4\beta\beta_0}{3\xi(1-\chi)}\frac{w}{w_0}\frac{\Delta p}{p}}, \quad (8)$$

where the first term describes the reverse emittance exchange and the second term takes into account the transverse scattering and energy straggling. Angle scattering and energy straggling will limit the achievable transverse emittance. The optimum maximum energy of REMEX is determined by energy straggling, which grows with energy. The following table 2 shows the expected REMEX change in the minimum $\varepsilon_\perp$ that can be found using equation (8) with appropriate choices of p, Δp, and ε_z.

Note that the integrated energy loss in such a fast process as REMEX is relatively small, as is the beam loss.

Table 2. Potential REMEX effect

Parameter	Unit	Initial	After 1st stage	After 2nd stage
Momentum	MeV/c	100	100	2500
Bunch length	cm	1	10	10
Momentum spread	%	3	3	3
Longitudinal norm. emittance	cm	3×10^{-2}	.15	7.5
Transverse norm. emittance	μm	25	8	2

COMPENSATION FOR ABERRATIONS

A principal challenge of the PIC and REMEX designs is to have all particles reach the minimum radial position at the absorber positions along the beam path. This synchronism is violated by the spread of betatron oscillation tunes, which is still quite large even after basic 6D cooling. In a linear focusing field, there are two fundamental mechanisms of optical aberration: 1) chromatic aberration due to tune dependence on particle energy and 2) spherical aberration due to particle path dependence on the square of transverse momentum. Compensation for both destructive factors is necessary in order to realize the maximum cooling effect. Compensation for chromaticity requires introduction of a large dispersion together with sextupole magnetic fields. Large dispersion, however, is detrimental to PIC and REMEX because of the energy straggling impact on transverse emittance discussed above. A resolution of this difficulty is to design the dispersion function with a period ½ of the betatron wavelength by applying alternating bends as shown in figure 2.

Compensation for spherical aberrations also seems achievable in this type of beam transport using an appropriate modulation of sextupole field along the beam path. An optimized design for resonance cooling requires a comprehensive analytical and simulation study of these possibilities which are on the way.

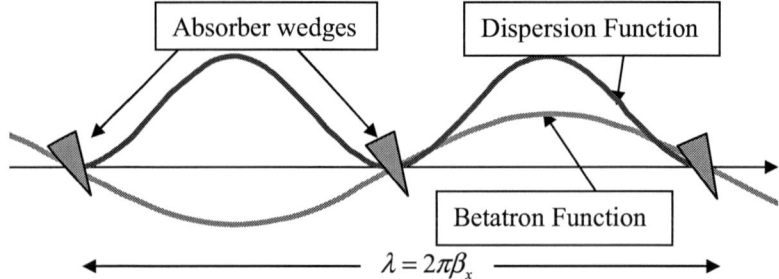

FIGURE 2. Schematic of an achromatic wiggler, where wedge absorbers are placed at symmetric locations relative to the dispersion function, which has a period half that of the betatron function.

CONCLUSIONS AND OUTLOOK

The transverse emittances of muon beams can be reduced to those normally associated with conventional electron or hadron colliders by implementing Parametric Resonance Ionization Cooling and Reverse Emittance Exchange techniques. This has extremely important consequences for any muon collider design, primarily in the number of muons required for high luminosity.

Some concepts of beam focusing and tune spread compensation for the best parametric resonance cooling and reverse emittance exchange have been proposed. The first analytic expressions shown here will be used for guidance of simulation efforts.

The compatibility of beam resonance focusing with effective regular and reverse emittance exchange is understood. The first results suggest some absorber technology issues for study and design.

REFERENCES

[1] Y. Derbenev and R. P. Johnson, *Phys.Rev.STAB* **8**, 041002, (2005)
[2] Y. Derbenev and R. P. Johnson, PAC05, http://snsapp1.sns.ornl.gov/pac05/TPPP014/TPPP014.PDF
[3] R. P. Johnson et al., PAC05, PAC05, http://snsapp1.sns.ornl.gov/pac05/ROAA005/ROAA005.PDF

MICE: The International Muon Ionization Cooling Experiment

Daniel M. Kaplan

Illinois Institute of Technology, Chicago, Illinois 60616, USA

(for the MICE Collaboration)

Abstract. Ionization cooling of a muon beam is a key technique for a Neutrino Factory or Muon Collider. An international collaboration is mounting an experiment to demonstrate muon ionization cooling at the Rutherford Appleton Laboratory. We aim to complete the experiment by 2010.

Keywords: muon cooling, muon, muon collider, neutrino, neutrino factory

INTRODUCTION

The experimental establishment of neutrino oscillations [1] has stimulated widespread interest in a muon storage ring-based Neutrino Factory [2], possibly the ultimate tool for studying the neutrino mixing matrix [3]. Two feasibility studies [4, 5] have shown that a high-performance Neutrino Factory can be built using available technology. However, some of the beam-manipulation techniques envisaged have yet to be applied in practice. Of these, ionization cooling of the muon beam [6, 7] is perhaps the most novel. In the longer term it holds the promise of s-channel Higgs Factories and multi-TeV muon-antimuon colliders, with potential for unique studies of matter and energy at the most fundamental level [8], complementing those at the Large Hadron Collider and the proposed International Linear Collider.

Ionization cooling contributes significantly to both the performance (up to a factor of 10 in intensity [9]) and cost (as much as 20% [5]) of a Neutrino Factory. This motivates the Muon Ionization Cooling Experiment (MICE). MICE is intended not only to demonstrate the *principle* of ionization cooling, but (and perhaps more importantly) to show how to build and operate a device with the performance required for a Neutrino Factory. The experience gained from MICE will provide input to the final design of the Neutrino Factory cooling channel and firm up its cost estimate. An important part of the MICE program is to study the cooling process by varying the relevant parameters, so that an extrapolation can be made to a different cooling-channel design, e.g., a ring [10] or helical cooling channel [11], should one of these be shown to be advantageous.

During 2001 and 2002, the international MICE Collaboration [12] was formed and developed a proposal [13] to carry out this program using a muon beam produced with the ISIS accelerator at Rutherford Appleton Laboratory (RAL). The proposal was approved in 2003. The MICE collaboration includes accelerator and experimental particle physicists from Europe, Japan, and the US. As of this writing, funding for the first phase of MICE has been provided in Italy, Japan, the Netherlands, Switzerland, the

UK, and the US. We aim for a definitive demonstration of ionization cooling by 2010.

DESIGN OF THE EXPERIMENT

The MICE design is presented in detail in the proposal [13] and Technical Reference Document (TRD) [14], and is briefly summarized here. The goals of MICE are

- to engineer and build a section of cooling channel (of a design that can give the desired performance for a Neutrino Factory) that is long enough to provide a measurable ($\approx$10%) cooling effect, but short enough to be moderate in cost; and
- to measure the resulting cooling effect with an absolute accuracy of 0.1% over a muon-beam momentum range of 140–240 MeV/c.

The layout of MICE is shown in Fig. 1. The tracks of single muons through the apparatus will be measured using standard particle-physics techniques, since bunched-beam diagnostics lack the needed precision. The 5.5 m-long cooling section, consisting of three absorbers and eight rf cavities encircled by lattice solenoids, is therefore surrounded at each end by tracking detectors, to measure beam emittance at the entrance and exit, and particle-ID detectors to reject particles other than muons. This requires the placement of tracking detectors close to the rf cavities and is therefore sensitive to backgrounds caused by dark-current electrons and their associated x-rays. Improving our understanding of such backgrounds is essential to the successful planning and execution of MICE and is a much-anticipated result from upcoming tests by the MuCool Collaboration [15].

The cooling section is one lattice cell of the "SFOFO" cooling channel developed in Feasibility Study II [5] (with minor modifications to reduce cost and comply with RAL safety requirements), with a full absorber at each end to protect the detectors from rf-cavity emissions. This arrangement provides considerable flexibility, as the solenoid polarities and currents can be varied to test a variety of lattices. Provision will be made for a variety of solid absorbers as well as liquid hydrogen and helium. (In principle, some other cooling cell could also be tested, perhaps in a subsequent MICE phase.)

Figure 2 shows the simulated effect of the cooling section on the normalized transverse beam emittance, as well as the beam transmission, vs. that emittance, for 200 MeV/c average beam momentum and nominal optics settings of the SFOFO lattice cell (3.2 T maximum on-axis field). For input emittance below the equilibrium value of 2π mm·rad the beam is heated; above 6π mm·rad scraping begins to deplete the beam. The detailed comparison of such measurements against Monte Carlo predictions, for a variety of beam momenta, emittances, and apparatus configurations, will serve to validate our Monte Carlo and design approach and allow extrapolation to the longer ($\sim$100 m) cooling channels typically used in Neutrino Factory designs.

Achieving 0.1% emittance resolution will require careful calibration and simulation. Scattering of the beam in the detectors causes a correctable bias, as illustrated (for input-beam emittance $\varepsilon_t = 2.5\pi$ mm·rad) in Fig. 3 [16]: before correction, the transverse emittance measured in each spectrometer is $\sim$1% larger than the "true" emittance. The goal of 0.1% emittance measurement will thus require that this bias be calibrated and corrected to $\sim$10% of itself (to be verified by calibration runs with no cooling section).

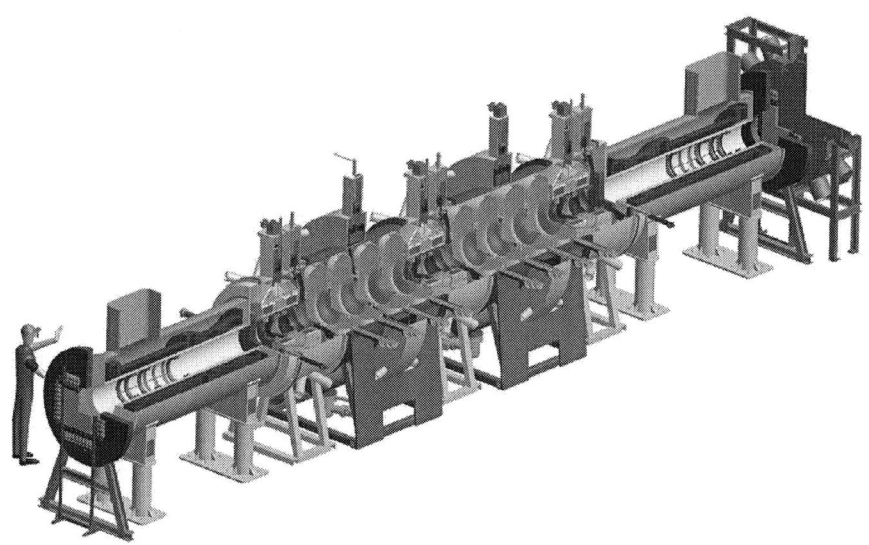

FIGURE 1. Three-dimensional cutaway rendering of the MICE apparatus. The muon beam enters from the lower left and is measured by time-of-flight (TOF) and Cherenkov detectors and a first solenoidal tracking spectrometer. It then enters the cooling section, where it is alternately slowed down in absorbers and reaccelerated by rf cavities, while being focused by a lattice of superconducting solenoids. Finally it is remeasured by a second solenoidal tracking spectrometer and its muon identity confirmed by Cherenkov and TOF detectors and a calorimeter.

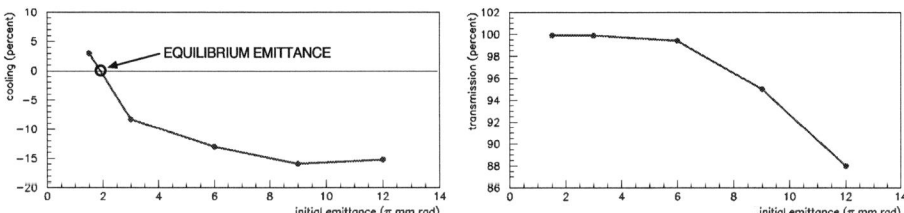

FIGURE 2. (Left) percent change of normalized transverse emittance and (right) beam transmission through cooling section, both vs. input emittance in π mm·rad.

Tracking Detectors

To minimize beam scattering and sensitivity to x-rays, the tracking detectors will be thin (350 μm-diameter) scintillating fibers, ganged by sevens to reduce the needed electronics channel count. Each group of seven adjacent fibers is mated to a 1 mm clear light-guide fiber that conveys the scintillation light to a VLPC photosensor. The >85% quantum efficiency of the VLPCs [17] results in an average of 11 photoelectrons per minimum-ionizing particle, as verified in cosmic-ray tests [16]. As in the D0 experiment [17], the use of two staggered layers per view ensures high efficiency. Each spectrometer will be made up of five detector stations, each with three views arranged in

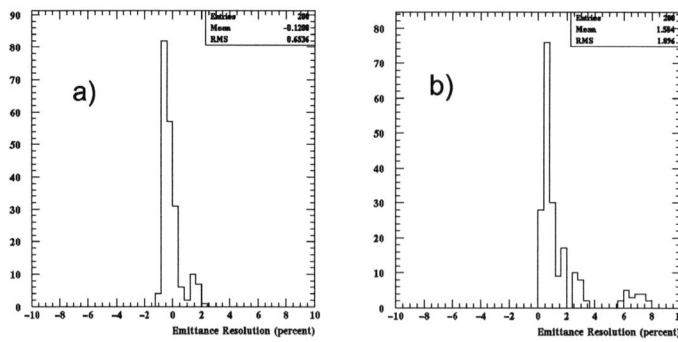

FIGURE 3. Resolution in uncorrected emittance (for 2.5π mm·rad input emittance) in a) upstream and b) downstream spectrometer.

120° stereo, deployed within a 1.1 m-long 4 T superconducting solenoid. A prototype 4-station detector is now undergoing beam tests in a 1 T solenoid at KEK.

Particle Identification

The muon beam may contain residual pions which are transported through the large momentum acceptance of the beamline, as well as electrons from the in-flight decay of muons. A three-plane time-of-flight system provides the precise time information needed for particle identification, emittance measurement, and off-line bunch construction and timing with respect to the rf phase. Additional particle identification is provided before and after the cooling channel by Cherenkov detectors and a calorimeter.

STAGING AND CURRENT STATUS

The need to carefully cross-calibrate the spectrometers as well as the cost of the cooling section suggest staging the installation and operation of MICE as indicated in Fig. 4. The currently funded first phase of MICE includes the detectors but not the cooling section; the second phase will be assembled in steps as funds allow. At present the first rf cavity is under test in the MuCool Test Area at Fermilab, the Absorber/Focus-Coil and RF-cavity/Coupling-Coil module designs are well advanced, and work is in progress on the spectrometers, beamline, and infrastructure. First beam is planned for April 2007.

ACKNOWLEDGEMENTS

The author gratefully acknowledges support of the US MICE institutions by the Department of Energy and the National Science Foundation.

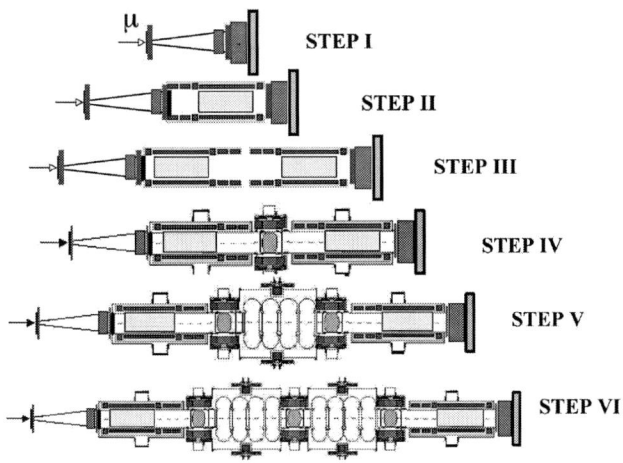

FIGURE 4. Six possible steps in the development of MICE.

REFERENCES

1. Y. Fukuda et al., Phys. Lett. B **433**, 9 (1998); ibid. **436**, 33 (1998); Phys. Rev. Lett. **81**, 1562 (1998); ibid. **82**, 2644 (1999); Q. R. Ahmad et al., ibid. **87**, 071301 (2001); ibid. **89**, 011301 (2002); ibid. **89**, 011302 (2002).
2. S. Geer, Phys. Rev. D **57**, 6989 (1998); A. Blondel et al., CERN Report 2004-002.
3. See, e.g., M. Lindner, in **Neutrino Mass**, Springer Tracts Mod. Phys. **190** (2003) 209, and C. Albright et al., report Fermilab-FN-692 (2000), available from http://arXiv.org/pdf/hep-ex/0008064
4. *Feasibility Study on a Neutrino Source Based on a Muon Storage Ring*, ed. D. Finley and N. Holtkamp, FERMILAB-PUB-00-108-E (2000), available from http://www.fnal.gov/projects/muon_collider/reports.html
5. *Feasibility Study-II of a Muon-Based Neutrino Source*, ed. S. Ozaki et al., BNL-52623 (2001), available from http://www.cap.bnl.gov/mumu/studyii/FS2-report.html
6. A. N. Skrinsky and V. V. Parkhomchuk, Sov. J. Part. Nucl. **12**, 223 (1981); D. Neuffer, Part. Acc. **14**, 75 (1983); E. A. Perevedentsev and A. N. Skrinsky, in Proc. 12th Int. Conf. on High Energy Accelerators, ed. F. T. Cole and R. Donaldson (Fermilab, 1984), p. 485.
7. D. Neuffer, in **Advanced Accelerator Concepts**, ed. F. E. Mills, AIP Conf. Proc. **156** (American Institute of Physics, New York, 1987), p. 201; R. C. Fernow and J. C. Gallardo, Phys. Rev. E **52**, 1039 (1995).
8. C. M. Ankenbrandt et al., Phys. Rev. ST Accel. Beams **2**, 081001 (1999).
9. K. Hanke, NuFact Note 59 (2000), available from http://slap.web.cern.ch/slap/NuFact/NuFact/NFNotes.html
10. R. B. Palmer, J. Phys. G: Nucl. Part. Phys. **29** (2003) 1577.
11. Ya. Derbenev and R. P. Johnson, Phys. Rev. ST Accel. Beams **8**, 041002 (2005).
12. See http://www.mice.iit.edu/
13. MICE proposal, available from http://mice.iit.edu/mnp/MICE0021.pdf
14. MICE TRD, available from http://www.isis.rl.ac.uk/accelerator/MICE/TR/MICE_Tech_ref.html
15. See talk by Y. Torun, this conference.
16. A. Khan et al., MICE-Note 90 (2005).
17. B. Baumbaugh et al., IEEE Trans. Nucl. Sci. **43**, 1146 (1996).

6D Muon Ionization Cooling with an Inverse Cyclotron

D. J. Summers*, S. B. Bracker*, L. M. Cremaldi*, R. Godang* and R. B. Palmer[†]

*Dept. of Physics and Astronomy, University of Mississippi-Oxford, University, MS 38677 USA
[†]Brookhaven National Laboratory, Upton, NY 11973 USA

Abstract. A large admittance sector cyclotron filled with LiH wedges surrounded by helium or hydrogen gas is explored. Muons are cooled as they spiral adiabatically into a central swarm. As momentum approaches zero, the momentum spread also approaches zero. Long bunch trains coalesce. Energy loss is used to inject the muons into the outer rim of the cyclotron. The density of material in the cyclotron decreases adiabatically with radius. The sector cyclotron magnetic fields are transformed into an azimuthally symmetric magnetic bottle in the center. Helium gas is used to inhibit muonium formation by positive muons. Deuterium gas is used to allow captured negative muons to escape via the muon catalyzed fusion process. The presence of ionized gas in the center may automatically neutralize space charge. When a bunch train has coalesced into a central swarm, it is ejected axially with an electric kicker pulse.

Keywords: beam cooling, cyclotron, muon
PACS: 13.66.Lm, 14.60.Ef, 14.60.Lm

INTRODUCTION

Cooling an ensemble of muons must be completed more rapidly than their 2.2 μs lifetime. Ionization cooling can help [1]. Random muon motion is removed by passage through a low Z material, such as hydrogen, and coherent motion is added with RF acceleration. Designs for 6D muon cooling using linear helical channels [2] at 100 MeV kinetic energies and using frictional cooling [3,4] at keV energies are under investigation. Muon cooling rings have been simulated at various levels [5]. In a ring, the same magnets and RF cavities may be reused each time a muon orbits. Transverse cooling can naturally be exchanged for longitudinal cooling by allowing higher momentum muons to pass through more material. Thus rings cool in all six dimensions.

Small emittance bunches of cold muons are useful to reduce the aperture of the acceleration system for a neutrino factory [6,7] and are required to provide adequate luminosity for a muon collider [8]. At a neutrino factory, accelerated muons are stored in a racetrack to produce neutrino beams ($\mu^- \to e^- \bar{\nu}_e \nu_\mu$ and $\mu^+ \to e^+ \nu_e \bar{\nu}_\mu$). Neutrino oscillations have been observed [9] and need more study. Further exploration at a neutrino factory could reveal *CP* violation in the lepton sector [10], and will be particularly useful if the ν_e to ν_τ coupling, θ_{13}, is small [7,11]. A muon collider can do s-channel scans to split the H^0/A^0 Higgs doublet [12]. Above the ILC's 800 GeV there are a large array of supersymmetric particles that might be produced [13] and, if large extra dimensions exist, so could mini black holes [14]. Note that the energy resolution of a 4 TeV muon collider is not smeared by beamstrahlung like CLIC.

OPERATION OF AN AZIMUTHALLY SYMMETRIC INVERSE CYCLOTRON AT LEAR (P-BAR) AND PSI (MU-)

An inverse cyclotron has been used to slow LEAR anti-protons at CERN [15,16]. An annular quasipotential well, $U(r,z)$, is formed which ferries anti-protons towards the center of an azimuthally symmetric cyclotron. The radius of the annulus decreases with the decreasing angular momentum of the $\bar{p}$.

$$U(r,z) = V(r,z) - (1/(2\eta r^2))(L_g/M + \eta rA_\theta)^2, \quad (1)$$

where $\eta = e/M$ and $L_g = L_z - erA_\theta$ is a generalized angular momentum. The radial well deepens with decreasing radius and the vertical well grows shallower (see Fig. 2 of Ref. 15). Particles must adiabatically spiral to the center. If dE/dx is too large, particles will not stay in the magnetic wells. The final $\bar{p}$ swarm has a radius of 1.5 cm, a height of 4 cm, and a kinetic energy of 2 keV. A long bunch train is coalesced into a single swarm, which is roughly the same diameter as the incoming beam. The spiral time is 20 μs with 0.3 mbar hydrogen and about 1 μs with 10 mbar hydrogen. Given the dependence of the cyclotron frequency on mass, $f = \omega/2\pi = qB/2\pi m$, the spiral time for a muon is nine time less than for a $\bar{p}$. The gas pressure in the center must be low, both to allow a particle to spiral all the way in before stopping, and to allow reasonable kicker voltages for axial extraction. An 80 ns electric kicker pulse rising to 500 V/cm in 20 ns is employed. The $\bar{p}$'s move 32 cm in 500 ns. Given that $F = ma$, muons will go nine times farther.

The cyclotron has now been moved from LEAR to PSI where it is used to slow negative muons to a few keV [17]. Three centimeter diameter beams with 30 000 μ^-/s below 50 keV and 0.8 cm diameter beams with 1000 μ^-/s in the 3 to 6 keV kinetic energy range are output for use. A static electric field continuously ejects the muons. The energy absorber and the negatively charged electrode consist of a single 30 μg/cm^2 Formvar foil (polyvinyl formal) with 3 nm of nickel produced by 30 minutes of sputtering.

SKETCH OF A SECTOR INVERSE CYCLOTRON WITH LARGE ADMITTANCE FOR MUONS

A scaling sector cyclotron would allow greater admittance [18] than the azimuthally symmetric cyclotron now running at PSI. For a given $\int \mathbf{B} \cdot d\ell$, the ratio of the fields in the hills and valleys can be adjusted to maximize acceptance. Only radial and neither spiral nor FFAG [19] sectors have been explored so far. The sector cyclotron may be able to function as a damped harmonic oscillator to lower the amplitude of horizontal and vertical betatron motion as a bunch train of muons spirals into a single central swarm.

$$F = \frac{\gamma m v^2}{r} = \frac{qQ}{4\pi\varepsilon_0 r^2} + qvB, \quad r = \frac{\gamma m v^2 \pm \sqrt{(\gamma m)^2 v^4 - 4(qvB)(qQ/4\pi\varepsilon_0)}}{2qvB} \quad (2)$$

With 10^{12} muons in a swarm, space charge is a concern. Table 1 and Eqn. 2 show the effect of space charge. Fortunately, the muons are swarming in an ionized gas which may be able to automatically neuralize the space charge [20]. Electrons experience 200 times

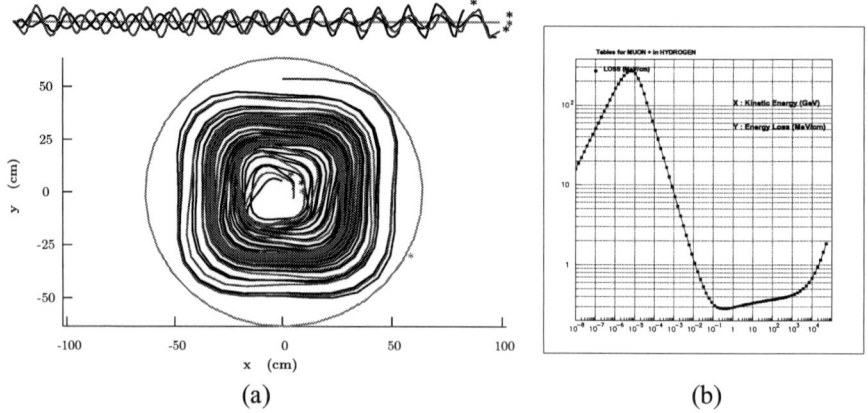

FIGURE 1. (a) ICOOL [21] simulation of single turn, energy loss injection. Three identical 172 MeV/c muons are injected into a 1.8 Tesla cyclotron with four sectors and soft edged magnetic fields. The inward spirals differ because of multiple scattering and straggling. The energy loss is caused by radial LiH wedges surrounded by hydrogen gas. The amount of matter encountered in a given orbit decreases adiabatically with radius to allow stable orbits. The upper left trace shows that vertical motion is completely contained within ± 5 cm along the 70 m spiral. The fractional energy loss required in the first turn for injection increases with the width of the muon beam and decreases as the cyclotron's magnetic field is lowered. The injection scaling relation is given by $\Delta p = .3 B \Delta r$. Units are GeV/c, Tesla, and meters, respectively. (b) Plot of μ^+ energy loss (MeV/cm) in liquid hydrogen versus kinetic energy (GeV) using GEANT3. The default value of "CUTMUO" was decreased from 10 MeV to 10 eV to propagate slow muons. The energy turnover at 8 keV corresponds to a momentum of 1.3 MeV/c. $p = \sqrt{2mE} = \sqrt{2 \times 105.7 \times 0.008}$. Aluminum, copper, iron, and liquid helium show similar results as does the PDG.

the acceleration of muons in an electric field. Movement of 10^{12} electrons in 100 ns requires 1.6 amps of current. A metallic grid might also be used for neutralization.

Muons must spiral in fast enough to minimize decay loss, but must not stop before reaching the central swarm. So the density of the absorber must decrease smoothly with radius. Radial LiH wedges immersed in a gas or high to low pressure gases

TABLE 1. The effect of space charge. Orbital radius of the last muon in millimeters is shown as a function of momentum, magnetic field, and central point charge. The radii come from Eqn. 2 using $v = pc^2/E = pc^2/\sqrt{p^2c^2 + m^2c^4}$. An "$i$" indicates that the radius is partly imaginary.

p	1 Tesla $Q=0$	1 Tesla $Q=10^{12}q$	1 Tesla $Q=4 \times 10^{12}q$	2 Tesla $Q=0$	2 Tesla $Q=10^{12}q$	2 Tesla $Q=4 \times 10^{12}q$
16 MeV/c	53	27+26	27+24	27	13+13	13+11
8 MeV/c	27	13+11	13+8.7i	13	6.7+3.6	6.7+9.1i
4 MeV/c	13	6.7+9.1i	6.7+22i	6.7	3.3+7.2i	3.3+16i
2 MeV/c	6.7	3.3+16i	3.3+32i	3.3	1.7+11i	1.7+22i
1 MeV/c	3.3	1.7+22i	1.7+45i	1.7	.83+16i	.83+32i
.5 MeV/c	1.7	.83+32i	.83+64i	.83	.42+23i	.42+45i
.25 MeV/c	.83	.42+45i	.42+90i	.42	.21+32i	.21+64i

TABLE 2. Emittance reduction goals for an inverse cyclotron. Emittance goes as $(\Delta p_x \Delta x)(\Delta p_y \Delta y)(\Delta p_z \Delta z)$. A muon collider needs a factor of 10^6 in cooling. The input assumes a factor of 10 in transverse cooling [7]. The output for Δp is from a study of what might be achieved with frictional muon cooling [3].

Δp_x	30 MeV/c → 0.3 MeV/c	Δx	70 mm → 50 mm
Δp_y	30 MeV/c → 0.3 MeV/c	Δy	70 mm → 50 mm
Δp_z	30 MeV/c → 0.3 MeV/c	Δz	10000 mm → 50 mm

separated by beam pipes might meet this criteria. The sector cyclotron geometry must transform into an azimuthally symmetric magnetic bottle as the muons approach the central swarm. Otherwise, as shown by GEANT3, muons will escape though the valleys. In the transition region the field might resemble a hexapole or octupole field as used in an Electron Cyclotron Resonance Ion Source (ECRIS) [22]. If 2×10^{12} 172 MeV/c muons (KE = 96 MeV) arrive at 30 Hz, they will deposit 920 watts of beam power.

Atoms can capture muons. Helium may be used to inhibit muonium (μ^+e^-) formation [3]. A possibility for negative muons is to use deuterium gas. Muons will catalyze fusion and be freed. The sticking factor is 10%. The reaction appears in Eqn. 3 [23]. 2×10^{12} fusions repeated at 30 Hz only generate 35 watts. The momentum of the freed muon ranges from 0 to 29 MeV/c. A negatively charged absorber foil might also prevent μ^- sticking and is used at PSI. The foil would have to dissipate roughly 100 watts.

$$d+d+\mu^- \rightarrow {}^3He+n+\mu^- + 3.3\,\text{MeV} \text{ or } t+p+\mu^- + 4.0\,\text{MeV} \qquad (3)$$

Busch's theorem (Eqn. 4) [24] has the effect of increasing the emittance as muons leave a magnetic field. A half Tesla field and a 50 mm radius give a 4 MeV/c azimuthal kick. One might be able to use radial iron fins in the exit port to alleviate this effect or reverse and increase the magnitude of the magnetic field to capture the unwanted angular momentum in an absorber after extraction. Using low fields with tall cylindrical swarms that have small diameters works for sure. An RF quadrupole is perhaps a natural choice for acceleration that would immediately follow the extraction electric kicker.

$$\dot{\phi} = [e/(2\pi \gamma m r^2(s))][\Phi(s) - \Phi_k], \quad L_z = xp_y - yp_x = r^2 \gamma m \dot{\phi} = -eBr^2/2 \qquad (4)$$

In summary, progress on a large admittance sector cyclotron is underway, including energy loss injection (see Fig. 1a), 6D muon cooling (see Table 2), and an axial electric kicker for extraction. Many thanks to Juan Gallardo and Franz Kottmann for useful suggestions. This work was supported by the U.S. Dept. of Energy, DE-FG02-91ER40622 and DE-AC02-98CH10886.

REFERENCES

1. A. Skrinsky and V. Parkhomchuk, *Sov. J. Part. Nucl.* **12**, 223 (1981);
    D. Neuffer, *Part. Accel.* **14**, 75 (1983); *Nucl. Instrum. Meth.* **A532**, 26 (2004);
    G. Penn and J. S. Wurtele, *Phys. Rev. Lett.* **85**, 764 (2000);
    K. Kim and C. Wang *Phys. Rev. Lett.* **85**, 760 (2000); *Phys. Rev. Lett.* **88**, 184801 (2002);

G. Franchetti, *Phys. Rev. ST Accel. Beams* **4**, 074001 (2001); G. Dugan, *ibid.,* **4**, 104001 (2001);
R. B. Palmer, COOL 05; Y. Derbenev, COOL 05; Y. Torun, COOL 05; M. Cummings, COOL 05;
C. Darve, COOL 05; K. Yonehara, COOL 05; K. Beard, COOL 05.
2. Y. Derbenev, R. Johnson, *Nucl. Instrum. Meth.* **A532**, 470 (2004); *AIP Conf. Proc.* **671**, 328 (2003); *Phys. Rev. ST Accel. Beams* **8**, 041002 (2005); R. Johnson, COOL 05.
3. H. Abramowicz et al., *Nucl. Instrum. Meth.* **A546**, 356 (2005).
4. R. Galea et al., *Nucl. Instrum. Meth.* **A524**, 27 (2004); *J. Phys.* **G29**, 1653 (2003);
A. Caldwell, *J. Phys.* **G29**, 1569 (2003); D. Taqqu, *AIP Conf. Proc.* **372**, 301 (1996);
M. Muhlbauer et al., *Nucl. Phys. Proc. Suppl.* **51A**, 135 (1996); *Hyperfine Interact.* **119**, 309 (1995).
5. V. I. Balbekov and A. Van Ginneken, *AIP Conf. Proc.* **441**, 310 (1998);
J. S. Berg, R. C. Fernow, and R. B. Palmer, *J. Phys.* **G29**, 1657 (2003);
R. B. Palmer, *J. Phys.* **G29**, 1577 (2003); *Nucl. Instrum. Meth.* **A532**, 255 (2004);
R. Palmer et al., *Phys. Rev. ST Accel. Beams* **8**, 061003 (2005).
6. A. Blondel et al., *Nucl. Instrum. Meth.* **A451**, 102 (2000); R. Palmer et al., *ibid.,* **A451**, 265 (2000);
D. Ayres et al., physics/9911009; N. Holtkamp et al., "A neutrino source based on a muon storage ring," Fermilab-Pub-00-108-E (2000); S. Ozaki et al., "Study II," BNL-52623 (2001);
M. Yoshida, *Nucl. Phys. Proc. Suppl.* **149**, 94 (2005); D. Kaplan, COOL 05.
7. C. Albright et al., physics/0411123.
8. G. Budker, *AIP Conf. Proc.* **352**, 4 (1996); *ibid.,* **352**, 5 (1996); A. Skrinsky, *ibid.,* **352**, 6 (1996);
D. Neuffer, *AIP Conf. Proc.* **156**, 201 (1987); *Nucl. Instrum. Meth.* **A350**, 27 (1996);
R. Palmer et al., *Nucl. Phys. Proc. Suppl.* **51A**, 61 (1996); *AIP Conf. Proc.* **372**, 3 (1996);
R. B. Palmer, J. C. Gallardo, A. Tollestrup, and A. Sessler, *Sci. Cult.* **13**, 39 (1998);
R. Raja, A. Tollestrup, *Phys. Rev.* **D58**, 013005 (1998); C. Ankenbrandt et al., *Phys. Rev. ST Accel. Beams* **2**, 081001 (1999); M. Alsharo'a et al., *Phys. Rev. ST Accel. Beams* **6**, 081001 (2003).
9. R. Davis et al. (Homestake), *Phys. Rev. Lett.* **20**, 1205 (1968); *Astrophys. J.* **496**, 505 (1998);
Y. Fukuda et al. (SuperK), *Phys. Rev. Lett.* **81**, 1562 (1998); *Phys. Rev. Lett,* **93**, 101801 (2004);
Q. R. Ahmad et al. (SNO), *Phys. Rev. Lett.* **89**, 011301 (2002); *Phys. Rev. Lett.,* **92**, 181301 (2004);
K. Eguchi et al. (KamLAND), *Phys. Rev. Lett.* **90**, 021802 (2003); *ibid.* **94**, 081801 (2005);
M. Ahn et al. (K2K), *Phy. Rev. Lett.* **90**, 041801 (2003); E. Aliu, *ibid.* **94**, 081802 (2005).
10. S. Geer, *Phys. Rev.* **D57**, 6989 (1998); C. Albright et al., hep-ex/0008064;
V. Barger et al., *Phys. Rev. Lett.* **45**, 2084 (1980); A. Cervera et al., *Nucl. Phys.* **B579**, 17 (2000).
11. M. Maltoni et al., *New J. Phys.* **6**, 122 (2004).
12. V. Barger et al., *Phys. Rev. Lett.* **75**, 1462 (1995); *Phys. Rept.* **286**, 1 (1997);
D. Atwood and A. Soni, *Phys. Rev.* **D52**, 6271 (1995); J. F. Gunion, hep-ph/9802258.
13. J. Ellis, LCWS 04, hep-ph/0409140.
14. R. Godang et al., hep-ph/0411248; M. Cavaglia and S. Das, *Class. Quant. Grav.* **21**, 4511 (2004).
15. J. Eades and L. M. Simons, *Nucl. Instrum. Meth.* **A278**, 368 (1989).
16. J. Eades et al., *Nucl.Phys.Proc.Suppl.* **8**, 457 (1989); E. Aschenauer et al., *Sov. J. Nucl. Phys.* **55**, 856 (1992); L. M. Simons et al., *Springer Proc. Phys.* **59**, 33 (1992); *Phys. Scripta* **T22**, 90 (1988); *Hyperfine Interact.* **81**, 253 (1993); *Phys. Bl.* **48**, 261 (1992); *Nucl. Instrum. Meth.* **B87**, 293 (1994);
D. Horváth et al., *Nucl. Instrum. Meth.* **B85**, 736 (1994).
17. P. DeCecco et al., *Nucl. Instrum. Meth.* **A394**, 287 (1997); Franz.Kottmann@psi.ch.
18. A. Garren et al., "Gas Filled Rings for Muon Beam Cooling," COOL 05; A. A. Garren et al., *AIP Conf. Proc.* **297**, 403 (1994); D. J. Summers et al., *Int. J. Mod. Phys.* **A20**, 3851 (2005).
19. M. Craddock, *CERN Cour.* **44N6**, 23 (2004); D. J. Summers et al., *Int. J. Mod. Phys.* **A20**, 3861 (2005);
K. Symon, D. Kerst, L. Jones, L. Laslett, and K. Terwilliger, *Phys. Rev.* **103**, 1837 (1956);
J. S. Berg, *AIP Conf. Proc.* **642**, 213 (2003); D. Trbojevic et al., *AIP Conf. Proc.* **530**, 333 (2000);
S. Koscielniak, C. Johnstone, *AIP Conf. Proc.* **721**, 467 (2004); *NIM* **A523**, 25 (2004);
E. Keil and A. Sessler, *NIM* **A538**, 159 (2005); Y. Mori, *ICFA Beam Dyn. Newslett.* **29**, 20 (2002).
20. Lloyd P. Smith, W. E. Parkins, and A. T. Forrester, "On Separation of Isotopes in Quantity by Electromagnetic Means," *Phys. Rev.* **72**, 989 (1947); C. Niemann et al., *J. Phys.* **D36**, 2102 (2003).
21. R. C. Fernow, *AIP Conf. Proc.* **721**, 90 (2004); PAC99, eConf C990329, THP31.
22. A. Girard et al., *Rev. Sci. Instrum.* **75**, 1381 (2004); A. Zelenski et al., *ibid.* **75**, 1535 (2004).
23. L. I. Ponomarev, *Contemp. Phys.* **31**, 219 (1990); P. Strasser et al., *Phys. Lett.* **B368**, 32 (1996).
24. H. Busch, *Annalen Phys.* **81**, 974 (1926);
A. W. Chao and M. Tigner, *Handbook of Accelerator Physics and Engineering,* page 101 (1999).

The Muon Cooling RF R&D Program

Y. Torun*, A. Bross†, D. Li**, A. Moretti†, J. Norem‡, Z. Qian†,
R. A. Rimmer§ and M. S. Zisman**

Illinois Institute of Technology, Chicago, Illinois 60616, USA
†*Fermi National Accelerator Laboratory, Batavia, Illinois 60510, USA*
***Lawrence Berkeley National Laboratory, Berkeley, California 94720, USA*
‡*Argonne National Laboratory, Argonne Illinois 60439, USA*
§*Jefferson Lab, Newport News, Virginia 23606, USA*

Abstract. Cooling muon beams in flight requires absorbers to reduce the muon momentum, accelerating fields to replace the lost momentum in the longitudinal direction, and static solenoidal magnetic fields to focus the muon beams. The process is most efficient if both the magnetic fields and accelerating fields are high and the rf frequency is low. We have conducted tests to determine the operating envelope of high-gradient accelerating cavities in strong static magnetic fields. These studies have already produced useful information on dark currents, magnetic fields and breakdown in cavities. In addition to continuing our program at 805 MHz, we are starting to test a 201 MHz cavity and are planning to look at a variety of appropriate geometries and materials. In parallel with these activities, we are supporting R&D on models and surface structure.

PACS: 29.17.+w, 52.80.Vp, 73.22.2f

IONIZATION COOLING

Neutrino Factories and Muon Colliders hold the promise of neutrino and muon beams of unprecedented intensity, energy and quality and there is a strong worldwide effort on design studies and hardware R&D toward the goal of building such facilities [1].

Ionization cooling [2] appears to be the only practical option for generating very high brightness muon beams [3] since the standard techniques (stochastic and electron cooling) are too slow for muons. Muons can be generated in large numbers only at low energies and therefore decay in a few microseconds. Transverse cooling involves simply passing the muon beam through an absorbing medium to shrink the momentum through ionization energy loss, and then restoring the longitudinal component of the momentum by rf acceleration, reducing the transverse emittance as a result. Efficient cooling using this method requires high-field solenoids for strong focusing, which rules out use of superconducting rf cavities.

The Neutrino Factory and Muon Collider Collaboration (NFMCC) [4] has been working to solve the technical challenges in Neutrino Factory and Muon Collider design [5]. Part of this effort is MuCool, the R&D program at Fermilab for developing components for muon ionization cooling, including liquid-hydrogen absorbers [6, 7], solenoid magnets and rf cavities. The MuCool program also covers instrumentation, high-power beam tests and support of MICE [8] (the international Muon Ionization Cooling Experiment), an experiment using MuCool-developed components that is now under construction at RAL [9].

MUCOOL RF R&D

The evolution of normalized transverse emittance, ε, of a muon beam as a function of distance s in a solenoidal channel is given by [10]

$$\frac{d\varepsilon}{ds} \simeq \frac{\left\langle \frac{dE}{ds} \right\rangle}{\beta^2 E} (\varepsilon - \varepsilon_0) \qquad (1)$$

where β and E are the average muon speed and energy and ε_0, equilibrium emittance, is given by

$$\varepsilon_0 \simeq -\frac{0.875 \text{MeV}}{\left\langle \frac{dE}{ds} \right\rangle X_0} \beta_\perp. \qquad (2)$$

The equilibrium emittance is directly proportional to the focusing length $\beta_\perp$, and inversely proportional to the energy loss per radiation length X_0. It is often referred to as the figure of merit for absorbers. In liquid hydrogen, which has the best figure of merit, energy loss is about 30 MeV/m. To replace lost momentum in a hydrogen absorber, one needs rf cavities with large enough aperture for the muon beam. Since muon beams can go through thin windows unperturbed, 201 MHz cavities with pillbox geometry have been proposed to make efficient use of rf power. They take up about four times the length of hydrogen for 16 MV/m peak accelerating gradient (which includes an extra factor of two for phase focusing) in the Feasibility Study II cooling channel [11] which is also the basis for MICE.

805 MHz Program

To gain experience with operation of rf cavities under these unusual conditions and to form a better understanding of the underlying physics and technical issues, initial tests were conducted at 805 MHz, in effect working with quarter-scale models, which were cheaper to build and easier to handle. The Lab-G facility at Fermilab was equipped with an rf system to operate a spare 805-MHz 12-MW klystron from the Fermilab Linac and a shielded cave for these tests. A superconducting magnet with 44 cm warm bore and two coils in a Helmholtz configuration was installed inside the cave. The coils can be powered individually, so it is possible to get various field distributions including a solenoidal mode for which the magnet is rated at 5 T as well as a gradient mode similar to that of the focus coils used in Study II and MICE.

First, a 1-m-long 805-MHz 6-cell open-iris cavity was installed in the magnet. This cavity had thin titanium windows at the ends which allowed dark current measurements on axis [12]. Dark currents from field emission appear to be precursors to damaging sparks in cavities [13], have a very steep dependence on gradient [14] and provide insight into the condition of the interior surfaces. A variety of diagnostics was used to characterize operating parameters and radiation around the cavity [15].

Next, a single-cell pillbox cavity with removable end plates was installed and used to map out the high-power conditioning behavior as a function of magnetic field. Although the cavity went up to 40 MV/m at zero field, the maximum stable operating gradient degraded quickly with the field turned on, dropping to about 35 MV/m at 0.25 T and 13.5 MV/m at 4 T [16].

Another approach to suppressing breakdown in cavities using high-pressure gas was also tested in Lab-G with an 805 MHz pillbox prototype [17, 18]. Such pressurized cavities have the potential to provide ionization cooling for muon beams in addition to acceleration [19].

Materials Research

Reliable operation of cavities in high magnetic fields is essential for a Neutrino Factory cooling channel and high dark currents would be a potential problem for MICE [20]. An important part of the rf program is to identify practical materials, coatings and surface processing techniques that will allow cavities to withstand high gradients. Any deviations from a perfectly smooth surface on the interior of a cavity left over from the manufacturing process can lead to local enhancement of the electric field and increased dark current. This can generate concentrated heating and very large forces on the surface causing melting [21], mechanical break-up [12] and field evaporation of clusters [22, 23]. Sparks can cause pitting of the Cu surface and deposition of molten Cu over the interior and the resulting rough features on the surface can act as new sources of emission. Analysis of dark current data from MuCool cavities seems to indicate that the observed gradient limits are consistent with the tensile strength of Cu. In addition, visual inspection of the cavities after operation in Lab-G has revealed useful hints on mitigating breakdown problems. When the Cu end plates that showed heavy pitting were replaced with TiN coated Be windows, the damage was confined to Cu surfaces. A special end flange has been built for the 805 MHz pillbox cavity that can accommodate buttons of different materials. Using buttons with curvature so that the surface field is highest in that region, we will systematically explore the limits of different materials and coatings.

MuCool Test Area

A dedicated long-term test facility with rf power, cryogenics infrastructure and high-power beam is required to carry out MuCool R&D on rf cavities, liquid-hydrogen absorbers and instrumentation. The south end of the Fermilab Linac was a suitable location to build such a facility [24, 25]. Civil construction was completed in 2003 and the rf hardware has now been moved there (see Fig. 1). Both 12 MW at 805 MHz and 5 MW at 201 MHz are available from the Linac. A beamline to bring the 400-MeV Linac H$^-$ beam into the area is designed and installation is expected in the near future.

201 MHz Cavity

The full-scale prototype cavity has a rounded cylindrical profile of about 1.2 m diameter and a 43 cm gap. It was built [26] by e-beam welding half-shells spun from

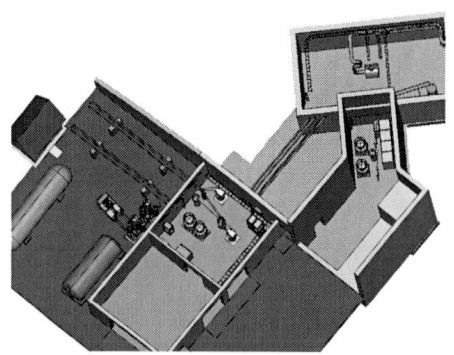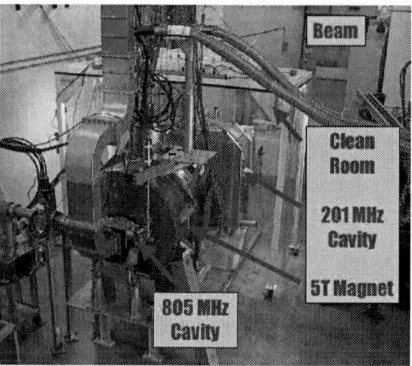

FIGURE 1. Layout of the MTA facility (left) and experimental hall (right)

6-mm-thick Cu sheet. Cooling tubes are welded to the outer surface and ports pulled on the equator for vacuum, rf power and diagnostic connections. Tuning and beam iris rings were also e-beam welded to each side. The interior was processed by mechanical polishing, light chemical cleaning and electropolishing for a clean and smooth finish to minimize field emission. In MICE, like in a Neutrino Factory cooling channel, the cavities would be in vacuum. Since this prototype is intended for testing in air, the cavity body is attached to thick aluminum plates to support it against atmospheric pressure. These plates also include attachment points for tuning rods to adjust the frequency through slight elastic deformation of the cavity shape and a frame to support its considerable weight. The cavity irises are terminated by 42-cm-diameter 0.38-mm-thick curved Be windows mounted with the concave/convex sides pointing in the same direction to minimize the frequency shift when they are distorted due to heating during high power operation. A detailed finite-element analysis has been performed [27] to establish the rf, thermal and structural behavior of the cavity. The design values [28] for the frequency, shunt impedance and quality factor are 201.25 MHz, 22 MΩ/m and 53500, respectively, and the cavity should provide a gradient of 8 MV/m at 1 MW and 16.2 MV/m at 4.6 MW input power. The cavity has been installed at the MTA and will be powered soon.

STATUS

The 805 MHz cavity is ready for high-power testing in a magnetic field with a pair of curved Be windows to establish stable operation. A program of systematic tests using buttons made of different materials and coatings will follow, to compare their suitability and to benchmark models of rf breakdown. Installation of the 201 MHz cavity should be complete soon and we intend to make the first measurements of radiation around the cavity at 8 MV/m to check previous estimates for MICE detector backgrounds. Successful conditioning and reliable operation of this full size prototype cavity is an important milestone in the MuCool rf program. It is desirable to operate this cavity in a large magnetic field as well. However, we do not yet have a magnet with large enough bore available for this purpose (a prototype for the coupling coil [29] in the cooling channel will be built in the future as funds allow). Since the effect of the magnetic field turns on rapidly at fields as low as 0.1 T which is the level of fringe field from the existing solenoid at one end of the cavity, we expect some guidance from operating this cavity with the magnet turned on.

ACKNOWLEDGMENTS

This work was supported by the U. S. Department of Energy, Office of High Energy Physics and the Illinois Board of Higher Education.

REFERENCES

1. Y. Torun, H. Haseroth, S. Machida and T. Yokoi, Nucl. Phys. B (Proc. Suppl.) 149, p. 251 (2005).
2. G. I. Budker and A. N. Skrinsky, Sov. Phys. Usp. 21, 277 (1978).
3. D. Neuffer, Particle Accelerators 14, 75 (1983).
4. http://www.cap.bnl.gov/mumu/
5. M. M. Alsharoa et al., Phys. Rev. ST Accel. Beams 6, 081001 (2003).
6. M. A. Cummings, these proceedings.
7. C. Darve, B. Norris and L. Pei, these proceedings.
8. http://mice.iit.edu/
9. D. M. Kaplan, these proceedings.
10. D. Neuffer, in Advanced Accelerator Concepts, ed. F. E. Mills, AIP Conf. Proc. 156 (AIP, New York, 1987), p. 201; R. C. Fernow and J. C. Gallardo, Phys. Rev. E 52, 1039 (1995).
11. *Feasibility Study-II of a Muon-Based Neutrino Source*, ed. S. Ozaki et al., BNL Report 52623 (2001).
12. J. Norem et al., Phys. Rev. ST Accel. Beams 6, 072001 (2003).
13. N. S. Xu and R. V. Latham, J. Phys. D: Appl. Phys. 27, 2547 (1994).
14. R. Fowler and L. Nordheim, Proc. R. Soc. A 119, 173 (1928).
15. P. Gruber and Y. Torun, Proceedings of the 2003 Particle Accelerator Conference (PAC03), Portland, OR, 2003 (IEEE, New York, 2003), p. 1413.
16. A. Moretti et al., Phys. Rev. ST Accel. Beams 8, 072001 (2005).
17. P. M. Hanlet et al., Proceedings of the 2005 Particle Accelerator Conference, Knoxville, TN, 2005.
18. K. Yonehara et al., Nucl. Phys. B (Proc. Suppl.) 149, pp. 286-288 (2005).
19. R. Johnson et al., these proceedings.
20. R. Sandström, Nucl. Phys. B (Proc. Suppl.) 149, pp. 301-302 (2005).
21. Proceedings of the 2003 Particle Accelerator Conference (PAC03), Portland, OR, 2003 (IEEE, New York, 2003), p. 1284.
22. J. Norem, Z. Insepov and I. Konkashbiev, Nucl. Instr. and Meth. In Phys. Res. A 537, 510 (2005).
23. Z. Insepov, J. Norem and A. Hassanein, Phys. Rev. ST Accel. Beams 7, 122001 (2004).
24. M. Popovic, Proceedings of the 2004 European Particle Accelerator Conference (EPAC2004), Lucerne, Switzerland, 2004.
25. C. Johnstone, A. Bross and I. Rakhno, Proceedings of the 2005 Particle Accelerator Conference, Knoxville, TN, 2005.
26. R. A. Rimmer et al., *ibid*.
27. S. Virostek and D. Li, *ibid*.
28. D. Li et al., Proceedings of the 2003 Particle Accelerator Conference (PAC03), Portland, OR, 2003 (IEEE, New York, 2003), p. 1243.
29. M. A. Green et al., Proceedings of the 2005 Particle Accelerator Conference, Knoxville, TN, 2005.

Mucool Hydrogen Absorber R & D

Mary Anne Cummings, for the Muon Collaboration

Muons, Inc., Batavia, IL 60510
Northern Illinois University, DeKalb, IL 60115

Abstract. The Mucool hydrogen absorber program will be presented. An update of current projects will be described, and the next year's plan will be reviewed, along with efforts in collaboration with the Muon International Cooling Experiment.

Keywords: Hydrogen, muon cooling, muon collider, neutrino factory

INTRODUCTION

The possibility of building a muon accelerator has been made more realizable by recent progress in the design of cooling channels. Muon beams at the required intensity for neutrino factories or muon colliders can only be produced into a large phase space and need to be focused fast enough to outlive the 2.2 μs muon lifetime. Ionization cooling, in which muons repeatedly traverse an energy absorbing medium, alternating with accelerating RF cavities within a strongly focusing magnetic lattice, addresses the latter challenge. The ionization energy loss, dE_μ/ds, decreases all three muon momentum components without affecting beam size, reducing the overall phase space by reducing the transverse momentum spread of the beam. The design of ionization cooling channels have been motivated by the following equation for the rate of change of the normalized transverse emittance [1]:

$$\frac{d\varepsilon_n}{ds} = -\frac{1}{\beta^2}\frac{dE_\mu}{ds}\frac{\varepsilon_n}{E_\mu} + \frac{1}{\beta^2}\frac{\beta_\perp (0.014)^2}{2E_\mu m_\mu L_R} \quad (1)$$

where s is the path length, E_μ is the beam energy in GeV, $\beta = v/c$, L_R is the radiation length of the absorber material and $\beta_\perp$ is the betatron function describing the size of the beam. The first term describes the cooling from ionization loss. The second term describes beam heating and is minimized when absorbers are placed in a strong focusing field (low $\beta_\perp$) and consist of material of a low atomic number (high L_R). Setting the cooling and heating terms equal defines the equilibrium emittance, the very smallest possible with the given parameters:

$$\varepsilon_n^{(equ)} = \frac{1}{2\beta}\frac{\beta_\perp (0.014)^2}{(dE_\mu/ds)m_\mu L_R} \quad (2)$$

A figure of merit can be defined as $[L_R dE\mu/ds)]^2$, the square coming from the two transverse directions. Figure 1 shows this number for several materials, with hydrogen, in its liquid or gaseous state, being optimal for the lowest achievable emittance.

TABLE 1. Cooling figures of merit for various light materials relative to liquid hydrogen.

Material	<dE/ds> (MeV g^{-1} cm^2)	L_R (g cm^{-2})	Merit
GH$_2$	4.103	61.28	1.03
LH$_2$	4.034	61.28	1
He	1.937	94.32	0.55
LiH	1.94	86.9	0.47
Li	1.639	82.76	0.30
GH$_4$	2.417	46.22	0.20
Be	1.594	65.19	0.18

THE MUCOOL COLLABORATION

The MuCool Collaboration consists of 18 institutions from the U.S., Europe and Japan. The mission is 1) to design and prototype cooling channel components; 2) perform a high-power beam test of a cooling cell; and 3) work with the Muon International Cooling Experiment (MICE), particularly with the design and building of cooling channel elements.

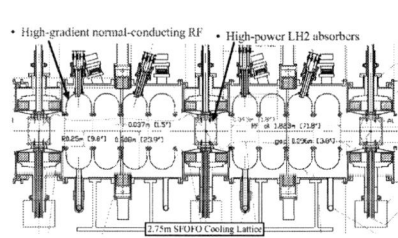
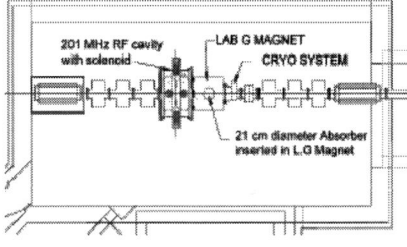

FIGURE 1. Cooling channel from Neutrino Factory the Study II.

FIGURE 2. Plan view of a complete cooling cell test at MTA.

The MuCool R&D efforts have been based on the SFOFO cooling lattice described in the Neutrino Factory Study II [2]. This lattice includes 201 MHz RF cavities {Figure 1). The MICE cooling channel is based on this design, and one goal of the MuCool program is to construct and test elements from this channel together in a high intensity proton beam in the newly completed MuCool Test Area (MTA). Additionally, an FNAL study group was formed to design the beamline that will be located in the old Linac access tunnel and brought into the MTA for a proposed high-power test of a complete cooling cell (Figure 2). The beamline has been designed, costed and approved, and a safety analysis and shielding assessment has been completed for the MTA. Beam is scheduled for sometime in 2006.

THE MUCOOL TEST AREA AT FNAL

In late 2003, construction of the MTA was completed at the site of the old access tunnel to the FNAL Linac, and is now the focus of MuCool activity at FNAL. The MTA is designed to accommodate the full Linac intensity of 1.6×10^{13} protons/pulse at 15Hz or 2.4×10^{14} protons/s, or 600W energy deposition into a 35 cm long liquid hydrogen absorber at 400 MeV. The test area will provide power from the Linac to operate 201 MHz and 805 MHz RF test cavities. A first LH_2 test has been completed with the KEK convection absorber, and RF testing (both 805 and 201 MHz) is planned. A cryogenic facility is scheduled for completion near the end of 2005 that will provide refrigeration power of up to 350W for the hydrogen absorber, as well as cooling for the superconducting solenoid.

LH_2 ABSORBERS

The issues that drive the absorber design and tests are: 1) the large amount of heat that needs to be extracted from high intensity beams; 2) the desire to minimize multiple scattering; and 3) the densely-packed and high radiation environment in which absorbers will be operating in a real cooling channel. Additionally, the combustive nature of hydrogen requires special safety considerations that drive much of the engineering and design.

Minimizing the multiple scattering has lead to novel window designs that depart from the standard shell profiles [3]. For efficient heat removal, two different absorber designs have been proposed: 1) an internal heat exchange design, where the LH_2 mixing is achieved by natural convection cells, driven by the beam-deposited heat and the cold walls of the heat exchanger (with cold He gas) and 2) an external heat exchange design where the heat exchanger is in an external loop of hydrogen and the LH_2 mixing is achieved with an external pump with nozzles oriented at various angles to establish turbulent flow. These are shown in Figure 4.

Thin Window Designs and Test

Designs for thin windows have matured, where windows of equivalent strength to standard designs have been realized with less than 30% their minimum thickness. Figure 3 shows the evolution of the window profiles. The current design has a profile with an inflected curvature where the thinnest section (around the center) is membrane-stress dominant: under the ultimate pressure (rupture) the greatest stress is experienced at the center.

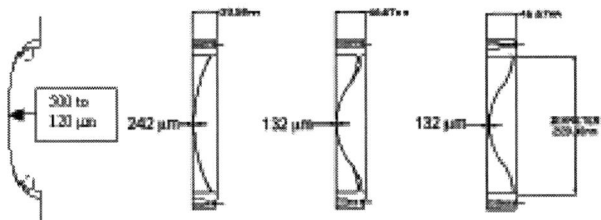

FIGURE 3. Evolution of thin window profiles. From left to right: standard torispherical shell with tapered ends to a normal surface, tapered torispherical to flange, inflected to flange, and thinned inflected.

Earlier burst tests on thin membrane-stress dominated windowed did not produce shards, and demonstrated a window strength that was consistent with predictions from finite element analysis (FEA) predictions. To test the window performance, photogrammetry was used as a non-contact, large sampling method of measuring window profile deformation under pressure. This technique uses points of light projected onto the window surface, whose space locations were determined by a camera programmed to make parallax calculations [4]. The system has been upgraded to include a new projector lens and new camera software. Additionally, new methods of optical coating, including vapor deposition, are under consideration, as the thinness of the windows makes the variations in the thickness of the TiO optical coating manually applied a potential source of error. A set of safety requirements for the window design have been established for the MICE and MuCool experiments, and the MuCool window approach has passed the initial MICE safety review. A certification procedure for windows used in the cooling cells is being developed.

Absorber Manifold Design

The program for the absorber manifolds is proceeding in parallel paths. The external heat exchange, or forced-flow prototype has been built, and is currently set up in Lab P8 at FNAL for room temperature flow tests using an infrared camera. A design for an absorber manifold, cryostat and external hydrogen loop with pump and heat exchange is near completion and will be the first absorber tested in a beam at the MTA (Figure 4, right). The cryostat is designed to slide into the solenoid from Lab G, with the absorber in the magnets center. This is the absorber planned for the first complete cooling cell test.

A prototype convection absorber design has been manufactured and tested by the KEK/Osaka group (Figure 4, left) [5]. This prototype was delivered to the MTA late in 2003. The absorber is cooled by cold helium gas (~15K) from dewars, circulated behind the wall of the absorber interior. Heating coils are installed at the absorber bottom and center using warm helium gas. In August 2004, after a lengthy safety review, the absorber was completely filled with liquid hydrogen, at a temperature of 18K maintained stably over a period of several hours. Several operational issues were successfully resolved and the safety and controls system refined with this first run. Initial heat loading tests indicated that heat absorption of 20W or more can be accomplished at a stable temperature. Plans for the next test in will include upgrades to the absorber instrumentation: new, better resolution temperature probes, a level sensor, and temperature probes placed inside the input and output cold helium for better determination of cooling efficiency.

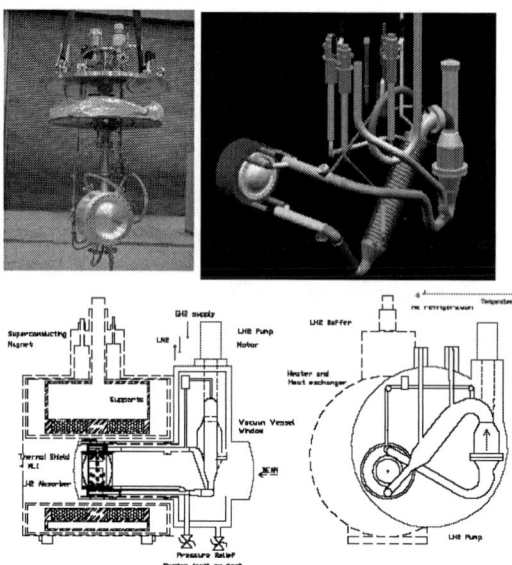

FIGURE 4. Two LH$_2$ absorber prototypes, convection type, internal heat exchange (left) and forced-flow, external heat exchange (right). The far right picture shows the forced-flow absorber inside of a solenoid for the MTA tests.

Results from both the KEK tests and the room temperature flow tests will be compared with flow simulations of both prototype designs to predict performance in a real beam.

HIGH-PRESSURE RF CAVITIES

A novel approach to ionization cooling is being developed that uses high-pressure hydrogen gas at liquid nitrogen temperatures, with a density 0.5 that of liquid hydrogen. This idea of filling RF cavities with gas is new for particle accelerators, and is only possible for muons since they neither scatter as do strongly interacting protons, nor shower as do less-massive electrons. Instead of discrete absorbers, the gaseous hydrogen would fill the accelerating RF cavities. Dark currents are suppressed due to Paschen's Law, a property first observed in 19th century when sparking in vacuum tubes was suppressed when the tubes where filled with a pressurized inert gas. Consequently, a much higher gradient is possible in a hydrogen-filled RF cavity than is needed to overcome the ionization energy loss, provided one can supply the required RF power. Hydrogen is also twice as effective as helium in ionization cooling effectiveness, viscosity, and heat capacity.

Measurements by Muons, Inc. and IIT at FNAL have demonstrated that hydrogen gas suppresses RF breakdown very well, about a factor six better than helium at the same temperature and pressure [6]. A small RF cavity test cell was built and run at

FNAL's Lab G using power from the Lab G klystron. The cell was run up to 80 MV/m after conditioning and before breakdown with a H_2 gas pressure of 31 atm at 80K. The next tests will be conducted at the MTA at FNAL. Current R & D plans include tests of pressurized RF cavities in magnetic fields and high radiation environments, and the use of new cavity construction materials, including beryllium RF windows for improved cavity performance. A beam test is scheduled for 2006 in the MTA.

SUMMARY

The MuCool collaboration has successfully resolved many outstanding technical and safety issues surrounding the use of hydrogen. The MTA is complete, on budget and on schedule. A major milestone was accomplished with the first absorber test with liquid hydrogen. The work involved in achieving safe, stable operation of the KEK convection type absorber builds a foundation for all subsequent reviews of absorber operation, including a high-powered run with a complete cooling cell. This test has also pushed the development of absorber instrumentation, and the work on safe operation at FNAL has expedited the successful completion of the first MICE safety review. The thin window design and testing have matured, and the window design has been adapted for the 201 MHz RF cell. An operational beamline is on schedule to run as early as 2006 in the MTA for tests on LH_2 and GH_2 absorber prototypes.

REFERENCES

1. D. Neuffer, μ^+-μ^- Collider, CERN 99-12 (1999).
2. M. Zisman and J. Gallardo, eds., "Feasibility Study-II of a Muon-based Neutrino Source" BNL-52623, June (2001). http://www.cap.bnl.gov/mumu/studyii/FS2-report.htm
3. M. A. C. Cummings et al, "The MuCool/MICE LH_2 Absorber Program", Nufact03 Conference Proceedings, (2003).
4. E. M. Mikhail, J. S. Bethel, and J. C. McGlone, Introduction to Modern Photogrammetry, New York, John Wiley & Sons, Inc. (2001).
5. M. A. C. Cummings and S. Ishimoto, "Progress on the Liquid Hydrogen Absorber for the MICE Cooling Channel", PAC 2005, paper WPAE022, (2005).
6. R. Johnson et al, "High Pressure, High gradient RF Cavities for Muon Beam Cooling", LINAC2004, Lubeck, Germany (2004).

Cryogenics for the MuCool Test Area (MTA)

Christine Darve, Barry Norris, Liujin Pei

Fermilab, Cryogenics department, MS347
Batavia, Illinois, 60510

Abstract. MuCool Test Area (MTA) is a complex of buildings at Fermi National Accelerator Laboratory, which are dedicated to operate components of a cooling cell to be used for Muon Collider and Neutrino Factory R&D. The long-term goal of this facility is to test ionization cooling principles by operating a 25-liter liquid hydrogen (LH_2) absorber embedded in a 5 Tesla superconducting solenoid magnet. The MTA solenoid magnet will be used with RF cavities exposed to a high intensity beam. Cryogens used at the MTA include LHe, LN_2 and LH_2. The latter dictates stringent system design for hazardous locations. The cryogenic plant is a modified Tevatron refrigerator based on the Claude cycle. The implementation of an in-house refrigerator system and two 300 kilowatt screw compressors is under development. The helium refrigeration capacity is 500 W at 14 K. In addition the MTA solenoid magnet will be batch-filled with LHe every 2 days using the same cryo-plant. This paper reviews cryogenic systems used to support the Muon Collider and Neutrino Factory R&D programs and emphasizes the feasibility of handling cryogenic equipment at MTA in a safe manner.

Keywords: Muon cooling, cryogenics, hydrogen absorbers, helium plant
PACS: 07.20.Mc

INTRODUCTION

Cooling channels are fundamental components for the development of a future Muon Collider or Neutrino Factory [1]. Because muons decay very quickly, ionization cooling appears as the preferred principle to reduce the normalized emittance of muon beams. These arguments drive the choice of a low-Z medium in order to reduce the Coulomb multiple scattering and maximize the ionization energy loss. The optimal candidates are thus hydrogen (as a fluid) and beryllium or aluminum (as metals). High gradient RF cavities are necessary to restore the beam's longitudinal momentum after cooling. The choices made in each step, cooling and reacceleration, will affect the final quality of the beam that can be achieved.

The cryo-engineering aspect of this research must address several challenges. The first is to design a test facility according to Fermilab hydrogen safety standards while housing absorbers filled with LH_2 contained in a manifold closed by two extremely thin aluminum windows [2]. Another technical challenge is the insertion of a LH_2 absorber into a high-magnetic field at the proximity of RF cavities without violating electrical safety standards.

In the present paper, we review the characteristics of the temporary and future MTA cryogenic systems to support MuCool research.

USER REQUIREMENTS AND CRYOGENIC FACILITIES

In terms of cryogenic fluid requirements, MTA facilities provide hydrogen, helium and nitrogen to the users. One of the purposes of the MTA building complex is to test two types of LH_2 absorbers: a convection-style one and a forced-flow one. In both cases, the ultimate goal of the testing is to deposit the maximum heat load to the LH_2 absorber volume, while limiting LH_2 density fluctuations in the absorber volume. The MTA test facility will also provide cooling to a 5 Tesla NbTi superconducting solenoid magnet.

Operations based on Dewar Fed LHe System

The mechanical, electrical and control approaches of this project were executed according to Fermilab design and review procedures. Hydrogen gas is stored in a separate building. An outdoor 21 m^3 vacuum buffer tank is connected to the cryostat vacuum vessel. The function of this vacuum buffer tank is to collect gas expanded to standard temperature and pressure in case of rupture from the hydrogen volume to the vacuum volume. This absorber was designed and built by KEK to validate the International Muon Ionization Cooling Experiment (MICE) absorber design [3]. The LH_2 absorber was instrumented with platinum-cobalt type temperature sensors and contained in a cryostat before being shipped to Fermilab. The beam energy deposition is simulated by warm helium flowing through a coil located inside of the LH_2 absorber volume. This dynamic heat load is transferred from the LH_2 to the cold helium flow through a manifold-integrated heat exchanger by means of convection heat transfer. Numerical simulations support the assumption that up to 100 W can be extracted from the LH_2 absorber by natural convection heat transfer. User requirements were to integrate the KEK cryostat in a system to operate at LH_2 conditions. Because the final refrigeration system is not operational yet, cold helium was transferred from a 500 liter LHe Dewar located outside of the experimental hall to the KEK cryostat through a 25-meter long transfer line. This first run has validated the feasibility of MTA control processes, instrumentation and safety equipments. Following the commissioning of the test, twenty different test points (different LH_2 temperature and pressure conditions) were investigated. Results are reported together with reflections to upgrade both the cryo-system and KEK absorber [2] [4-5].

Similar cryogenic installation is under construction to feed the superconducting solenoid magnet. The solenoid magnet liquid helium and nitrogen vessels are refilled every 2 days from the Dewar located outdoors.

A Future Continuous Helium Supply

Work is presently underway to install the required refrigeration system for long-term MuCool research. User requirements are to design the cryostat and cryo-loop integrating the forced-flow LH_2 absorber in the bore of the superconducting solenoid magnet and, simultaneously, to provide cryogenic facilities capable of delivering appropriate refrigeration. A continuous flow of 27 g/s, 14 K helium is necessary to cool the 25 liters of sub-cooled LH_2 flowing in a closed loop composed of a 6.9-liter absorber, a LH$_2$ pump, a He/H$_2$ counter-flow heat exchanger and a heater. Although a conceptual design for the forced-flow system was completed in 2002 [6-7] an alternative solution using the existing cryostat from the SAMPLE experiment is being considered [8]. Numerical simulations were necessary to size the cryo-system and to optimize the LH$_2$ flow hydro-dynamics [7] so that up to 300 W can be deposited by the dummy beam to the LH$_2$ absorber.

A refrigerator and compressor system are being installed to fulfill this specification. The compressor room houses rotating machinery, i.e. two (2) Sullair 300 kilowatt 2-stage oil injected screw compressors with associated after-coolers and coaleasers. Each compressor is hooked up to a purification skid to remove oil vapor contamination from helium gas. Each purification skid includes a coalescing filter, a charcoal adsorber, and a final filter. The compressor system is connected to a Tevatron satellite refrigerator style, which was modified to match the MTA requirements. One helium and one nitrogen storage tank are located outdoors. A transfer line connects the refrigerator complex to the experimental hall and will be used to feed the superconducting solenoid magnet and the forced-flow cryostat.

The full capacity of the MTA future refrigerator complex will permit two operational temperature ranges: 14 K to support the forced-flow LH$_2$ absorber system and 5 K to "batch fill" the superconducting solenoid magnet. Therefore the cryo-plant design approach is to allow the refrigeration system to switch from 14 K to 5 K mode. Figure 1 shows the flow schematic of the MTA refrigeration system.

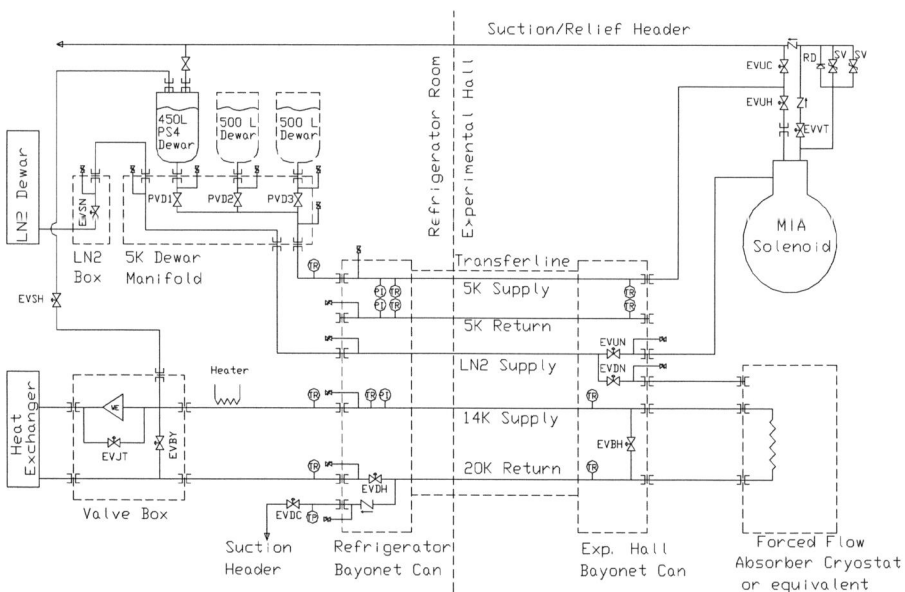

FIGURE 1. Overall scheme of MTA refrigeration including support of the 5 K helium needs for the superconducting solenoid magnet and the 14 K system for absorber experiments.

ELECTRICAL COMPONENT SOLUTIONS

The design of the different cryo-systems (forced-flow LH_2 absorber, KEK LH_2 absorber and superconducting solenoid magnet) utilizes the expertise of existing US hydrogen experiments [8-9] and Fermilab LH_2 target guidelines. Due to the presence of hydrogen in the experimental hall, all ignition sources must be rated Class I (areas with flammable vapors), Division 2 (hazard normally not present), group B (hazard is hydrogen) in accordance with the National Electrical Code (NEC). If these conditions cannot be met then devices must be engineered intrinsically safe or placed in a shunt trip circuit. For this purpose a nitrogen purge box is located in the experimental hall to house non-code devices and electronics, which are in proximity of devices located inside the LH_2 absorbers. For instance, Cernox®, platinum-cobalt, platinum and carbon temperature sensors are excited from electronics located in the experimental hall purged box. Cernox® is the temperature sensor of choice due to its reliable performances at cryogenic temperatures and under magnetic field. Hydrogen safety requirements also drive the use of diffusion pump, capacitance-type vacuum gauges, MC type cable and PLTC cables. The experimental hall ventilation system is monitored by a flow switch mounted at the end of the ventilation duct, which would become a hazardous area in the presence of hydrogen in the experimental hall. Therefore, an intrinsically safe barrier is required to connect the electrical cable to the I/O box providing read-out. Interlocks set the system in fail-safe mode if hydrogen or helium were to leak in the experimental hall. Further, flammable gas detectors, ODH detectors and purged-system pressure transducers are continuously monitoring the MTA building conditions through the QUADLOG® safety PLC. Alarms are activated if the read-out is beyond the chosen set values. Depending on the degree of emergency, audio and visual warnings, automatic fluid venting, or activation of the FIRUS system (Fermilab Fire Department is alerted) can occur. Additional engineering solutions were considered throughout the MTA projects demanding developments [5].

Hydrogen safety issues can trigger a broad variety of answers [8-10]. For instance, the NFPA panel is delegated to generate H_2 standards and create a Proposed Hydrogen Technology Correlating Committee. Although safety standards vary according to laboratory policy, many guidelines exist to run a safe but realistic hydrogen project.

PROCESS AND CONTROL SYSTEM SOLUTIONS

Two types of programmable logic controller (PLC) technology are used for the operation of LH_2 absorber and superconducting solenoid magnet. In the case of the superconducting solenoid magnet operation, the data acquisition (DAQ) and control system are relaxed from explosion proof criteria and a conventional PLC was implemented to automatically supply helium and nitrogen to the solenoid magnet. The implementation of the instrumentation and controls for the MTA helium cryo-plant in the compressor and refrigerator rooms are based on Tevatron standards.

Finally, the data-acquisition system and control must accurately monitor a hazardous environment like flammable gas, and oxygen deficiency hazard. The Fermilab safety committee requires the use of a safety PLC when operating a test with hydrogen presence. Siemens-Moore QUADLOG® safety PLC is designed for this type of application and permits us to operate in a safe manner using dual architecture. QUADLOG® makes use of device redundancy. It provides a very reliable system with the fault tolerance and special self-testing software. QUADLOG® safety PLC is used for continuous operations or batch processing and is fail-safe. QUADLOG® is a programmable control system designed specifically for critical applications, such as emergency shutdown systems, fire and gas detection systems. QUADLOG® combines the beneficial features of a PLC (such as modularity, ladder logic and sequential programming) with high safety, high availability, and extensive diagnostics. It also incorporates continuous PID control, analog I/O, and a variety of operator interface options not typically available from a PLC.

CONCLUDING COMMENTS

The Fermilab MTA facility is an operational hydrogen facility equipped with cryogenic capacity. The cryogenic utilities and developments supporting the MuCool research were described. The MTA was fully designed to comply with hydrogen project guidance. A first LH_2 absorber was successfully tested using a temporary liquid helium supply system. This first hydrogen test permitted a validation of the reliability of the process system operating under a hazardous environment.

ACKNOWLEDGMENTS

The authors want to thank the Cryogenic Department for the donation of equipment and for its technical availability to support the work. Fermilab is operated by Universities Research Association Inc. under contract No. DE-AC02-76CH03000 with the U. S. Department of Energy.

REFERENCES

1. D. Kaplan et. al., "Muon-cooling research and development", *Nucl. Instr. and Meth*, 241-248 (2004).
2. M. Cummings, "Progress in the Liquid Hydrogen Absorber for the MICE Cooling Channel", submitted to PAC'05 proceeding (2005).
3. S. Ishimoto et. al., "Convection-type LH_2 absorber R&D for muon ionization cooling", *Nucl. Instr. and Meth.* **A 503,** 2003, pp. 396.
4. B. Norris et. Al, "Reflections on Initial MTA Tests", FNAL/Cryogenic Dept. Iinternal Note (2004).
5. A. Bross et. al, "An upgrade for the MuCool Test Area", submitted to CEC'05 proceeding (2005).
6. C. Darve et al., "Cryogenic Design for a Liquid Hydrogen Absorber System", Proceeding of ICEC19, edited by G. Gistau Baguer and P.Seyfert, Narosa, Grenoble, France, 2002, pp. 593-596.
7. C. Darve et al., "The Liquid Hydrogen System for the MuCool Test Area," in *Advances in Cryogenic Engineering* **49A**, edited by J. Waynert et al., Plenum, New York, 2003, pp. 48-55.
8. E.J. Beise et al., "A high power liquid hydrogen target for parity violation experiments", *Nucl. Instr.and Meth.* **A378**, 1996, pp. 383-391.
9. J. G. Weisend II et. al, "Safety Aspects Of The E158 Liquid Hydrogen Target System", Proceeding of ICEC19, edited by G. Gistau Baguer and P.Seyfert, Narosa, Grenoble, France, 2002, pp. 605-608.
10. Mike Green, "Hydrogen safety issues compared to safety issues with methane and propane", submitted to CEC'05 proceeding (2005).

G4BEAMLINE Simulations of Parametric Resonance Ionization Cooling of Muon Beams[*]

Kevin Beard[#], S. Alex Bogacz[#], Yaroslav Derbenev[#], Katsuya Yonehara[†], Rolland P. Johnson[‡], Kevin Paul[‡], Thomas J. Roberts[‡]

[#]*Jefferson Lab, Newport News, Virginia*
[†]*Illinois Institute of Technology, Chicago, Illinois*
[‡]*Muons, Inc, Batavia*

Abstract. The technique of using a parametric resonance to allow better ionization cooling is being developed to create small emittance beams so that high collider luminosity can be achieved with fewer muons. While parametric resonance ionization cooling (PIC) of muons has been shown to work in matrix-based simulations using OptiM [1] when the system is properly tuned, doing the same using a much more detailed GEANT-based g4beamline [2] simulation has been more difficult.

INTRODUCTION

The starting point for this work is a the linear channel; a half integer resonance is induced such that the normal elliptical motion of particles in x-x' phase space becomes hyperbolic, with particles moving to smaller x and larger x' as they pass down the channel. Thin absorbers placed at the focal points of the channel cool the angular divergence of the beam by the ionization cooling mechanism where each absorber is followed by RF cavities. Thus the phase space of the beam is compressed in transverse position by the dynamics of the resonance and its angular divergence is compressed by the ionization cooling mechanism.

The concepts and basic equations for PIC have been described previously [3]. An essential practical question for PIC is how well detuning effects can be compensated. Chromatic aberration, the detuning effect we first considered, is where the momentum-dependent betatron frequency causes off- momentum particles to be out of resonance with the focusing lattice. Analytic studies [4] and matrix based simulations [5] using the OptiM program showed that by choosing suitable synchrotron motion parameters, the resonance condition can be maintained. This paper reports the extension of that work using a much more detailed GEANT-based simulation program g4beamline.

SOLENOID TRIPLET CELL

Our studies of optimum lattice configurations indicated that an ideal cell for PIC consists of an alternating solenoid triplet. This provides strong focusing, so that the horizontal and vertical betatron phases advance by 3π across the cell and by

π between the absorbers. This configuration provides a periodic half-integer and integer resonant lattice in a very compact cell.

For p = 286.8 MeV/c, the length of the cell is 7.2 m. While OptiM uses the three "soft edge" solenoids with fields of B_0= -34.1, 32.4, and -34.1 kG, g4beamline models the solenoids with cylindrical current sheets. These currents were adjusted empirically, giving fields of B_0= -33.1, 34.8, and -33.1 kG. In both cases the solenoid lengths are L=80, 130, and 80 cm, respectively, and radius a=20 cm.

The periodic solution of the cell is illustrated in terms of Twiss functions below:

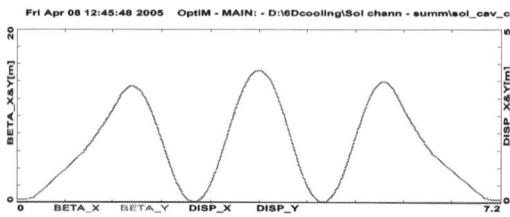

FIGURE 1. Beta functions for the solenoid triplet cell. Thin absorbers are placed at the two central focal points. In the OptiM simulations for FIG 2 the lost energy is simply replaced at the absorbers.

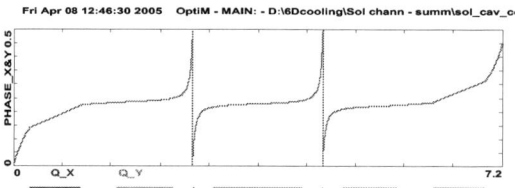

FIGURE 2. The phase advance over a single cell; 400 MHz RF cavities at each end provide synchrotron motion.

FIGURE 3. g4beamline view of one cell; the Be absorbers are shown in yellow, RF cavities as dark red.

THIN ABSORBER AND RF

The OptiM model considered the ionization cooling as due to energy loss (-Δp) in a thin absorber followed by immediate re-acceleration (Δp):

$$\Delta\theta_\perp = -\theta_\perp \frac{\Delta p}{p}$$

The corresponding canonical transfer matrix can be written as

$$M_{abs} = K \begin{bmatrix} 1 & 0 & 0 & 0 \\ 0 & 1-\frac{\Delta p}{p} & 0 & 0 \\ 0 & 0 & 1 & 0 \\ 0 & 0 & 0 & 1-\frac{\Delta p}{p} \end{bmatrix} K^{-1}$$

$$K = \begin{bmatrix} 1 & 0 & 0 & 0 \\ 0 & 1 & -k/2 & 0 \\ 0 & 0 & 1 & 0 \\ k/2 & 0 & 0 & 1 \end{bmatrix} \quad \hat{x} \equiv \begin{bmatrix} x \\ p_x \\ y \\ p_y \end{bmatrix} \quad x \equiv \begin{bmatrix} x \\ \theta_x \\ y \\ \theta_y \end{bmatrix}$$

where $k = eB_z/pc$ and $\hat{x} = Kx$.

TABLE 1. Initial parameters for simulation studies.

normalized emittance: $\varepsilon_x/\varepsilon_y$	mm	30
longitudinal emittance: ε_l ($\varepsilon_l = \sigma_{\Delta p}\sigma_z/m_\mu c$)	mm	0.8
momentum spread: $\sigma_{\Delta p/p}$		0.01
bunch length: σ_z	mm	30
momentum	MeV/c	286.8

In contrast, the g4beamline model placed a 400 MHz pillbox RF cavity just after the (not necessarily thin) absorber; this cavity provides on-crest acceleration and some RF focusing. The time offset was simply set as the time-of-flight at the nominal energy. The energy loss compensating voltage, Vacc, was iteratively adjusted to replace the energy loss of the absorber (57.3 MV/m for a 32.8mm Be absorber).

To induce synchrotron oscillation into the channel dynamics two 400 MHz RF cavities at zero crossing were added symmetrically to each cell. Their time offset was also set by a the nominal time-of-flight, and their voltage (Vsyn) iteratively adjusted to give the smallest final 6D emittance, which for a 32.8mm Be absorber Vsyn was 13.08 MV/m. Since the RF cavities in the simulations provide some focusing, it was necessary to readjust the solenoids.

SIMULATIONS

A single text input file described the model for g4beamline. Typical runs used from 100 to 1000 particles and the same starting distribution as used in the OptiM model.

It was quite difficult to analytically determine the correct operating parameters for the g4beamline simulation. A program, kmimf, using the CERNLIB-based MINUIT package, ran an iterative search in parameter space to minimize a specified quantity (such as the final r_{RMS} or ε_{6D}); that quantity was extracted from the analysis of each g4beamline run. This was very helpful in tuning the system. Another utility program, retrack, was used to gather statistics, convert output formats, and manipulate the output of g4beamline.

The first goal was to compare with the previous OptiM simulations. kmimf found a reasonable tune within a few iterations. The emittance was calculated using a slightly modified ecalc9 [6] for the cases with the synchrotron cavities turned off (Vsyn=0) and with an optimum value (Vsyn=13.08 MV/m). To compare to OptiM in these simulations, all stochastic processes (multiple scattering, energy straggling, muon decay, etc.) were disabled.

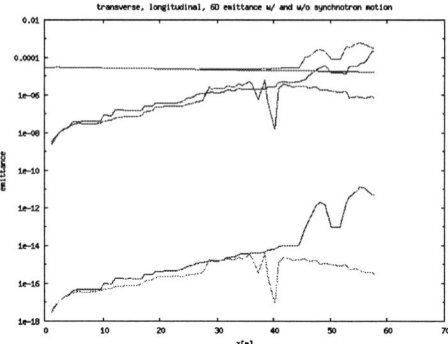

FIGURE 4. The transverse (ε_T [m-rad]), longitudinal (ε_L [m]), and 6-dimensional emittance (ε_{6D} [m^3]) in the absence of all stochastic processes is plotted as a function of distance along the channel for the cases without and with synchrotron motion. In this example L_{Be}=32.8 mm, Vacc=57.29 MV/m, and Vsyn=0 or 13.08 MV/m. In each case, the curve ending higher is the case without synchrotron motion.

The next step was to turn on all the stochastic processes to see what would happen. It was quite clear that multiple scattering and energy straggling are very important effects (FIG 5.). Optimization, not shown here, using kmimf resulted in an addition 10x reduction in the final 6-dimensional emittance.

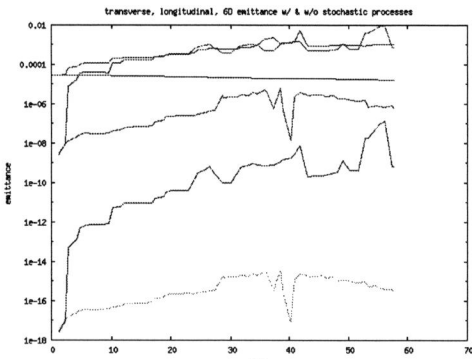

FIGURE 5. The transverse (ε_T [m-rad]), longitudinal (ε_L [m]), and 6-dimensional emittance (ε_{6D} [m^3]) is plotted as a function of distance along the channel for the cases without and with stochastic processes enabled. In this case, L_{Be}=32.8 mm, Vacc=57.29 MV/m, and Vsyn=13.08 MV/m. In each case, the curve ending higher is the case with stochastic processes enabled.

CONCLUSIONS

To realistically explore compensation schemes for PIC, much more realistic g4beamline simulations, which many import effects lacking in the earlier OptiM simulations, are being conducted.. While both programs show the importance of synchrotron motion, the new simulations make it clear that, in addition to optimization, new methods should be considered to make PIC work in the presence of physical stochastic processes.

REFERENCES

[1] V. Lebedev, http://www.bdnew.fnal.gov/pbar/organizationalchart/lebedev/OptiM/optim.htm
[2] T.J.Roberts, http://www.muonsinc.com/g4beamline.html
[3] Yaroslav Derbenev and Rolland P. Johnson, Ionization Cooling Using a Parametric Resonance, PAC05
[4] Y. Derbenev and R. P. Johnson, Phys. Rev. ST Accel. Beams **8**, 041002 (2005)
[5] Kevin B. Beard, et al, Simulations of Parametric Resonance Ionization Cooling of Muon Beams, PAC05
[6] R.C.Fenrow, Physics Analysis Performed by ECALC9, MUC-NOTE-COOL_THEORY-280

Simulations of MANX, A Practical Six Dimensional Muon Beam Cooling Experiment

Katsuya Yonehara[a,b], Kevin Beard[c], Alex Bogacz[c], Yaroslav Derbenev[c]
Rolland P. Johnson[b], Daniel Kaplan[a], Kevin Paul[b], Thomas Roberts[b],

[a]*Illinois Institute of Technology, Chicago, Illinois, U.S.A.*
[b]*Muons, Inc., Batavia, Illinois, U.S.A.*
[c]*Thomas Jefferson National Accelerator Facility, Newport News, Virginia, U.S.A.*

Abstract. A helical cooling channel (HCC) has been proposed to quickly reduce the six-dimensional phase space of muon beams for muon colliders, neutrino factories, and intense muon sources. Simulation studies of the HCC have already verified the use of a channel with solenoidal, and helical magnetic fields of constant amplitude where, by moving to a rotating frame, a z or time-independent Hamiltonian can be obtained for detailed analytic treatment. In the discussion below, the HCC concept has been extended to have momentum-dependent magnetic field strengths for a six-dimensional Muon collider And Neutrino factory muon beam cooling demonstration eXperiment (MANX). The simulation studies reported here for this experiment have shown that liquid helium can be used as an energy absorber and coolant for superconducting magnetic coils and that the HCC parameters can be varied to reduce the maximum required field magnitudes. These developments make the experiment more practical in that safety requirements are relaxed and the required fields can be achieved with existing technology.

Keywords: Muon Beam Cooling, Helical Magnet, Muon Collider
PACS: 29.27.-a, 29.20.-c, 14.60.Ef, 41.85.Lc

INTRODUCTION

Helical magnets have been used for some time in the control of spin precession in polarized beam devices. One advantage of helical magnets is that, in conjunction with a solenoid field, the fields can provide continuous focusing, dispersion, and correction of chromatic aberration. By using a low Z material like hydrogen or helium as an absorber in this helical cooling channel (HCC), the six-dimensional phase space of muon beams can be continuously and effectively cooled [1]. The concept of the HCC has been verified by earlier simulation studies [2].

The HCC concept has been extended to have momentum-dependent magnetic field strengths for two new applications. The first is a six-dimensional precooler which would follow a pion decay channel and the second, discussed here, is a six-dimensional Muon Collider And Neutrino Factory muon beam cooling demonstration experiment, MANX [3], which is being designed to follow MICE [4]. We discuss simulations of practical design concepts of the HCC for the MANX experiment in this article.

A PRACTICAL HCC

By filling a HCC with liquid helium (LHe), a muon beam can be decelerated and cooled in six-dimensional phase space by ionization energy loss. The HCC magnet parameters must be varied to match the momentum of the beam as it slows down. A muon beam with initial beam momentum of 300 MeV/c is injected into the HCC. It is possible to produce a large number of muons with this momentum by an 8-GeV proton driver. Approximately 200 MeV/c of beam momentum is lost in the HCC in along a 6.4 meters path length through ionization energy loss in LHe, such that the beam momentum is 100 MeV/c at the end of the HCC. The final beam momentum must be more than 100 MeV/c to maintain particle stability.

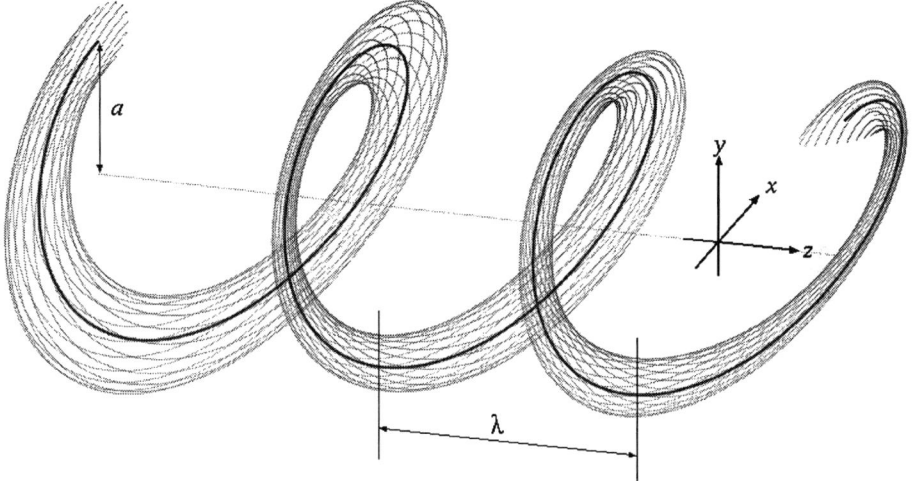

FIGURE 1. Simulated particle tracking in the HCC. The red line is the reference orbit and the blue lines correspond to particles with positions offset from the reference orbit. Here, a is 25 cm.

HCC Magnet Parameters

From reference [1], the solenoidal and helical magnet parameters for helix period λ and helix pitch angle $\arctan(\kappa)$ are given by

$$B_z = \frac{pk(1+q)}{\sqrt{1+\kappa^2}} = p_z k(1+q), \qquad (1)$$

$$b = \frac{kB_z\left(1 - 1/(1+q)\right)}{1+\kappa^2}, \qquad (2)$$

$$b' = -\frac{gpk^2}{(1+\kappa^2)^{3/2}}, \qquad (3)$$

$$b'' = -\frac{b'}{2\hat{D}}, \qquad (4)$$

where B_z is the solenoidal magnet, b, b', and b'' are the helical dipole, quadrupole, and sextupole field components at the reference orbit, p and p_z are the reference momentum and its z component, k is the helix wave number ($=2\pi/\lambda$), and $\hat{D}$ is the dispersion factor ($= p/a \cdot da/dp$, where a is the radius of reference orbit), respectively. At the periodic reference orbit, the q ($= k_c/k - 1$, where k_c is the cyclotron wave number) and the effective field index g are given by

$$q = \sqrt{\frac{1 + \kappa^2 - \kappa^2 \hat{D}/2}{1 + \frac{\kappa^2 \hat{D}/2}{1 + \kappa^2}}}, \quad (5)$$

$$g = \frac{1}{\hat{D}} - \frac{\kappa^2 + (1 - \kappa^2)q}{1 + \kappa^2}, \quad (6)$$

where we have assumed that the cooling decrement in all directions is equal. When λ and κ are constant at the reference orbit, all magnet parameters are determined by the reference momentum, p.

The solution of Maxwell's equations for the helical magnet in a cylindrical coordinate system is given by [5]

$$b^{(m)} = \left\{ \sum_{n=1}^{\infty} n! \left(\frac{2}{nka} \right)^n \frac{I_n(nk\rho)}{\rho} \tilde{b}_n \cos n\Psi \right\}^{(m)}, \quad (7)$$

where I_n is the n-th order modified Bessel function, (ρ, $\Psi = \varphi + 2\pi z/\lambda + \varphi_0$) is the position in cylindrical coordinates, and $\tilde{b}_n$ is the coefficient of the n-th field component. For simplicity, $\tilde{b}_n$ (n=1,2,3) are replaced by *bd*, *bq*, and *bs* when the magnet field map is calculated.

The parameters λ and κ are variables. They can be determined by optimizing the cooling factor in the HCC, constrained by the fact that the field strength of the HCC is determined by them. Therefore, the determination of λ and κ is essential for the practical design of a HCC. Longer λ requires lower field strength. However, this makes for a larger reference radius, and a correspondingly bigger beam pipe radius. Smaller κ also lowers the field strength. However, this implies smaller cooling decrements. By tuning both parameters by hand, we found optimum values for λ and κ, which are 2.0 meters and 0.8, respectively. Figure 2 shows the calculated magnet parameters (Bz, *bd*, *bq*, and *bs*) as a function of z.

SIMULATION RESULTS

Simulation studies of the HCC for MANX have been done by using two Monte Carlo simulation programs; ICOOL which is developed and maintained by R. Fernow [6], and G4Beamline which is developed and maintained by T. Roberts [7]. Each program gives the same results. Beam tracking is converted in terms of the beam emittance as calculated by the ICOOL post-processing tool ECALC9 [8].

Figure 3 shows the G4Beamline simulation results for the MANX experiment. The emittances are calculated for all particles that survive to z = 5. The cooling factors in transverse and longitudinal emittances are about 1.5 at z = 4 meters, respectively. Therefore, the six-dimensional cooling factor in this simulation is more than 3.

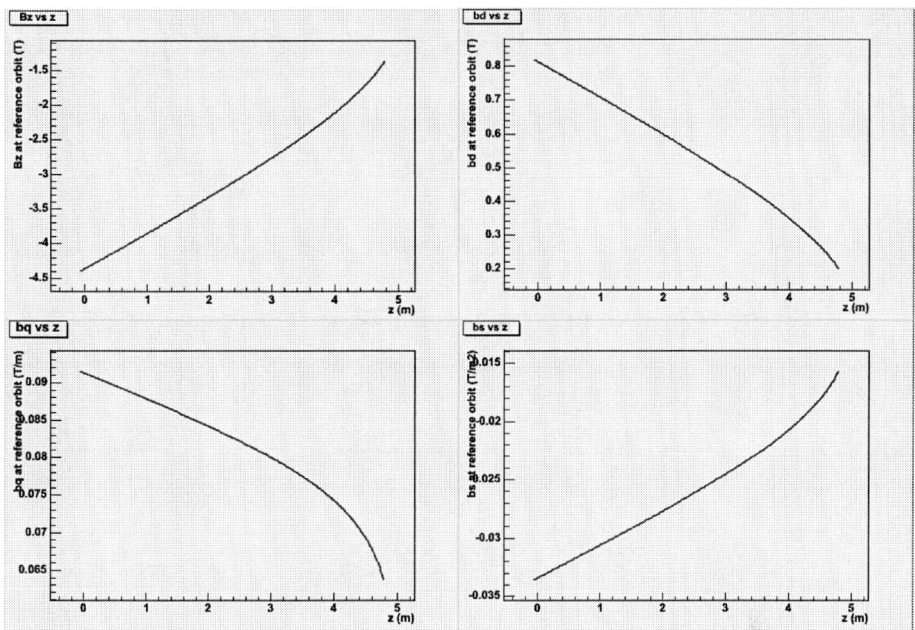

FIGURE 2. Calculated magnet parameters (Bz, bd, bq, and bs) as a function of z. The initial beam momentum is 300 MeV/c, the helix period is 2.0 meters, and the helix pitch angle is 0.8, respectively. The LHe is used as the absorber.

CONCLUSION

The first simulation studies in the HCC for the MANX experiment using LHe have shown encouraging results. The use of LHe is only slightly less effective than using LH2 [9]. However, the use of LHe compared to LH2 provides relaxed safety constraints and provides a natural means to cool superconducting magnet coils.

The lower initial momentum, smaller helix pitch angle, and longer helix period compared to those used in earlier studies reduce the maximum field strength to less than 5.5 T at the superconductor coils. This reduction in the maximum field should be more easily realized in a practical magnet design.

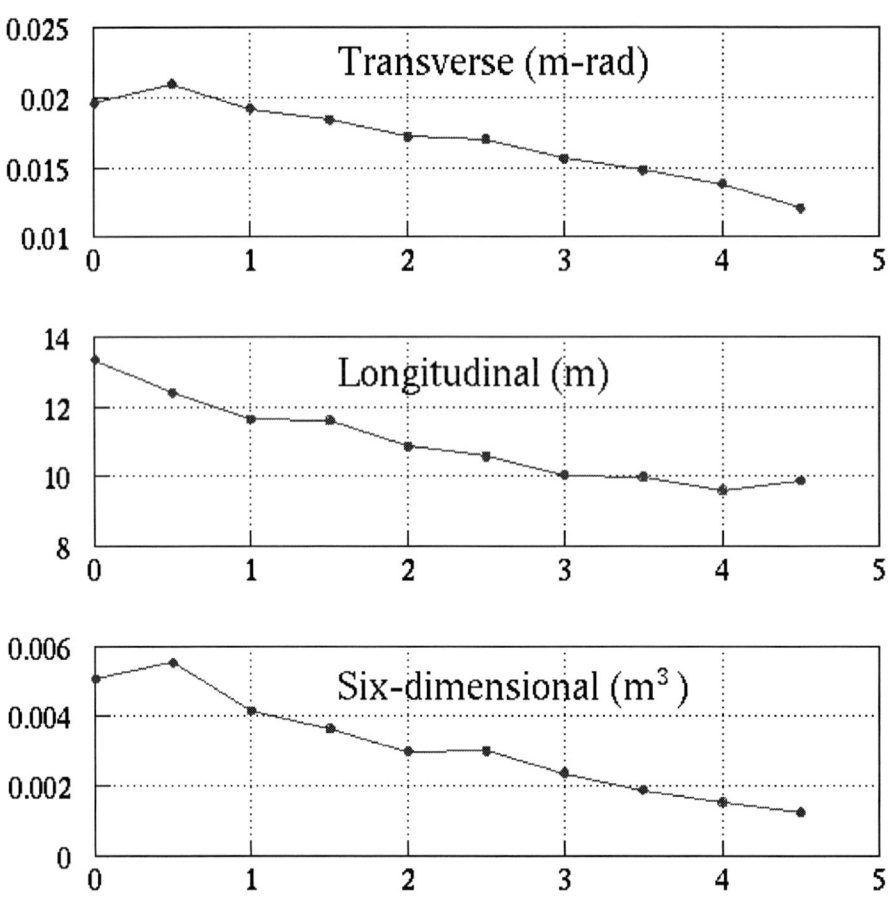

FIGURE 3. Evolution of transverse, longitudinal and six-dimensional emittances as a function of z as calculated by the G4Beamline simulation program.

REFERENCES

1. Y. Derbenev and R. P. Johnson, *Phys.Rev.STAB* **8**, 041002, (2005).
2. K. Yonehara et al., *to be published in PAC05 proceedings*.
3. T. Roberts et al., to be published in PAC05 proceedings.
4. M. A. Cummings et al., to be published in PAC05 proceedings.
5. T. Tominaka et al, *Nucl. Instr. and Meth.*, **A459**:398, (2001).
6. http://pubweb.bnl.gov/people/fernow/icool/
7. http://www.muonsinc.com/g4beamline.html/
8. http://www-mucool.fnal.gov/mcnotes/public/pdf/muc0280/muc0280.pdf
9. http://snsapp1.sns.ornl.gov/pac05/ROAA005/ROAA005.PDF

ELECTROSTATIC RINGS

DESIREE – A Double Electrostatic Storage Ring for Merged-Beam Experiments

H. Danared[*], L. Liljeby[*], G. Andler[*], L. Bagge[*], M. Blom[*], A. Källberg[*], S. Leontein[*], P. Löfgren[*], A. Paál[*], K.-G. Rensfelt[*], A. Simonsson[*], H. T. Schmidt[†], H. Cederquist[†], M. Larsson[†], S. Rosén[†] and K. Schmidt[†]

[*]*Manne Siegbahn Laboratory, Frescativägen 28, S-104 05 Stockholm, Sweden*
[†]*Department of Physics, Stockholm Universtity, S-106 91 Stockholm, Sweden.*

Abstract. DESIREE is a double electrostatic storage ring cooled to cryogenic temperatures. It is built at the Manne Siegbahn Laboratory for merged-beam and single-beam experiments in atomic and molecular physics as well as biophysics. This paper describes the present status of the design of DESIREE.

Keywords: Storage rings, electrostatic rings, merged beams, ion-ion collisions
PACS: 29.27.-a, 41.75.-i, 82.30.Fi

INTRODUCTION

Electrostatic storage rings were introduced [1] a little less than a ten years ago as a tool for studies in atomic and molecular physics and related disciplines. Since then, several other electrostatic rings have been or are being built, and yet others are being planned. At the Manne Siegbahn Laboratory, a double electrostatic storage ring, DESIREE [2], is being designed, as reported in this paper.

There are several reasons for building electrostatic rather than magnetic storage rings when high particle energies are not required. Perhaps the most important factor, at least in the present case, is cost. An electrostatic bending element, such as a pair of deflection plates, is much cheaper to manufacture than a dipole magnet. It is also lighter and smaller, allowing a more compact design of the ring (or rings) as a whole.

Another factor, often quoted, is that the electrical force is stronger than the magnetic force for low particle velocities, since the magnetic force is proportional to the velocity. As a result, heavy particles in low charge states can have higher velocities in electrostatic rings than in magnetic ones. Assuming electrical field strengths similar to those in the existing electrostatic rings and magnetic fields similar to those in small magnetic storage rings like CRYRING, one finds a cross-over at a mass of about 100 for singly charged ions. Heavier particles will move faster in an electrostatic ring and vice versa. Whether high particle velocities is an advantage is another issue, and the answer depends on what type of experiments one wants to perform.

A third advantage is that, for an ion source on a given electrical potential, ions with all charge-to-mass ratios can be stored in an electrostatic ring without change of the voltages of the ion-optical elements. The charge-to-mass ratio used in the ring is then selected by an analyzing magnet in the injection line. The ring can thus be set up with an ion species where a high current is available, and a second species can then be stored by just changing the setting of the analyzing magnet, even if the current is too low for the available ring diagnostics.

Some difficulties in building electrostatic storage rings are due to the fact that there exists less experience from such devices, and there are fewer design tools available. Also, electrostatic beam-optical elements tend to have larger aberrations than magnetic ones.

DESIREE (Double ElectroStatic Ion Ring ExpEriment), see fig. 1, is a double electrostatic ring designed for experiments with merged beams of positive and negative ions. It can also be used for experiments with only one ring, such as measurement of atomic lifetimes or studies of the physical properties of biomolecules. The entire device will be cooled to a temperature of 5–8 K in order to achieve a good vacuum and allow the storage of molecular ions in their lowest vibrational and rotational states.

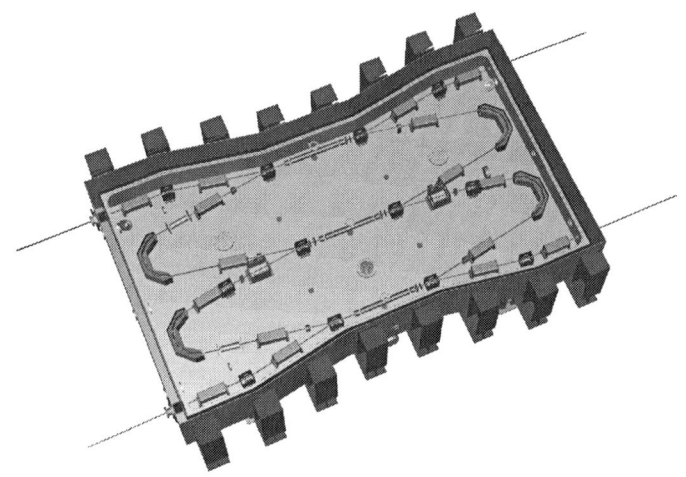

FIGURE 1. Schematic diagram of DESIREE with the lids of the vacuum vessels removed.

The fact that DESIREE will be cooled to cryogenic temperatures motivates the use of electrostatic technology, since the device then can be made sufficiently compact so that both rings can be built inside a single cryostat, as seen in the figure.

DESIREE is now being designed, and construction of the main components will begin during 2006. It is expected that commissioning can start at the end of 2007.

TECHNICAL DESIGN

Rings

DESIREE consists of two rings, each one with a circumference of approximately 9.3 m. The rings have one common straight section where the beams of the two rings are merged. Both rings have 160-degree cylindrical bends and 10-degree parallel-plate deflectors. The upper ring in fig. 1 (referred to as the light-ion ring) always has a 10-degree deflection at the ends of the common straight section. The lower ring (the heavy-ion ring) can have ions with different energy or charge compared to the light-ion ring, and these will then deflect by an angle that is different from 10 degrees. Due to geometric constraints, the ions in the two rings must have charges with opposite sign, and the deflection angle at the end of the straight section can range between 0.5 and 10 degrees for ions in the heavy-ion ring. Additional deflectors are required in this ring to compensate for the varying deflection angle. All deflecting and bending elements have a focussing action in the horizontal plane, and, in addition, there are four quadrupole doublets in each ring and a few horizontal and vertical correction elements.

Injection is made by a rapid switching of the respective 10-degree deflectors. A settling time of 1 μs is foreseen for the deflector voltage supply, which will be sufficient for the injection of H^+ or H^- at 100 kV. Possibly, extraction will also be implemented in a similar way.

The rings are designed for a maximum energy of 100 keV times the charge state of the ion, i.e., for injection from a platform on 100 kV. In this case, the maximum voltage on the electrodes of the optical elements is 16 kV.

An additional geometric constraint is due to the fact that neutral reaction products from the common straight section must be able to reach detectors located at the extension of this section, rather than being intercepted by deflection plates. It will also be possible to detect neutral particles produced on the two injection straight sections. Since both optical elements and detectors are mounted directly on the bottom of the inner vacuum vessel, as described below, it will be possible to put detectors in additional positions if there will be experiments requiring these.

Vacuum and Cryogenics

The two rings will be housed in a common vacuum chamber/cryostat with double walls. The outer wall or vacuum chamber will be built from 5 mm thick steel, and it will have to be reinforced by external beams to support the atmospheric pressure. Its size is approximately 4.9 m × 2.6 m × 0.7 m plus reinforcement beams which add 0.24 m on all sides. The inner chamber will be made from 5 mm aluminium, but it will have a thicker bottom on which optical elements and detectors will be mounted as mentioned below. It will have less reinforcements, and it will not be built to withstand atmospheric pressure. Between the two vessels, there will be a thermal screen and multilayer insulation between the screen and the outer vessel. The vessels will have large lids, almost as large as the vacuum vessels themselves. Feedthroughs will mainly be mounted at the bottom of the vessels, such that the lids can be opened without the need to disconnect all feedthroughs.

The cooling will be achieved through the use of cryogenerators. According to calculations, two cryogenerators (Sumitomo CSW71/RDK-515D) will be needed to reach the desired temperature. The first stage of the cryogenerators will be connected to the screen, and the second stage to the inner vacuum vessel. It is estimated that the heat load on the screen will be approximately 60 W, which will result in a temperature below 60 K. The inner vessel will receive a heat load of around 3 W, which should give a temperature of 5 K on the rings. If the heat loads will become bigger, it is possible to add one or two more cryogenerators.

It will also be possible to run DESIREE at higher temperatures, up to room temperature. The inner vacuum vessel thus has to be bakeable to 150°C, putting restrictions on the choice of materials, etc. It is then also necessary to have a high pumping capacity on the inner vessel. This will be achieved through titanium sublimation pumps in combination with turbo pumps, and it is expected that a pressure of 1×10^{-11} mbar can be reached at room temperature. At low temperatures, these pumps will remove hydrogen and helium while all other gases have negligible vapour pressure at 5 K.

Fig. 2 shows results of tests of a cryogenerator of the above model. These tests were made in order to verify the specifications from the manufacturer and to measure the cooling power at higher temperatures than given in the specifications. It can be seen from the figure that the agreement between measurements and specifications is very good in the range where specifications exist.

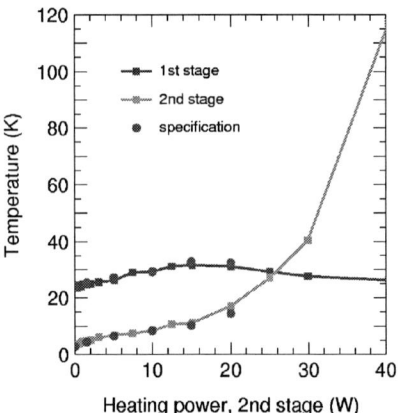

FIGURE 2. Measurements of temperatures on the first and second stage of a Sumitomo CSW71/RDK-515D cryogenerator when a heat load is applied to the second stage. The curves show measurements and the filled circles represent specifications.

An efficient heat transfer from all components seen by the beams to the cryogenerators is important in order to get reasonably short cooldown times and low final temperatures. For this reason, and also for the sake of mechanical stability, the bottom of the inner vacuum vessel will be made from 12 mm thick aluminium, and the beam-optical elements, detectors, etc., will be mounted directly on this aluminium plate. Flexible copper braids will connect this plate to the cryogenerator heads. Tests have indicated that a temperature difference between the aluminium plate and the cryogenerator head lower than 1 K can be obtained.

Injectors

DESIREE will be provided with at least two injectors, one for each ring, although both injectors will be able to supply ions to both rings. The injectors consist of a high-voltage platform, an ion source and an analyzing magnet. The injector mainly supplying the heavy-ion ring will be tailored to heavier ions in that it will be built for a higher platform voltage, 100 kV, and it will have a larger magnet with higher resolving power. The magnet itself, with a bending power of 1.9 Tm, will be on earth potential, but the vacuum chamber will be insulated for 100 kV. The injector mainly used for the light-ion ring will have a 25 kV platform voltage.

Several kinds of ion sources are foreseen, such as a standard plasmatron ('Nielsen' type) source for singly charged atomic and molecular ions, an expansion source for molecules in low rotational states, an electron-impact source producing molecules in known vibrational states, a sputter ion source for negative ions, an electrospray source for complex molecules and biomolecules, and possibly an ECR ion source for multiply charged ions. The use of a cooled ion trap after the electrospray source is being investigated. With such a trap the ions can be accumulated and buffer-gas cooled in order to increase the particle intensity in the beam pulse and to decrease the beam emittance.

ION OPTICS

The light-ion ring has two symmetry planes, and if the voltages on the ion-optical elements are to obey that symmetry, the optics is defined by only two parameters: the focusing and defocusing quadrupole voltages. Using standard linear transfer matrices for ideal quadrupoles and cylindrical deflectors, and approximating the parallel-plate deflectors also with cylindrical deflectors, a stability diagram like the one in fig. 3 is obtained simply from matrix multiplication and standard procedures for calculating the stability criterion and Q values.

It is seen that the stable islands are relatively small, but that it is possible to find several working points with, e.g., round well-focused beams in the common straight section. The smallness of the islands is related to the rather long straight sections without focusing, which is an inevitable consequence of the merged-beams design.

The existence of stable orbits was initially proven using COSY INFINITY [3], and later on also with SIMION [4]. The latter code evaluates the electric fields numerically and in this sense gives more realistic results, at the expense of much longer execution times, so that much less detail can be obtained. The results from both the COSY INFINITY and SIMION calculations agree well with those from the simple matrix multiplications discussed here.

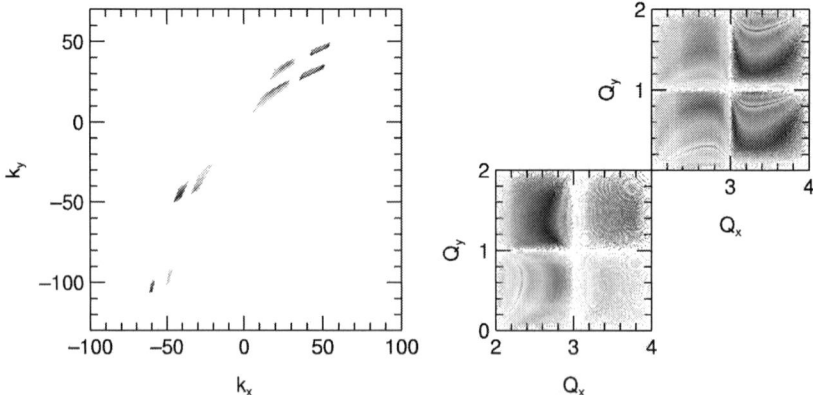

FIGURE 3. The left part of the figure shows stable islands for the light-ion ring as a function of the horizontal and vertical quadrupole strengths (where the strength multiplied by the 0.10 m electrode length is the inverse of the focal length for a single quadrupole). The right part shows the Q values corresponding to the islands to the left, and the upper right Q diagram corresponds to the upper right islands and vice versa. The colours are such that red represents round beams on the common straight section while blue and violet represent flat beams. Saturated colours represent small, well-focused beams in the common straight section while unsaturated colours represent unfocused beams.

The heavy-ion ring has four more parallel-plate deflectors, which are required in order to merge ions with different charge-to-mass ratios or energies with the ions stored in the other ring. As a result, the heavy-ion ring only has one symmetry plane, and, since the amount of deflection depends on the charge-to-mass ratios in the two rings (or, in the general case of different ion velocities, on the ratio of particle energy per unit charge in the two rings), the optics and the lattice functions will depend on this ratio. More specifically, the three deflectors on each side of the common straight section will contribute to focusing in the horizontal plane, such that a larger deflection angle gives stronger focusing. The stable islands for the heavy-ion ring will thus be larger if the deflection in the deflectors closest to the common straight section is large, which is the case when the energy per unit charge is the same in both rings.

Aberrations tend to be larger with electrostatic optics than with magnets. For instance, an electrostatic quadrupole gets aberrations of octupole character because the particles change speed in the electrical fields. Such aberrations can limit the dynamic aperture of an electrostatic ring, as illustrated in fig. 4. Here, particles were traced through the same lattice as used for the calculations of fig. 3, but the octupole effect of the quadrupoles were taken into account in a simple approximation. Particles were traced up to 100 turns through light-ion ring with the same parameters as were used for fig. 3, except that the range of focusing strengths is smaller. The largest, purple areas represent k and Q values where particles with an emittance of 0.1 mm mrad in each plane survive 100 turns. Blue and green indicate larger emittances, and red shows k and Q values where particles with an emittance of 100 mm mrad survive at least 100 turns.

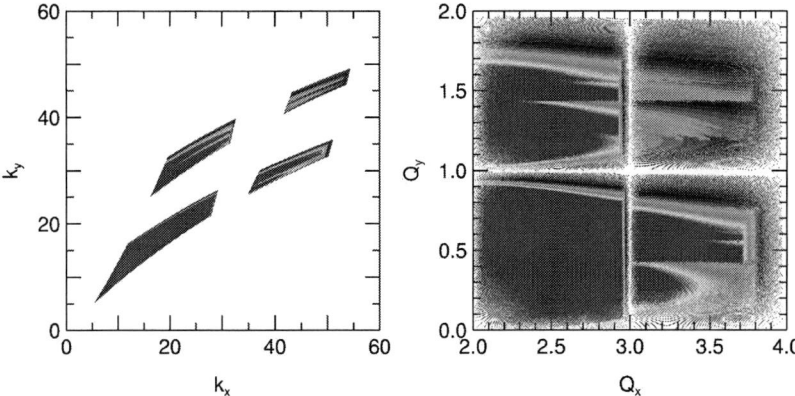

FIGURE 4. The different colours show k and Q values where particles with different emittances survive at least 100 turns in the light-ion ring. Purple indicates survival of particles with 0.1 mm mrad, and the emittance increases in steps of 10 mm mrad up to 100.1 mm mrad which is indicated with red.

Fig. 4 is not an exact representation of the dynamic aperture in DESIREE since aberrations in the deflecting elements are not included, nor higher-order effects in the quadrupoles. Also, fringe fields are not included. A complete result again requires a more realistic evaluation of the fields, using a code like SIMION. Such calculations have been performed for DESIREE, although a map as detailed in fig. 4 would be far too time-consuming to produce using such a technique.

APPLICATIONS

As mentioned above several plans exist for the use of DESIREE in single-ring configuration for lifetime measurements and interactions with laser fields ranging from high-precision spectroscopy with continuous wave lasers to experiments involving ultra-short pulses in the fs range. However, as the single most original feature of DESIREE is the possibility to investigate interactions between positive and negative ions at low relative velocity, we will focus on this aspect here.

The process of mutual neutralization (MN) between two singly charged – one positive and one negative – ions to form two neutral products is an ideal case for DESIREE. The neutral products will continue straight in the field where the two ion beams are separated and can readily be detected. By applying a detector with multihit capability and position sensitivity it is even possible to determine the amount of kinetic energy released in the process. Another advantage of such a scheme will be the direct detection of more than two neutral fragments in coincidence. This is of importance when molecular ions are taking part in the collisions and the question of whether the electron transfer is accompanied by dissociation is addressed. The study of molecular ions will further benefit from one of the other advanced DESIREE features namely the very low temperature, which in turn leads to very good vacuum and presumably thereby also very long storage lifetimes. This combination means that infrared-active ions will be found to reach internal temperatures of similarly low values, thus resulting in population of only one or a few quantum states.

The interest in the MN process is both in the fundamental mechanisms and in total cross section determinations for related fields such as the chemistry of interstellar clouds and in planetary atmospheres (including our own). As an example of the fundamental interest we mention the fact that it is a matter of current debate whether or not the neutral hydrogen molecule formed in the MN of H^- with H_2^+ (or HD^+) will dissociate [5]. An even simpler system to consider is the p-H^- system. Here the obvious objection to performing such an experiment in DESIREE would be that no use is made of the low temperatures and the beam storage. This is, however, exactly the point. In the near future we will engage in an effort to measure this process in a single-pass merged-beams experiment lead by Prof. Urbain of the Catholique University of Louvain-la-Neuve in Belgium. The later repetition of the exact same experiment in DESIREE is expected to provide very useful calibration information needed for the determination of absolute cross sections for mutual neutralization and other merged-beams experiments involving molecular ions.

REFERENCES

1. S. P. Møller, *Nucl. Instr. Meth. A* **394**, 281 (1997).
2. K.-G. Rensfelt et al., in *Proc. EPAC 2004,* edited by J. Poole, J. Chrin, Ch. Petit-Jean-Genaz, C. Prior and H.-A. Synal, Lucerne 2004, p. 1425.
3. M. Berz, COSY INFINTY Version 8.1 User's Guide and Reference Manual, Michigan State Univ. 2002.
4. D. A. Dahl, SIMION 3D Version 7.0 User's Manual, INEEL, Idaho Falls, USA.
5. M. J. J. Eerden et al., *Phys. Rev. A* **51**, 3362 (1995).

The Heidelberg CSR: Stored Ion Beams in a Cryogenic Environment

A. Wolf[*], R. von Hahn[*], M. Grieser[*], D. A. Orlov[*], H. Fadil[*], C. P. Welsch[*], V. Andrianarijaona[*], A. Diehl[*], C. D. Schröter[*], J. R. Crespo López-Urrutia[*], M. Rappaport[†], X. Urbain[**], T. Weber[*], V. Mallinger[*], Ch. Haberstroh[‡], H. Quack[‡], D. Schwalm[*], J. Ullrich[*] and D. Zajfman[†,*]

[*]*Max-Planck Institute for Nuclear Physics, Heidelberg*
[†]*Weizmann Institute of Science, Rehovot, Israel*
[**]*Université Catholique de Louvain, Louvain-La-Neuve, Belgium*
[‡]*Technische Universität Dresden, Germany*

Abstract. A cryogenic electrostatic ion storage ring CSR is under development at the Max-Planck Institute for Nuclear Physics in Heidelberg, Germany. Cooling of the ultrahigh vacuum chamber is envisaged to lead to extremely low pressures as demonstrated by cryogenic ion traps. The ring will apply electron cooling with electron beams of a few eV up to 200 eV. Through long storage times of 1000 s as well as through the low wall temperature, internal cooling of infrared-active molecular ions to their rotational ground state will be possible and their collisions with merged collinear beams of electrons and neutral atoms can be detected with high energy resolution. In addition storage of slow highly charged ions is foreseen. Using a fixed in-ring gas target and a reaction microscope, collisions of the stored ions at a spead of the order of the atomic unit can be kinematically reconstructed. The layout and the cryogenic concept are introduced.

Keywords: Electrostatic storage ring, molecular beams, cryogenic ring
PACS: 29.20.Dh

INTRODUCTION

Collision experiments with stored slow ion beams at energies in the range of ~ 100 keV are still relatively rare, although they have a large potential. Thus, a number of experiments were performed with stored, but uncooled heavy ion beams, including clusters and biological species, with heavy ion beams in electrostatic storage devices [1]; however, they mainly concerned unimolecular relaxation and dissociation after their production or radiative excitation. Only a few recent examples exist for studies using electron impact on stored molecular ion beams [2, 3, 4].

Electrostatic storage rings [5] can with a reasonable technological effort be constructed for ion beams below a few hundred keV, using cylinder-capacitor-like deflectors with ~ 10 kV/cm field and a bending radius of the order of 1 m. In addition, fast-beam ion traps have been established as physics instruments [6]. With diameters of ~ 10 m it can be considered to cool electrostatic storage rings and traps entirely to low cryogenic temperatures (~ 2 K) sufficient to freeze out all relevant gases and potentially to reach gas pressures of $\sim 10^{-16}$ mbar, similar as in penning traps for antiprotons [7]. The development of a fully equipped cooler storage ring for ~ 20–300 keV ion beams over a large mass range is pursued since 2004 in the Cryogenic Storage Ring (CSR) project at

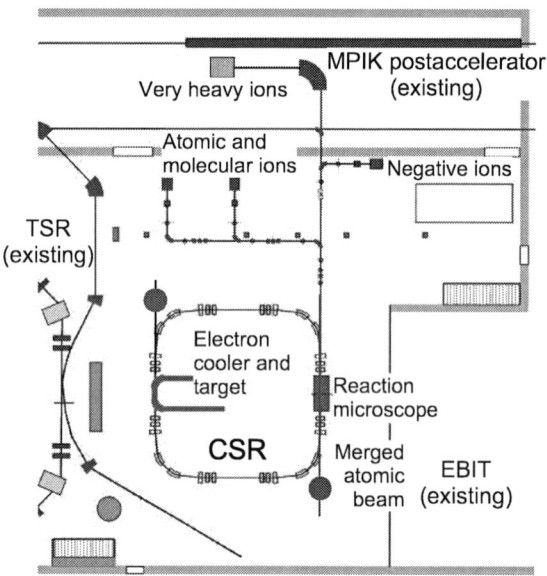

FIGURE 1. View of the CSR project, Heidelberg, close to the existing accelerator and the electron beam ion trap (EBIT). Detector regions for the experimental zones are marked with green dots.

the Max-Planck Institute for Nuclear Physics, Heidelberg.

MOTIVATION

Low-energy inelastic electron collisions with MeV molecular ion beams are successfully studied since 1992 at magnetic storage rings focusing in particular on the process of dissociative recombination (DR) with electrons [8, 9]. This process is one of the main loss processes for molecular ions in thin ionized media where also electrons are present. In general, reactions of molecular ions both with electrons and with other heavy species are of utmost importance for the chemical composition of thin media such as interstellar space [10] and their evolution, such as star formation [11]. DR runs exothermically even at meV interaction energies and sensitively depends on the properties of neutral molecular resonance states, produced by binding the electron, in the vicinity of the equilibrium geometry of the molecular ions.

While molecular vibrations could be well controlled and the effect of vibrational excitation eliminated in previous experiments [8], a similar research effort now is underway regarding the control of rotational excitation. In contrast to vibration, only few-meV excitation energies are involved and, correspondingly, temperatures of the order of 10 K are required to suppress rotational excitation. Presently, in measurements of the energy dependence of the dissociation cross section, contributions of various excited rotational levels are usually mixed and, in spite of the achieved high energy resolution of $\sim$0.5 meV, can hardly be disentangled [12].

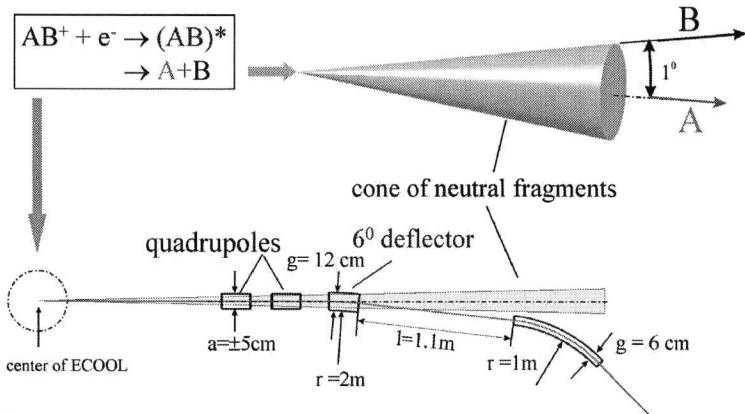

FIGURE 2. Detector arrangement for neutral recombination products.

Performed with stored ion beams at cryogenic temperatures (<10 K), electron impact measurements of this type would assume the character of pure state-to-state experiments. It should be noted that the use of energetic particle beams together with suitable multi-hit imaging detectors makes it possible to detect and reconstruct the full dissociation geometry of the interacting molecule, including both charged and neutral fragments [12].

The performance of molecular collision experiments with high energy resolution and including the analysis of the dissociation channels is one of the main motivations of the Heidelberg CSR. Owing to the moderate beam energy of <300 keV, it is planned to use also stationary gas targets and a merged atomic beam.

LAYOUT

The CSR is a fourfold-symmetry electrostatic storage ring for ion beams of 20–300 keV, consisting of four pairs of separated 39° deflectors and four pairs of small 6° deflectors at the start and end of each bend. It has four straight sections with a length of 4 m, each of which includes also a pair of electrostatic focusing doublets. The total circumference is 34 m; further accelerator properties are discussed by Fadil et al. [14]. Two straight sections will be equipped with detectors for neutral and charged fragments and used for collision experiments; in one of them an electron cooler and target will be installed, whereas the other section will house a fixed target and an associated reaction microscope as well as the merged atomic beam. The planned emplacement and the basic experimental arrangement are shown in Fig. 1.

The electron cooling system uses the cryogenic photocathode system run already at the TSR [12]. At low beam energies in the range <10 eV this system is expected to provide large advantages regarding the obtainable electron density at a transverse temperature below ~1 meV; thus, the emission current limit of presently ~400 A of the photocathode becomes irrelevant as the space-charge limited currents for, e.g., 3 Perv are clearly lower than this limit (< 100 A) [15]. Moreover, the small (<10 meV) initial energy spread of the photoelectrons makes it possible to reduce the beam expansion required for a given transverse temperature by a factor of ~ 10 and also yields lower longitudinal energy spread at low energy [15]. The reaction microscope [16] will use a well collimated gas jet (primarily He) of a thickness of $<10^{11}$ cm^{-2} and kinematically reconstruct collisions with the circulating keV ions (in particular also highly charged ions) by completely detecting all released collision products (ions, electrons and the deflected projectile). In the same ring section, it is also planned to produce a neutral atomic beam (e.g., H$^-$, D$^-$, O$^-$) of $>10^3$ cm^{-3} velocity-matched to and collinearly overlapped with the circulating molecular ion beam. The beam will be produced by photodetaching a negative ion beam (see Fig. 1). Heavy particle reactions such as deuteration (the exchange of a hydrogen atom against deuterium) should become observable at an estimated energy resolution of a few meV. From the zero-energy rate coefficient of $\sim 10^{-9}$ cm^3 s^{-1} reaction rates, observable essentially background-free, of ~ 10 s^{-1} are estimated for 1 A of ion current (e.g., CH$^+$). All experiments will use the detector regions set up in two regions of the ring (see Fig. 1). Molecular fragmentation leads to quite large transverse fragment momenta, so that a free drift has to be ensured for neutral fragments with an opening angle of 1° over several meters. This is accomplished by the deflector layout and positioning as shown in Fig. 2. In addition to the neutral fragment detectors, several positions are foreseen for charged atomic and molecular fragments having experienced up- or downcharging as well as pick-up and loss of heavy particles. The operation of the detectors at cryogenic temperatures is under study.

CRYOGENIC CONCEPT AND PROTOTYPE

Starting with a room-temperature ultrahigh vacuum, the cooling of the storage ring to cryogenic temperatures, down to 2 K, can be expected to reduce the residual gas density to extremely low values ($\sim 10^3$ cm^3), corresponding at this temperature to pressures of the order of a few 10^{-16} mbar). To our knowledge, the CSR will be the first large-scale machine exploring such extreme vacua. In addition, the thermal blackbody radiation field will be virtually eliminated, offering the experimental possibilities discussed above.

The beam pipe and the ion-optical components will be situated in an inner, stainless-steel vacuum chamber designed to reach a base pressure below 10^{-11} mbar at room temperature. This chamber will will be pumped mainly by volume getters (NEG) after bakeout to $\sim 300°$C. During the subsequent cooling down to cryogenic temperatures, most of the remaining gas components will be cryo-sorbed on the chamber walls; in contrast to practically all other gases, cryo-sorption of hydrogen requires very low temperatures [17] and the attainment of wall temperatures near or slightly below 2 K can be considered the safest prerequisite to ensure the stable adsorption even of this gas component. Liquid helium will be evaporated under low pressure inside cooling units mounted at the inner vacuum chamber and interconnected by a pipe system. The cooling lines, the suspension, and two thermal screens will be located in the insulation vacuum ($\sim 10^{-6}$ mbar) of the outer chamber, acting as a cryostat interconnected by 600 mm diam. flanges between machine modules. As a critical elememt the suspension of the inner vacuum chamber, which has to ensure stable positioning of the ion optical elements, will be carefully designed and tested. The specifications of the cryogenic supply system have been considered iteratively also under the aspect of economical optimization.

A number of technological issues, such as the indispensable connections to room-temperature equipment (at, e.g., the ion injection lines), which risk to be the main critical load of external gas and blackbody radiation input, will be investigated experimentally at a test setup presently under construction. This prototype contains an electrostatic ion beam trap [6] designed for electrode voltages of 30 keV, corresponding to the deflector voltage of the CSR, with cryogenic sections for detectors and for observations on the stored ions. At the date of the Conference, this system is being manufactured, while the detailed design of the CSR machine elements is being started on the basis of the completed prototype design.

REFERENCES

1. L. H. Andersen, O. Heber and D. Zajfman, *J. Phys. B* **37** R57–R88 (2004)
2. T. Tanabe, K. Noda, M. Saito, S. Lee, Y. Ito, and H. Takagi, *Phys. Rev. Lett.* **90**, 193201 (2003)
3. A. Diner, Y. Toker, D. Strasser, O. Heber, I. Ben-Itzhak, P. D. Witte, A. Wolf, D. Schwalm, M. L. Rappaport, K. G. Bhushan, and D. Zajfman, *Phys. Rev. Lett.* **93**, 063402 (2004); O. Heber, P. D. Witte, A. Diner, K. G. Bhushan, D. Strasser, Y. Toker, M. L. Rappaport, I. Ben-Itzhak, N. Altstein, D. Schwalm, A. Wolf, and D. Zajfman, *Rev. Sci. Instrum.* **76**, 013104 (2005)
4. M. O. A. El Ghazaly, A. Svendsen, H. Bluhme, A. B. Nielsen, S. B. Nielsen, and L. H. Andersen, *Phys. Rev. Lett.* **93**, 203201 (2004)
5. S. P. Møller, *Nucl. Instrum. Methods A* **394**, 281 (1997)
6. M. Dahan, R. Fishman, O. Heber, M. Rappaport, N. Altstein, D. Zajfman, and W. J. van der Zande, *Rev. Sci. Instrum.* **69**, 76 (1998)
7. G. Gabrielse et al., Phys. Rev. Lett. **65**, 1317 (1990); ibid. **74**, 3544 (1995)
8. M. Larsson, *Annu. Rev. Phys. Chem.* **48**, 151 (1997)
9. A. Wolf, S. Krohn, H. Kreckel, L. Lammich, M. Lange, D. Strasser, M. Grieser, D. Schwalm, D. Zajfman, *Nucl. Instrum. Methods A* **532**, 69–78 (2004)
10. E. Herbst and W. Klemperer, *Astrophys. J.* **185**, 505–534 (1973)
11. R. B. Larson, *Rep. Prog. Phys.* **66**, 1651–1697 (2003)
12. D. A. Orlov, F. Sprenger, M. Lestinsky, U. Weigel, A. S. Terekhov, D. Schwalm, and A. Wolf, *J. Phys.: Conf. Ser.* **4**, 290–295 (2005)
13. A. Wolf, D. Schwalm, and D. Zajfman, Chapter 26 of *Many-Particle Quantum Dynamics in Atomic and Molecular Fragmentation*, ed. by J. Ullrich and V. P. Shevelko (Springer, Berlin, 2003)
14. H. Fadil et al., these proceedings
15. D. Orlov et al., these proceedings
16. J. Ullrich, R. Moshammer, A. Dorn, R. Dörner, L. Ph. H. Schmidt, and H. Schmidt-Böcking, *Rep. Prog. Phys.* **66**, 1463 (2003)
17. C. Benvenuti, R. S. Calder, and G. Passardi, *J. Vac. Sci. Technol.* **13**, 1172 (1976)

Ultra-Cold Electron Beams for the Heidelberg TSR and CSR

D. A. Orlov*, M. Lestinsky*, F. Sprenger*, D. Schwalm*, A. S. Terekhov[†]
and A. Wolf*

*Max-Planck-Institut für Kernphysik, 69117 Heidelberg, Germany
[†]Institute of Semiconductor Physics, 630090 Novosibirsk, Russia

Abstract. A cold electron target with a cryogenic GaAs-photocathode electron source was developed for the Heidelberg Test Storage Ring. Two independent electron facilities (cooler and target) allow to separate cooling of the ion beam from target operation improving the quality of electron and ion beams. In addition a strong gain in the resolution was achieved with a help of a cryogenic photoelecron source providing dc electron currents up to 0.5 mA with an emission energy spread of about 10 meV. In first recombination measurements at the target, performed on HD^+, H_3^+ and Sc^{18+}, low energy resonant structures at milli-eV collision energies revealed unprecedented low transverse and longitudinal electron temperatures of about 0.5 meV and 0.025 meV, respectively. The photocathode source will be also used to provide cold beams for electron cooling of low-energy ions stored at the electrostatic Cryogenic Storage Ring which will be built at MPIK. The perspectives of photocathode-driven electron coolers operating at very low laboratory energies are discussed.

Keywords: Cold electron beams, electron cooling, electron target, GaAs photocathode.
PACS: 29.25.Bx, 29.20.Dh, 41.75.Fr, 79.60.-i, 79.60.Jv

INTRODUCTION

A magnetically guided electron beam, overlapped in some part of the ring with stored ions moving with the same average velocity, acts as an effective cooler for hot ions [1]. Moreover, for the electron-ion collision experiments the electron cooler can be used as a target by detuning the electron velocity to a desired relative energy. Despite high energy resolution of a few milli-eV obtained in such experiments [2] the resolution limit of this technique is still far away. To improve the collision energy resolution an electron target was developed for the Heidelberg TSR in addition to the electron cooler. Two independent electron beams allow us to perform electron collision experiments at variable relative energies avoiding the resolution loss caused by the interruption of electron cooling [3]. Furthermore, a strong gain in energy resolution was obtained by using a cryogenic photocathode source [4] releasing electrons within energy spreads 10 times narrower than conventional thermocathodes [5].

In this paper the performance of the TSR ultra-cold electron target is presented. In addition, the perspectives of photocathode-driven electron coolers operating at very low laboratory energies are discussed with a connection to a new project (low-energy electrostatic Cryogenic Storage Ring) being developed in the Institute.

BACKGROUND

In this section we will recall the background of high-resolution merged beam experiments at storage rings for typical electron energies of a few keV. (The topic of low-energy electron cooling will be covered at the end of the paper.) In the next part of this section we will consider the principles of GaAs-photocathodes which can provide electrons with laboratory energy spreads below 10 meV.

Electron Beam Temperatures

While the typical laboratory energies of the stored ions and electrons are about a few MeV/u and keV, respectively, collision energies below a milli-eV can be realized by small longitudinal velocity detuning between the ion beam and the magnetically guided electron beam. In such experiments the ion velocity usually is well defined and the collision energy spread is mainly given by the electron energy distribution, being strongly anisotropic in the co-moving reference frame. For strongly magnetized electron beams, when transverse-longitudinal relaxation (TLR) is suppressed, the longitudinal temperature $kT_\parallel$ of electrons in the co-moving frame at zero detuning energy is given by

$$kT_\parallel = \frac{(kT_c)^2}{2E_{kin}} + C \frac{e^2}{4\pi\varepsilon_0} n_e^{1/3} \qquad (1)$$

where kT_c is a cathode temperature, E_{kin} is the electron energy in the laboratory frame, and n_e is the electron density. The parameter C describes the quality of the acceleration and is about 1.9 for fast acceleration [6]. It can be reduced by slow adiabatic acceleration suppressing a potential energy relaxation when the relaxation time, defined by the plasma frequency, is short compared to the acceleration time [6]. For typical electron energies E_{kin} of a few keV and densities $n_e=10^{-5}$–10^{-7} cm^{-3} the first kinematic term (Eq. (1)) is very small and the longitudinal temperature is mainly defined by the density term (the second one in Eq. (1)). It is typically below 0.1 meV and always much smaller than the transverse temperature $kT_\perp$ which is about 5-30 meV. In recombination experiments, the electron temperatures lead to an asymmetric broadening δE of the observed cross-section resonances. The low-energy side of the resonance is broadened by $kT_\perp$, while $kT_\parallel$ broadens the resonance symmetrically by $4(E_r kT_\parallel ln2)^{1/2}$, with E_r being the resonance energy [7]. The resolution (FWHM) in merged beam experiments can be approximated by:

$$\delta E \approx \sqrt{(kT_\perp ln2)^2 + 16\, ln2\, E_r\, kT_\parallel} \qquad (2)$$

Hence the highest energy resolution can be obtained for resonances in the vicinity of zero energy. For E_r below ≈ 0.1 eV, δE is limited by $kT_\perp$. To improve δE in this energy range, $kT_\perp$ has to be decreased. Initially $kT_\perp$ is given by the cathode temperature and can be reduced by adiabatic magnetic expansion to $kT_\perp = kT_c/\alpha$, where α is the expansion coefficient [8]. While for thermocathodes kT_c is above 100 meV, semiconductor photocathodes were considered as promising candidates for cold electron sources.

Cold Electrons from GaAs-Photocathodes

Conventionally used thermocathodes with the work function for electrons of about 2 eV must be heated to 1300-1500 K to release electrons into the vacuum, resulting in emission electron energy spreads of about 110–120 meV. In the case of GaAs photocathodes (Fig. 1) the situation is different: GaAs with a thin layer of cesium and oxygen produces a state with effective Negative Electron Affinity (NEA), when the vacuum level lies below the position of the conduction band in the bulk [9].

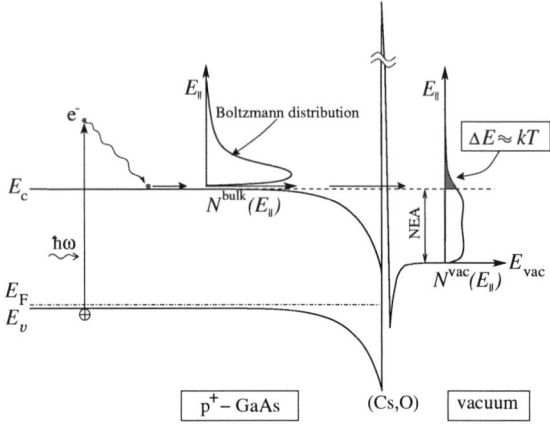

FIGURE 1. Band diagram of NEA-GaAs showing schematically the photoemission process.

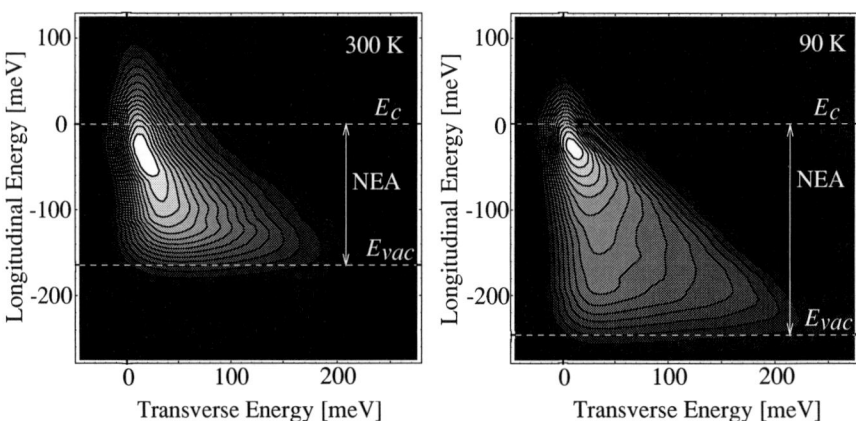

FIGURE 2. Energy distributions of photoemitted electrons from GaAs at 300 K (QY=23%) and at 90 K (QY=29%) [10]. The density of electrons is shown as contours (15 in total, linear scale). The maximum density corresponds white contour, whereas black contour reflects area of zero density.

Electrons photoexited from the valence band to the conduction band are rapidly thermalized at the bottom of the conduction band and reach the surface with energy spreads defined by the temperature of the bulk. Due to the NEA condition, a large fraction of these electrons can escape into the vacuum. However, during the escape process electrons undergo strong energy and momentum relaxation [5]. As a result they occupy the complete available phase space, enlarging the transverse and longitudinal energy spreads to about the value of the NEA (see Fig. 2), which is typically about 150-250 meV. Moreover, in contrast to thermocathodes a decrease of the temperature causes a rise of the energy spreads due to further NEA increase [10]. Hence the overall energy distribution from a photocathode is about the same or even worse as compared to thermocathodes. However, there is a way to reduce the energy spreads. It is seen that electrons emitted above a suitable longitudinal energy ($\geq E_c$) have transverse and longitudinal spreads of about the bulk temperature [5, 11]. For these electrons energy spreads of about 5–7 meV and 25 meV were measured at 90 K and 300 K, respectively [5, 14]. Effective quantum yield QY_{eff} for the cold electrons is found to be 1-1.5% at 90 K [12]. In the case of the conventional space charge operation of the electron source the required longitudinal barrier is naturally built by the space charge. In current mode operation of the photocathode source the laser power of about 100 mW will produce about 20 mA of emission current (with $QY \approx 1-1.5\%$). Reducing the extraction voltage allows running the cathode in space charge mode and bringing electron currents down to about required value (<1 mA) that orresponds to $QY_{eff} \approx 1-1.5\%$ and consequantly to the right position of the longitudinal space charge barrier. To be on safety side (with the barrier above E_c) the laser power up to 1 W is usually applied.

EXPERIMENTAL

The Heidelberg TSR is equipped with two electron beam facilities in different sections of the ring (see Fig. 3). One is used for phase space cooling of the stored ion beam and the other as a target [3]. This separation allows independent control of the parameters of both electron beams, so that the quality of the electron and ion beams, respectively, can be optimized independently in each of the device. For the used electron densities in the target (a few times smaller than in the cooler) no velocity changes of the ion beam, circulating under stationary conditions, were found even at very low (<0.1 meV) relative energies. Moreover, the cryogenic photocathode electron source developed for the TSR target allowed us to reduce the transverse temperature of the electron beam down to 0.5 meV, which finally strongly improved the collision energy resolution.

Electron Target

The target consists of an acceleration section (including the electron gun), an interaction section, a collector and two toroid sections. Several valves allow the separation of the acceleration/collector section from the toroid sections. The maximum magnetic field at the cathode position is up to 3.5 T and the guiding magnetic field amounts to 0.04–0.1 T. Hence, magnetic expansion ratios up to 90 are possible, expanding low-energy electrons adiabatically on a distance of about 0.3–0.4 m. Voltages applied to the Pierce and to the extraction electrode control the beam profile and the electron current from the cathode, respectively. After the extraction (gun perveance is 1μPerv) the electron beam enters a drift tube with a length of about 0.2 m, where the electron energy is

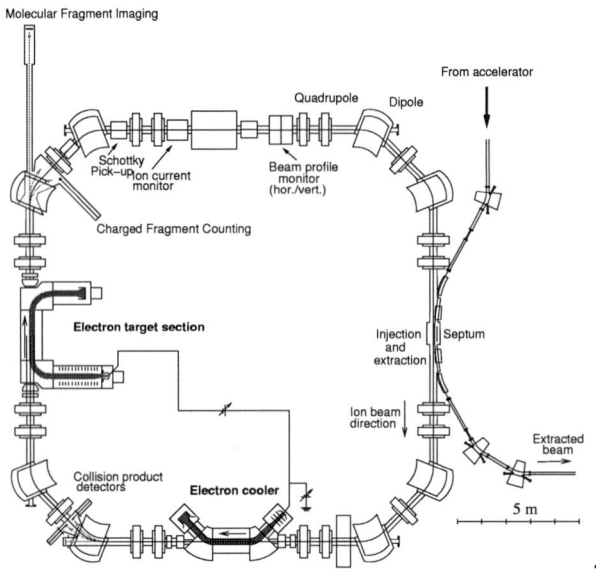

FIGURE 3. Layout of the Heidelberg Test Storage Ring equipped with two electron facilities.

kept low (100–250 V) during magnetic expansion. Then the beam reaches the adiabatic acceleration section which is about 1.5 m long and includes 77 independent electrodes. After the acceleration (maximum energy is 20 keV) the electrons are bent in a 90° toroid with a radius of 0.6 m and overlap with the ion beam in the interaction section over a distance of about 1.5 m. The interaction section is designed for magnetic-field $\Delta B_\perp/B$ angles to be smaller than 10^{-4} in oder to achieve an energy resolution below 0.1 meV in recombination experiments. At the second toroid the electrons are separated from the ion beam and collected by a Faraday cup. The recombination processes resulting from electron-ion interactions at the target are observed downstream by direct counting of charge-changed products, which are separated from the stored ion beam at the next storage ring bending magnet. Details of the target setup are described elsewhere [3].

Setup and Photocathode Handling

We use transmission mode GaAs photocathodes consisting of a two layer GaAs/AlGaAs heterostructure bonded to a sapphire substrate. The emitting p^+-GaAs(100) layer was $\approx 1.5\,\mu$m thick and doped with Zn to the hole concentration of about $5 \times 10^{18}\,\text{cm}^{-3}$. The vacuum setup includes a loading (with an attached atomic hydrogen chamber), a preparation and a gun chamber, separated by all-metal gate valves, with base pressures of 10^{-10}, 10^{-12} and 3×10^{-11} mbar, respectively. Inside the vacuum the samples are transferred by magnetically coupled manipulators. In the preparation chamber samples are fixed on a carousel capable of keeping four cathodes. The carousel can be rotated into different positions for thermal cleaning and activation of

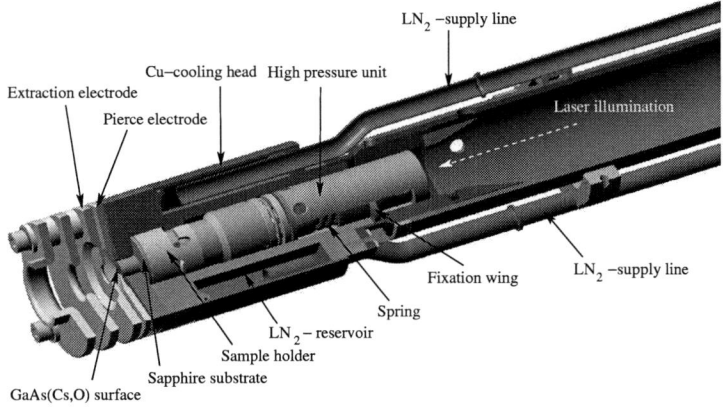

FIGURE 4. Part of the electron gun, showing the photocathode sample in the mounted position.

the photocathodes with cesium and oxygen. The quantum yield QY of a fully activated cathode is about 20-25%. In the electron gun (see Fig. 4) the sapphire substrate of the cathode is pressed to the polished axial surface of a copper cold head by a spring force of 100 N in order to obtain a good thermocontact [15]. The photocathode is illuminated in transmission mode with a 800 nm diode laser capable to apply a power of 1.7 W. By flooding the cold head with liquid nitrogen the typical operating temperature of about 100 K is presently used with laser illumination of about 1 W.

At this temperature, the photocathodes are typically run at emission currents of about 0.2 mA during 10-15 hours with a current stability of about 5%. Then the electron current rapidly falls down due to the rise of the work function so that the vacuum level reaches the position of the conduction band in the bulk (see Fig. 1). A further improvement of the vacuum conditions (the pressure is estimated to be about few 10^{-10} mbar at the photocathode region) is required to obtain the currents and lifetimes beyond the present limitation. The degraded cathode can be changed against another one stored in the preparation chamber. It takes about 20-30 min to change the sample and to cool it down. The samples were usually re-used 3–5 times by radiative heating and re-activation in the preparation chamber before they loose their performance. Then an atomic hydrogen cleaning is used to recover the properties of the degraded cathode by removing contaminations, as well as As- and Ga-oxides and old (Cs,O) layers [13]. In total, we had 4 samples which were used for all measurements without even removing them from the vacuum chamber.

HIGH-RESOLUTION MEASUREMENTS AT THE TSR TARGET

In first recombination measurements at the target, performed on HD^+, H_3^+ and Sc^{18+}, using the photocathode source, low energy resonant structures at milli-eV collision energies revealed unprecedented low transverse and longitudinal electron temperatures. To demonstrate high-energy resolution the data on dissociative recombination (DR) of

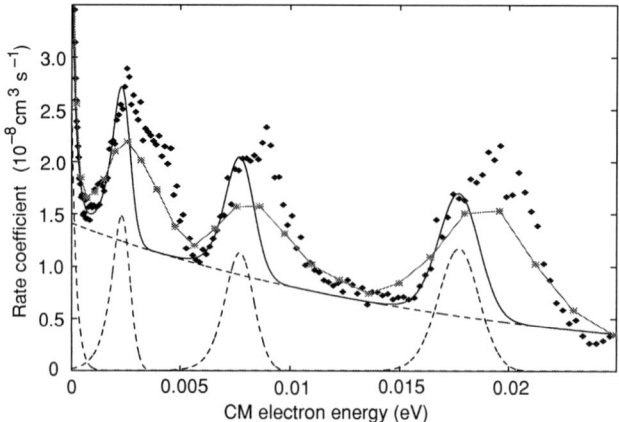

FIGURE 5. Dissociative recombination rate coefficient of stored HD$^+$ ions at the TSR observed with cryogenic photocathode electron gun (dots) [4, 16] compared to the previous high-resolution data from CRYRING [7] (olid line with stars). To estimate the transverse electron energy spread the experimental shape at the low-energy sides of the three peaks was fitted with a model rate coefficient (——) as obtained from the sum of several components (- - - -), representing a smoothly varying base contribution and three delta resonances, broadened with the electron energy distribution for $kT_\perp$=0.5 meV and $kT_\parallel$=0.02 meV.

HD$^+$ are presented. The spectrum was measured with a magnetic expansion of 20, electron currents of about 0.1 mA ($n_e \approx 5 \times 10^5$ cm^{-3}) and an average ion current of 1 μA. The ion energy was 1.7 MeV/u (corresponding electron energy was 1000 eV). The diameters of the electron beam and the stored ion beam were about 13 mm and 1 mm, respectively. The recombined neutral products (H, D and H+D) were detected downstream with a surface barrier detector. Fig. 5 shows the DR rate coefficient for HD$^+$ (dots) at energies below 25 meV (in the CM frame), where transverse energy spreads play a crucial role for the resolution. The best spectrum obtained before at CRYRING in Stockholm [7] with a resolution of 2 meV is also shown. Aside from a zero energy peak, both the previous and new spectra have three peaks at about the same positions. It is also seen, that the data obtained at TSR target have a sharper structure and clearly demonstrate the presence of an even finer structure within each peak.

While an a-priory the theoretical prediction of the DR peaks is presently not available, we have used a model of three delta peaks asymmetrically broadened according to electron velocity distributions to fit the low energy side of the peaks under the assumption that there is also a smoothly varying base contribution to the DR rate coefficient whose strength was chosen to fit the valley between the peaks. The steepness of the low energy side delivers $kT_\perp \approx 0.5$ meV, while that of the high energy sides yields $kT_\parallel$ of about 0.02 meV. It is seen that the fit describes very well the low energy side of each peak. It should be also mentioned that the same procedure applied to the Stockholm data gives the previously quoted [7] value of $kT_\perp$=2 meV. The measurements of dielectronic recombination on Sc^{18+} (not shown here) and the following fitting with a theoretical spectrum convoluted with a thermal electron velocity distribution allowed us to confirm that transverse energy of the electron beam stays below 1 meV [17].

LOW-ENERGY ELECTRON COOLING

Our interest in this field stems from a development of a novel cryogenic electrostatic ring (CSR at MPIK) used to store low-energy ions (20-300 keV per charge) [18, 19]. An electron cooler, based on the photocathode technology developed for the TSR, will serve as a major tool to cool the ions and to study electron-ion interactions. A strong anisotropy of an electron velocity distribution with $kT_\parallel \ll kT_\perp$ results in different amplitudes of friction forces in longitudinal and transverse directions. Due to a strong mass difference the velocity spread of injected ions is expected to be smaller than the longitudinal velocity spread of cooling electrons, so that the linear range of the friction force has to be considered. To calculate cooling times usually two limited case are considered of the nonmagnetic or strongly magnetized electron beams. However, the used guiding magnetic fields are typically in between. The dependence of cooling times on electron temperatures and on electron densities in the case of nonmagnetic is given by Eq. .(3) [21]. In the limiting case of complete magnetization (Eq. .(4)) the friction forces depends only on longitudinal temperature of electron beam [21].

$$\tau_{cool,\perp}^{nomag} \propto \frac{(kT_\perp)^{3/2}}{n_e} \qquad \tau_{cool,\parallel}^{nomag} \propto \frac{kT_\perp (kT_\parallel)^{1/2}}{n_e} \qquad (3)$$

$$\tau_{cool}^{mag} \propto \frac{(kT_\parallel)^{3/2}}{n_e} \qquad (4)$$

In this section the influence of emission electron energy spreads from a cathode on electron beam temperatures and on cooling times will be considered for the strong and weak magnetic fields in the case of very small laboratory electron energies.

Electron Beams Guided by Very Low Magnetic Field

Low-energy light ions (e.g. protons) are very sensitive to the magnetic field of the cooler and can be strongly deflected by the toroid magnetic fields. It limits the maximum guiding magnetic fields down to 15-30 Gauss for protons at the Heidelberg CSR [20]. This field is enough to keep confined the electron beam but it is too weak to suppress the transverse-longitudinal relaxation, so that the longitudinal temperature rises up to the transverse one resulting in an isotropic velocity distribution of electrons in the beam with $kT_\parallel \approx kT_\perp \approx kT_c/\alpha$. Raising of the magnetic expansion decreases both $kT_\perp$ and electron densities, so that cooling time will be improved as $\alpha^{-1/2}$. However, increasing the α more than 100 does not appear to lower $kT_\perp$ further. The lowest transverse energy of about 2 meV was obtained at CRYRING with a thermocathode at $\alpha=90$ and of about 0.5 meV at the TSR target with cryogenic photocathode at $\alpha=20$. According to Eq.(3) the cooling time will depend on the cathode temperature as $\tau_{cool} \propto (kT_c)^{3/2}$. For the same magnetic expansion, the τ_{cool} can be by a factor of 30-40 shorter for the photocathode electron source ($kT_c \approx 10$ meV) in comparison to the thermocathode one ($kT_c \approx 110-120$ meV).

Electron Beam Guided by Strong Magnetic Field

The higher the ion mass the less sensitive they are to the toroid magnetic fields at the cooler. For the Heidelberg CSR it was found [20] that guiding magnetic fields up to 150-200 Gs can be used at the cooler without a strong deflection effect for stored ions with masses of A>12. At these magnetic fields and for the used electron densities the transverse longitudinal relaxation is suppressed and the longitudinal temperature is described by Eq. (1).

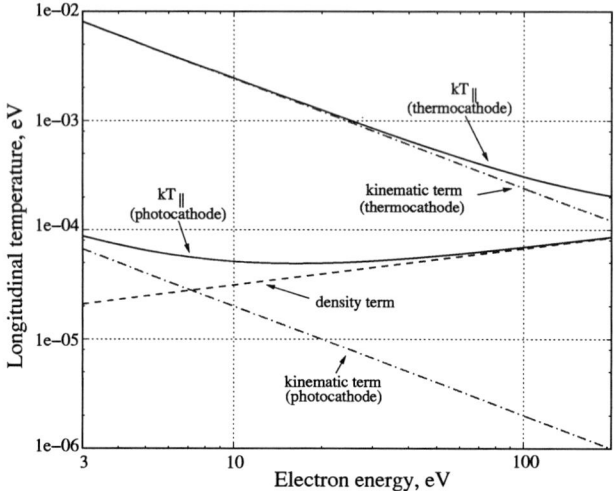

FIGURE 6. Longitudinal temperature, as well as the kinematic and density contributions to $kT_\parallel$, as a function of electron kinetic energy for expanded beams ($\alpha = 20$).

For kinetic energies in keV range the $kT_\parallel$ depends mainly on the density term (see Eq. (1)) and, as a result, cooling times are independent on cathode temperatures in the case of strong magnetization (see Eq. (4)). However, the situation is completely different for low-energy beams when the kinematic contribution to the $kT_\parallel$ plays a crucial role. Fig. 6 shows $kT_\parallel$, as well as the kinematic and the density contributions to $kT_\parallel$, as a function of the electron kinetic energy E_{kin} in the range of 3-200 eV. For the calculation of the electron density n_e we assumed a cathode diameter of 3 mm, a gun perveance of 2μPerv and magnetic expansion of 20. It corresponds to currents of $\sim$0.01 mA ($n_e \approx$4.5 10^5 cm^{-3}) and $\sim$ 5.7 mA ($n_e \approx$3 10^7 cm^{-3}) at 3 V and 200 V, respectively. The cathode temperature was 110 meV and 10 meV for thermocathodes and photocathodes, respectively. It is important to mention that for the calculations of the kinematic contribution using Eq. (1) the cathode temperature was doubled, assuming that during magnetic expansion almost all transverse energy spread will be transferred to the longitudinal one. It is seen (Fig. 6) that for thermocathodes the kinematic term dominates over all energy range, approaching the density contribution at 200 eV. For photocathodes the kinematic term can be strongly suppressed, that finally reduces longitudinal temperatures of electron beams by a factor of $\sim$ 100 at low energies and by a factor of about 3 at 200 eV. For the Heidelberg CSR with the ion energies of

300 keV and for molecules with A>12, the corresponding cooling electron energies will be below 15 eV. It means that it covers the region (see Fig. 6) where the kinematic term gives the main contribution and cooling times ($\tau_{cool} \propto T_c^3$) can be strongly improved by a factor of 1000 with a help of cryogenic photocathode source. Of cause it is a limited case and for the used magnetic fields the effect will be smaller. As a low limit the formula (3) can be used. Compare to thermocathode, the photocathode can reduce transverse cooling time by a factor of 30-40 and longitudinal times by a factor of 100.

CONCLUSIONS

A cold electron target with a cryogenic photocathode source was developed for electron-ion merged experiments at the Heidelberg TSR. The source provides electrons with energy spreads below 10 meV at currents up to 0.5 mA. After subsequent adiabatic magnetic expansion and acceleration transverse and longitudinal temperatures of the electron beam below $kT_\perp$<1.0 meV (down to 0.5 meV) and $kT_\parallel$<0.03 meV (down to 0.02 meV) were obtained at the TSR target. The developed photocathode source will be also used at the Heidelberg CSR. A photocathode-driven electron cooler is expected to be an excellent tool for cooling low-energy ion beams.

REFERENCES

1. I. N. Meshkov, *Phys. Part. Nucl.*, **25**, 631–661 (1994)
2. G. Gwinner et al, *Phys. Rev. Lett.*, **84**, 4822–4825 (2000)
3. F. Sprenger et al, *NIM A*, **532**, 298–302 (2004)
4. D. A. Orlov et al, *J. Phys.: Conf. Ser.*, **4**, 290–295 (2005)
5. D. A. Orlov et al, *Appl. Phys. Lett.*, **78**, 2721–2724 (2001)
6. N. S. Dikansky et al, *Ultimate possibilities of electron cooling*, Preprint 88-61, Institute of Nuclear Physics, Novosibirsk, 1988
7. A. Al-Khalili et.al., *Phys. Rev. A*, **68**, 42702 (2003)
8. T. M. O'Neil and P. G. Hjorth, *Phys Fluids*, **28**, 3241–3252 (1985)
9. J. .J. Scheer and J. van Laar, *Solid State Commun.*, **3**, 189–193 (1965)
10. D. A. Orlov et al, *NIM A*, **532**, 418–421 (2004)
11. S. Pastuszka et al, *J. Appl. Phys.*, **88**, 6788-6800 (2000)
12. D. A. Orlov et al, *Proceedings of the 9-th International Workshop on Polarized Sources and Targets*, World scientific, New Jersey, 2002, pp. 151–155.
13. D. A. Orlov et el, *Proceedings of Polarized Electron Sources and Polarimeters*, 2004, to be published.
14. D. A. Orlov at al, *Hyperfine interactions*, **146/147**, 215–218 (2003)
15. U. Weigel et al, *NIM A*, **536**, 323–328 (2005)
16. H. Buhr et al, to be published
17. M. Lestinsky et al, to be published
18. D. Zajfman et al, *J. Phys.: Conf. Ser.*, **4**, 296–299 (2005)
19. A. Wolf et al, this proceedings
20. H. Fadil et al, this proceedings
21. Yu. S. Derbenev and A. N. Scrinski, *Sov. Phys. Rev.*, **1**, 165–237 (1981)

LASER COOLING

Laser Cooling for 3-D Crystalline State at S-LSR

Akira Noda[1], Shinji Fujimoto[1], Masahiro Ikegami[1], Toshiyuki Shirai[1], Hikaru Souda[1], Mikio Tanabe[1], Hiromu Tongu[1], Koji Noda[2], Satoru Yamada[2], Shinji Shibuya[3], Takeshi Takeuchi[3], Hiromi Okamoto[4] and Manfred Grieser[5]

[1] *Advanced Research Center for Beam Science, Institute for Chemical Research, Kyoto University, Gokano-sho, Uji-city, 611-0011, Kyoto, Japan*
[2] *National Institute of Radiological Sciences, Anagawa 4-9-1, Inage-ku, Chiba-city, 263-8555, Chiba, Japan*
[3] *Accelerator Engineering Co. Ltd., Anagawa 4-9-1, Inage-ku, Chiba-city, 263-8555, Chiba, Japan*
[4]. *Graduate School of Advanced Science of Matter, Hiroshima University, Higashi Hiroshima, Hiroshima, Japan*
[5] *Max-Planck-Institut für Kernphysik, Postfach 103980, D-69029, Heidelberg, Germany*

Abstract. At ICR, Kyoto University, an ion storage and cooler ring, S-LSR has been constructed. Its mean radius and maximum magnetic rigidity are 3.6 m and 1.0 Tm, respectively. $^{24}Mg^+$ ions with the kinetic energy of 35 keV are to be laser-cooled by the frequency doubled ring dye laser with the wavelength of 280 nm. In order to avoid the shear heating, dispersion compensation is planned by the overlap of the electric field with the dipole magnetic field in all 6 deflection elements. Intermediate electrodes, which can be potential adjusted, are to be utilized so as to realize a uniform electric field radial direction within a rather limited vertical gap, 70 mm of the dipole magnet. Synchro-betatron coupling needed for 3-dimensional laser cooling is to be realized by placing the RF cavity at the siraight section with finite dispersion for the normal mode lattice, which is expected to realize 1 dimensional string. For the case of dispersion compensated lattice to suppress the shear heating, possibility of realizing "tapered cooling" with use of an Wien Filter combined with the laser cooling is being investigated in order to avoid the usage of the coupling cavity, which seems to be difficult to fabricate. With the presence of such a tapered cooling, formation of a 1 shell crystalline structure is expected.

Keywords: 3-D Crystalline Beam, Shear Heating, Tapered Cooling, Dispersion Suppressor.
PACS: 41.75Ak, 41.85.Lc

INTRODUCTION

At the workshop ECOOL84 in Karlsruhe, the ordering of proton beam by electron beam cooling when the beam intensity goes down to lower value was reported based on the experiments at NAPM in Novosibirsk [1]. Recently similar ordering was reported from ESR at GSI [2] and CRYRING [3] for electron cooled heavy ion beams. The crystal beam with 3 dimensional extension, however, has not yet realized because of "shear heating" except for the success with PALLACE at very low energy [4]. Possible suppression of "shear heating" by dispersion compensated lattice [5] is to be applied for S-LSR [6] or "tapered cooling" with use of Wien Filter [7] is to be experimentally studied by 3 dimensional laser cooling of ^{24}Mg$^+$ at S-LSR. In the present paper, the experimental program to approach the 3-dimensional crystalline beam is shown together with the preliminary indication from the computer simulation with use of molecular dynamics.

ION STORAGE AND COOLER RING, S-LSR

At Institute for Chemical Research, Kyoto University, an ion storage and cooler ring, S-LSR has been under construction for the purpose of demonstrating the capability of generation of high quality laser-produced ion beam by the combination of phase rotation and electron beam cooling of hot ion beam [8]. At Advanced Research Center for Beam Science of ICR, Kyoto University, an approach to crystalline beam by 3-dimensional laser cooling has also been started utilizing S-LSR. In Fig.1, the layout of the experimental area of S-LSR is shown.

In this section, the lattice structure of S-LSR is described in connection with the requirements from beam dynamics related to beam crystallization and efforts to suppress the "shear heating" is described. Then the experimental plan to attain crystalline beam is shown together with preliminary result of computer simulation with molecular dynamics.

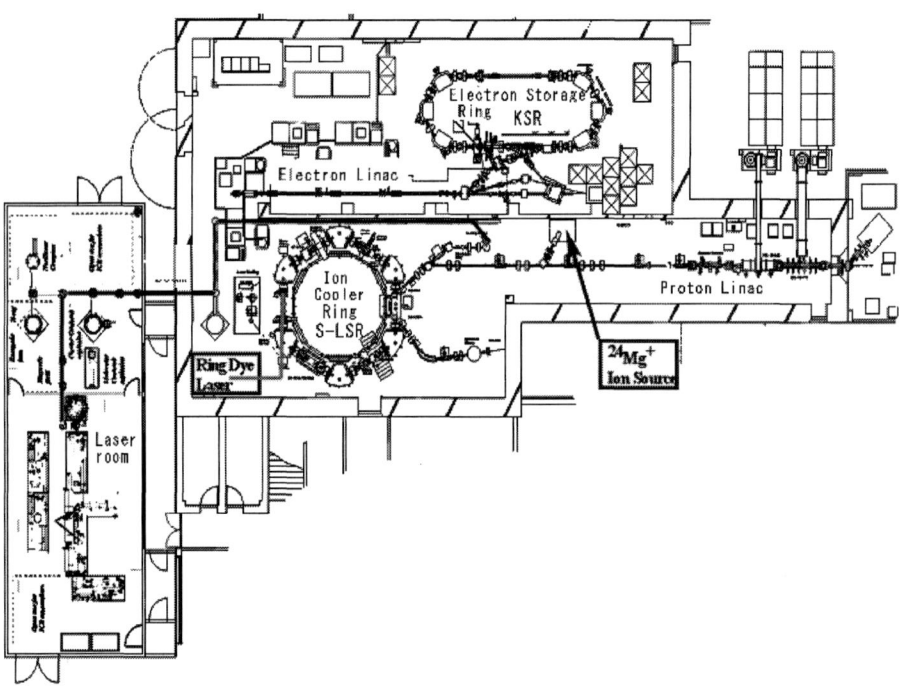

FIGURE 1. Layout of Experimental Area of S-LSR and its laser cooling system.

Lattice Structure of S-LSR

Main magnet system of S-LSR is composed of 6 dipole and 12 quadrupole magnets and has 6-fold symmetry. As the criteria to keep the crystalline structure of the beam, the following so-called "*crystal formation and maintenance conditions*" [9] are well known,

$$\gamma \leq \gamma_t \qquad (1)$$

$$\nu_{H,V} \leq \frac{N_s}{2\sqrt{2}}, \qquad (2)$$

where γ_t, $\nu_{H,V}$, N_s are transition γ, betatron tune in horizontal, vertical directions and ring superperiodicity, respectively. In the cooling experiment of $^{24}Mg^+$ ion beam with the kinetic energy of 35 keV, the first condition is satisfied at all the operating points of S-LSR. S-LSR has been designed so as to have the operating points satisfying also the second condition. The geometrical structure of S-LSR, main parameters and the operating points of S-LSR are given in Fig.2 and table 1, 2, respectively.

Normal Lattice with Finite Dispersion

With the normal lattice where the deflection of the ion beam is performed with only the magnetic fields, the finite size (>1 m for the present case) momentum dispersion appears as shown in Fig. 3 for the operation tune of (2.07, 1.07). The stable region of S-LSR for the lattice with only magnetic fields is given in Fig. 4 together with typical operating points. This suffers "shear heating" [10] as described below and makes the creation of the multi-dimensional crystal difficult. Even with this mode, however, formation of one dimensional string is expected, if enough strong and 3-dimensional cooling force is available[11]. For this purpose, syncho-betatron coupling

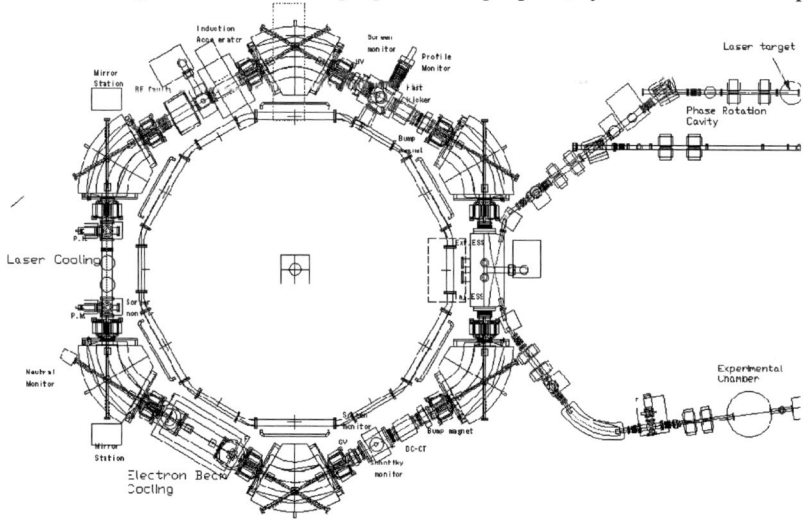

FIGURE 2. Geometrical structure of S-LSR.

TABLE 1. Main parameters of S-LSR.

Parameter	Value
Circumference	22.557 m
Average Radius	3.59 m
Length of Straight Section	2.66 m
Number of Period	6
Bending Magnet	H-Type
Maximum Field	0.95 T
Radius of Curvature	1.05 m
Gap Height	70 mm
Pole End Cut	Rogowski Cut + Filed Clamp
Deflection Angle	60°
Weight	4.5 tons
Quadrupole Magnet	
Maximum Field Gradient	5 T/m
Core Length	0.20 m
Bore Radius	70 mm

TABLE 2. Main operating points of S-LSR

Operating mode	betatron tune	transition gamma
3D cooling using a coupling resonance method	(2.07, 1.07)	1.754
Tapered cooling of coasting beams	(1.44, 1.44)	1.231
Dispersion free mode	(2.06, 2.06)	∞

induced through dispersion at an rf cavity is assumed in order to realize the transverse cooling force [12]. In order to maximize the coupling effect, the betatron tunes and the synchrotron tune are required to satisfy the following resonance conditions,

$$v_s - v_H = n \quad (n: \text{integer}, v_s: \text{synchrotron tune}) \quad (3)$$

$$v_H - v_V = m \quad (m: \text{integer}) . \quad (4)$$

The coupling of the horizontal and the vertical motions is to be induced by a solenoid field of the electron cooler. In addition, an RF cavity with the main parameters given

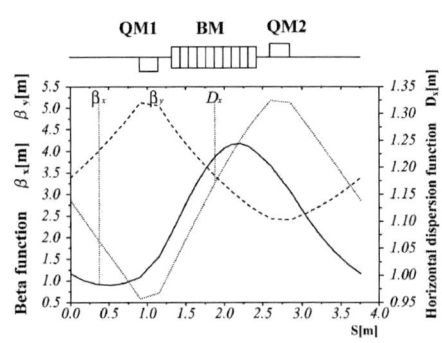

FIGURE 3. Beta and dispersion functions of S-LSR with normal mode without electric field with the operation tune at (2.07, 1.07)

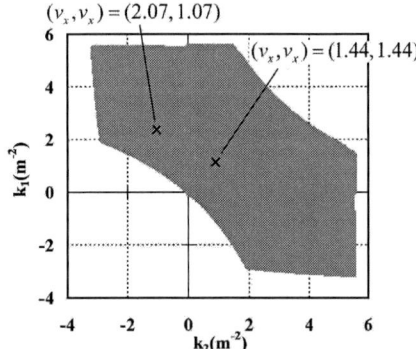

FIGURE 4. Operating points of S-LSR and its stable region.

in table 3 located at the finite dispersion position or coupling cavity whose longitudinal acceleration voltage has horizontal (or vertical) position dependence [13] is required for longitudinal and transverse coupling. With this scheme, 3-dimensional cooling is considered to be possible for bunched beam.

From the computer simulation by molecular dynamics assuming the cooling force approximated as a linear friction force with its coefficients of fx=fy=0.1, fz=0.15, one dimensional string is expected also for the operating point at (2.06, 2.06) as shown in Fig. 5, although the typical operating point at (2.07, 1.07) for the normal lattice is considered to be much stable because its phase advance per cell is smaller. The multi-dimensional crystal is, however, difficult to be realized for this mode because of the shear heating.

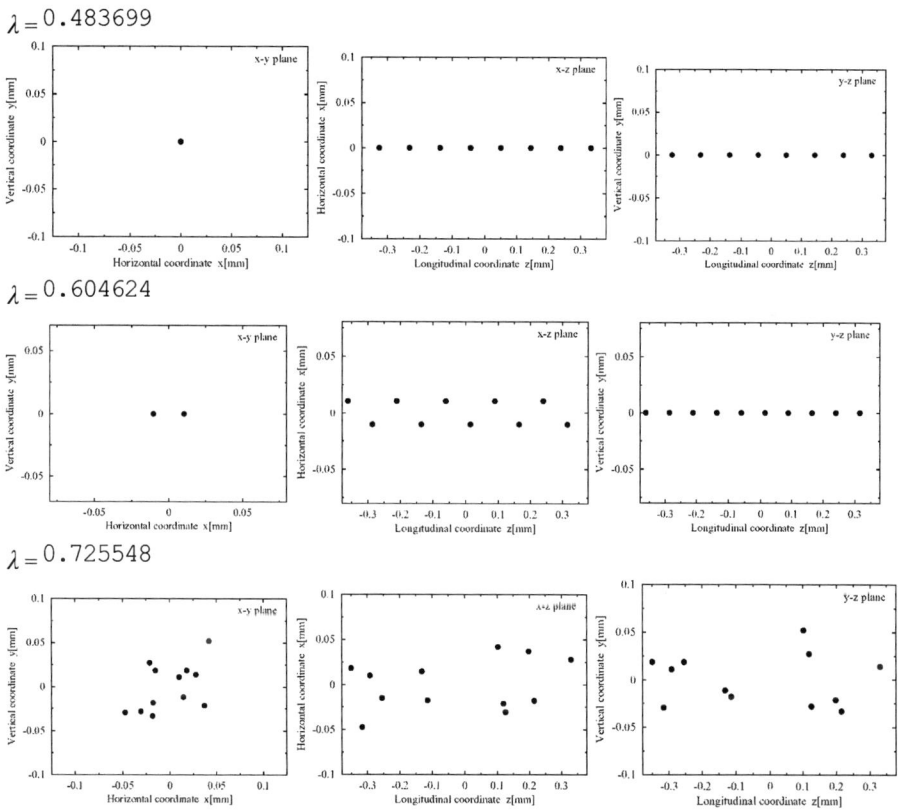

FIGURE 5. Particle distribution in the real space at the final state of the cooling. The cooling force was given as a linear friction force of the three directions. λ means the normalized line density. Low-line-density zigzag structure was formed for the case of normalized line density, λ=0.604.

Tapered Cooling Mode

A tapered cooling force [14] enables us to avoid the "shear heating" of the cooled beam, which is due to angular velocity difference among ions with different energies at the deflection elements with finite momentum dispersion. Therefore, if the tapered

TABLE 3. Main parameters of the rf cavity for 3D cooling at the operation point (2.07, 1.07).

Quantity	Value
Ions to be laser cooled	$^{24}Mg^+$
Total kinetic energy	35 keV
Betatron tune	(2.07, 1.07)
Synchrotron tune	0.07
rf voltage	30.7 V
rf frequency	2.32 MHz
rf harmonics	100
Momentum compaction factor	0.325

cooling is realized, the multi-dimensional crystalline beam may be obtained in conventional storage rings with finite dispersion. It is also known that the tapered cooling force which acts on the beam at the position with finite dispersion, the cooling force also acts on the horizontal motion. Therefore, it enables the three-dimensional cooling for coasting beams. We are planning the tapered cooling at the operating point (1.44, 1.44). This operating point is best for generating the multi-dimensional crystal, since the phase advances of the betatron motions in both horizontal and vertical directions are kept below 90 degrees [15].

The tapered cooling method using the Wien filter has been proposed [7]. The ion beam orbit is partially shifted to overlap with the laser only inside the Wien filter, where the $^{24}Mg^+$ ions are laser cooled (Fig.6). Due to electrostatic potential created by the Wien filter, the ions have energy difference according to their radial positions while all the ions are cooled down to the same energy by the laser cooling. So in the straight section after coming out from the Wien filter, the ions are cooled down to the energies dependent on their radial positions, which is expected to attain the so called "tapered cooling". The magnetic and electrostatic fields of the Wien filter and magnets to create orbit chicane, however, might reduce the superperiodicity, which needs further careful study with molecular dynamics simulation..

Lattice without Linear Dispersion Utilizing Dispersion-Suppresser

In order to suppress the shear heating, the possibility of adopting the lattice without the linear dispersion throughout the whole circumference has been studied with use of the dispersion-suppresser [6]. The dispersion-suppresser has a *cross-field* composed of

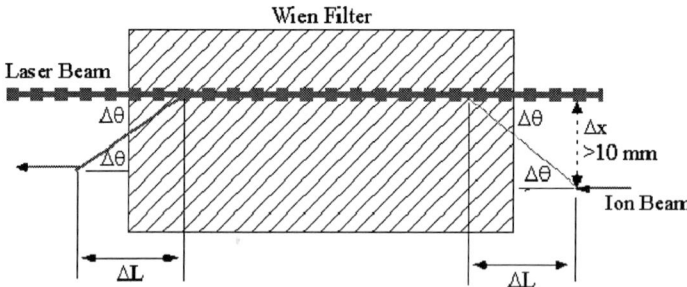

FIGURE 6. Proposed scheme for tapered cooling by combination of the Wien Filter with the laser cooling. By application of orbit chicane to the ion beam, the laser is overlapped only inside of the Wien filter, where the ion beam has the different energy depending on its radial position due to electrostatic potential of the Wien filter.

magnetic and electric fields [16]. In ref.6, we considered the electric field generated by a cylindrical electrostatic deflector. When this electric field is superposed with the magnetic field, the radial focusing of the dispersion-suppresser is enhanced. Due to this effect, in S-LSR, the operating point satisfying the maintenance condition has been lost in the dispersion-free mode.

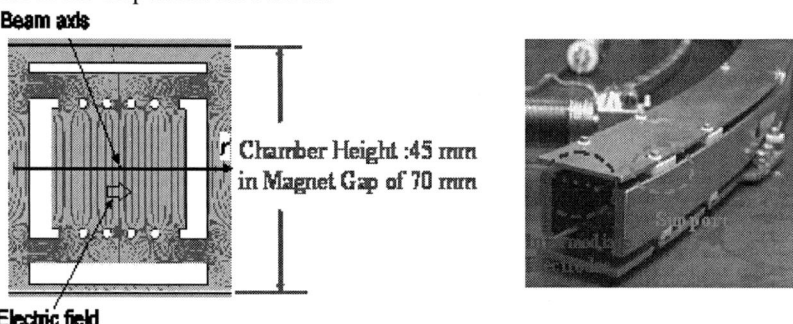

(a) Cross-sectional shape (b) Photograph of the fabiricated electrodes

FIGURE 7. Fabricated electrodes installed into the vacuum chamber inside the dipole magnet of S-LSR.

If the field distribution of the electric field is changed, the strength of the radial focusing can be controlled, in the same way that the focusing of the combined function magnet changes with the field index. The ideal electric field distribution for our purpose has been found to be uniform in the radial direction, which is generated by the electrode with a complex curved line shape [17]. The real fabricated electrodes have a cylindrical shape and 4 pairs of intermediate electrodes as shown in Fig. 7 [18]. It is found that by adjusting the potentials of these intermediate electrodes, almost uniform electric field in the radial direction is realized. The betafunctions for this case is shown in Fig. 8 for the operating point of (2.06, 2.06). The stable region for this mode is given in Fig. 9.

With this mode, the linear dispersion is zero throughout the whole circumference and it is not possible to attain synchro-betatron coupling by using an ordinary RF acceleration cavity. So a coupling cavity above mentioned becomes inevitable for this

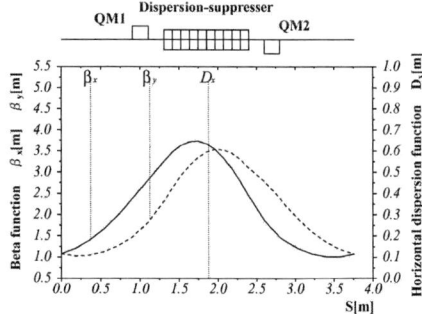
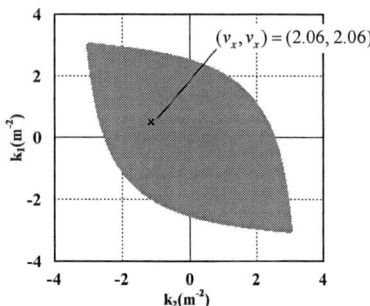

FIGURE 8. Betafunctions for the operating point of (2.06,2.06) with dispersion free lattice.

FIGURE 9. The operation point and stable region of S-LSR with dispersion free mode created by superposition of electric field with the magnetic field.

case to attain 3-D laser cooling for bunched beam. We have found the fact that the ideal tapered cooling will realize the 3 dimensional laser cooling for coasting beams, which is expected to be effective for the case of the dispersion free ring [19].

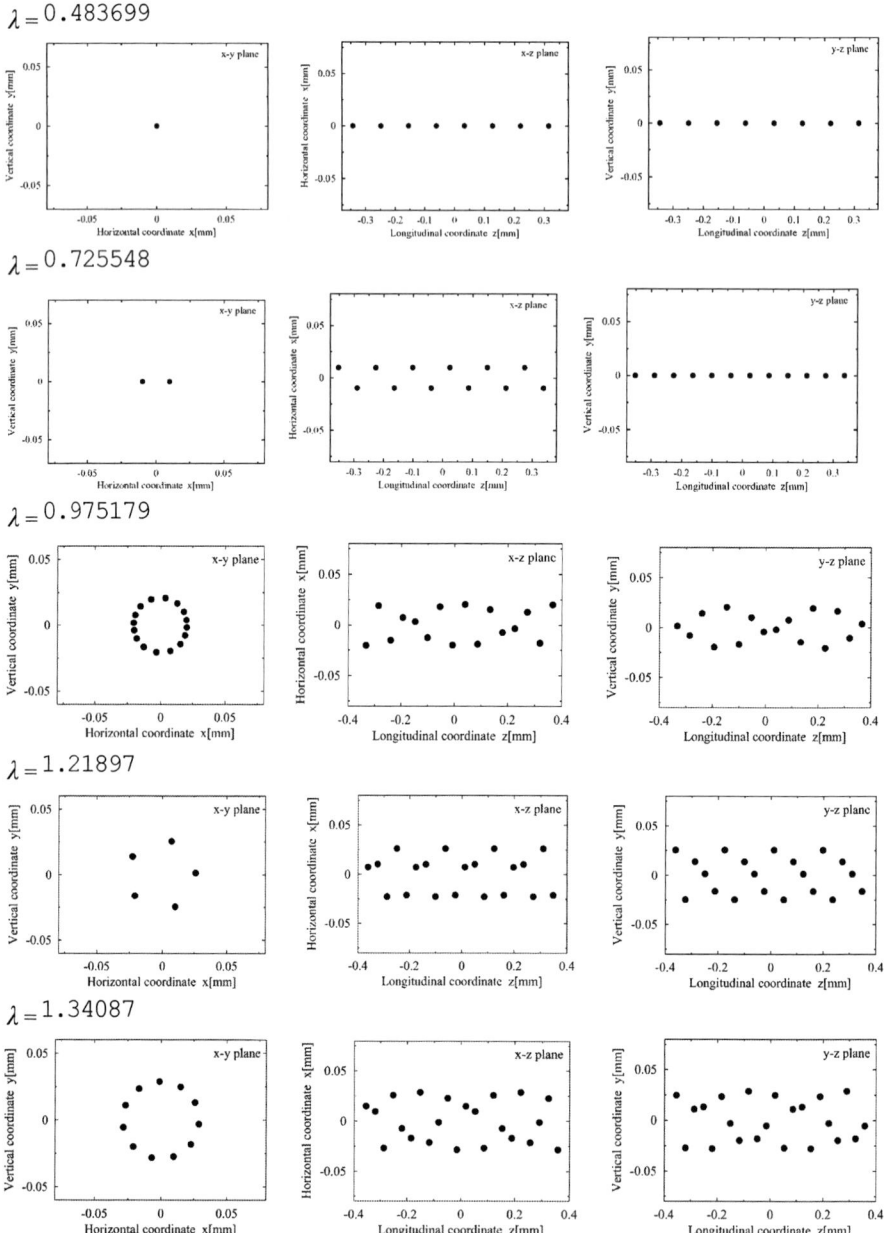

FIGURE 10. Particle distribution in the real space at the final state of the cooling. Three-dimensional crystalline beam

The molecular dynamics simulation performed for the dispersion-free mode of the betatron tune (2.06, 2.06) also assuming the linear friction coefficients of the cooling forces as fx=fy=0.1, fz=0.15, shows the generation of 1-shell crystalline structure (Fig.10).

LASER COOLING SYSTEM AT S-LSR

In order to realize ultra-cold ion beam, laser cooling is to be applied for $^{24}Mg^+$ beam with the kinetic energy of 35 keV, which is extracted from the ion source by the high voltage of 35kV and is directly transferred and injected into the ring, S-LSR as shown in Fig. 1. From the energy levels of this atom, the laser with the wave length 280 nm is required, which is provided by a ring-dye laser utilizing Rhodamine 110 pumped with a solid green laser with the wavelength 532 nm and following second harmonics generator. In Fig. 11, the laser system under tuning is shown. At the beginning, a single laser is to be utilized sweeping the energy of the stored ions by an induction accelerator, which leads to beam loss in case of intra beam scattering is dominant. Its cooling force is shown in Fig. 12(a). We are planning to improve the system to the one utilizing two lasers counter and co propagating with the circulating ion beam, the cooling force of which, is shown in Fig. 12(b).

FIGURE 11. Photograph of the Ring-Dye laser pumped with Solid Green laser.

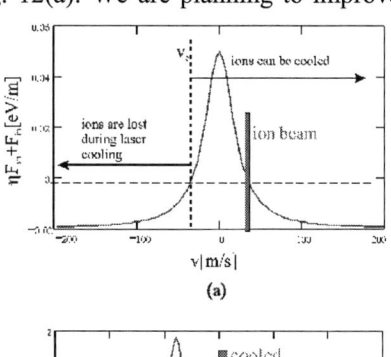

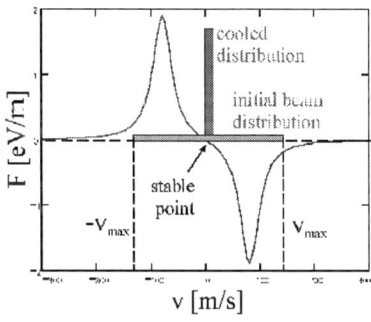

FIGURE 12. Laser cooling force for the case of (a) a single laser, (b) a couple of laser counter propagating each other.

SUMMARY

S-LSR lattice adopts rather high super-periodicity as 6, in order to make the so-called maintenance conditions to be satisfied. By the ordinary lattice with finite dispersion, one dimensional bunched ordered state is expected to be realized at the operating point (2.07, 1.07) with the synchrotron tune of 0.07 by the coupling resonance method using ordinary RF accelerating cavity. Its extension to multi-dimensional structure, however, needs, tapered cooling. The 1-shell crystalline structure may be realized by application of tapered laser cooling.

ACKNOWLEDGMENTS

The work presented here has been performed as the Advanced Compact Accelerator Development financed by Ministry of Education, Culture, Sports, Science and Technology of Japanese Government. This project is also supported from the 21st Century Center of Excellence program -Center for Diversity and Universality in Physics- at Kyoto University.

REFERENCES

1. V.V. Parkhomchuk, Physics of Fast Electron Cooling, Proc. of ECOOL1984, Karhlsruhe, Germany, pp.71-83 (1984).
2. M. Steck et al., Phys. Rev. Lett., **77**, pp3803-3806 (1996)
3. H. Danared, A. Källberg, K.-G. Rensfelt, and A. Simonsson, Phys. Rev. Lett. **88**, 174801 (2002)
4. T. Schaetz, U. Schramm and D. Habs, Crystalline ion beams, Nature, **412**, 717(2001)
5. R.E. Pollock, Zeitschrift fur Physik A, Hadrons and Nuclei, **341**, 95 - 99, (1991)
6. M. Ikegami, A. Noda, M. Tanabe, M. Grieser, H. Okamoto, Phys. Rev. ST Accel. Beams **7**, 120101 (2004).
7. A. Noda and M. Grieser, Beam Science and Technology, **9**, pp12-15 (2004).
8. A. Noda et al., Ion production with a high-power short-pulse laser for application to cancer therapy, Proc. of EPAC2002, 2748-2750 (2002).
9. J. Wei, H. Okamoto and A. M. Sessler, "Necessary conditions for attaining a crystalline beam", Phys. Rev. Lett. **80**, 1998, pp2606-2609.
   J. Wei, X.-P. Li, and A. M. Sessler, in Advanced Accelerator Concepts, Proceedings of the Sixth Advanced Accelerator Concepts Workshop, Fontana, 1994, AIP Conf. Proc. 335, edited by P. Schoessow (AIP, New York, 1995)
10. H. Okamoto, Nucl. Instrum. Methods. A **532**, 32 (2004)
11. Y. Yuri and H. Okamoto, Phys. Rev. Lett. **93**, 204801 (2004)
12. H. Okamoto. Phys. Rev. E **50**, 4982-4996 (1994)
13. H. Okamoto, A. M. Sessler, D. Möhl, Phys. Rev. Lett. **72**, 3977 (1994)
    T. Kihara, H. Okamoto, Y. Iwashita, K. Oide, G. Lamanna, and J. Wei, Phys. Rev. E **59**, 3594-3604 (1999)
14. H. Okamoto and J. Wei, Phys. Rev. E **58**, 3817-3825 (1998)
15. H. Okamoto, AIP Conf. Proc., in press.
16. W. Henneberg, Annalen der Physik. 19, 335 (1934)
    W. E. Millet, Phys. Rev. 74, 1058 (1948)
17. M. Ikegami, Y. Iwashita, M. Tanabe, A. Noda, submitted to Phys. Rev. ST Accel. Beams
18. M. Tanabe et al., , Proc. of this workshop.
19. M. Ikegami, et al. to be submitted to Phys. Rev.

Combined Laser and Electron Cooling of Bunched C3+ Ion Beams at the Storage Ring ESR

U. Schramm*, M. Bussmann*, D. Habs*, T. Kühl[†], P. Beller[†], B. Franzke[†], F. Nolden[†], M. Steck[†], G. Saathoff**, S. Reinhardt** and S. Karpuk[‡]

*Department für Physik, LMU München, 85748 Garching, Germany
[†]Gesellschaft für Schwerionenforschung (GSI), Darmstadt, Germany
**Max-Planck-Institut für Kernphysik, Heidelberg, Germany
[‡]Institut für Physik, Universität Mainz, Mainz, Germany

Abstract. We report on first laser cooling studies of bunched beams of triply charged carbon ions stored at an energy of 1.46 GeV at the ESR (GSI). Despite for the high beam energy and charge state laser cooling provided a reduction of the momentum spread of one order of magnitude in space-charge dominated bunches as compared to electron cooling. For ion currents exceeding $10\,\mu A$ intra-beam-scattering losses could not be compensated by the narrow band laser system presently in use. Yet, no unexpected problems occurred encouraging the envisaged extension of the laser cooling to highly relativistic beams. At ESR, especially the combination with modest electron cooling provided three-dimensionally cold beams in the plasma parameter range of unity, where ordering effects can be expected and a still unexplained signal reduction of the Schottky signal is observed.

Keywords: laser cooling, heavy ion beam cooling, storage rings, Coulomb ordering
PACS: 29.20.Dh, 41.75.-i, 42.50.Vk, 52.27.Gr

LASER COOLING OF STORED ION BEAMS

At heavy ion storage rings electron cooling represents the prominent technique for the reduction of the momentum spread as well as of the transverse emittance of stored ion beams until equilibrium with competing processes like intra-beam-scattering (IBS) is reached. Heating due to IBS increases with the phase space density of the beam implying that the equilibrium momentum spread for rf-bunched harmonically confined ion ensembles scales with the number of stored particles N as $N^{1/6}$. Thus, a further reduction of the momentum spread of an electron-cooled beam can be achieved by a reduction of the ion beam current or by the application of an additional cooling method increasing the cooling rate. Concerning the momentum spread, such an increase can be provided by laser cooling, relying on the resonant momentum transfer originating from the repeated scattering of photons out of a laser beam merged with the ion beam. Yet, the extremely steep momentum gradient of the laser force comes at the expense of a narrow momentum acceptance range, predominantly collinear action, and, at present machines, the limited number of ions with suitable optical transitions.

A strong motivation for the development of the laser cooling technique at the storage rings TSR in Heidelberg [1] and ASTRID in Aarhus [2] was the principle capability

---
[1] internet: www.ha.physik.uni-muenchen.de/uschramm/

of the method to reach ion beam temperatures, describing the energy spread in the co-moving frame, far below the mutual Coulomb energy of neighboring ions. In this regime a phase transition into a Coulomb-ordered or crystalline beam is expected [3]. Such a state is characterized by an almost complete vanishing of collision dominated heating mechanisms like IBS. Unfortunately, it turned out that this suppression of collisional heating also meant a reduction of the vital coupling between the transverse ion motion, experiencing little yet too much stochastic heating due to the randomness of the scattering process, and the longitudinal ion motion, directly laser cooled [4]. Thus, though techniques were developed that provide direct transverse laser cooling making use of storage ring dispersion[5], no stable crystalline beams could be observed in these two machines [4, 6], benchmarking and present activities being summarized in [7].

In the very low energy regime this phase transition could recently be demonstrated with laser-cooled $^{24}Mg^+$ ion beams in the rf quadrupole (RFQ) storage ring PALLAS [8] for coasting as well as for bunched beams [9]. The strong IBS heating of the cold beam could be overcome by first reducing the tune of the RFQ storage ring, which leads to a well-defined increase of the transverse beam size and thus to a reduced IBS rate, and increasing it again after the beam is longitudinally sufficiently cold. Here, the advantage of the RFQ is the adjustable tune whereas in conventional machines transverse heating might help [10]. In the crystalline regime transverse laser heating, as discussed for TSR and ASTRID was directly observed [11]. Yet, the online tuning capability of the focusing strength was sufficient for the compensation of the loss of indirect cooling by stronger transverse confinement, as again reviewed in [7].

At the heavy ion storage ring ESR (GSI) electron cooling of highly charged heavy ions has developed into a routine tool for the cooling of highly charged ions independent from their internal atomic properties [12]. As the cooling force (as well as the inter-ion coupling) roughly increases with the square of the ion charge, one dimensional beam ordering effects of highly charged ions were observable with electron cooling at extremely low beam currents [13, 14], where density dependent heating mechanisms become negligible.

Combining the experience from ESR, TSR and PALLAS the idea for the experiment presented in this paper is the demonstration of combined laser and electron cooling of C^{3+} ion beams. Laser cooling then should provide lowest momentum spread, electron cooling the transverse cooling and a larger momentum acceptance, and the relatively high charge state increases the ion-ion coupling.

BUNCHED BEAM (PURE) LASER COOLING AT THE ESR

At the ESR combined electron and laser cooling of C^{3+} ion beams can be performed at an energy of 122 MeV/u ($\beta = 0.47$, $\gamma = 1.13$). At this energy the closed optical $2S_{1/2} - 2P_{3/2}$ transition ($\lambda_0 = 154.82$ nm, $\tau = 3.8$ ns [15]) of the Li-like carbon ions is Doppler-shifted into resonance with the UV-laser line at $\lambda_{laser}/2 = 257.34$ nm [16] when counterpropagating laser and ion beams are used. The decelerating laser force is counteracted by the restoring force of a bucket when the beam is bunched. This established technique [17, 9, 7], sketched in the cartoons in Fig. 1, provides the momentum dependent friction force required for cooling without the need of a copropagating laser beam.

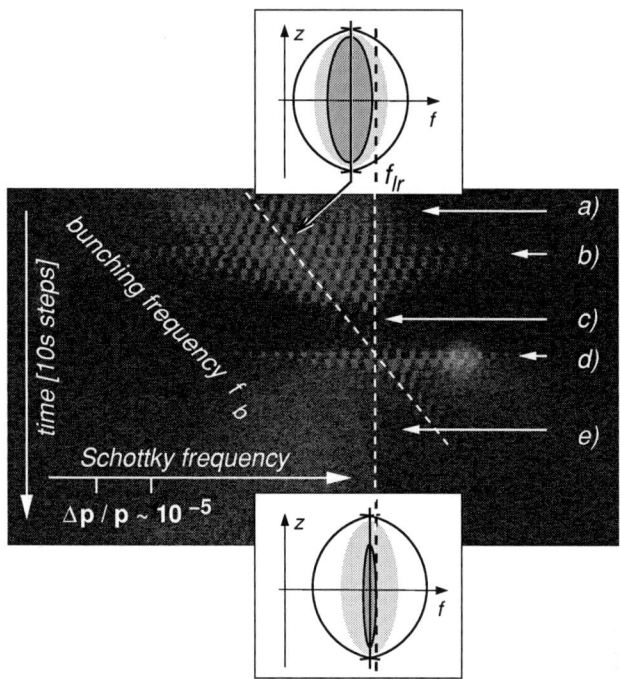

FIGURE 1. Schottky noise spectrum (47th harmonic, log. intensity coding) of a laser-cooled bunched C^{3+} beam recorded at scanning bunching frequency f_b (diagonal dashed line). The decelerating laser force is always resonant with ions at f_{lr} (vertical dashed line). For illustration $f_b = hf_{rev} = 10 \times 1.295\,\text{MHz}$ is increased in steps of 10 Hz every 10 s. Starting at a low bunching frequency ions close to the separatrix of the bucket come into resonance with the laser beam and are cooled into the bucket (a, and upper cartoon). Synchrotron sidebands appear (b), indicating a longitudinally laser-cooled bunch where the ions perform 'incoherent' synchrotron oscillations with $f_{sync} \sim 100\,Hz$. The initial envelope corresponds to the rms momentum acceptance of the bucket $\Delta p/p \sim 2 \times 10^{-5}$ and the cooling can be followed. Closer to resonance most sidebands vanish and the signal intensity decreases (c) when the beam enters the 'space-charge dominated' regime. Crossing the resonance (d) the synchrotron motion is driven instead of damped. Ions are decelerated (e) out of the bucket until the cycle restarts.

Moderate voltages for bunching of only few volts were applied at the 10th as well as at the 20th harmonic of the revolution frequency $f_{rev} = 1.295\,\text{MHz}$. The bucket depth was determined by the measurement of the synchrotron frequency $f_{sync} \sim 100\,\text{Hz}$ ($h = 10$) and $f_{sync} \sim 170\,\text{Hz}$ ($h = 20$) and corresponds to a momentum acceptance of the order of $\Delta p/p \sim 2 \times 10^{-5}$. For a purely electron-cooled beam this equilibrium momentum spread is reached for a total ion number of few 10^7 (few $10\,\mu A$) [7]. The storage-time of the beam, electron pre-cooled for few seconds after injection, amounted to $\tau \sim 450\,\text{s}$, avoiding recombination losses in the electron cooler.

Bunched beam laser cooling, invented at ASTRID [17] now means damping of the synchrotron motion by the strong but narrow-band resonant laser force. For the cw-lasers used in the experiment [16], the band-width of the force is determined by the

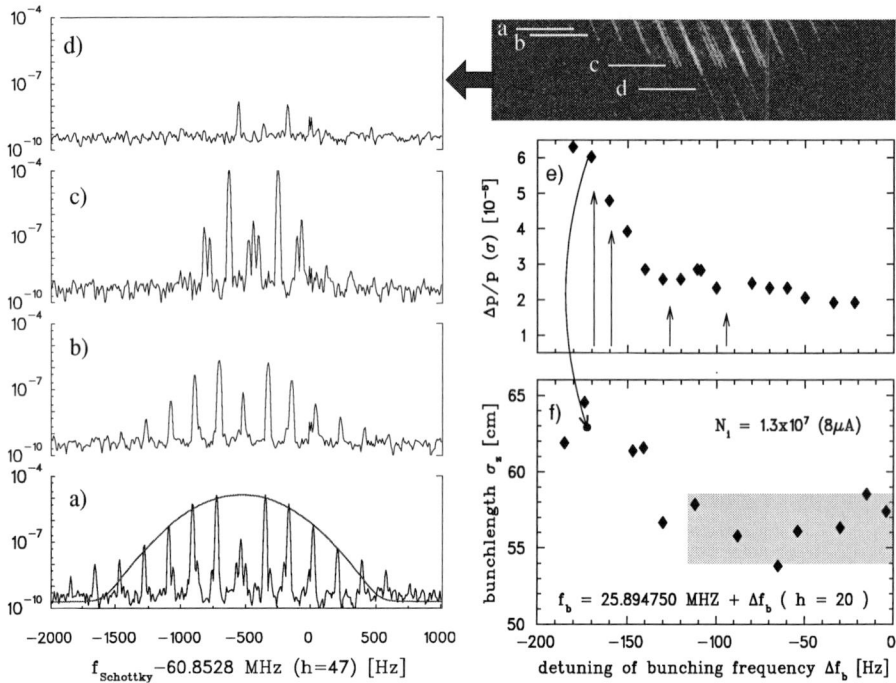

FIGURE 2. Left: Schottky-noise spectra corresponding to the detuning positions indicated in the upper right image with decreasing detuning from a) to d). The envelope in a) represents the Gaussian distribution used for the estimation of the momentum spread. Middle right: Momentum spread e) deduced from the width of the envelope of the 'incoherent' Schottky side-band spectrum as a function of the detuning Δf_b ($h = 20$). Arrows indicate the position of the spectra displayed in a)-d). Bottom right f): Bunch length (pick-up measurement) limited by the equilibrium length for constant ion density. The curved arrow indicates the correspondence between momentum spread and bunch length in the emittance (or IBS) dominated regime.

line-width of the transition corresponding to a momentum spread of $\Delta p/p \approx 5 \times 10^{-8}$. Two schemes exist to overcome this tremendous mismatch of the width of the cooling force and the initial momentum distribution besides the non-straightforward increase of the band-width of the force and, possibly, the additional application of momentum-matched electron cooling. In the first, the laser– or equally well the bunching–frequency is continuously tuned from a value where the laser is resonant with ions at the edge of the bucket to a final frequency close to its center [17]. Thereby, all ions are subsequently decelerated to the center of the bucket in momentum space and slightly shifted out of center in real space for the compensation of the decelerating laser force. However, note, that binary collisions may kick ions out of the narrow momentum acceptance range and a second broader distribution may form in the bucket. This scheme is illustrated in Fig. 1.

The development of the momentum distribution with continuously increasing cooling strength is depicted in Fig. 2e) as a function of the detuning Δf_b of the bucket center with respect to the laser resonance. The tuning rate is slow compared to the longitudinal cooling rate so that the situation can be regarded as equilibrated. For large detuning the momentum spread can be deduced from the envelope of the 'incoherent' side-band spectrum [19] as indicated in Fig. 2a). The development of the spatial distribution, independently measured for a similar ion current using capacitive pick-up devices, is shown in Fig. 2f). Starting at large detuning, the momentum spread as well as the bunch length both decrease with increasing cooling strength or momentum compression of the bunch. For less detuning than $-100\,\text{Hz}$, corresponding to $\Delta p/p = 6 \times 10^{-6}$, the bunch length remains constant and can be reproduced under the assumption of longitudinally 'space-charge dominated' bunches [20] of constant linear density (gray area in the graph). Note, that the integrated pick-up signal and the beam current monitor do not show any unexpected ion losses at and beyond this point.

In the left Fig. 2 individual Schottky spectra that represent the different situations discussed above are shown in detail. The initial distributions (a,b) show a symmetric distribution of 'incoherent' synchrotron sidebands with reduced signal strength at the carrier. This reduction does not correspond to the Bessel-function description of the modulated spectrum for the given momentum spread [19] and might be attributed to prior laser heating. Close to the point where space-charge becomes dominant (c), presently unexplained satellites become observable on one side of the even side-bands at a frequency separation of $\sim 40\,\text{Hz}$. Well in the 'space-charge dominated' regime (d), identified by the behavior of the spatial distribution, most sidebands as well as the satellites have vanished, and the integrated signal intensity is unexpectedly reduced, leaving only the carrier and two distinct sidebands at the unaltered spacing of $f_{sync} = 188\,\text{Hz}$. Again, note, that the number of ions in the bunch remains constant.

In this regime the momentum spread cannot be derived from the envelope of the Schottky signal any more. However, laser cooling itself provides a unique diagnostic. Tuning the laser frequency across the Doppler-broadened transition, the momentum distribution can be directly observed via the laser fluorescence signal as described in [21].

COMBINED LASER AND ELECTRON COOLING

The second scheme for bunched beam laser cooling relies on the repeated interaction of the ions oscillating in the bucket with the laser beam tuned to optimum cooling slightly above to the center of the bucket. The advantage of this scheme, first used at TSR [18, 9], is that all momentum classes frequently interact with the laser beam, and that it is stationary so that electron and laser cooling can be easily synchronized. Systematic studies were performed at the ESR simultaneously recording the Schottky-spectra, the momentum dependent fluorescence signal [21], and the longitudinal and the transverse spatial profiles, for the latter using a residual gas ionization beam profile monitor close to its spatial resolution and at integration times of about a minute. For comparison continuous electron cooling was applied at different electron currents. Results are displayed in Figs. 3 and 4, filled symbols representing electron cooling and open ones laser cooling at fixed detuning for all the following graphs.

For continuous electron cooling of the C^{3+} beam, momentum matched to the center of the bucket, the momentum spread as well as the longitudinal and transverse spatial

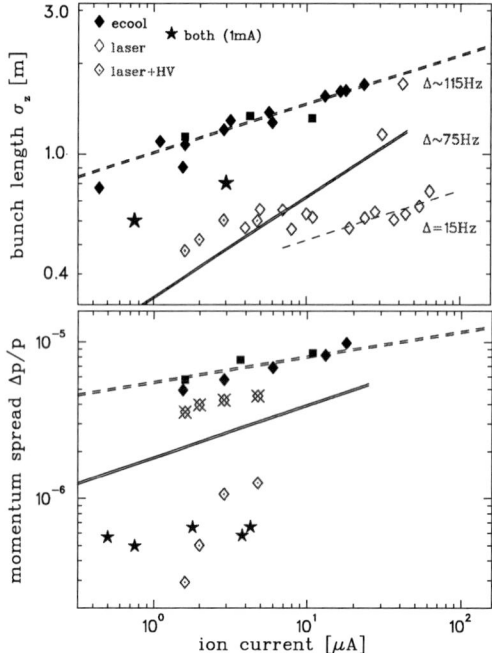

FIGURE 3. Upper graph: Spatial bunch length as a function of the ion current, measured with a current transformer. Filled symbols stand for electron cooling with different electron current, open ones for laser cooling at given detuning and stars for combined cooling, details consistently given in the legend of the following three graphs. The dashed lines indicate the $N^{1/6}$ scaling expected from IBS, the solid the equilibrium length for space charge dominated bunches ($\propto N^{1/3}$). Lower graph: Momentum spread deduced from the Schottky spectra for the electron-cooled case and only giving upper limits for the laser cooled one and deduced from the fluorescence scan (marked HV).

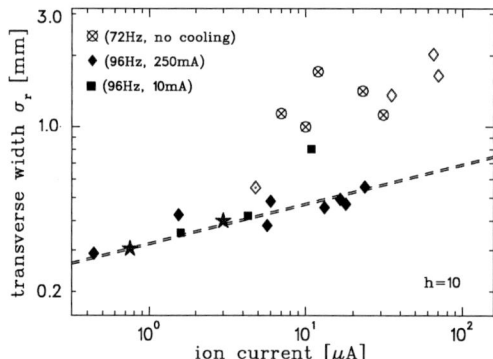

FIGURE 4. Transverse beam size as a function of the ion current for the beams presented in Fig. 3. A size of 0.4 mm corresponds to an rms emittance of $\varepsilon = 4 \times 10^{-3} \pi$ mm mrad.

profiles are well in agreement with standard IBS theory ($\propto N^{1/6}$) and existing ESR data [12] and will serve as a reference for the laser cooling measurements. No dependence in equilibrium beam properties was observable on the electron beam current, varied between 10, 50 and 250 mA, however, for 10 mA transverse cooling became less effective at higher ion currents.

Regarding first the spatial bunch length for ion currents below $10\,\mu A$, laser cooling almost at resonance ($\Delta f_b = 15\,\text{Hz}$ corresponds to $\Delta p/p \sim 10^{-6}$ at $h = 10$) leads to a reduction of the length of a factor of two as compared to electron cooling. The solid line in the upper Fig. 3 depicts the absolute value for the equilibrium length for space-charge dominated bunches of corresponding ion number ($\propto N^{1/3}$) indicating that all ions in the bunch are efficiently laser-cooled. This seems to be different for higher ion currents. Starting from higher currents after injection, the decrease almost proportional to $N^{1/6}$ suggests that, as not enough phase space is available in the region defined by the laser detuning, a hot IBS dominated fraction of the beam exists that could not be resolved due to large noise-dominated base-line fluctuations in the pick-up measurement. Increasing the detuning in this situation and thereby opening momentum space for laser-cooled ions around the center of the bucket leads to an increase in the bunch length and in the integrated pick-up signal, corresponding to a larger number of laser-cooled ions. The same general behavior is observed when different rf voltages are applied (not shown).

Related to the bunch length, the momentum spread is shown in the lower Fig. 3. For the electron-cooled beams it is derived from the Schottky spectra. The solid line is meant to guide the eye indicating at which momentum spread space charge becomes relevant in the upper graph. Evidently, the true momentum spread measured via the laser fluorescence method decreases far beyond this limit into the few times 10^{-7} range, while the bunch length remains constant.

Regarding the little data recorded for the transverse beam profiles the laser-cooled beams above $10\,\mu A$ do not show any sign of transverse cooling, while for low currents, an onset might be visible. This completely changes when simultaneously electron cooling at an electron current of only 1 mA is applied (stars in the graphs). Though the momentum spread is slightly increased, the beam becomes transversally as cold as for strong electron cooling. Given the low momentum spread and the low emittance, the plasma parameter of these beams is of the order of unity.

SUMMARY AND FUTURE PERSPECTIVES

Summarizing the latter, this first combination of laser and electron cooling provided three-dimensionally cold beams of unprecedented momentum spread for ion currents below $10\,\mu A$. In the longitudinal degree-of-freedom, laser cooling lead to a reduction of a factor of two in the bunch length and of one order of magnitude in momentum spread as compared to the purely electron-cooled beam. Although only an upper limit of the transverse beam temperature can be given as for the case of electron cooling, stating a plasma parameter of about $\Gamma \gtrsim 1$ seems justified. As the interesting beam currents correspond to linear densities where one-dimensional ordering is possible, one might speculate about such effects being involved in the observation of the unexpected intensity drop of the Schottky signal, shown in Fig. 2d), yet, further experimental as well

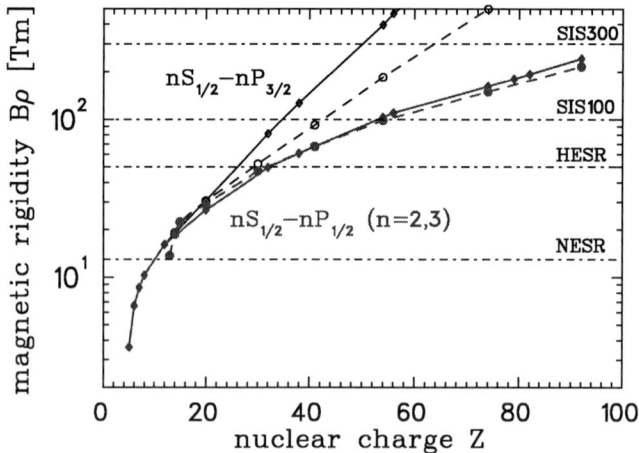

FIGURE 5. Magnetic rigidity required for the storage of ions at an energy where the transition wavelength of the ground state transitions of Li-like (solid lines, rhombs, $n = 2$) and Na-like ions (dashed lines, circles, $n = 3$) are Doppler-shifted into resonance with a counterpropagating laser beam of wavelength $\lambda_{uv} = 257$ nm.

as theoretical studies are mandatory.

Longitudinal cooling times were of the order of seconds and strongly dependent on the cooling scheme, especially when scanning is involved. Consequently, the introduction of an additional broad-band laser system should improve the cooling time. Besides the direct addressing of the whole initial momentum distribution, a broad-band system also has the advantage of efficiently recycling those ions into the cooling process that are lost out of resonance in an IBS event and thus also the control of higher currents should become possible. Ideally, the broad-band laser system can be realized by conventional pulsed laser systems with pulse length in the ns-range. At present, an additional scanning laser system is prepared for first tests instead of a true broad-band system.

The prominent drawback for the practical use of laser cooling as a general cooling method is the lack of suitable optical transitions in most ions of interest. This drawback, however, can be overcome in future heavy ion synchrotrons like the SIS 300 envisaged within FAIR, as the high magnetic rigidity of such machines allows for beam energies Doppler-shifting the ground state excitation of all Li-like heavy ions into an accessible laser frequency range [23], as shown in Fig. 4. Moreover, in this highly relativistic regime, where electron cooling cannot be readily applied any more, the laser force principally increases with the third power of the ion energy [7, 22]. On the one hand, this gain in efficiency is due to the relativistic Doppler-shift, increasing the momentum transfer while, on the other hand, optical transitions in the highly charged ions of interest become faster with the nuclear charge of the ion, for details see [7, 22]. Assuming that broad-band laser systems can be used, cooling times of the order of only few seconds seem possible at energies of $\gamma \sim 30$ for Li-like uranium ions, clearly warranting the further investigation of the method while, for an estimation of equilibrium temperatures, a profound analysis of the competing mechanisms is required.

ACKNOWLEDGMENTS

Work supported by the German BMBF (06ML183).

REFERENCES

1. S. Schröder, et al., Phys. Rev. Lett. **64**, 2901 (1990)
2. J.S. Hangst, et al., Phys. Rev. Lett **67**, 1238 (1991)
3. J.P. Schiffer and P. Kienle, Z. Phys. **A 321**, 181 (1985)
4. N. Madsen, et al., Phys. Rev. Lett. **87**, 274801 (2001)
5. I. Lauer, et al., Phys. Rev. Lett. **81**, 2052 (1998)
6. U. Eisenbarth, et al., Hyperfine Interactions **127**, 223 (2000)
7. U. Schramm, D. Habs, Progress in Particle and Nuclear Physics **53**, 583 (2004)
8. T. Schätz, et al., Nature (London) **412**, 717 (2001), U. Schramm, et al., Phys. Rev. **E 66**, 036501 (2002)
9. U. Schramm, et al., Phys. Rev. Lett. **87**, 184801 (2001)
10. A. Smirnov, et al., these proceedings (COOL05)
11. U. Schramm, et al., Journal of Physics **36**, 561 (2003)
12. M. Steck, J. Opt. Soc. Am. B **20**, 1016 (2003)
13. M. Steck, et al., J. Phys. B **36**, 991 (2003) and Phys. Rev. Lett. **77**, 3803 (1996)
14. H. Danared, et al., J. Phys. B **36**, 1003 (2003)
15. W.R. Johnson, et al., At. Data Nucl. Data Tab. **64**, 279 (1996)
16. U. Schramm, et al., Hyperfine Int. 115 (1998) 57
17. J.S. Hangst, et al., Phys. Rev. Lett. **74**, 4432 (1995)
18. H.-J. Miesner, et al., Nucl. Instr. Meth. **A 383**, 634 (1996)
19. D. Boussard, CAS, CERN-87/3, 416 (1987), O. Boine-Frankenheim, T. Shukla, Phys. Rev. ST AB **8** 034201 (2005)
20. T.J.P. Ellison et al., Phys. Rev. Lett. **70**, 790 (1993)
21. U. Schramm. et al., Proc. PAC 2005, Jacow, FOAD004 (2005)
22. U. Schramm, et al., Nucl. Instr. Meth. **A 532**, 348 (2004)
23. U. Schramm, et al., LoI#18 FAIR APPA–PAC (2004)

TRAPS

Electron Cooling of Ions and Antiprotons in Traps

Günter Zwicknagel

Institut für Theoretische Physik, Universität Erlangen-Nürnberg,
Staudtstr. 7, D - 91058 Erlangen, Germany

Abstract. For a theoretical description of electron cooling of ions or antiprotons in traps we have investigated the energy loss and cooling force in a strongly magnetized electron plasma employing both perturbation approaches and more complete numerical simulations. Some characteristic features for cooling under conditions prevailing in Penning traps are presented. One particular feature is, that the energy loss in strongly magnetized electrons, which tend to move along the field lines like beads on a wire, strongly depends on the sign of the interaction. The energy loss can be significantly larger for antiprotons than for protons. Special attention is paid to the cooling of highly charged ions, here bare Uranium, in HITRAP. The time evolution of the energy distribution of the trapped ions is studied within a simplified model which takes into account the related heating of the electrons. The feedback of this heating on the energy loss results in an intricate dependency of the cooling times on the density of the electrons and the ratio of the number of ions to the number of electrons in the trap. From this analysis we find that cooling times less than about a second are feasible for electron cooling of bare Uranium in HITRAP.

INTRODUCTION

Electron cooling is a well-established method to improve the phase space quality of ion beams in storage rings by the energy loss of the ions to a superimposed electron beam [1, 2]. More recently ions and antiprotons have also been cooled by electrons or positrons in traps, as e.g. in the Penning traps employed in the recent and planned experiments for the production of antihydrogen at CERN [3, 4] or the generation of slow highly charged ions at GSI [5, 6]. Although electron cooling is routinely used in these applications a lot of observations are not yet satisfactorily explained and understood. A key issue is here an improvement of the theoretical understanding of the energy loss of ions in a magnetized plasma as the fundamental process of electron or positron cooling. Compared to the case of the energy loss of ions in unmagnetized electrons, the presence of a (strong) external magnetic field considerably complicates the description of the energy loss and imposes a formidable challenge to the theoretical description. In the next section we will give a brief overview of different theoretical approaches for this task.

The main subject, however, will be the cooling under the specific conditions in Penning traps, that is, for a rather strong magnetic field and, in contrast to electron cooling in storage rings, an isotropic velocity distribution of the electrons. We will discuss some typical features of the energy loss under such conditions. As one specific result we find that the energy loss in strongly magnetized electrons, which tend to move along the field lines like beads on a wire, strongly depends on the sign of the interaction. At low velocities it can be more than 50 percent larger for antiprotons than for protons.

In the second part we then focus on the cooling of highly charged ions, namely bare Uranium, in HITRAP [5, 6] and will present results for the time evolution of the energy distribution of the trapped ions. This is done in framework of a simplified model taking into account the related heating of the electrons. The feedback between the electron heating and the ion energy loss turns out to be rather intricate and important for the cooling rates, which strongly depend on the density of the electrons and the ratio of the number of ions to the number of electrons in the trap. From these studies we conclude that cooling times of about a second or less are feasible for electron cooling of bare Uranium in HITRAP.

THEORETICAL TREATMENTS OF THE ENERGY LOSS OF IONS BY COLLISIONS WITH MAGNETIZED ELECTRONS

Whereas the energy loss of ions in unmagnetized electron plasmas has already been studied extensively [7, 8], a qualified and comprehensive description of the interaction of ions with magnetized electrons is a rather formidable task [2,9-26] which is still in progress. The presence of a magnetic field considerably complicates the description of the energy loss mainly because of the loss of symmetries as compared to the case of nonmagnetized electrons where the energy transfer in a collision is a function only of the relative velocity. With magnetic field, additional dependencies show up, the ion-electron motion no more separates in a center-of-mass and relative part [22, 25], and the cooling force strongly depends on the direction of the ion velocity $\vec{V}$ relative to the magnetic field $\vec{B}$, i.e. on the angle $\alpha = \sphericalangle(\vec{B},\vec{V})$.

As for the nonmagnetized case there are basically two complementary approaches: In the dielectric theory (DT) the decelerating force on the ion is due to the polarization which it creates in its wake. This can be either calculated in linear response (LR) [17, 18, 19, 20] or numerically by a particle-in-cell (PIC) simulation of the underlying nonlinear Vlasov-Poisson equation [8, 20, 21]. In the complementary binary collision (BC) approximation the drag force is accumulated from the velocity transfers in individual collisions. This can be calculated by scattering ensembles of magnetized electrons from the ions in a classical trajectory Monte-Carlo method (CTMC) [23, 22], and by treating the Coulomb interaction as a perturbation in $O(Z^2)$ to the helical motion of the electrons, see [24, 25, 18] for details. The resulting force $\vec{F}(\vec{V})$ on an individual ion is then obtained by integrating with respect to the impact parameter and the velocity distribution of the electrons. A detailed discussion of the conformity between this perturbation treatment of binary collisions and the linear response description of DT is given in [18].

Using these approaches we have studied the energy loss and drag force of ions in a magnetized electron plasma for various situations. Some explicit results have been presented for the conditions prevailing in the Heidelberg test storage ring TSR [22, 24] and for the LEIR at CERN [25]. Some results for typical trap conditions have been given in [26], more will be presented and discussed in the next section. A critical comparison of the various treatments allows the following conclusions: The numerically expensive solution of the Vlasov-Poisson equation by PIC simulations is, despite its numerical noise, the method of choice at strong coupling, that is, high ion charges and low ion velocities, and at large magnetic fields. It accounts for both the nonlinearity and the collectivity of the response. In many cases the collectivity is, however, not so important. Then the CTMC implementation of binary collisions comes next. The $O(Z^2)$ perturbation treatment of BC is acceptable as long as the ion-electron coupling is not too large and large velocity transfers in the binary collisions are rare. For weak coupling also the dielectric linear response treatment is acceptable as long as the magnetic field is not to strong. For strong magnetic fields it predicts unrealistic large drag forces for ions moving with small angles α to the magnetic field lines.

ELECTRON COOLING OF IONS AND ANTIPROTONS IN PENNING TRAPS

Here we are specifically interested in the energy loss and cooling forces of ions and antiprotons at the typical conditions prevailing in Penning traps, in particular the large magnetic field of $B \approx 6$ T. In addition, like in HITRAP, high ion charges may occur. This usually results in a strong ion-electron coupling [7, 8] which then prevents the use of weak coupling approaches. A comparison for the energy loss obtained by the different treatments outlined in the last section is presented in the left part of Fig. 1 for a strong coupling situation, i.e for bare Uranium at low velocities. Here significant deviations between the different approaches can be seen. Compared to the most complete PIC simulation results, the linear response (LR) and binary collision in $O(Z^2)$ (BC) both fail at low velocities and strongly disagree among one another at medium velocities. At sufficiently large velocity, that is, when we proceed toward the weak coupling regime, all shown approaches tend to merge. The non-perturbative CTMC calculations are rather close to the PIC results except for the velocity region around the maximum of the energy loss. This deviation is due to dynamic screening contributions which are accounted for in the nonlinear Vlasov-Poisson treatment underlying the PIC simulations. In the CTMC

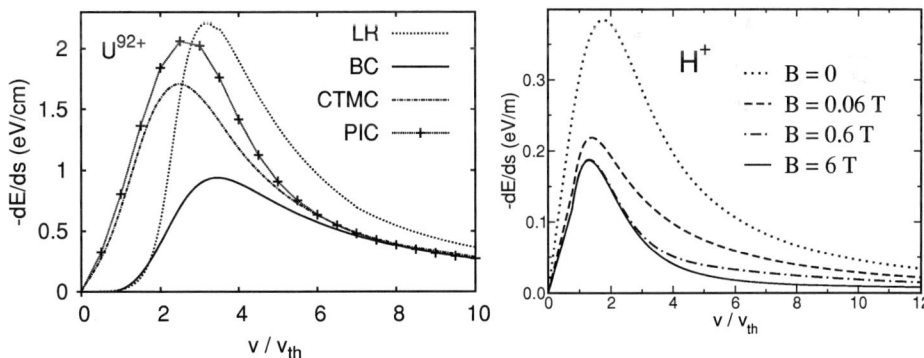

FIGURE 1. Energy loss dE/ds as a function of the ion velocity V in units of $v_{th} = (k_B T_e/m_e)^{1/2}$ in an electron plasma with $n_e = 10^7 \text{cm}^{-3}$ and $T_e = 4$ K. Left: Results of different methods for calculating the energy loss for the case of bare Uranium and $B = 6$ T. Right: Energy loss of protons obtained by the CTMC method at different magnetic field strengths. The direction of the ion motion is $\alpha = 30^o$ in all cases.

description they are only included in an approximate manner by choosing a velocity dependent screening length $\lambda(v) = \lambda_D (V^2 + v_{th}^2)^{1/2}$ in the employed Yukawa like screened interaction [22], where λ_D is the Debye length. The CTMC method nevertheless represents a very reasonable approximation in particular in view of the much less numerical expense compared to the PIC treatment.

The dependence on the strength of the magnetic field B is given in the right part of Fig. 1, here for a proton. The same qualitative behavior, that is, a overall decrease of the energy loss with increasing magnetic field also applies to higher charged ions. This is distinct from the dependency on the magnetic field of the cooling force for electron cooling in a storage ring. There we have, in contrast to the present case, an anisotropic electron distribution and the energy loss may be enhanced or reduced with increasing B depending on the ion velocity with respect to the thermal velocities ($v_{th,\perp}$ and $v_{th,\parallel}$) of the electrons.

Although we have an isotropic electron distribution here, the (strong) magnetic field nevertheless introduces are rather pronounced dependency of the energy loss on the direction of the ionic motion relative to the magnetic field, i.e. on $\alpha = \sphericalangle(\vec{B}, \vec{V})$, as shown in the left part of Fig. 2. In the right part of Fig. 2 the dependence on the ion charge state Z is discussed. The energy loss dE/ds divided by Z^2 is given here for different ion species and demonstrates that the energy loss grows weaker with the charge state than Z^2. This is in qualitative agreement with findings obtained for the nonmagnetized case [7, 8] and cooling force measurements at the TSR [27] and the ESR [28].

A particular remarkable feature is the strong sensitivity of the cooling force and the energy loss on the sign of the ion charge i.e. if the ion-electron interaction is attractive or repulsive. Explicit results for a comparison of the energy loss of protons with that of antiprotons under typical trap conditions ($B = 6$ T, $T = 4$ K) are given in Fig. 3. At low velocities the energy loss of antiprotons is significantly larger than for protons in particular for small angles of the (anti)proton motion, i.e. of $\vec{V}$, with the magnetic field $\vec{B}$. This behavior can be understood in a binary collision picture. For large magnetic fields the electrons tend to move like beads on a wire along the magnetic field lines.

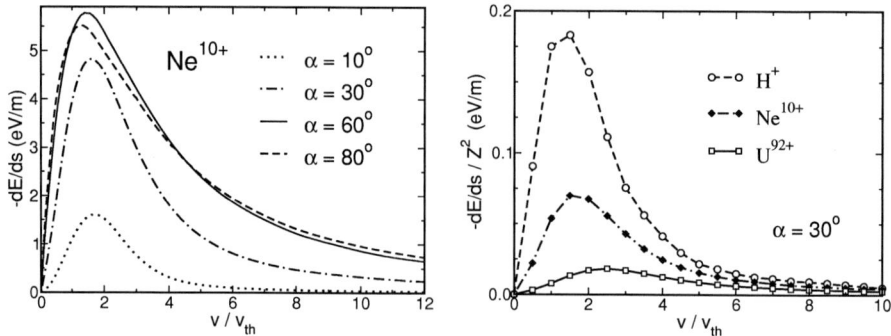

FIGURE 2. Energy loss dE/ds from CTMC calculations as function of the ion velocity V for plasma parameters as in Fig. 1 and $B = 6$ T. Left: Energy loss of Ne^{10+} for various angles α of the ion motion with respect to $\vec{B}$. Right: Energy loss scaled by Z^2 for different ion species and $\alpha = 30°$.

The velocity- and energy transfer between ion and electrons thus vanishes for ions also moving along the magnetic field lines for symmetry reasons. But this symmetry argument yields a vanishing velocity transfer only in the case of an attractive electron-ion potential. For a repulsive potential there occur large velocity transfers when the particles are reflected from each other. This can be well seen in the resulting energy loss for $\alpha = 0$. For nonzero angles this symmetry argument of course does not apply any longer, but some effect of this completely different scattering behavior persists and results in a considerable difference between attractive and repulsive interaction which only slowly diminishes with increasing angle. The shown results have been obtained by the CTMC method which accordingly shows a vanishing energy loss for $\alpha = 0$ as expected in a binary collision treatment. With the dynamic screening contribution due to the nonlinear response of the electron plasma (as e.g. present in the PIC approach) the above symmetry argument does not apply, and there will be a finite cooling force on a proton for $\alpha = 0$. It will be, however, still small compared to that on an antiproton. From the strong dependency on the sign of the potential we can immediately conclude that any perturbative treatment like the linear response or binary collision to $O(Z^2)$ which yields results independent of the sign of the potential runs into difficulties and become doubtful. There exists apparently no suitable parameter of smallness in case

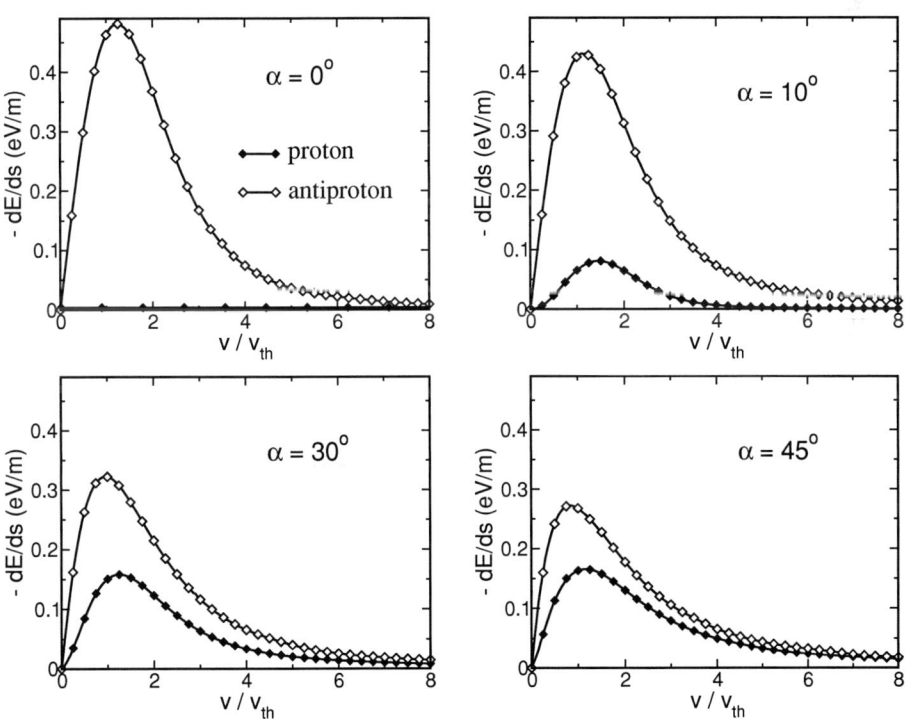

FIGURE 3. Energy loss dE/ds (CTMC) of antiprotons and protons in a magnetized electron plasma ($n_e = 10^7 \text{cm}^{-3}$, $T_e = 4$ K, $B = 6$ T) for different directions α of the ion motion.

of this quasi-one-dimensional electron motion. This asymmetry in the cooling between protons and antiproton concerns as well the cooling of (highly charged) ions by electrons or positrons. That is, positron cooling of positively charges ions is more effective at large magnetic fields than electron cooling, provided that we have, of course, the same density of electrons and positrons and the (initial) energy of the ions in not to large. For large velocities the differences due to the different signs of the potential strongly diminish.

ELECTRON COOLING OF HIGHLY CHARGED IONS IN HITRAP

We now focus on the whole cooling process in the specific case of bare Uranium in HITRAP. To study the time evolution of the ion energies and the electron temperature we use a simplified description which takes into account:
(i) the deceleration of the N_i ions ($\mu = 1, \ldots, N_i$) by collisions with magnetized electrons

$$M\frac{d\vec{V}_\mu}{dt} = \vec{F}_\mu = \vec{F}[n_e, T_e, B, \vec{V}_\mu(t)], \qquad (1)$$

which results in the Energy loss of the ions $\frac{dE_\mu}{dt} = M\vec{V}_\mu \cdot \frac{d\vec{V}_\mu}{dt} = \vec{V}_\mu \cdot \vec{F}_\mu.$
(ii) the transfer of the ionic energy to the trapped electrons

$$\langle\frac{dE}{dt}\rangle(t) = \frac{1}{N_i}\sum_\mu^{N_i}\frac{dE_\mu}{dt} = -\frac{dE_e}{dt} \stackrel{!}{=} \frac{3}{2}\frac{N_e}{N_i}k_B\frac{dT_e}{dt}. \qquad (2)$$

(iii) the related heating of the electrons and their cooling by synchrotron radiation

$$\frac{dT_e}{dt}(t) = -\frac{2}{3k_B}\frac{N_i}{N_e}\langle\frac{dE}{dt}\rangle(t) - \frac{1}{\tau}(T_e(t) - T_0); \qquad \frac{1}{\tau} = \frac{1}{3\pi\varepsilon_0}\frac{e^4 B^2}{m_e^3 c^3}, \qquad (3)$$

where $\tau \approx 0.1$ s for $B = 6$ T, and $T_0 = 4$ K is the ambient temperature supplied by the cryostat. A similar treatment of the cooling process in terms of simple coupled differential equations has already been proposed and used in some early studies [29] and also more recently in [30, 31]. But in these approaches the energy loss and heat transfer has been implemented by rate equations for the electronic and ionic temperatures. Here we use a more detailed description of the energy loss by considering the cooling force $\vec{F}[T_e(t), \vec{V}_\mu(t)]$ for individual ions followed by an ensemble average over the ion distribution in the trap for the total energy loss and heat transfer to the electrons [see Eq. (2)]. As in the earlier models several assumptions and simplifications are still made. One is the instantaneous conversion of the ion energy into an electron temperature, that is, a sufficiently short equilibration time of the electron plasmas. Furthermore, the actual geometry and the electrostatic fields in the trap are neglected. The ions just move in an infinitely extended strongly magnetized electron plasma. To obtain, however, a correct heating rate the corresponding ratio of the finite number of electrons N_e and ions N_i which are confined together in the trap during the cooling process is used in Eqs.(2) and (3). And since the ions only interact in the trap when they travel trough the electron plasma a correction factor κ is introduced. It represents the ratio of the extension of the electron cloud(s) to the total path length of the ions between their turning points in the trap. For the present calculations we used $\kappa = 0.4$ which enters the above description by multiplying the cooling force $\vec{F}$ in Eq.(1) by κ. The scheme (1)-(3) has been applied to an ion distribution of 500 U^{92+} ions representing the ion ensemble of roughly 10^5 ions which are expected to be trapped during a typical cooling cycle at HITRAP [6]. The initial ion distribution, i.e. at time $t = 0$, was obtained from a preceding simulation of the injection of an ion bunch into the cooler trap for the HITRAP setup [32]. Due to the high charge of the U^{92+} ions and the correspondingly large energy loss a strong heating of the electrons takes place. It thus happens that the ion velocity is almost all the time about or less than the thermal velocity of the electrons and we have to consider the energy loss in the low velocity regime, like e.g. in Figs. 1-3. But since the electron temperature is large (see below) we are nevertheless in a weak coupling situation. To reduce the computational effort we thus used for the cooling force $\vec{F}$ in Eq. (1) the expression for an infinitely strong magnetic field given in [17] as a reasonable approximation for the present situation. Compared to the use of the more advanced methods (CTMC, PIC) we estimate the error made in the final global observables (cooling times, electron temperatures) of at most 20%. Fig. 4 shows the resulting time evolution of the average ion energy and some snapshots of the energy

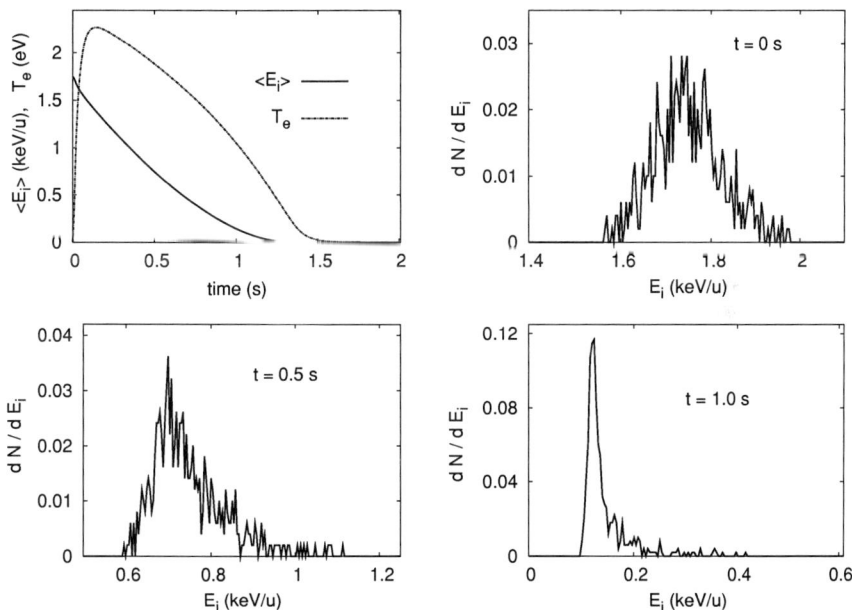

FIGURE 4. Time evolution of an ensemble of bare Uranium ions in a magnetized electron plasma with $n_e = 10^7 \text{cm}^{-3}$, $N_i/N_e = 10^{-4}$, $T_0 = 4$ K and $B = 6$ T. The average energy of the ions and the electron temperature as function of time (left top) are shown together with the energy distribution of the ions at some instants in time.

distribution for U^{92+} ions with an initial energy around 1.7 keV/u. The average ion energy and the maximum of the ionic energy distribution goes down within roughly one second, whereas the energy distribution develops an increasingly pronounced peak. In the beginning of the cooling process a strong and rapid heating of the electrons from $T_0 \approx 0.34$ meV to about 2.2 eV takes place. Then the energy and the energy loss of the ions continuously decreases and thus the heating of the electrons. The cooling of the electrons by synchrotron radiation now slightly prevails and the electrons start to cool down again. At the final stage of the cooling (not resolved in the shown figure) the ion energy decays exponentially, the heating of the electrons stops, and the electron temperature falls off $\propto exp(-t/\tau)$ towards T_0 due to synchrotron radiation.

To get an idea about the optimal conditions for the cooling process we also varied some of the parameters. Namely the density n_e of the electron plasma, which can be varied to some extend by varying the electrostatic potential in the trap, and the ratio N_i/N_e of the number of ions to electrons filled in. Some examples for the found dependencies are shown in Fig. 5. An increasing number of ions (upper part of Fig. 5) at

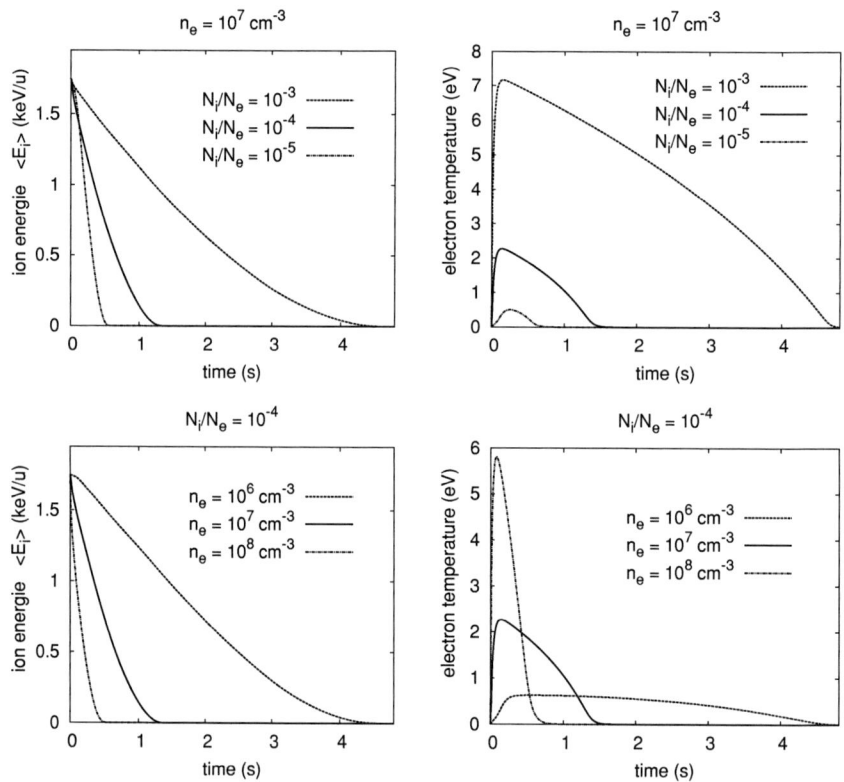

FIGURE 5. Average ion energy (left) and electron temperature (right) as function of time for an ensemble of bare Uranium ions in a magnetized electron plasma with $T_0 = 4$ K, $B = 6$ T and different n_e and N_i/N_e.

a given number of electrons (i.e. an increasing N_i/N_e) results as expected in a stronger heating of the electrons. This increases the cooling time as the cooling force is smaller here at a larger electron temperature. On the other hand a larger electron temperature will help to reduce ion-electron radiative recombination. And for a much lower number of ions than shown here the overall trend is reversed and the cooling time will start to grow again. Now the ion velocities V are almost all the time large compared to the thermal velocity v_{th} of the electrons, and the cooling force is considerably smaller than for ion velocities $V \approx v_{th}$. Due to this feedback of the electron temperature on the energy loss, the dependence on the electron density is also more involved than expected. The cooling force scales in first order linear in n_e which would result in a cooling time going like $1/n_e$. While the trends for the cooling time and the heating of the electrons for a varying electron density are as expected, the amount of change of the cooling time is much weaker, see the lower part of Fig. 5.

Summing up we found that the cooling time strongly depends on the ratio N_i/N_e and the density n_e, and is governed by a nonlinear feedback between electron heating and cooling force. This prevents simple estimates and scaling rules for the cooling process in a trap, offers, however, various options for optimization. But our results also clearly indicate that cooling times of about a second or less are feasible for electron cooling of bare Uranium in HITRAP.

To further improve these predictions and for a more detailed description of the cooling process in a Penning trap work is in progress by taking into account both the full electrostatic potential and the spatial electron distribution in the trap, as well as recombination processes. Also the gyration of the ions due to the magnetic field, which was neglected so far, needs further attention, as does the energy transfer from the ions to the electrons. Here the major questions are: Is the instantaneous thermalization and the assumption of an isotropic electron distribution justified? Or are at least two different electron temperatures $T_{e,\perp}, T_{e,\|}$ needed for a proper description?

ACKNOWLEDGMENTS

This work was supported by the Bundesministerium für Bildung und Forschung (BMBF, 06ER128) and by the Gesellschaft für Schwerionenforschung (GSI, ER/TOE).

REFERENCES

1. G.I. Budker, Atomnaya Energiya **22** (1967), 346.
2. H. Poth, Phys.Rep. **196** (1990,) 135.
3. M. Amoretti et al., Nature **419** (2002) 456.
4. G. Gabrielse et al., Phys.Rev.Lett. **89** (2002) 213401.
5. W. Quint et al., Hyp.Int. **132** (2001) 457.
6. Th. Beier et al.,*HITRAP, Technical Design report*, GSI Darmstadt, 2003.
7. G. Zwicknagel, C. Toepffer and P.-G. Reinhard, Phys.Rep. **309** (1999) 117.
8. G. Zwicknagel, Nucl.Instr. and Meth. **B197** (2002) 22.
9. Ya.S. Derbenev, A.N. Skrinsky, Part.Accel. **8** (1978), 235.
10. Ya.S. Derbenev, A.N. Skrinsky, Soviet Physical Reviews **3** (1981), 165.
11. A.H. Sorensen, E. Bonderup, Nucl.Instr. and Meth. **215** (1983) 27.

12. V.V. Parkhomchuk, A.N. Skrinsky, Rep.Prog.Phys. **54** (1991) 919.
13. I.N. Meshkov, Phys.Part.Nucl. **25** (1994) 631.
14. D.K. Geller and J.C. Weisheit, Phys.Plasmas **4** (1997) 4258.
15. V.V. Parkhomchuk, Nucl.Instr. and Meth. **A441** (2000) 9.
16. D.L. Bruhwiler et al., eds. I. Hofmann, J.-M. Lagniel, R. Hasse, AIP Conference Proceedings **773**, 2005, p. 394.
17. H. B. Nersisyan, M. Walter, G. Zwicknagel, Phys.Rev. **E61** (2000) 7022.
18. H.B. Nersisyan, G. Zwicknagel, C. Toepffer, Phys.Rev. **E67** (2003) 026411.
19. M. Walter, thesis, Universität Erlangen, 2002.
20. M. Walter, G. Zwicknagel, and C. Toepffer, Eur.Phys.J. D **35** (2005) 539.
21. M. Walter, C. Toepffer, G. Zwicknagel, Nucl.Instr. and Meth.**B168** (2000) 347.
22. G. Zwicknagel and C. Toepffer, in *Non-neutral Plasma Physics IV*, eds. F. Anderegg, L. Scheikhard and C.F. Driscoll, AIP Conference Proceedings **606**, 2002, p. 499.
23. G. Zwicknagel, in *Non-neutral Plasma Physics III*, eds. J.J. Bollinger, R.L. Spencer and R.C. Davidson, AIP Conference Proceedings **498**, 1999, p. 469.
24. C. Toepffer, Phys.Rev. **A66** (2002) 022714.
25. B. Möllers, M. Walter, G. Zwicknagel, C. Carli, C. Toepffer, Nucl.Instr. and Meth. **B207** (2003) 462.
26. B. Möllers, C. Toepffer, M. Walter, G. Zwicknagel, C. Carli, H. B. Nersisyan, Nucl.Instr. and Meth. **A532** (2004) 279.
27. M. Beutelspacher, PhD Thesis, Universität Heidelberg, 2000.
28. Th. Winkler et al., Hyp.Int. **99** (1996) 277.
29. S.L. Rolston, G. Gabrielse, Hyp.Int. **44** (1988) 233.
30. H. Higaki et al., Phys.Rev. **E65** (2002) 046410.
31. J. Bernard et al., Nucl.Instr. and Meth. **A532** (2004) 224.
32. F. Herfurth, private communications

Aspects of Cooling at the TRIμP Facility

L. Willmann, G.P. Berg, U. Dammalapati, S. De, P. Dendooven, O. Dermois, K. Jungmann, A. Mol, C.J.G. Onderwater, A. Rogachevskiy, M. Sohani, E. Traykov, and H.W. Wilschut

Kernfysisch Versneller Instituut, Rijksuniversiteit Groningen, Zernikelaan 25, NL 9747 AA Groningen, The Netherlands

Abstract. The TRIμP facility at KVI is dedicated to provide short lived radioactive isotopes at low kinetic energies to users. It comprised different cooling schemes for a variety of energy ranges, from GeV down to the neV scale. The isotopes are produced using beam of the AGOR cyclotron at KVI. They are separated from the primary beam by a magnetic separator. A crucial part of such a facility is the ability to stop and extract isotopes into a low energy beamline which guides them to the experiment. In particular we are investigating stopping in matter and buffer gases. After the extraction the isotopes can be stored in neutral atoms or ion traps for experiments. Our research includes precision studies of nuclear β-decay through β-ν momentum correlations as well as searches for permanent electric dipole moments in heavy atomic systems like radium. Such experiments offer a large potential for discovering new physics.

Keywords: Fundamental Symmetries, Radioactive Beams, Atom Trapping, Weak interactions, Permanent Electric Dipole Moments
PACS: 11.30.Er, 32.80.Pj, 34.70.+e

Short lived radioactive isotopes are interesting for several promising lines of research and currently a number of facilities to provide the isotopes are planned or are being set up. The research topics cover a wide range in atomic, nuclear and particle physics [1]. At the Kernfysisch Versneller Instituut (KVI), Groningen, The Netherlands, we are commissioning the TRIμP Facility (Trapped Radioactive Isotopes: μicrolaboratories for fundamental Physics) [2], which is open to outside users.

The physics interest of the TRIμP group are tests of discrete fundamental symmetries, i.e. charge conjugation (C), space inversion (P), and time reversal (T). In standard theory the structure of weak interactions is V-A, which means that vector (V) and axial vector (A) currents with opposite relative sign causing a left handed structure and thus parity violation [3]. Other possible currents like scalar, pseudo-scalar or tensor like are signs for new physic. They can be tested by searching for β-ν correlations in weak interactions [4]. In order to determine the neutrino momentum, the recoil to the nucleus needs to be measured. Since the recoil energy is on the order of 100 eV, a precision measurement of the recoil momentum can only performed when the nuclei are suspended in a very shallow potential, which can be provided by confining atoms by light forces [5]. Particularly good candidates are ^{20,21}Na and ^{18,19}Ne.

Another research direction is searching for permanent electric dipole moments (edm), which violated C and P simultaneously [6]. Any observation of an edm would be an indication of physics beyond the standard model. Currently the most sensitive experiment on a nuclear edm, was performed with ^{199}Hg [7], which gives a limit of 2.1×10^{-28} ecm. Recently Flambaum and collaborators pointed out that radium offers a large sensitivity to edm's due to its nuclear as well as atomic level structure [8]. Currently we are investigating the feasibility of a search for an edm using radium. Both experiments require to produce the isotopes and to store them subsequently in a neutral atom traps. We will describe the setup of the facility.

Isotopes of interest are produced in fragmentation or fusion-evaporation reactions utilizing heavy ion beams from the AGOR cyclotron on fixed targets, which are chosen for optimum production rates. The production mechanism favors proton rich isotopes. The goal is to provide a clean beam of the requested isotopes with very low background radiations. This requires dedicated separation and isotopes selective extraction stages. The primary heavy ion beam is separated from the reactions product in a magnetic device, which is designed to cover a wide range from light to heavy isotopes. Through unique magnet design we achieved a compact device. This magnetic separator has been successfully commissioned in the fall of 2004 [9]. After the separator the reaction products are stopped in matter, which will be discussed in below. The extraction time from the stopping device sets the lower limit for the lifetime of the isotopes which can be provided by the facility. We are aiming at times shorter than 1 s. After the extraction from the stopping device the isotopes are cooled and trapped in an radio frequency buncher cooler (RFQ-cooler). From the buncher they are transported to the experiment in an electrostatic beamline. After neutralisation the atoms can be stored in neutral atom traps by laser cooling and trapping methods.

A central role in the design of a facility for radioactive beams takes the cooling from the energies of several MeV/u at the productions to the eV range. For short lived isotopes the stopping and extraction should be minimal. Reaction products stopped in matter will be spread over a distance given by the range straggling and initial momentum uncertainty. There are two options.

- Stopping ions in a buffer gas, preferentially helium. The main common argument is that the ionization potential for helium (24.5eV) is much larger than for any other elements. Thus the neutralization of the incoming ion is energetically forbidden, at least at low enough energies. The ionic isotope can be extracted from the buffer gas by electrostatic guiding fields. A technical aspect is that the helium gas has to be extremely clean. A main question is the survival of ions in a buffer gas while they are cooled by collisions especially at high rates of incoming ions [10, 11]. This approach is followed by groups at several accelerators because it is less dependent on the specific element.
- Implanting in a solid which can be heated to a high temperature at which the isotopes diffuse out of the material [12]. At high temperature the diffusion and effusion time can be less than 1s if the stopping foils are sufficiently thin ($\approx 1\mu$m). Such a thermoionizer could provide high efficiencies of order 1 for alkaline and alkaline earth metals.

We have investigated the possibility of using a gas cell. While an ion traverses through the buffer gas it changes its charge state very rapidly as it undergoes many neutralization-reionization cycles. The neutralization has a kinematic cutoff because all isotopes have a lower ionization potential then helium. The neutralization cross section for a singly charged ion is maximal around an energy of about 25 keV/nucleon (Fig. 1). Below

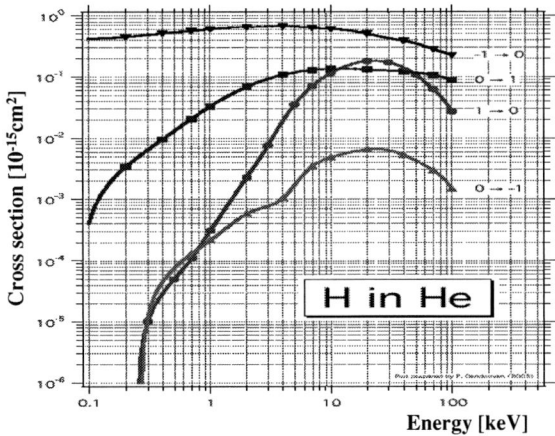

FIGURE 1. Charge exchange cross section for H on He [13].

that energy the cross section drops rapidly. The fraction of charged isotopes at thermal energies is expected to be determined by the ratio of the ionization to neutralization cross section. These charge exchange cross sections at low energies are important input for the development of devices like radio frequency coolers as well as gas catcher cells for stopping high energetic ions.

We have measured cross sections for neutralization at well as stripping cross sections at energies below 1keV/u for different buffer gases. The survival rate strongly depends on the composition of the buffer gas. The preferred choices for a buffer gas is highly pure inert gases where helium stands out because of its high ionization potential of 24.5eV.

We measured the change of the charge state of multiply charged ions after passing through a differentially pumped He gas target. Typically less than 10 collisions are sufficient to reach an equilibrium charge state distribution, while the energy loss is small compared to the total energy. Because of the small number of collisions such a setup is less sensitive to the purity of the gas than in a measurement where we completely stop the ions. In Fig.2 our results for Xe on helium is plotted in addition to data at higher energies and for different isotopes. We could extend these measurements to lower momentum of 15×10^8 cm u/s, respectively 80 eV/u of energy. The average charge state is decreasing with decreasing momentum of the ions. Measurements at lower momenta were not possible in our apparatus because of the increasing scattering angle.

For the TRIμP facility we are currently commissioning a stopping device of the second type, since it is ideal for alkaline and alkaline earth isotopes. We plan to stopping the ions in thin tungsten foils, which can be heated to temperatures of 2500 K were the expected diffusion times are less than 1 s for 1μm thick foils.

After the stopping device the isotopes are extracted as ions, which allows for easy manipulation. The ions are guided by electrostatic means into a gas filled radio-frequency buncher cooler system. It consists out of two identical segmented rfq's of 330 mm length, which are separated by small apertures for differential pumping purposes. The rods of the quadrupoles are 10 mm apart and are segmented in order to apply axial field gradient.

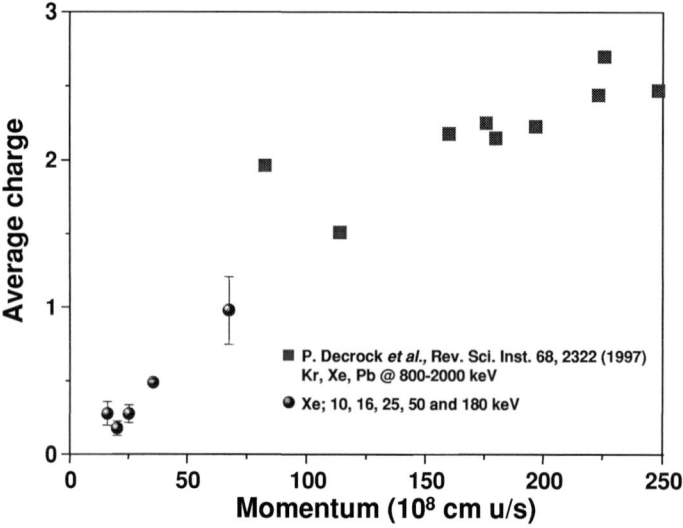

FIGURE 2. Average charge state of ions passing through helium at different momenta. The Xe data are from our measurements.

The segments are connected by a dc resistor chain, while the rf is capacitively coupled to the segments. This reduces the number of electrical feedthroughs significantly. We use frequencies from 0.5-2 MHz and a voltage V_{pp} of up to 200 V, which is sufficient for the isotopes of interest. The device is housed in standard UHV double crosses (Fig. 3).

The system was commissioned using ions with energies of 10-60 eV. In the first section they are slowed and transversely cooled by collisions with He buffer gas of about 3×10^{-2} mbar. A small drag potential of 0.5 V/cm moves the ions along the axis. The second rfq operates at a pressure ten times lower. The axial potential has is shaped to allow for trapping near to the exit. The potential depth is on the order of 5 V The ions can be ejected into a pulsed drift tube accelerator by switching the last electrode by several ten volt. Preliminary measurements indicate that we find more than 60% of the ions entering the device are transferred into the drift tube. The drift tube is pulsed and the ion pulse is detected by a micro channel plate in the low energy beamline. The pulse width is in the order of several hundredths of ns in agreement with simulations. The ion pulse can then be transported in an electrostatic beamline to the experiments.

The ultimate cooling of the radioactive isotopes is provided in neutral atom traps. Here, atoms are well localized at typical temperatures of the order of μK. Storage times in a MOT depends on the particular atom and the background pressure and Na trapping times of more than 100 s have been achieved. Recently the first Na MOT for TRIμP has been brought into operation. An advantage of optical trapping is that it allows to manipulate the state of the systems. The limitation of laser cooling are that the forces are rather small and atomic level scheme has to be suitable. Thus we are developing new laser cooling schemes for atoms like radium, extending the list of trapable atoms.

Research with trapped rare isotopes offer unique possibilities for testing fundamental

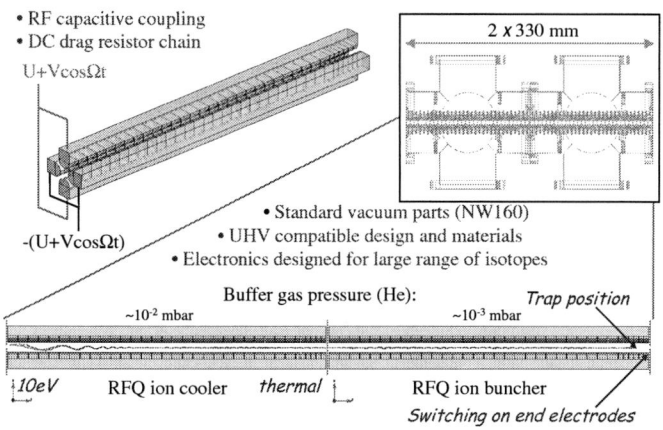

FIGURE 3. The Radio Frequency Cooler Buncher for the TRIμP Facility.

interactions in a complementary way to high energy physics. Atomic physics techniques allow for precision measurements which can test extensions to the standard model very sensitively [7, 14]. The upcoming facilities at KVI and other places are on their way to enable promising experimental test in the near future.

REFERENCES

1. Atomic Physics at Accelerators: Laser Spectroscopy and Applications, L. Schweikhard and H.J. Kluge (eds.), Hyperfine Interactions 127.
2. For more informations in the TRIμP Facility: http://www.kvi.nl/ trimp/web/html/trimp.html.
3. P. Herczeg, Precision Tests of the Standard Electroweak Model (World Scientific, Singapore, 1995).
4. N. D. Scielzo et al., Phys. Rev. Lett. 93, 102501 (2004); A. Gorelov et al., Phys. Rev. Lett. 94, 142501 (2005).
5. J.W. Turkstra, H.W. Wilschut, D. Meyer, R. Hoekstra, R. Morgenstern, Hyperfine Interactions 127, 533-536 (2000).
6. "CP Violation without Strangeness", I.B. Krhiplovich, S.K. Lamoreaux, Springer, Berlin (1997).
7. M. V. Romalis et al. Phys. Rev. Lett. 86, 2505 (2001).
8. V.V. Flambaum, Phys. Rev. A 60, R2611 (1999); V. A. Dzuba et al. Phys. Rev. A61 062509 (2000).
9. G.P.A. Berg et al., accepted for publication Nucl. Inst. Meth. A, xxx.lanl.gov:nucl-ex/0509013
10. L. Weissman, P.A. Lofy, D. A. Davies, D.J. Morrissey, P. Schury, S. Schwarz, and G. Bollen, Nucl. Phys. A746c 655 (2004).
11. M. Huyse, M. Facina, Yu.Kudryavtsev, P.Van Duppen, Nucl. Instr. Meth. B187, 535 (2002).
12. R. Kirchner, Rev. Sci. Instrum. 67 928 (1996).
13. Atomic Data for Fusion, Volume 1, C.F. Barnett editor, ORNL 6086. and from R.Hoekstra, H.P. Summers and F.J. de Heer, Nucl. Fusion Suppl. 3, 63 (1992).
14. C.S. Wood et al., Science 275, 1759 (1997).

Workshop Program

Presentations are available online at: http://conferences.fnal.gov/cool05/

Overview – S. Nagaitsev, Chair

Welcome
(*Piermaria Oddone – FNAL*)
The International Year of Physics — Remembering Albert Einstein: His Impact on Accelerators; His Impact on the World
(*Andrew Sessler – LBNL*)
The Reason for Beam Cooling: Some of the Physics that Cooling Allows
(*Walter Oelert – FZJ*)
Overview of Recent Trends in Beam Cooling Methods and Technology
(*Igor Meshkov – JINR*)
Future Directions in Accelerator Physics
(*David Sutter – DOE, retired*)

Reports from Labs – I. Ben-Zvi, Chair

The FAIR Project
(*Markus Steck – GSI*)
Antiproton Cooling in the Fermilab Recycler
(*Sergei Nagaitsev – FNAL*)
Report on Operations of Antiproton Decelerator
(*Pavel Belochitskii – CERN*)

General Topics – D. Rcistad, Chair

Recent and Future Cooling Experiments at COSY
(*Dieter Prasuhn – FZJ/IKP*)
Transverse and Longitudinal Phase-Space Manipulations
(*Kwang–Je Kim – ANL*)
Transverse-Longitudinal Correlations: FEL Perfomance and Emittance Exchange
(*Andrew Sessler – LBNL*)
Optics of Electron Beam in the Recycler: Analysis of First Results
(*Alexey Burov – FNAL*)
Experimental Study of Dispersion Control Utilizing Both Magnetic and Electric Fields
(*Mikio Tanabe – Kyoto University*)
Transverse Echo Measurements in RHIC
(*Wolfram Fischer – BNL*)
Simulation of Beam Dynamics in Cooler Rings
(*Alexander Smirnov – JINR*)

Stochastic Cooling – R. Pasquinelli, Chair

Stochastic Cooling Developments at GSI
(*Fritz Nolden – GSI*)
Antiproton Production Rate Increase
(*David McGinnis – FNAL*)
Bunched-Beam Stochastic Cooling for RHIC
(*J.Michael Brennan – BNL*)
Cooling Scenario for the HESR Complex
(*Hans Stockhorst – FZJ/IKP*)
Stacking of 3 GeV Antiprotons with Moving Barrier Bucket Method at GSI–RESR
(*Takeshi Katayama – GSI*)
Bunched Beam Stochastic Cooling and Coherent Lines
(*Michael Blaskiewicz – BNL*)
Applications of Schottky Spectroscopy at the Storage Ring ESR of GSI
(*Fritz Nolden – GSI*)

Muon Cooling – A. Sessler, Chair

Review of Muon Cooling Development
(*Robert Palmer – BNL*)
Recent Innovations in Muon Beam Cooling
(*Rolland P. Johnson – Muons, Inc*)
6D Cooling of a Circulating Muon Beam
(*Al Garren – UCLA*)
Parametric Resonance Ionization Cooling and Reverse Emittance Exchange for Muon Collider
(*Yaroslav Derbenev – Jefferson Lab*)

Muon Cooling – I. Meshkov, Chair

MICE: The International Muon Ionization Cooling Experiment
(*Daniel Kaplan – IIT*)
6D Muon Ionization Cooling with an Inverse Cyclotron
(*Don Summers – Mississippi University*)
The Muon Cooling RF R&D Program
(*Yagmur Torun – IIT*)
Mucool Hydrogen Absorber R&D
(*Mary Anne C. Cummings – NIU*)
Cryogenics for the MuCool Test Area (MTA)
(*Christine Darve – FNAL*)

Electrostatic Rings – A. Noda, Chair

An LN2-Cooled Electrostatic Ring
(*Toshiyuki Azuma – Tokyo Metrolopitan University*)

DESIREE – A Double Electrostatic Storage Ring for Merged-Beam Experiments
(*Håkan Danared – MSL*)
The Heidelberg CSR: Low-Energy Ion Beams in a Cryogenic Electrostatic Storage Ring
(*Andreas Wolf – MPI–K*)
Ultra-Cold Electron Target for the Heidelberg TSR
(*Dmitry Orlov – MPI–K*)
FLAIR Project Proposal at GSI
(*Carsten P Welsch – MPI–K*)

Electron Cooling – M. Steck, Chair

Detailed Studies of Electron Cooling Friction Force
(*Alexei Fedotov – BNL*)
Simulations of Dynamical Friction Including Spatially-Varying Magnetic Fields
(*David Bruhwiler – Tech–X Corporation*)
Development of a New Generation of Coolers with a Hollow Electron Beam and Electrostatic Bending
(*Vasily Parkhomchuk – BINP*)
LEIR Cooler Status
(*Gerard Tranquille – CERN*)
Commissioning of HIRFL-CSR and Its Electron Coolers
(*Xiaodong Yang – IMP*)

Electron Cooling – H. Danared, Chair

Longitudinal Cooling Force Measurements at CELSIUS
(*Björn Gålnander – The Svedberg Laboratory*)
Experimental Benchmarking of the Magnetized Friction Force
(*Alexei Fedotov – BNL*)
Electron Cooling of Intensive Ion Beam
(*Igor Meshkov – JINR*)

High Energy Electron Cooling – D. Prasuhn, Chair

High-Current ERL-Based Electron Cooling System for RHIC
(*Ilan Ben–Zvi – BNL*)
Attainment of a High-Quality Electron Beam for Fermilab's 4.3-MV Cooler
(*Alexander Shemyakin – FNAL*)
HESR Electron Cooling System Proposal
(*Dag Reistad – The Svedberg Laboratory*)
COSY 2-MeV Cooling System Proposal
(*Juergen Dietrich – FZJ/IKP*)
Budker INP proposals for HESR and COSY Electron Cooling System
(*Vladimir Reva – BINP*)

Laser Cooling, Traps – K.-J. Kim, Chair

Laser Cooling for 3-D Crystalline State at S-LSR
(*Akira Noda – Kyoto University/ICR*)

Laser Cooling (and Stopping) of Relativistic Heavy Ion Beams
(*Ulrich Schramm – Munich University*)

Status of LEPTA Project
(*Igor Seleznev – JINR*)

Laser Cooling, Traps – Y. Derbenev, Chair

Cooling Techniques for Trapped Particles and New Trends in Physics with Trapped Particles
(*Yasunori Yamazaki – RIKEN*)

Electron Cooling of Highly Charged Ions in Traps
(*Guenter Zwicknagel – Erlangen University*)

A Radio Frequency Quadrupole Cooler/Buncher System for TRImP
(*Lorenz Willmann – KVI*)

List of Participants

Toshiyuki Azuma
Tokyo Metropolitan University
azuma@phys.metro-u.ac.jp

Kevin Beard
Thomas Jefferson National Accelerator Facility
beard@jlab.org

Ilan Ben-Zvi
Brookhaven National Laboratory
ilan@bnl.gov

Chandra Bhat
Fermi National Accelerator Laboratory
cbhat@fnal.gov

Mike Blaskiewicz
Brookhaven National Laboratory
blaskiewicz@bnl.gov

J. Michael Brennan
Brookhaven National Laboratory
brennan@bnl.gov

Daniel Broemmelsiek
Fermi National Accelerator Laboratory
broemmel@fnal.gov

David Bruhwiler
Tech-X Corporation
bruhwile@txcorp.com

Alexander Bubley
BINP SB RAS, Novosibirsk
bubley@inp.nsk.su

Alexey Burov
Fermi National Accelerator Laboratory
burov@fnal.gov

Kermit Carlson
Fermi National Accelerator Laboratory
kermit@fnal.gov

Fritz Caspers
CERN
fritz.caspers@cern.ch

Mary Anne Cummings
Northern Illinois University
macc@fnal.gov

Hakan Danared
Manne Siegbahn Laboratory
danared@msl.se

Christine Darve
Fermi National Accelerator Laboratory
darve@fnal.gov

Yaroslav Derbenev
Thomas Jefferson National Accelerator Facility
derbenev@jlab.org

Paul Derwent
Fermi National Accelerator Laboratory
derwent@fnal.gov

Jürgen Dietrich
Forschungszentrum Jülich
j.dietrich@fz-juelich.de

Brian Drendel
Fermi National Accelerator Laboratory
drendel@fnal.gov

Alexei Fedotov
Brookhaven National Laboratory
fedotov@bnl.gov

Wolfram Fischer
Brookhaven National Laboratory
wolfram.fischer@bnl.gov

Björn Gålnander
Uppsala University
bjorn.galnander@tsl.uu.se

Al Garren
University of California
aagarren@lbl.gov

Consoalto Gattuso
Fermi National Accelerator Laboratory
gattuso@fnal.gov

Romulus Godang
University of Mississippi
godang@phy.olemiss.edu

Elvin Harms
Fermi National Accelerator Laboratory
harms@fnal.gov

Martin Hu
Fermi National Accelerator Laboratory
martinhu@fnal.gov

Gerald Jackson
Hbar Technologies, LLC
gjackson@hbartech.com

Rolland Johnson
Muons, Inc.
rol@muonsinc.com

Stephen Kahn
Muons Inc.
kahn@bnl.gov

Vsevolod Kamerdzhiev
Fermi National Accelerator Laboratory
vsevolod@fnal.gov

Daniel Kaplan
Illinois Institute of Technology
kaplan@iit.edu

Takeshi Katayama
GSI
takeshi-katayama@nifty.com

Kwang-Je Kim
Argonne National Laboratory
kwangje@aps.anl.gov

Thomas Kroc
Fermi National Accelerator Laboratory
kroc@fnal.gov

Valeri Lebedev
Fermi National Accelerator Laboratory
val@fnal.gov

Jerry Leibfritz
Fermi National Accelerator Laboratory
leibfritz@fnal.gov

Vladimir Litvinenko
Brookhaven National Laboratory
vl@bnl.gov

David McGinnis
Fermi National Accelerator Laboratory
mcginnis@fnal.gov

Igor Meshkov
JINR, Dubna
meshkov@jinr.ru

Fred Mills
Fermi National Accelerator Laboratory (Retired)
fmills@fnal.gov

Sergei Nagaitsev
Fermi National Accelerator Laboratory
nsergei@fnal.gov

David Neuffer
Fermi National Accelerator Laboratory
neuffer@fnal.gov

Akira Noda
Institute for Chemical Research Kyoto University
noda@kyticr.kuicr.kyoto-u.ac.jp

Fritz Nolden
Gesellschaft für Schwerionenforschung
f.nolden@gsi.de

Gunnar Norman
Uppsala University
gunnar.norman@tsl.uu.se

Piermaria Oddone
Fermi National Accelerator Laboratory
pjoddone@fnal.gov and michelle@fnal.gov

Walter Oelert
Forschungszentrum Jülich
w.oelert@fz-juelich.de

Dmitry Orlov
Max-Planck-Institut für Kernphysik, Heidelberg
orlov@mpi-hd.mpg.de

Robert Palmer
Brookhaven National Laboratory
palmer@bnl.gov

Vasily Parkhomchuk
BINP RAN, Novosibirsk
parkhomchuk@inp.nsk.su

Ralph Pasquinelli
Fermi National Accelerator Laboratory
pasquin@fnal.gov

Kevin Paul
Muons, Inc.
kpaul@muonsinc.com

Claudius Peschke
Gesellschaft für Schwerionenforschung
c.peschke@gsi.de

Dieter Prasuhn
Forschungszentrum Jülich
d.prasuhn@fz-juelich.de

Lionel Prost
Fermi National Accelerator Laboratory
lprost@fnal.gov

Dag Reistad
Uppsala University
dag.reistad@tsl.uu.se

Vladimir Reva
BINP RAN, Novosibirsk
v.b.reva@inp.nsk.su

Ulrich Schramm
Universität Munich
ulrich.schramm@physik.uni-muenchen.de

Sergei Seletskiy
University of Rochester
smsvm@fnal.gov

Igor Seleznev
JINR, Dubna
seleznev@jinr.ru

Andrew Sessler
Lawrence Berkeley National Laboratory
amsessler@lbl.gov

Alexander Shemyakin
Fermi National Accelerator Laboratory
shemyakin@fnal.gov

Vladimir Shiltsev
Fermi National Accelerator Laboratory
shiltsev@fnal.gov

Toshiyuki Shirai
Kyoto University
shirai@kyticr.kuicr.kyoto-u.ac.jp

Alexander Smirnov
JINR, Dubna
smirnov@jinr.ru

Al Sondgeroth
Fermi National Accelerator Laboratory
sondgeroth@fnal.gov

Markus Steck
Gesellschaft für Schwerionenforschung
m.steck@gsi.de

Hans Stockhorst
Forschungszentrum Jülich
h.stockhorst@fz-juelich.de

Donald Summers
University of Mississippi, Oxford
summers@phy.olemiss.edu

Yin-e Sun
Argonne National Laboratory
yinesun@aps.anl.gov

David Sutter
U. S. Department of Energy
accelphys@aol.com

Mikio Tanabe
ICR, Kyoto
tanabe@kyticr.kuicr.kyoto-u.ac.jp

Yagmur Torun
Illinois Institute of Technology
torun@iit.edu

Gerard Tranquille
CERN
gerard.tranquille@cern.ch

Vitali Tupikov
Fermi National Accelerator Laboratory
tupikov@fnal.gov

Tomonori Uesugi
National Institute of Radioilogical Sciences (NIRS)
touesugi@nirs.go.jp

Gang Wang
State University of New York Stony Brook
gawang@bnl.gov

Arden Warner
Fermi National Accelerator Laboratory
warner@fnal.gov

Carsten Welsch
MPI-K, Heidelberg & CERN
carsten.welsch@cern.ch

Lorenz Willmann
Kernfysisch Versneller Instituut
willmann@kvi.nl

Andreas Wolf
Max-Planck Institute for Nuclear Physics
a.wolf@mpi-hd.mpg.de

Yasunori Yamazaki
University of Tokyo & RIKEN
yasunori@phys.c.u-tokyo.ac.jp

Xiaodong Yang
Institute of Modern Physics
yangxd@impcas.ac.cn

Katsuya Yonehara
Muons, Inc
yonehara@fnal.gov

Max Zolotorev
Lawrence Berkeley National Laboratory
max_zolotorev@lbl.gov

Günter Zwicknagel
Universität Erlangen
zwicknagel@theorie2.physik.uni-erlangen.de

AUTHOR INDEX

A

Abell, D. T., 319, 329
Alsharo'a, M., 405
Andler, G., 465
Andrianarijaona, V., 473
Ankenbrandt, C., 405

B

Bagge, L., 465
Barzi, E., 405
Beard, K., 405, 453, 458
Beckert, K., 177, 211
Behtenev, E., 334
Bell, G. I., 329
Beller, P., 108, 177, 196, 211, 501
Belochitskii, P., 48
Ben-Zvi, I., 75
Berg, G. P., 523
Blaskiewicz, M., 185, 206
Blom, M., 465
Bocharov, V., 308, 334, 355, 360
Bogacz, A., 458
Bogacz, S. A., 405, 453
Bolshakov, A., 39
Bolshakov, T. B., 370
Bracker, S. B., 432
Brennan, J. M., 185, 206
Brizgunov, M., 355
Broemmelsiek, D., 39, 226
Bross, A., 437
Bruhwiler, D. L., 319, 329
Bryzgunov, M., 308, 341
Bubley, A., 308, 355, 360
Bubley, V., 334
Burov, A. V., 39, 139, 159, 280, 370
Busby, R., 329
Bussmann, M., 501

C

Carlson, K., 39, 280, 380
Cary, J. R., 329
Caspers, F., 177

Cederquist, H., 465
Cline, D., 415
Cremaldi, L. M., 432
Crespo López-Urrutia, J. R., 473
Cullerton, E., 242
Cummings, M. A., 442

D

Dammalapati, U., 523
Danared, H., 465
Darve, C., 448
De, S., 523
Del Frate, L., 405
Dendooven, P., 523
Derbenev, Y., 405, 420, 453, 458
Dermois, O., 523
Derwent, P. F., 237, 242
Diehl, A., 473
Dietrich, J., 154, 270, 299
Dolinskii, A., 177

E

Ershov, V., 355
Evtushenko, Yu., 334

F

Fadil, H., 346, 473
Fedotov, A. V., 259, 265, 319
Fischer, W., 149
Franzke, B., 108, 177, 196, 211, 501
Fujimoto, S., 103, 491
Fujimoto, T., 103

G

Gålnander, B., 259, 265
Garren, A., 415
Gattuso, C., 39, 226
Godang, R., 432
Goncharov, A., 334, 355

Gonin, I., 405
Gosteev, V., 308
Gostishchev, V., 211
Grieser, M., 85, 103, 346, 473, 491
Grishanov, B., 365
Gusachenko, I., 159

H

Haberstroh, Ch., 473
Habs, D., 501
Hanlet, P. M., 405
Hartline, R., 405
Hu, M., 39, 280

I

Ikegami, M., 103, 144, 491
Ivanov, A., 334
Iwata, S., 103

J

Jandewerth, U., 177
Johnson, R. P., 405, 420, 453, 458
Jungmann, K., 523

K

Kahn, S., 415
Källberg, A., 465
Kamerdjiev, V., 270
Kaplan, D. M., 405, 427
Kaplan, R. P., 458
Karpuk, S., 501
Katayama, T., 196
Kazakevich, G., 39, 139, 280, 375, 380
Kazarezov, I., 308
Kienle, P., 108
Kikuchi, T., 196
Kim, K.-J., 115
Kirk, H., 415
Kobets, A., 95
Kokoulin, V., 334
Kolmogorov, V., 334
Kondaurov, M., 334

Konstantinov, S., 334, 355, 360
Koop, I., 108
Korotaev, Yu., 95, 270
Kozak, V., 334
Kozhuhzrov, C., 211
Krainov, G., 334
Kramper, B., 39, 280
Kroc, T. K., 39, 139, 280, 370, 375
Kruchkov, A., 334
Kruecken, R., 108
Kryuchkov, A., 308
Kuchnir, M., 405
Kühl, T., 501
Kumada, M., 365
Kuper, E., 334

L

Larsson, M., 465
Lebedev, V., 139, 231
Leibfritz, J., 39, 280
Leontein, S., 465
Lestinsky, M., 478
Li, D., 437
Li, J., 65, 334
Liljeby, L., 465
Litvinenko, V. N., 259, 265, 329
Litvinov, Y. A., 211
Löfgren, P., 465
Lofnes, T., 259, 265
Lomakin, A., 355
Lorentz, B., 190
Lu, W., 65, 334

M

Maier, R., 190, 270
Malakhov, V., 95
Mallinger, V., 473
Mao, L. J., 65, 334
McGinnis, D., 237, 242
Medvedko, A., 334
Meshkov, I., 16, 95, 154, 270, 317, 351
Messmer, P., 329
Mills, F., 415
Mironenko, L., 334
Möhl, D., 16, 177, 196

Mol, A., 523
Moretti, A., 405, 437

N

Nagaitsev, S., 39, 139, 159, 280, 317, 375, 380
Nesmiyan, I., 177, 196
Neuffer, D., 405
Noda, A., 103, 144, 491
Noda, K., 103, 144, 351, 491
Nolden, F., 177, 196, 211, 221, 501
Norem, J., 437
Norris, B., 448

O

Oelert, W., 6
Okamoto, H., 103, 491
Onderwater, C. J. G., 523
Orlov, D. A., 346, 473, 478

P

Paál, A., 465
Palmer, R. B., 432
Panasyuk, V., 308, 334, 341, 355, 360
Parkhomchuk, V. V., 65, 108, 249, 299, 308, 334, 341, 355, 360, 365
Pasquinelli, R., 237, 242
Paul, K., 405, 453, 458
Pavlov, V., 95, 308
Pei, L., 448
Peschke, C., 177, 221
Pestrikov, D., 308
Petri, P., 177
Pivin, R., 95
Polukhin, V., 355
Popovic, M., 405
Prasuhn, D., 190, 270
Prost, L. R., 39, 139, 280, 375, 380, 391
Pruss, S., 39, 139, 280

Q

Qian, Z., 437
Quack, H., 473

R

Rappaport, M., 473
Rastigeev, S., 365
Reinhardt, S., 501
Reistad, D., 289
Rensfelt, K.-G., 465
Reva, V. B., 169, 299, 308, 334, 341, 355, 365
Rimmer, R. A., 437
Roberts, T. J., 405, 453, 458
Rogachevskiy, A., 523
Romanov, G., 405
Rosén, S., 465

S

Saathoff, G., 501
Saewert, G., 39, 280
Schmidt, C. W., 39, 280, 375
Schmidt, H. T., 465
Schmidt, K., 465
Schramm, U., 501
Schröter, C. D., 473
Schwalm, D., 346, 473, 478
Schwinn, A., 211
Seletskiy, S. M., 39, 280, 370, 386
Seleznev, I., 95
Sessler, A. M., 3, 115
Shamovskij, V., 308
Shatunov, Y., 108
Shemyakin, A., 39, 139, 159, 280, 370, 375, 391
Shibuya, S., 103, 144, 351, 491
Shirai, T., 103, 144, 491
Shrainer, K., 334
Sidorin, A. O., 95, 154, 259, 265, 270, 319
Simonsson, A., 465
Skarbo, B., 334, 355
Skrinsky, A., 108, 308, 334
Smirnov, A. V., 95, 154, 259, 265, 270, 397
Smirnov, B., 334
Sohani, M., 523
Song, M., 65
Souda, H., 103, 144, 491
Sprenger, F., 478
Steck, M., 29, 177, 196, 211, 501

Stein, J., 154, 270
Stockhorst, H., 190, 270
Sukhina, B., 308, 334, 355
Summers, D. J., 432
Sun, D., 242
Sutherland, M., 39, 139, 280, 375, 380
Syresin, E. M., 103, 351

T

Takeuchi, T., 103, 491
Takubo, A., 103
Tanabe, M., 103, 144, 491
Terekhov, A. S., 478
Thorndal, L., 177, 221
Tinsley, D., 242
Tiunov, M., 139, 380
Tongu, H., 103, 144, 491
Torun, Y., 437
Tranquille, G., 57
Traykov, E., 523
Trubnikov, G., 95
Tupikov, V., 39, 280, 375, 386
Turrioni, D., 405

U

Uesugi, T., 351
Ullrich, J., 85, 473
Urbain, X., 473

V

Vander Meulen, D., 237
Vedenev, M. A., 299, 308, 334, 341, 355
Veitzer, S., 329
von Hahn, R., 346, 473
Voskoboinikov, R., 334
Vostrikov, V., 108, 308, 365

W

Wang, G., 164
Wang, Z. X., 65, 334

Warner, A., 39, 139, 280, 375, 380
Weber, T., 473
Welsch, C. P., 85, 397
Werkema, S., 237
Widmann, E., 108
Willmann, L., 523
Wilschut, H. W., 523
Wolf, A., 346, 473, 478

X

Xia, J., 65

Y

Yakovenko, S., 95
Yamada, S., 491
Yan, H. B., 334
Yang, X., 65
Yang, X. D., 334
Yarba, V., 405
Yonehara, K., 405, 453, 458
Yuan, Y., 65
Yuri, Y., 103

Z

Zajfman, D., 346, 473
Zakhvatkin, M., 334, 355
Zapiatkin, N., 334, 355
Zenkevich, P., 39
Zhan, W., 65
Zhang, J. H., 334
Zhang, W., 334
Zhao, H., 65
Zhao, H. W., 334
Ziemann, A. V., 259
Ziemann, V., 265
Zisman, M. S., 437
Zwicknagel, G., 513